Animal Feeds, Feeding and Nutrition, and Ration Evaluation with CD-ROM

ANIMAL FEEDS, FEEDING AND NUTRITION, AND RATION EVALUATION WITH CD-ROM

David A. Tisch

Australia • Brazil • Japan • Korea • Mexico • Singapore • Spain • United Kingdom • United States

Animal Feeds, Feeding and Nutrition, and Ration Evaluation with CD-ROM
David A. Tisch

Vice President, Career Education Strategic Business Unit: Dawn Gerrain

Director of Editorial: Sherry Gomoll

Acquisitions Editor: David Rosenbaum

Developmental Editor: Gerald O'Malley

Editorial Assistant: Christina Gifford

Director of Production: Wendy A. Troeger

Production Manager: J. P. Henkel

Production Editor: Kathryn B. Kucharek

Technology Project Manager: Sandy Charette

Director of Marketing: Wendy Mapstone

Marketing Specialist: Gerard McAvey

Cover Images: Getty Images

Cover Design: TDB Publishing Services

Library of Congress Control Number: 2005051182

ISBN-13: 978-1-4018-2640-6

ISBN-10: 1-4018-2640-7

Delmar
Executive Woods
5 Maxwell Drive
Clifton Park, NY 12065
USA

Cengage Learning is a leading provider of customized learning solutions with office locations around the globe, including Singapore, the United Kingdom, Australia, Mexico, Brazil, and Japan. Locate your local office at **international.cengage.com/region**

Cengage Learning products are represented in Canada by Nelson Education, Ltd.

For your lifelong learning solutions, visit **www.cengage.com/delmar**

Visit our corporate website at **www.cengage.com**

Printed in the United States of America
2 3 4 5 6 7 11 10 09

To my wife, Pam.

CONTENTS

PREFACE

The animal feeding and nutrition industries are as computerized as any industry in the world. Computers are used extensively to prepare the complex mixtures from hundreds of available feedstuffs that provide the bioavailable levels of the dozens of nutrients required by the tissues of domestic animals. Ration formulation cannot be accomplished without the aid of computer software. There is a need, therefore, to integrate the textbook's coverage of the subjects of animal feeding and nutrition with computer software used during ration formulation. This is the need that this project attempts to meet. For each of the eleven species of animals discussed in the textbook, an application for ration formulation is included.

Though the applications provided in this project have similar functionality, they are designed differently than most software used in the feed and animal nutrition industries:

1. Like industry software, this project's applications contain a feed table from which feedstuffs may be chosen for inclusion in a developing ration. Unlike industry software, the applications included with this project are designed as learning tools. On the feed table, the nutrient content of feedstuffs is easily viewed, and numerous comments are inserted to provide information about feedstuffs, including AFIA definitions and limitations as ration components.
2. Like industry software, this project's applications contain an animal input section in which the animal and aspects of its management and environment that would affect nutrient requirements are described. Unlike most industry software, this project's applications show the consequences of changing inputs on predicted animal nutrient requirements.
3. Like industry software, this project's applications contain ration-balancing sections in which feedstuff amounts are entered in the attempt to bring the ration nutrient levels in balance with the animal's predicted nutrient requirements. Also like industry software, this project's applications provide guidance to the user, informing her or him of the effect on nutrient balance as feedstuffs are added or removed from the developing ration. This project's applications highlight in red those nutrient levels and performance parameters that need adjustment. Using this project's applications, the work of ration balancing takes on a video-game aspect as students monitor the highlights going off and on with varying feedstuff selections and pounds fed.
4. Like industry software, this project's applications are capable of blending feedstuffs into mixtures that would be handled during the feeding activity. Also like most industry software, the formula and analysis of these blends are stored and available for viewing or printing.

Because a single author produced each application, all ration formulation applications are operated in the same way—once a student has mastered the buttons necessary to run the cat software, he or she will be able to begin productive work immediately on a ration for a lactating dairy cow, a growing rabbit, a working horse, or any of the eleven species covered in the textbook. Such compatibility and continuity has not been available heretofore.

Because the author of these applications is also the author of the textbook, the applications are thoroughly integrated into the text and vice versa. This is important because ration formulation applications are used to apply the knowledge presented in an animal nutrition course. As an example, the applications make predictions of rumen pH for goat, sheep, beef, and dairy rations. The rumen pH prediction is highlighted in red if it falls below a certain value. The text explains how the rumen pH prediction is made, providing the knowledge necessary to make appropriate ration adjustments.

Other benefits resulting from the integration of applications and textbook involve making best use of the strengths of the two types of learning instruments. Textbooks are best suited to explanations and descriptions. Applications are better suited to handling large amounts of information presented in tabular form, images, real-time updates, and anything that is best learned through interaction. Specifically, the integration of textbook and applications in this project has resulted in the following benefits and opportunities:

1. The text and applications are complementary, enhancing the learning process. For example, any text can make the statement that the protein of soybean meal is superior in quality to that of any other oilseed meal. The veracity of this statement becomes apparent when the student actually tries replacing the soybean meal in a balanced pig ration with canola meal or peanut meal. The uniform blue display of a well-balanced ration explodes with red highlights, indicating amino-acid deficiencies. The impact of this "Whoa!" effect on learning and retention is undoubtedly significant.

2. Students are spared the difficulty of reading tables of feed composition and nutrient requirements that span several pages. These are replaced by the easily scanned feed- and nutrient-requirement sections of the ration formulation applications.
3. The way images have been used in previous textbooks has not resulted in optimal educational value. Images are not contained in this project's textbook, but rather on a PowerPoint file on the accompanying online companion. Included are color images of feedstuffs along with their analysis values, feed processing equipment, livestock digestive tracts, and more.
4. Finally, the applications provide links to current sources of information related to topics discussed in the text. The project's ration formulation applications also allow the instructor to add web links.

One of the challenges in the development of the ration formulation applications was to establish some degree of standardization. For example, a discussion of feed nutrient densities and animal requirement concentrations may be made on either an as fed basis or a dry matter basis for any species. However, it would impede the learning process if a single standard was not adopted. Likewise, the approaches taken to deal with the dietary need for energy vis-à-vis feed intake have been particularly disparate among the various specialists in animal nutrition. And, of course, in the United States, there is the debilitating disconnect between how feed weights are expressed on the farm (avoirdupois) and how they are expressed in research (metric). The choice of standard was based on what is most conducive to learning. It has been my experience that by applying a standard format from pig to trout, the graduate is better able to apply the principles of animal nutrition to any species in any format.

IMPORTANT: since the CD accompanying this text went into production, the website address of the FDA database of approved animal drug projects has changed. As a result, the FDA link contained in the ration formulation workbooks is no longer functional. The new address to the FDA website is:

http://dil.vetmed.vt.edu/NADA/default.cfm

Access to the FDA website is needed to complete the ration formulation assignments in Chapters 15, 23, and 27. The updated link may be added to the ration formulation workbooks at the *Relevant Hyperlinks* portion of the opening display.

CHAPTER 1

INTRODUCTION

As America became the center of the planetary food system, trade routes were transformed, new economic relationships took shape, and grain became one of the foundations of the postwar American Empire.

MORGAN, 1980

INTRODUCTION

The waste products generated during the processing of food and fiber for human use are often rich in nutrients. The feed industry began as a recycler of these nutrients through livestock diets, thereby turning waste nutrients into edible products. Today, feed costs amount to between 20 and 70 percent of the cost of raising livestock, and these by-products of food processing industries play an important role in managing feed costs. The profitability of livestock operations will depend on how efficient the animals are in converting feed nutrients into saleable products. This efficiency declines when animals are raised in stressful environments. The production of livestock is closely related to many other fields of science and business including management, quantitative analysis, agronomy, computer science, animal behavior, environmental policy, agriculture engineering, animal health, and microbiology.

TERMINOLOGY

In discussing the subject of feeding and animal nutrition, a few terms must be defined at the outset. *Feedstuffs* contain the substances that are the *nutrient requirements* of animals. A feedstuff is an ingredient in a *ration.* A ration is a mixture of feedstuffs formulated to meet the daily nutrient requirements for the target animal. The ration is a subset of the *diet.* The diet refers to all feedstuffs consumed by the animal over time. In intensely managed animal operations, the composition of the diet is described by the ration. *Feed* can mean the same as feedstuff, but feed can also mean a mixture of feedstuffs as a "finished feed" or "complete feed."

Grain is a classification of feedstuffs. Examples of grains include corn, sorghum, wheat, oats, and barley. What these feedstuffs have in common is energy: they all consist of the seed of the plant, and the seed is rich in high-energy compounds. Grains are essential components of most animal diets because the need for energy is what drives animals' appetites. The most important grain is corn grain. Corn is such a dominant grain that the term *grain* is sometimes used to refer to corn grain specifically. To the feed company, "grain" may refer to the feed product shipped to the farm, regardless of what is contained in the feed product. In this case, grain is used synonymously with feed.

Like grain, *forage* is a classification of feedstuffs. Examples of forage include pasture, green chop, hay, hay crop silage, corn silage, forbes, and browse. What these feedstuffs have in common is fiber: they all include the portions or stage of the plant when fiber is highest. Because of their relatively high fiber content, forages are sometimes called *roughages*. The term *roughage*, however, lost favor among many nutritionists because it has, for some, a negative connotation. The reality is that some diets suffer for lack of roughage. Although there will be exceptions, forages are generally homegrown feedstuffs and grains are generally purchased from feed mills.

Although the term *nutrient requirement* is not unique to the feed and animal nutrition industries, application of the term to different species of livestock presents some challenges. In animal nutrition, a nutrient requirement is the level of a specific chemical or general chemical category that must be consumed each day if the animal is to meet specific performance criteria. Depending on the animal and the situation, these performance criteria may include body weight status, reproductive performance, and health and/or production criteria. In animal nutrition, it is becoming increasingly evident that the level of nutrition necessary will vary depending on the performance criteria used. For example, there may be different levels of nutrients required if the animal is to attain maximum productivity compared to maximum health or maximum longevity.

Given the complexity with which animals absorb, metabolize, modify, mobilize, and excrete nutrients, not to mention the interactions that occur between nutrients, it may never be possible to establish precise optimal levels of nutrients in animal diets. Because of this, each nutrient requirement value established by researchers is taken from a range of acceptable nutrient content. The required value may be the minimum within the acceptable range or it may be the minimum augmented with some degree of overage. The companion application to this text represents the nutrient requirement as a range and the user is cautioned when the nutrient level of the developing ration falls outside either end of the acceptable range. Use of a nutrient range rather than a nutrient value necessitates replacement of the term *requirement*. In the companion application to this text, the term *requirement* has been replaced by the term *target*.

RECYCLING IN THE FEED AND ANIMAL NUTRITION INDUSTRIES

Few people realize that the businesses involved in feed and animal nutrition are some of the most active and effective recycling businesses in existence. Nutrient-rich by-products (also referred to as *coproducts*) of food manufacturing processes such as brewing, distilling, bread making, milk processing, edible nut processing, sugar refining, citrus processing, and meat processing are potentially powerful pollutants. These by-products could be disposed of in a landfill or drain off the site of origin into a stream. In the stream, microorganisms could use available oxygen to oxidize the nutrients contained in the by-products, impacting the stream as a habitat for other species.

Alternatively, these by-products can be used as ingredients in livestock feeds. In fact, the feed industry originated as a recycler of the by-products of food processing, and these by-products continue to be an important feed industry product today. Laboratories analyze the nutrient content of by-products and the results are compared to the nutrient requirements of the animals to be fed. Supplements are added to the formula to make good the deficiencies of the by-product(s) and a balanced ration is developed.

In addition to by-products, food products that are wholesome, yet do not meet quality standards for human consumption, are routinely used in livestock feeds. Bakery waste, candy, cookies, and instant breakfast drink mixes are all examples of products that would enter the waste system if they were not used in livestock diets.

STRESS, STRAIN, ANIMAL WELFARE, AND ANIMAL NUTRITION

Animals partition feed nutrients to different functions with specific priorities. The function of maintenance is the top priority. If the diet does not supply enough nutrients to meet maintenance needs, none of the other functions will be supported. On the other hand, if the diet supplies more than enough nutrients to support maintenance, the excess nutrients can go to support reproductive and productive functions.

Stress is the environmental situation that provokes an adaptive response by the animal. The adaptive response is called *strain*. Strain is a maintenance function. If an animal is in a stressful environment, the strain will cause nutrients to be diverted away from potential reproductive and productive functions. The profitability of farms and ranches, therefore, is dependent on the quality of animal welfare at the operation.

The neuroendocrine system is responsible for the repartitioning of nutrients during animal strain. The nervous system detects the environmental stress and communicates the information to the hypothalamus of the brain. The hypothalamus directs the endocrine system's response to the environmental stress. Norepinephrine from the adrenal medulla and glucocorticoids from the adrenal cortex are secreted in increasing amounts in an animal experiencing strain, and these are apparently responsible for the repartitioning of available nutrients.

Figure 1–1 illustrates qualitatively how three factors have affected the milk production potential of the average dairy cow over time. Actual milk production over time is less than animal genetic potential for milk production and less than what should be possible through application of our knowledge of nutrition. Actual milk production by the dairy cow and productivity for all species of livestock is probably limited by our understanding of how to manage animal environments to minimize stress and strain.

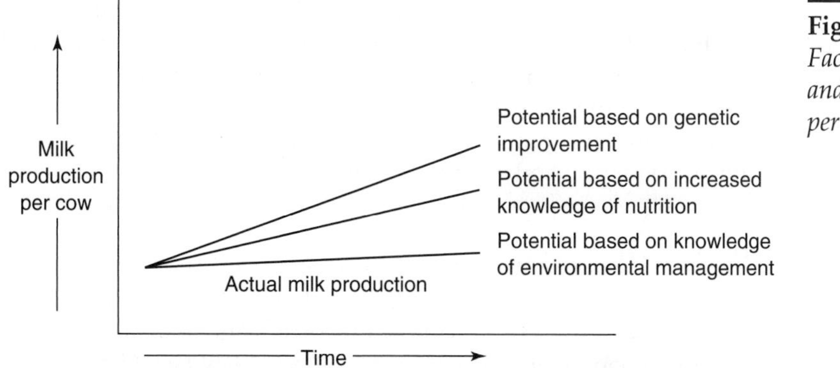

Figure 1–1
Factors affecting potential and actual milk production per cow

RELATIONSHIP OF THE FEED AND ANIMAL NUTRITION INDUSTRIES TO OTHER FIELDS OF BUSINESS AND STUDY

Management

As farms increase in size, the management functions separate into areas of specific expertise. Areas of expertise in the management function include technical activities, commercial activities, and financial accounting activities. The establishment and operation of a feeding program involve all aspects of management. On most farms and ranches, a nutritionist deals with the technical aspects of the feeding program. The recommendations of the nutritionist are communicated to the individual responsible for making feed purchases who, in turn, may need the assistance of the financial person to investigate the acquisition of credit.

Quantitative Analysis

Laboratories are available to perform numerous tests on feed samples. These laboratories perform quantitative analyses to measure nutrient content and pH, and to test for the presence of toxins. Most laboratories offer more than one method of analysis. The goal of the laboratories is to give the nutritionist a sense of the value of the feedstuff in terms of meeting animal nutrient requirements. But the laboratory processes are not identical to the digestion and absorption processes. As a consequence, quantitative analysis equipment and techniques are continually upgraded and improved to better predict the nutritive value of feedstuffs.

Agronomy

In many cases, animal diets are built around homegrown crops. That is, the foundation of the diets fed to animals on a farm or ranch will be what is grown on the farm or ranch. Additional feedstuffs will be purchased to supply nutrients lacking in homegrown feedstuffs. If the farm or ranch can produce feedstuffs that closely match the requirements of the animals, the need to purchase additional feedstuffs will be minimized and this will improve profitability. Agronomy, the science of crop production, is a closely allied field to the feed and animal nutrition industries.

Computer Science

To develop livestock rations that meet requirements for dozens of nutrients, farmers and ranchers need more than feedstuff tables, nutrient requirement tables, and a calculator. Computer applications are essential to utilize the hundreds of values involved in the formulation of just one ration. However, it is probable that a greater number of individuals are trained in software development than are trained in animal production and nutrition. A computer expert who has no knowledge of animal production will be unable to develop a usable ration formulation product, so there is a need for individuals trained in both animal production and computer science.

Animal Behavior

Farmers and ranchers must have a knowledge of animal behavior if they are to feed livestock profitably. When computer feeders first became available to provide supplemental feed to cattle, for example, they lacked chutes leading to the feeder. Without the chutes, an animal that had begun to consume feed could be bumped out of the way in a demonstration of natural competitive behavior. In a fish farm example, the feed was delivered to ponds via a truck that drove beside them. Fish were conditioned to rise to the pond surface as they sensed the feed truck approaching. But when an engine problem resulted in a change in the sound and vibrations made by the truck, the fish did not rise to receive their feed. This situation created two problems for the fish feeder. First, the feeder was deprived of an important management tool: watching the fish eating was useful in assessing their health and the suitability of the feed. Second, if the fish were primarily surface feeders, much of the feed could sink before the fish ate it.

Environmental Policy

The Environmental Protection Agency (EPA) and the U.S. Department of Agriculture (USDA) are establishing regulations for confined animal feeding operations (CAFOs) to ensure that medium to large farms account for the manure nutrients they generate. These regulations have fostered new enthusiasm for processing manure into usable products including bedding, fuel, and compost, as well as fertilizer. They have also highlighted the value of balanced diets that maximize the capture of feed nutrients in animal products, thereby minimizing the amount excreted.

Agriculture Engineering

Before the dawn of agriculture, there were hunters and gatherers. The hunters harvested wild animals living in an environment of their own choosing. Today, livestock are increasingly being raised in confinement in a structure of some type. The animal in confinement has only those environmental choices offered to it by the designers of the structure. This limitation is important because animals seek the environment that presents the least stress. If the structure is not properly designed by the agricultural engineer, the animals living in it may experience excessive strain, thereby reducing profitability of the operation.

Animal Health

Long ago, those involved in animal husbandry recognized the connection between the diet and health of the animal. The connection exists in two parts. First, mismanagement of the feeding program can make animals ill. Excessive dietary vitamin A can result in skin and bone lesions. Too little energy can result in excessive body weight loss, leading to a metabolic disease called ketosis. Second, animals resist infectious disease with immune factors built with feed nutrients. Animals that do get sick will have the best chances for recovery if their immune system is adequately supplied with nutrients. Many medications are intended to help the immune system help itself, but if the nutrients are not available for the immune system to work optimally, the medications will not work. Curing sick animals is complicated by the fact that they usually have reduced appetites.

Genetics

Efforts are currently underway to identify, locate, and determine the role of the genes present in the DNA of livestock cells. With a knowledge of the genome sequence, it will be possible for veterinarians and nutritionists to identify unusual genetic sequences characteristic of certain breeds and even specific animals. These unusual sequences are referred to as *genetic polymorphisms.* Some genetic polymorphisms are responsible for inoperative metabolic pathways that cause disease by affecting the animal's nutrient metabolism and requirements. Applying a knowledge of the genome sequence will make it possible to develop a diet to treat or prevent a specific disease for which an animal has a genetic predisposition. Other genetic polymorphisms may explain observed differences in feed intake, energy balance, milk production, milk components, marbling scores, fertility, and immune functions. Determining the genome of livestock and identifying the roles of each component gene will allow producers to select for more productive traits, and to more effectively feed and manage livestock. As of early 2004, the human genome has been completely sequenced; the dog, cow, and chicken genomes are being sequenced, and genome sequencing of the pig will be commenced shortly (Swanson, 2004).

Microbiology

The digestive tract of all animals considered in this text (and probably all animals not considered as well) hosts a population of microbes. These microbes vary in species and activity. Beneficial microbes aid digestion and absorption and help to suppress pathogenic microbes; pathogenic microbes disrupt digestive activity and may produce toxins leading to diarrhea. Numerous feed products have been developed to influence the activity of the microbes living in the digestive systems of domestic animals, including yeast products, antibiotics, and direct-fed microbials (also called *probiotics*).

Other aspects of microbiology that are important in the feed industry have to do with feed preservation. Products containing live microbes are added to ensiled feedstuffs to encourage the fermentation that generates the preserving acid. When feed is not well preserved, its deterioration is caused by the activity of other types of microbes. There are many feed additives designed to control the growth of bacteria and molds in feed that has begun to deteriorate.

CAREERS IN THE FEED AND ANIMAL NUTRITION INDUSTRIES

In addition to employment in the many allied fields, the student of feeds and animal nutrition may work as a professional animal nutritionist. Table 1–1 (Lobo, 2003) gives animal numbers and feed tonnage potential nationwide. The feed that is sold is fed to animals in rations that have been developed, for the most part, by professional nutritionists. A professional nutritionist may work as an independent consultant or be employed by a feed mill/feed manufacturer. The nutritionist may develop diets for the manager of a farm or ranch or for the shopper in various outlets (grocery store, pet specialty store, feed store, vet/kennel).

Most professional nutritionists are employed by feed manufacturers and work out of the feed mills they operate, developing diets for their customers. These nutritionists must consider the uniqueness of each operation: any grouping strategies must be based on the available facilities and labor. Performance targets must be established based on the owner's goals. Homegrown feedstuffs must be analyzed for nutrient content and inventories must be determined. This data is then used to develop diets that will meet short- and long-term performance goals.

Nutritionists also work to develop pet-food formulations to be marketed at grocery stores, pet specialty stores, feed stores, veterinary hospitals, and kennels. Many more will be involved in the marketing and sales of these products. Increasingly, the salespeople are expected to be familiar with the nutritional content and performance characteristics of the products they sell.

Table 1–1
Livestock operations and feed potential, 2002

Region	Swine Operations	Total Swine Feed Potential (tons)
North Atlantic	6,140	965,000
North Central	42,600	27,926,000
South Atlantic	9,040	10,027,000
South Central	12,280	5,298,000
West	5,290	2,449,000

Region	Beef Operations	Total Beef Feed Potential (tons)
North Atlantic	22,270	30,000
North Central	240,900	19,101,000
South Atlantic	103,430	311,000
South Central	352,000	12,064,000
West	86,480	8,478,000

Region	Dairy Cow Numbers	Total Dairy Feed Potential (tons)
North Atlantic	1,533,000	5,275,000
North Central	3,209,000	12,139,000
South Atlantic	537,000	2,166,000
South Central	735,000	2,944,000
West	3,137,000	13,621,000

Region	Poultry—Layers, Broilers, and Turkey Numbers	Total Poultry Feed Potential (tons)
North Atlantic	183,579,000	2,694,000
North Central	366,661,000	11,375,000
South Atlantic	3,374,603,000	22,463,000
South Central	4,375,646,000	25,540,000
West	42,162,000	2,480,000

Region	Sheep—Ewes and Lamb Numbers	Total Sheep Feed Potential (tons)
North Atlantic	237,000	21,000
North Central	2,206,000	229,000
South Atlantic	126,000	12,000
South Central	1,250,000	103,000
West	4,632,000	521,000

Fish type	Operations in the Top 15 States	Pounds of Food-Size Fish Sold
Catfish	1,155	673,743,000
Trout	378	54,451,000

Dog Food Type	Sales Value
Dry	$5,201,000,000
Canned	$1,380,000,000
Semi-moist	$1,475,000,000

THE FEED AND ANIMAL NUTRITION INDUSTRIES AND HUMAN WELFARE

Human welfare is dependent on a diverse planet. Diversity is possible only if we manage the land devoted to food production efficiently. A consequence of efficient use is that a minimal amount of land will need to be devoted to agriculture. This will give us the opportunity to preserve forests and the wildlands that make our planet diverse.

From a nutrient-transfer perspective, it is wasteful for humans to consume the meat of animals fed grain. This is because of the inefficiencies involved in transferring the nutrients contained in grain to meat before they are consumed by humans. In facing concerns over this inefficiency, it is important that meat animal producers create conditions in which these transfers are made as efficiently as possible. Animal health must be optimized because healthier animals make more efficient transfers than do sick animals.

The grass-fed beef alternative has the advantage of not using potential human food—grain—to produce meat. Animals fed primarily grass transfer the nutrients in a product that is indigestible for humans into meat. However, there are inherent inefficiencies in the grass-fed beef feeding system. The nutrient analysis of grass is generally higher in protein and lower in energy than what would support optimal growth. As a result, the animals on pasture are unable to transfer much of the grass protein to meat, and the excess protein is processed and excreted as urea in urine. Because grass-fed beef grow less efficiently and more slowly, land used to produce grass-fed beef represents a less efficient use of agricultural resources than land used for animals fed diets balanced with grasses and grains and whatever else is required for optimal growth.

To minimize the acreage of land that must be devoted to agriculture to feed the human population, it is important that human health be optimized. This is because a considerable number of calories are wasted by fevers caused by infections. Table 1–2 shows how prevalent these infections are worldwide. Consider the following example. The Recommended Dietary Allowance (RDA) for energy in kilocalories is 2900 for men between the ages of 25 and 50 and 2200 for women in the same age bracket (NRC, 1989a). The average value is 2550 kilocalories. Resting metabolic rate is that portion of dietary energy that goes to maintain the basic body functions of heartbeat, respiration, nervous function, muscle tone, body temperature, and the like. Approximately 60 to 65 percent of total energy expenditure in humans goes toward the resting metabolic rate. The portion of the energy RDA that represents the resting metabolic rate is, therefore, $0.60 \times 2550 = 1530$ kilocalories.

Table 1–2
The Prevalence of Human Infections

Disease Category	Human Infections
Worm parasites	4.5 billion
Tuberculosis	1.7 billion
Malaria	300–500 million
HIV/AIDS	40 million
Entamoeba histolytica (a protozoan parasite usually found in the digestive tract)	50 million

Helminth data from Chan, 1997, and Hopkins, 1992; tuberculosis data from WHO, 1994; malaria data from CDC, 2000; AIDS data from UNAIDS, 2003; protozoan data from Walsh, 1998.

Resting metabolic rate increases by about 7.2 percent for each degree Fahrenheit rise in body temperature. With a low-grade fever of 99.6, the resting metabolic rate of the body would increase by 7.2 percent or $.072 \times 1530 = 110.2$ kilocalories, the approximate kilocalorie content of one-half cup of brown rice. To fight their infections, these people need the energy equivalent of an additional half cup of rice daily. A society with a million people suffering this low-grade fever would have an additional energy demand equivalent to 31,939 tons of rice per year (Figure 1–2). Clearly, the quality of human health has an impact on how much of our planet must be devoted to food production and preservation of land.

1,000,000 persons/day × 365 days/year × 110.2 kcal/person ×
0.5 cup rice/110.2 kcal × 1 ton (5,714 cups) rice = 31,939 tons of rice/year

Figure 1–2
The impact of fever on rice production needs

As the population of the planet grows toward a forecasted peak of 8.5 billion in 2040, the farms of the world will need to produce about three times as much food (Avery, 1997). Societies will be faced with three choices of how to feed their people.

1. Clear additional wildlands and put them into agricultural production.
2. Make society become vegan or vegetarian.
3. Increase the productivity of natural resources already used in agriculture.

Although the first option would result in increased food production, it would be detrimental to the quality of life on this planet and is to be avoided. The vegan/vegetarian option would result in improved efficiencies and would mean that more people could be fed using existing agricultural land. These are lifestyle choices, however, and the political reality in most societies makes a vegan or vegetarian requirement unenforceable. The third option, to increase the productivity of natural resources already used in agriculture, is one that has a proven track record. Improved accuracy in predicting nutrient requirements and new products in animal nutrition have allowed us to produce the human food needed on the fewest possible acres. Consider the fact that if U.S. dairy farms had not advanced beyond the technology of 1960, when nutrients were acquired primarily through grazing pasture, the clearing of another 178 million acres for pasture would have been necessary to produce the milk we consumed between 1960 and 1997 (Avery, 1997).

THE NATIONAL RESEARCH COUNCIL

The National Research Council (NRC) is the primary source of nutrient requirement information used by animal nutritionists working in North America. A history and description of the NRC follows.

The National Academy of Sciences was given a mandate by Congress in 1863 that requires it to advise the federal government on scientific and technical matters. The National Academy of Sciences has become a private, nonprofit, self-perpetuating society of research scientists, dedicated to the furtherance of science and technology, and to their use for the general welfare (http://www.nas.edu/nrc).

In 1916, the National Academy of Sciences organized the National Research Council to create a linkage between the research scientists and the academy's purposes of furthering knowledge and advising the federal government.

Within the National Research Council is a Board on Agriculture, and within this Board is a Committee on Animal Nutrition. It is the subcommittees of the Committee on Animal Nutrition that produce the publications referenced in this text.

The National Academy Press (NAP) was created by the National Academies to publish the reports issued by the National Research Council and other councils under the authority of the National Academy of Sciences. The NAP operates under a charter granted by the Congress of the United States. NAP publishes over 200 books a year on a wide range of topics in science, engineering, and health (http://www.nap.edu/about.html).

The nutrient requirement information in the application associated with this text is taken primarily from the information produced by the National Research Council's Board on Agriculture, Committee on Animal Nutrition, and published by the National Academy Press.

The committees and their publications include:

Committee on Animal Nutrition, Subcommittee on Beef Cattle Nutrition. (2000). *Nutrient Requirements of Beef Cattle* (7th rev. ed.).
Committee on Animal Nutrition, Subcommittee on Dairy Cattle Nutrition. (2001). *Nutrient Requirements of Dairy Cattle* (7th rev. ed.).
Committee on Animal Nutrition. (1985). *Nutrient Requirements of Dogs.*
Committee on Animal Nutrition. (1986). *Nutrient Requirements of Cats.*
Committee on Animal Nutrition. (1993). *Nutrient Requirements of Fish.*
Committee on Animal Nutrition, Subcommittee on Goat Nutrition. (1981). *Nutrient Requirements of Goats: Angora, Dairy, and Meat Goats in Temperate and Tropical Countries.*
Committee on Animal Nutrition, Subcommittee on Horse Nutrition. (1989b). *Nutrient Requirements of Horses* (5th rev. ed.).
Committee on Animal Nutrition, Subcommittee on Poultry Nutrition. (1994). *Nutrient Requirements of Poultry* (9th rev. ed.).
Committee on Animal Nutrition, Subcommittee on Sheep Nutrition. (1985). *Nutrient Requirements of Sheep* (6th rev. ed.).
Committee on Animal Nutrition, Subcommittee on Swine Nutrition. (1998). *Nutrient Requirements of Swine* (10th rev. ed.).
Committee on Animal Nutrition, Subcommittee on Rabbit Nutrition. (1977). *Nutrient Requirements of Rabbits.*

When one reads the NRC publications, it becomes obvious that the level of knowledge in animal nutrition is not equal among the various species of livestock. It is necessary, therefore, that some values and conclusions be extrapolated to nonstudy species. For this reason, the student of animal nutrition must be aware of work that is being done outside the area of primary interest.

THE ASSOCIATION OF AMERICAN FEED CONTROL OFFICIALS

The Association of American Feed Control Officials (AAFCO) is an association of state employees involved in regulating the production, labeling, distribution, and sale of animal feeds. Until 1990, the AAFCO had relied on the NRC as its recognized authority on pet nutrition. The NRC publications served as the basis on which pet food manufacturers formulated their products, and regulatory officials could validate nutritional adequacy claims based on the information from the NRC.

In the NRC publications for dogs and cats, nutrient requirement recommendations are based on bioavailable nutrients. Accurate nutrient requirements can only be determined using bioavailable or absorbed nutrients, and the science of nutrition is moving in this direction. In addition to the dog and cat NRC publications, the latest NRC publications from the swine, poultry, dairy, and fish nutrition committees present at least some nutrient requirements in terms of bioavailable or absorbed nutrients.

Formulating diets by using nutrient requirement values based on what is absorbed by the animal necessitates knowing the bioavailabilities of nutrients in ingredients. This knowledge is acquired only through expensive and time-consuming research. The AAFCO judged that the NRC publications for dogs and cats could not be practically applied to the laws regulating the pet food industry, so the AAFCO elected to develop an alternative to the NRC.

In 1990, the AAFCO established a Canine Nutrition Expert Subcommittee and in 1991, the AAFCO established a Feline Nutrition Expert Subcommittee. The charge of these two subcommittees is to develop alternatives to the NRC approach of expressing nutrient requirements on a bioavailable basis. The work of these two subcommittees has resulted in rules that allow companies to choose between two options to substantiate claims of nutrition adequacy in their pet foods. In the first option, the company may formulate a pet food so that a laboratory analysis of the food gives nutrient concentrations that are similar to AAFCO guidelines. In the second option, the company may conduct feeding trials for a prescribed length of time and the animals tested must pass an examination. There are additional performance standards for foods intended for growing, gestating, and lactating animals. The AAFCO is now the nutrition authority most often referenced on dog and cat food labels.

In 2003, the NRC made available a prepublication manuscript to replace the 1985 dog and 1986 cat publications. The dog and cat NRC committee is accepting comments and suggestions in advance of publication (Cook, 2003).

In the dog ration application associated with this text, nutrient requirements are taken primarily from the NRC (1985) publication and AAFCO requirements are referenced. In the cat ration application associated with this text, nutrient requirement information from the AAFCO committee is used.

SUMMARY

Chapter 1 sets the stage for a study of the feed and animal nutrition industries. Getting the most out of the nutrients in the feeds used involves managing livestock and their environments to minimize stress and strain. There are many potential career paths that may lead from a study of feed and animal nutrition, not the least of which is the professional animal nutritionist. The NRC publications are the primary source of information regarding most livestock nutrient requirements. The AAFCO is the most frequently referenced source of nutrient requirement information for dogs and cats.

END-OF-CHAPTER QUESTIONS

1. Nutrient requirements are predicted based on specific performance criteria. Give four performance criteria that may be used in animal nutrition.
2. What is the consequence of the fact that the quantitative analysis procedures used in the laboratory to measure nutrient content are not identical to the animal's digestion and absorption processes?
3. Explain the reasoning behind replacement of the term *requirement* with the term *target* in reference to ration nutrient levels in the companion application to this text.
4. What are the meanings of the terms *stress* and *strain*, and what is their relationship to animal nutrition?
5. In the utilization of nutrients from digested feed, what function is the body's top priority?

6. Describe the role of the National Research Council as it relates to the nutrition of domestic animals.
7. Describe the role played by the Association of American Feed Control Officials in dog and cat nutrition.
8. Discuss the impact on land use if increasing amounts of beef are produced using grass-fed feeding management.
9. Discuss the importance of the concept of nutrient bioavailability.
10. Discuss the importance of each of the following as a factor that may limit animal performance on profitable farms and ranches: animal genetics, animal nutrition, and management of animal environments. Which, in your opinion, is first limiting? Explain why.

REFERENCES

Avery, D. T. (1997, October 21–23). Saving the planet with high feed efficiency. In *Proceedings Cornell Nutrition Conference for Feed Manufacturers* (pp. 77–87). Rochester, NY.

CDC. (2000). Centers for Disease Control and Prevention, National Center for Infectious Diseases, Travelers' Health. Retrieved 12/1/2003 from: www.cdc.gov/travel/malinfo.htm.

Chan, M. S. (1997). The global burden of intestinal nematode infections—fifty years on. *Parasitology Today, 13,* 438–443.

Cook, N. K. (2003, November). Whoa: Pet food institute urges changes to the new NRC report. *Petfood Industry, 45,* 50–51.

Hopkins, D. R. (1992). Homing in on helminthes. *American Journal of Tropical Medicine and Hygiene. 45,* 626–634.

Lobo, P. (2003, October). North America market data reports. *Feed Management, 54,* 7–14.

Morgan, D. (1980). *Merchants of grain.* New York: Penguin Books.

National Academies Press Web site: http://www.nap.edu/about.html.

National Academies. The National Research Council Web site: http://www.nas.edu/nrc/

National Research Council. (1977). *Nutrient requirements of rabbits.* Washington, DC: National Academy Press.

National Research Council. (1981). *Nutrient requirements of goats: Angora, dairy, and meat goats in temperate and tropical countries.* Washington, DC: National Academy Press.

National Research Council. (1985). *Nutrient requirements of dogs.* Washington, DC: National Academy Press.

National Research Council. (1985). *Nutrient requirements of sheep,* (6th rev. ed.). Washington, DC: National Academy Press.

National Research Council. (1986). *Nutrient requirements of cats.* Washington, DC: National Academy Press.

National Research Council. (1989a). *Recommended dietary allowances* (10th rev. ed.). Washington, DC: National Academy Press.

National Research Council. (1989b). *Nutrient requirements of horses* (5th rev. ed.). Washington, DC: National Academy Press.

National Research Council. (1993). *Nutrient requirements of fish.* Washington, DC: National Academy Press.

National Research Council. (1994). *Nutrient requirements of poultry* (9th rev. ed.). Washington, DC: National Academy Press.

National Research Council. (1998). *Nutrient requirements of swine* (10th rev. ed.). Washington, DC: National Academy Press.

National Research Council. (2000). *Nutrient requirements of beef cattle* (7th rev. ed.). Washington, DC: National Academy Press.

National Research Council. (2001). *Nutrient requirements of dairy cattle* (7th rev. ed.). Washington, DC: National Academy Press.

Swanson, K. S. (2004). Nutritional genomics, mapping the future of petfood. *Petfood Industry,* 46, 2.

UNAIDS, United Nations Acquired Immune Deficiency Syndrome. (2003). Retrieved 12/1/2003 from: http://www.unaids.org/en/resources/epidemiology.asp.

Walsh, J. A. 1998. Prevalence of *Entamoeba histolytica* infection. In J. I. Ravdin (Editor), *Amebiasis: Human infection by* Entamoeba histolytica. New York: Wiley & Sons.

WHO, World Health Organization. (1994). TB deaths increasing in eastern Europe. Retrieved 12/1/2002 from: http://www.who.int/archives/inf-pr-1994/pr94-48.html.

CHAPTER 2

DIGESTION AND ABSORPTION

Through painstaking research, chemists and physiologists have been able to gain much information on the various steps in the digestion of food. When the nutrients leave the digestive tract and enter the body, the difficulties of learning what becomes of them are much greater.

F. B. MORRISON, 1949

INTRODUCTION

The digestive systems of livestock are diverse in anatomy but the function of the digestive system of all animals is essentially the same: to extract and absorb the nutrients in ingested feedstuffs. To this end, digestive organs of livestock reduce feed particle size through physical and chemical action. Populations of microbes live within the digestive systems of all animals and play some role in digestion in all livestock, but especially in the ruminant, where a digestive organ has evolved specifically to accommodate the microbial population.

THE DIGESTIVE SYSTEM

The digestive system of an animal connects the animal's diet with the metabolic activities that make life possible. It is essentially a muscular tube that runs from the mouth to the anus. Animals with molar teeth begin the process of digestion by grinding the feed material. Muscles of the tube grind, mix, and move the feed. At various locations along the tube are glands that manufacture products secreted into the tube. In ruminant animals, an expanded area along the digestive tube harbors a population of microbes that assist in digestion through the fermentation of feed materials. The activities of the microbes in the rumen also include the synthesis of essential nutrients for their host. All animals have microbial populations living in their digestive tubes, and most if not all animals can potentially benefit from the activity of these microbes. In the dog, for example, the microbes inhabiting the small intestine make a contribution to overall starch utilization through their fermentation activities (Murray et al., 2001).

DIGESTION

The actions and secretions of the digestive system work to degrade (digest/ferment) feed material in the tube into absorbable compounds and to synthesize essential nutrients from ingested feed. Absorbable products of the digestion include monosaccharides, amino acids, peptides, fatty acids, monoglycerides, glycerol, vitamins, and salts. Only if feed is reduced to these compounds will the animal benefit from being fed.

ABSORPTION

Only if the compounds produced by actions of the digestive system are actually absorbed will the consumed ration be successful in meeting the animal's nutrient requirements. Absorption involves moving these compounds from inside the lumen of the digestive tube across the wall of the digestive tube, and making them available to the body's tissues. Nutrients may be absorbed by active transport, facilitated diffusion, or diffusion. In active transport, energy is required to absorb the nutrient. In facilitated diffusion, energy is not needed, but absorption will not occur without a specific carrier molecule. In diffusion, energy is not needed and the nutrient moves from the area of high concentration (inside the digestive tube) to the area of lower concentration (the cytoplasm of the cells lining the digestive tube).

Most nutrient absorption takes place across the wall of the small intestine. However, some nutrients may be absorbed before reaching the small intestine, as in the ruminant (Figure 2–1), and some nutrients may be absorbed beyond

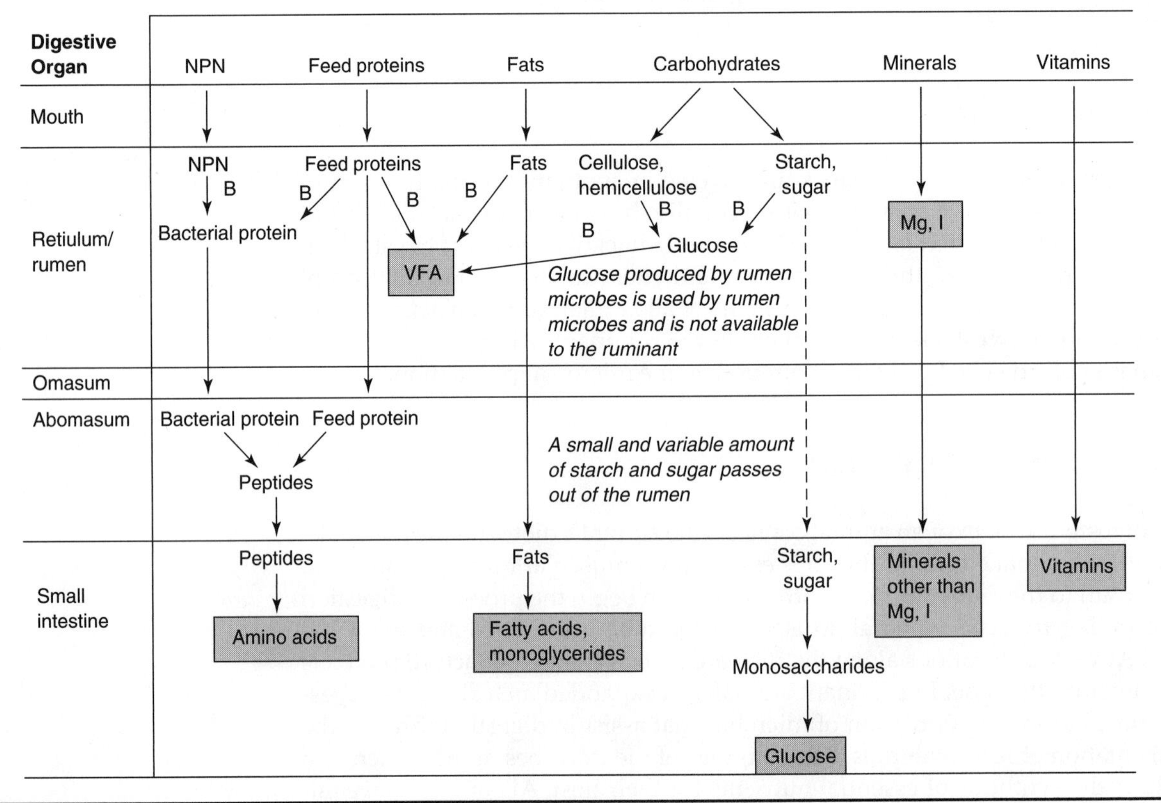

Figure 2–1
Generalized fate of feed components consumed by the ruminant. B indicates bacterial action; each component is boxed and shaded at its primary site of absorption. VFA: Volatile fatty acids.
Adapted from Ishler, V., Heinrichs, J., & Varga, G. (1996). From feed to milk: Understanding rumen function. Extension Circular 422. http://www.das.psu.edu/dcn/catnut/PDF/rumen.pdf

the small intestine. The latter situation is particularly important in some monogastrics (Figure 2–2).

During absorption in the small intestine, nutrients move inside the cells lining the lumen of the tube. The opportunity for absorption is maximized through the arrangement of these cells in finger-like projections called *villi* (singular *villus*). Figure 2–3a is a diagram representing the villi. From inside the cells of the villus, nonfat nutrients then pass into blood capillaries of the villus and are transported via the hepatic portal system to the liver for possible modification before entering the blood circulation. Figure 2–3b illustrates this activity.

Most digested fat, however, is not absorbed in this way. Most fatty acid and monoglyceride molecules move into the epithelial cells of the villus, as do carbohydrates and amino acids. As illustrated in Figure 2–3b, instead of being picked up in the blood circulation serving the villus, these products of fat digestion enter a lacteal within the villus. The lacteal is part of the lymphatic system. Fat absorbed into the lacteal is carried in the lymphatic circulation from the digestive system directly to a major vein, bypassing the liver. The fact that absorbed fat does not pass through the liver is important when considering the nutrition-related diseases of ketosis and fatty liver.

The efficiency with which animals absorb feed nutrients is an important economic issue for livestock feeders. As the passage rate of feed material through the digestive tube increases, the efficiency of absorption declines, and more of the nutrients in the tube will pass out in manure. Given an increased feed passage rate, the various methods of feed processing become increasingly

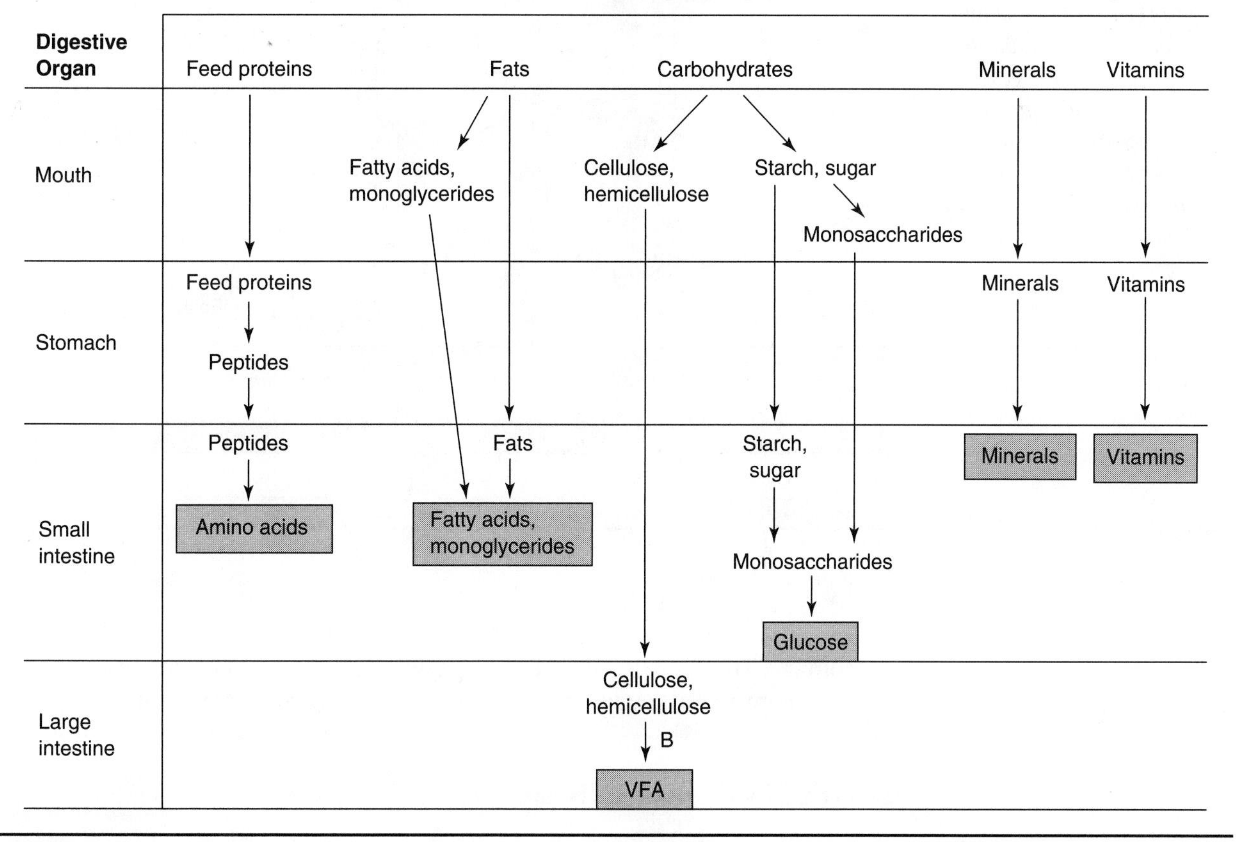

Figure 2–2
Generalized fate of feed components consumed by the monogastric. B *indicates bacterial action; each component is boxed and shaded at its primary site of absorption. VFA: Volatile fatty acids.*

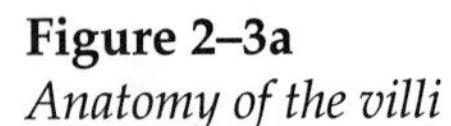

Figure 2–3a
Anatomy of the villi

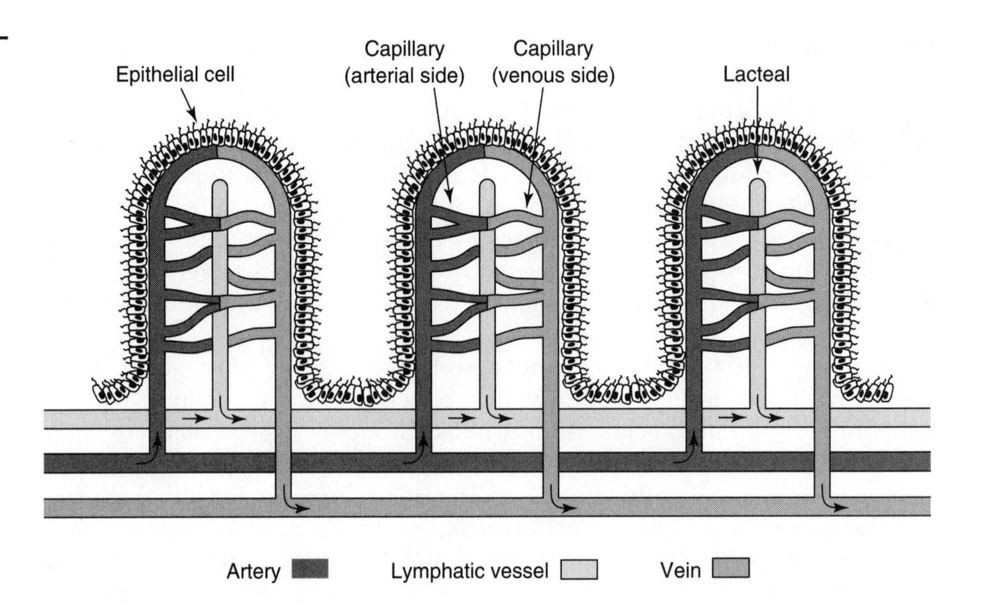

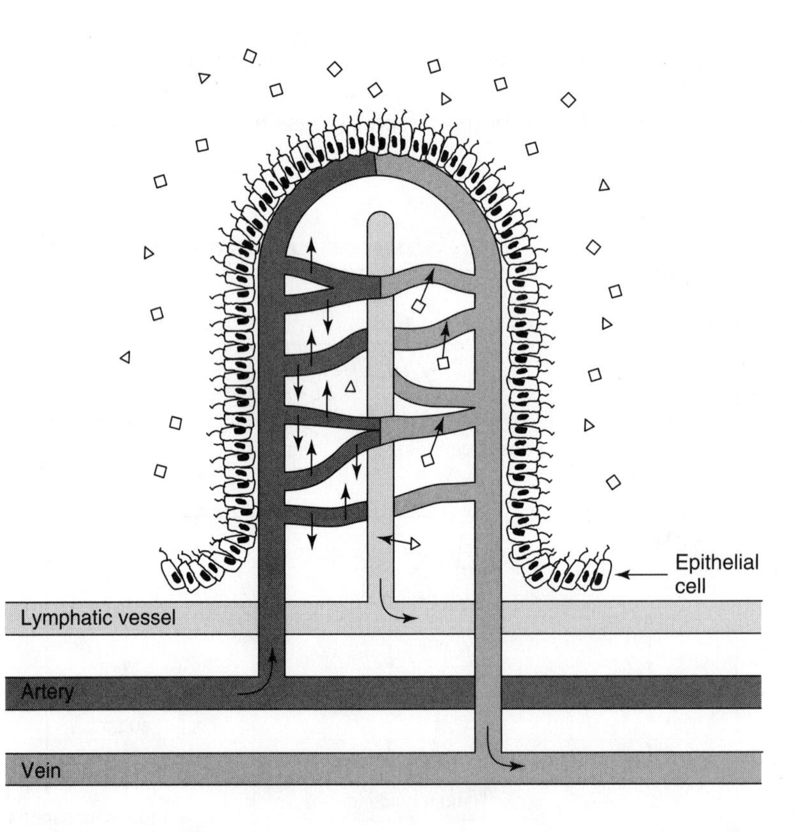

Figure 2–3b
Absorption of fat versus non-fat type nutrients at the intestinal villus (except poultry).

- □ *Nonfat-type nutrients are absorbed into the venous blood serving the villus.*
- △ *Products of fat digestion including most fats, oils, and fat-soluble vitamins are absorbed into the lymphatic vessels serving the villus.*

economical. These processes aid the digestive system in degrading feed into absorbable compounds so less resident time in the digestive tract is needed.

DIGESTIVE ORGANS

The anatomy of the digestive tract determines the type of feed that is nutritionally useful for a particular species. Meat-eating animals (carnivores) have

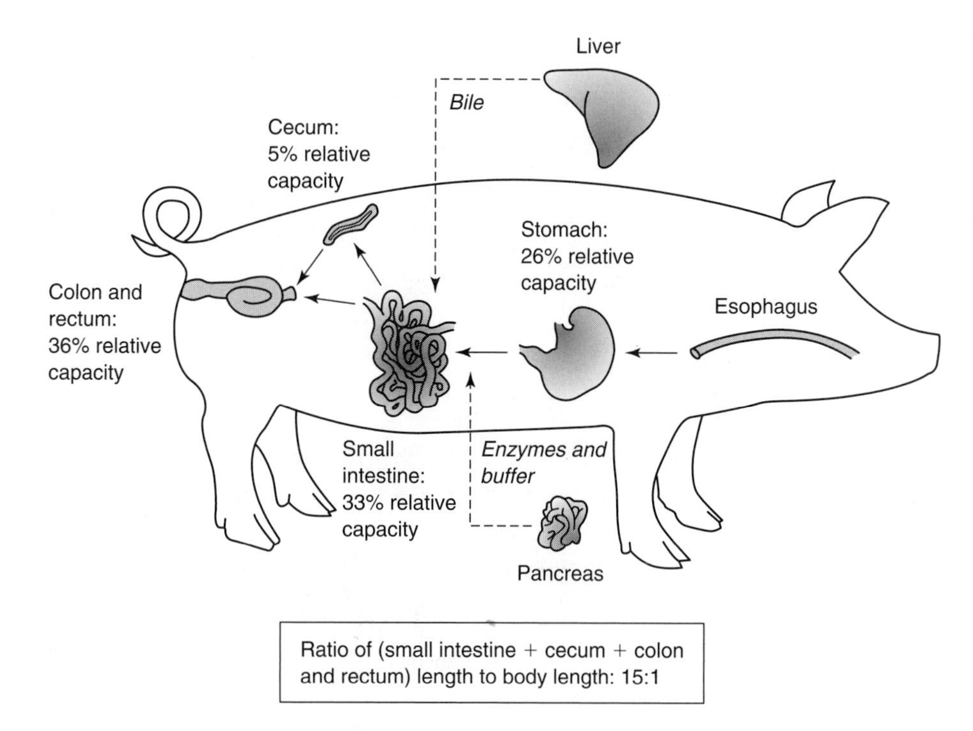

Figure 2–4a
Arrangement and relative capacities of the organs in the digestive system of the pig
Capacities data for swine digestive anatomy from Gillespie, J. R. (1998). *Animal science.* Clifton Park, NY: Thomson Delmar Learning.

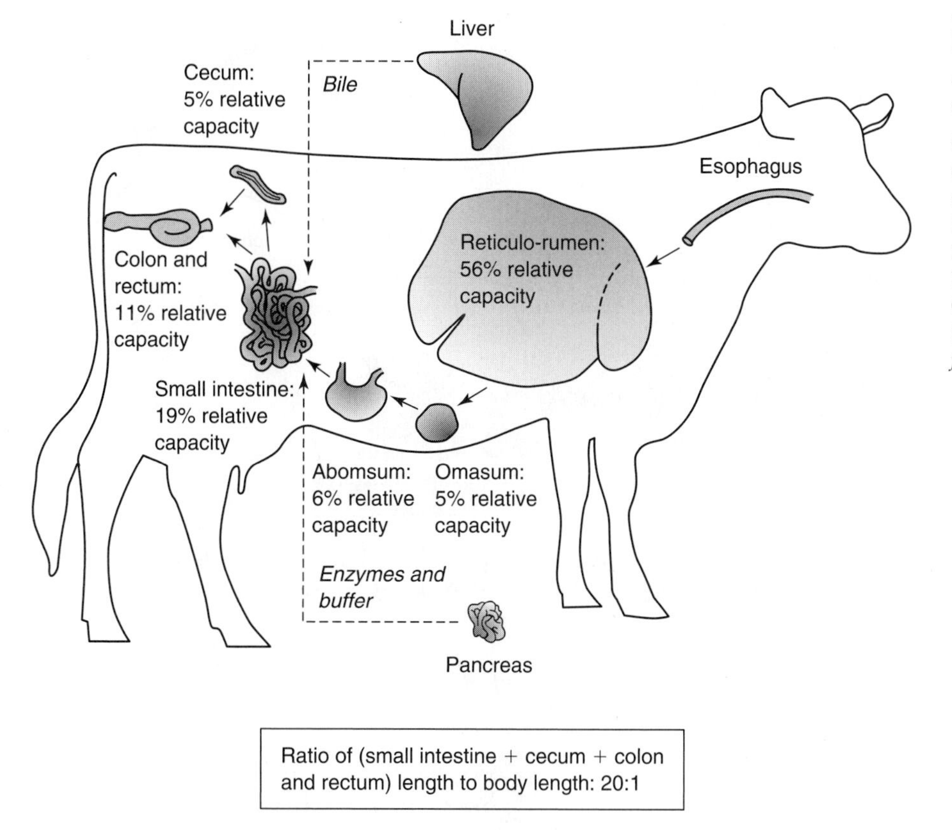

Figure 2–4b
Arrangement and relative capacities of the organs in the digestive system of cattle
Capacities data for ruminant digestive anatomy from Gillespie, J. R. (1998). *Animal science.* Clifton Park, NY: Thomson Delmar Learning. Ratio data from Pond, W. G., Church, D. C., & Pond, K. R. (1995). *Basic animal nutrition and feeding.* NY: John Wiley & Sons.

digestive tracts that are relatively short and low in volume, whereas in plant-eating species (herbivores), the digestive tract volume is relatively large. Diagrams of livestock digestive tracts with capacity data are shown in Figures 2–4a through 2–4i. Images and supplemental information on livestock digestive tracts are found in the file titled Images.ppt on this text's companion CD.

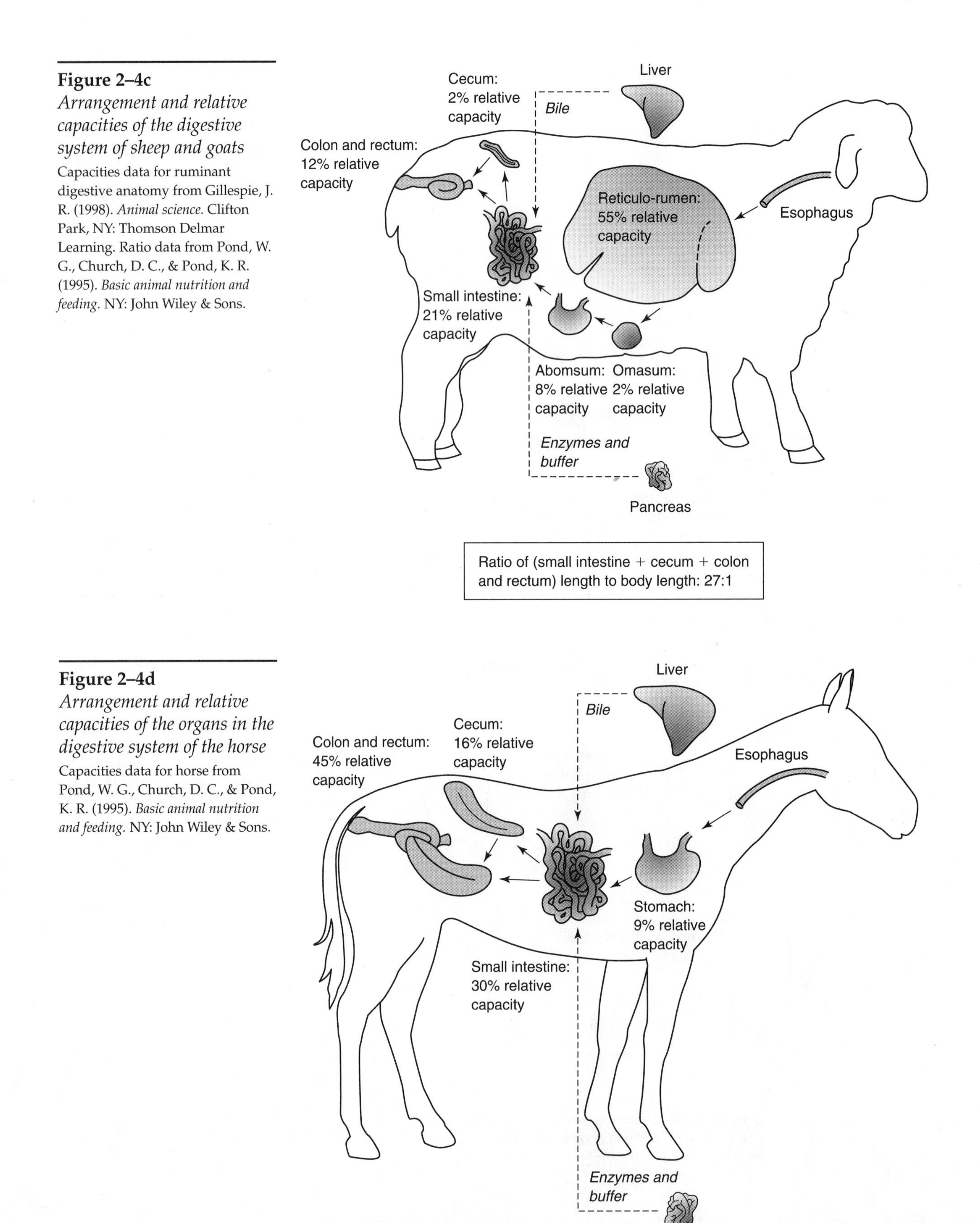

Figure 2–4c
Arrangement and relative capacities of the digestive system of sheep and goats
Capacities data for ruminant digestive anatomy from Gillespie, J. R. (1998). *Animal science.* Clifton Park, NY: Thomson Delmar Learning. Ratio data from Pond, W. G., Church, D. C., & Pond, K. R. (1995). *Basic animal nutrition and feeding.* NY: John Wiley & Sons.

Figure 2–4d
Arrangement and relative capacities of the organs in the digestive system of the horse
Capacities data for horse from Pond, W. G., Church, D. C., & Pond, K. R. (1995). *Basic animal nutrition and feeding.* NY: John Wiley & Sons.

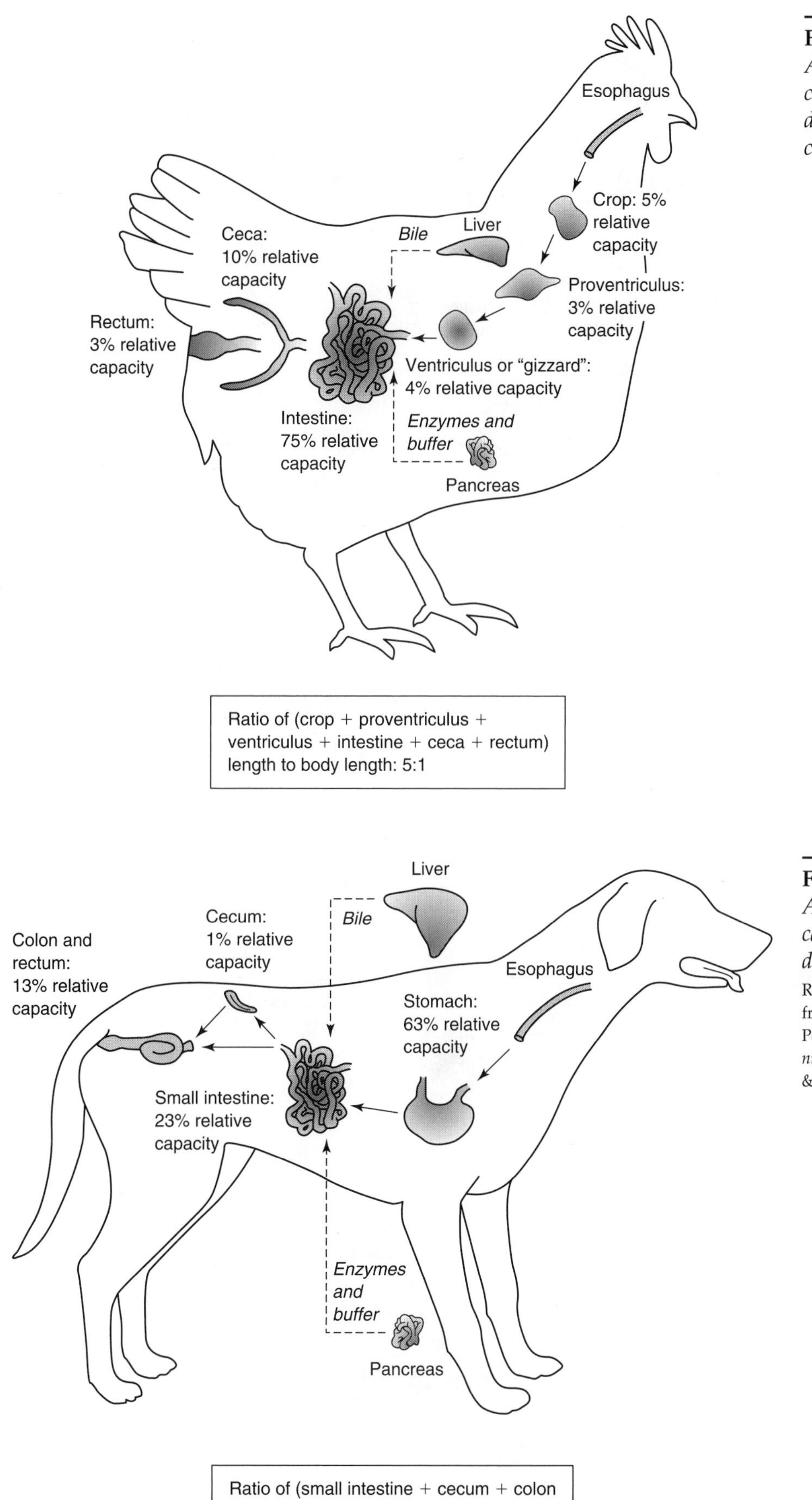

Figure 2–4e
Arrangement and relative capacities of organs in the digestive system of the chicken

Figure 2–4f
Arrangement and relative capacities of organs in the digestive system of the dog
Ration and capacities data for dog from Pond, W. G., Church, D. C., & Pond, K. R. (1995). *Basic animal nutrition and feeding*. NY: John Wiley & Sons.

Figure 2–4g
Arrangement and relative capacities of organs in the digestive system of the cat
Ration and capacities data for cat from Pond, W. G., Church, D. C., & Pond, K. R. (1995). *Basic animal nutrition and feeding*. NY: John Wiley & Sons.

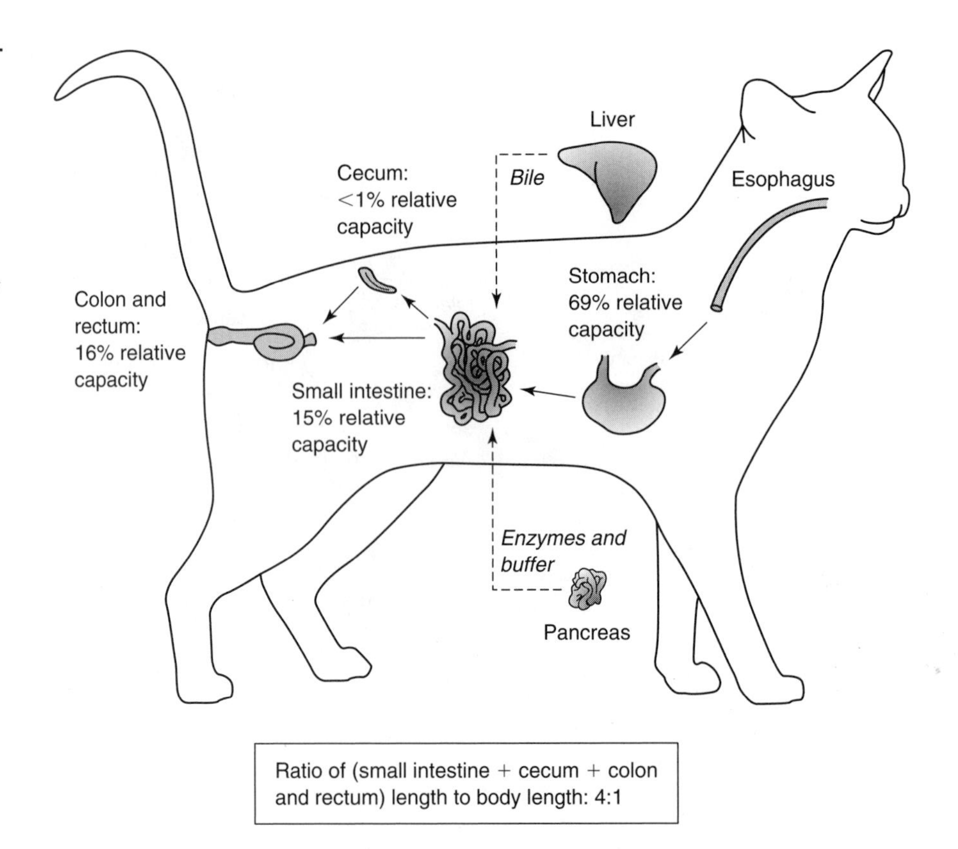

Figure 2–4h
Arrangement and relative capacities of organs in the digestive system of the rabbit

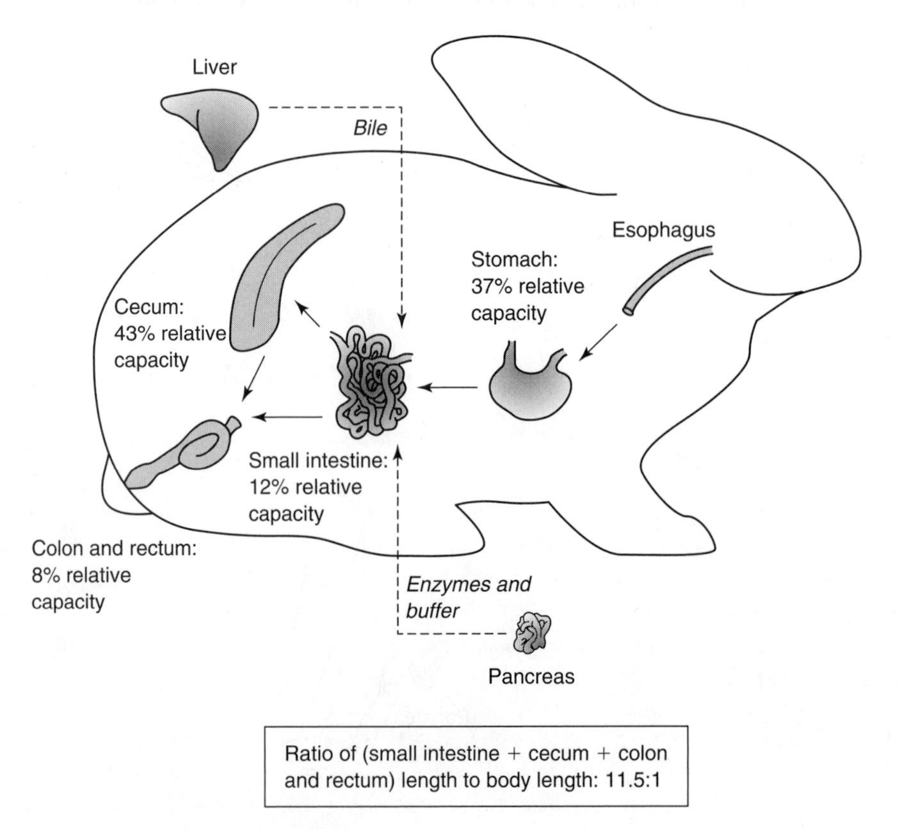

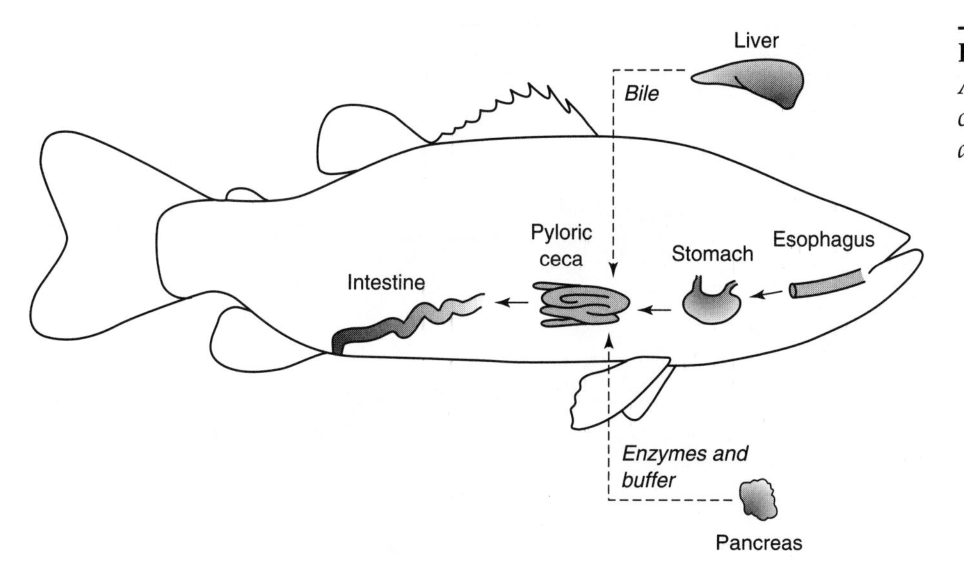

Figure 2–4i
Arrangement and relative capacities of organs in the digestive system of the fish

The Mouth, Saliva Glands, and Esophagus

For those animals that chew, the initial grinding (mastication or chewing) of feed takes place in the mouth. Mastication functions to reduce feed particle size and to mix feed with saliva to form a bolus for swallowing. In ruminant animals, complete mastication of regurgitated ruminal contents (rumination) occurs while ruminants are resting.

Saliva is produced by three pairs of glands in most livestock. These include the parotid, the mandibular, and the sublingual. Saliva plays many important roles in the process of digestion. A summary of saliva functions follows:

- Moistens the feedstuffs for chewing and bolus formation
- Lubricates the bolus for easy swallowing and passage down the esophagus
- Helps provide digestive action. In some monogastric species, the saliva contains enzymes that begin the process of enzymatically digesting the carbohydrate and fat in feed.
- Helps maintain water balance. A mature dairy cow will produce upwards of 50 gallons of saliva each day, about 98.7 percent of which is water (Church, 1988).
- Contains buffering compounds. For ruminant animals, these are essential in maintaining rumen health.
- Contains recycled nitrogen, sodium, potassium, calcium, magnesium, phosphorus, and chlorine (Table 2–1) (McDougall, 1948). For ruminant animals, these serve as a source of nutrients for bacteria and protozoa in the rumen.

There are significant differences in mouth anatomy and physiology among livestock. Chickens do not have teeth. Trout have teeth but lack molar teeth for chewing. Ruminants lack incisor teeth on the upper jaw.

The lips, tongue, teeth, and beak are organs of prehension for animals when eating. They are used to pick up feed to be eaten. The lips and tongue are responsible for animals being able to sort through a mix of feed and eat only selected portions. Among all livestock, the lips of goats are the most prehensile and enable them to select only the most palatable portions of plants when grazing. In the chicken, lips and teeth are replaced by a horny mandible on each jaw, forming the beak.

Table 2–1
Saliva composition of sheep

Constituent	Sheep Saliva (mg/100 ml)
N	20
Na	370–462
K	16–46
Ca	1.6–3.0
Mg	0.6–1.0
P	37–72
Cl	25–43
pH	8.4–8.7

From McDougall, E. I. (1948). Studies on ruminant saliva. *Biochemical Journal. 43,* 99.

As food is chewed and mixed with saliva, a wad of moist feed material called a *bolus* is formed. The tongue is used to move the bolus to the back of the mouth and down the esophagus during swallowing. The muscle in the wall of the esophagus contracts reflexively in response to the bolus of feed material. Peristalsis involves alternate relaxation and contraction of rings of muscle in the wall of the esophagus, coupled with contraction of longitudinal muscles in the area of the bolus.

The Forestomachs of the Ruminant and the Crop of Poultry

The reticulum, rumen, and omasum are called stomachs because of their location, not because of their function; they lack the secretory function that characterizes a true stomach. Their function is similar to that of the crop of a chicken; both are feed storage organs, and while in storage, feed is exposed to bacterial activity.

The Crop

The crop of poultry is a pouch, formed as a specialized area of the esophagus. The primary function of the crop is storage of ingested feed material. While in storage, feed begins to be degraded (fermented) by bacteria that inhabit the crop. In terms of both qualitative and quantitative output, the bacterial activity occurring in the crop is not as productive as that occurring in the rumen, and it is usually ignored as a source of nutrients for the bird.

The Reticulum or Honeycomb

The reticulum is a compartment of the ruminant stomach. There is a reticulated or honeycomb-like pattern on the tissue of the inner wall that distinguishes the reticulum from the other compartments of the ruminant stomach. The reticulum is separated from the rumen by a low pillar of tissue. The reticulum is much smaller than the rumen and aids in rumination and particularly regurgitation. During regurgitation, a bolus of coarse feed material is sent up the esophagus to the mouth for rechewing. As a human food item, the reticulum of cattle is called *tripe*.

The reticulum is located where heavy materials, such as pieces of metal, often become lodged. Ingested bits of wire and nails may penetrate the wall of the reticulum and slowly move through the diaphragm into the pericardial sac causing a condition described as pericarditis or "hardware disease." Feed mills pass feed ingredients over powerful magnets during processing so that stray

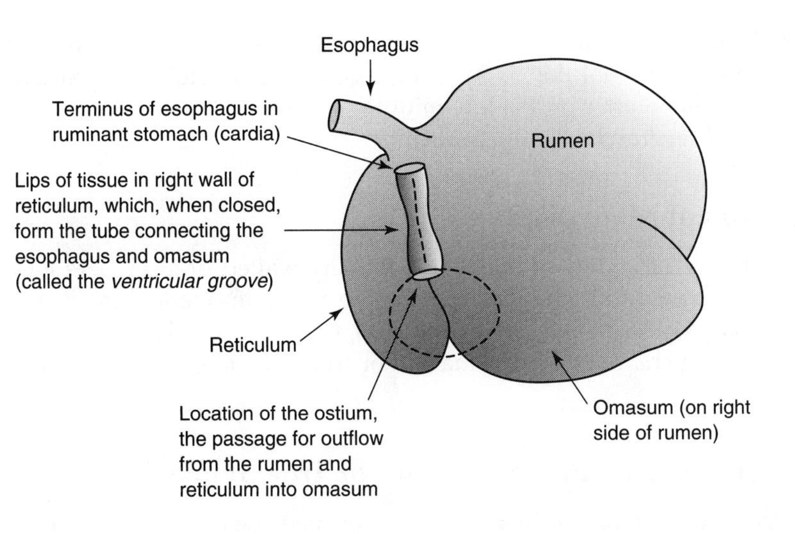

Figure 2–5
Reticulo-rumen structures, mature ruminant, transverse view

bits of metal are removed from mixes. Nonmagnetic metals like aluminum will not be removed by magnets, though these metals are not in widespread use in feed manufacturing.

The exit point for feed material leaving the reticulum and rumen is located low on the wall of the reticulum where there is a connection with the omasum (Figure 2–5). The opening from the reticulum to the omasum is small, and thus larger particles in the rumen-reticulum are retained for regurgitation, chewing, and more complete fermentation.

Rumen or Paunch

Ingested feed arriving at the rumen is mixed to establish good contact with the microorganisms inhabiting the rumen and reticulum. These microorganisms are primarily bacteria, some protozoa, and a relatively small number of fungi. The bacteria colonize particles of feed and produce digestive enzymes. These enzymes contribute to fermentation of feed material, including cellulose, thereby releasing nutrients on which the microbes grow and reproduce. Fermentation also results in the synthesis of nutrients utilized by the ruminant. Bacterial enzymes break down the complex carbohydrates (such as cellulose and starch) by fermenting them into short-chain fatty acids. Acetic, propionic, and butyric acid are the primary short-chain fatty acids produced, and as a group, these are referred to as *volatile fatty acids.* The volatile fatty acids are absorbed directly from the rumen and reticulum and used as energy sources and for milk fat synthesis.

Similarly, protein in feed is broken down into peptides, amino acids, ammonia, and amines. The microorganisms use these substances as building blocks for their own cells. Eventually, the microorganisms are passed down the intestinal tract, digested, and used as a protein source by the ruminant.

Other end products of rumen fermentation are B vitamins and vitamin K. Thanks to the activities of the microbes inhabiting the rumen, ruminant animals do not require these nutrients in their diets.

The gases carbon dioxide (CO_2) and methane (CH_4) are also produced during rumen fermentation. Gas production accounts for 13 to 15 percent of the carbon consumed. The motility of the reticulum and rumen ensures that the gases produced make their way into the thoracic esophagus so that they can be belched or eructated out the mouth. Interference with the process of eructation results in a serious buildup of pressure in the reticulo-rumen and is called *bloat.*

The exit of undigested feed residue from the reticulo-rumen leads to the omasum; this location is termed the *ostium.* Feed particles too large to pass through the ostium are repeatedly returned to the mouth during regurgitation for rechewing to reduce particle size and make them more accessible to the microbes. Feed material that is fed and ingested as small, dense particles may leave the rumen after only a few minutes, whereas larger, less dense particles may spend several days in the rumen.

Omasum or Manyplies

The omasum of ruminant animals is a many-walled organ where part of the water is removed from the mass leaving the rumen and reticulum. By removing water and passing on a drier feed material, the omasum allows the secretions of the following chamber, the abomasum or true stomach, to be more effective in digestion.

The Abomasum, Proventriculus, or True Stomach

The true stomach of ruminant animals is called the *abomasum*. In poultry, the true stomach is called the *proventriculus*. Channel catfish and rainbow trout have a true stomach, but this is not the case for all commercially important fish species.

The walls of the true stomach contain glands made up of cells that manufacture and secrete substances that aid in the digestion of the different feed components. The walls of the true stomach contain parietal cells that secrete hydrochloric acid. The chief cells of the stomach wall secrete at least two enzymes that function in digestion. In young mammals, the chief cells secrete gastric lipase, which functions in fat digestion. The production of gastric lipase is less important in the adult. The chief cells also secrete enzymes to aid protein digestion. Pepsinogen is secreted by the chief cells of the true stomach. Pepsinogen is the *zymogen* (precursor) of the protein-digesting enzyme, pepsin. Pepsin is not secreted directly (to avoid the enzyme from digesting the protein tissues of the true stomach itself). The conversion of pepsinogen to pepsin is initiated by the presence of acid and active pepsin in the stomach contents. In young ruminants, stomach glands are responsible for the secretion of another protein-digesting enzyme, rennin (converted from the zymogen, prorennin). The semifluid mixture of digestive secretions and partially digested feed material in the true stomach is called *chyme*.

The Ventricular Groove

In ruminant animals, the wall of the reticulum participates in a shunt that permits liquids to bypass the rumen and reticulum. This shunt is described as the *ventricular groove* (formerly esophageal groove). It appears to be most important in preweaned ruminants, allowing milk to avoid the immature reticulum and rumen and move directly to the omasum and on to the abomasum (Figure 2–5). Glands in the wall of the abomasum are responsible for the production of enzymes (rennin and pepsin) and acid (hydrochloric acid) that cause the coagulation of milk proteins. This coagulation results in the formation of a curd in the abomasum. The curd contains protein and embedded fat. The curd is retained in the abomasum and digested over a 12- to 18-hour period (Merchen, 1988). Most ruminant nutritionists believe that curd formation in the abomasum is necessary for optimal digestion of liquid feed consumed by the preweaned calf.

The Gizzard

The poultry stomach is actually made up of two parts: the glandular proventriculus and the muscular ventriculus or gizzard. The gizzard connects the proventriculus and the small intestine. Its function is to grind coarse feed. This process may be aided by the presence of grit consumed in the diet. Grit is unnecessary if poultry are fed uniformly ground diets.

The Pancreas

The beginning of the small intestine is folded into a loop called the *duodenum*. Inside this loop lies the pancreas. The pancreas secretes digestive juices into the duodenum. These juices contain buffers to neutralize the hydrochloric acid added to the chyme in the true stomach. Pancreatic secretions also contain enzymes that degrade proteins (proteases), starches (amylases), and fats (lipases). Note that the pancreas does not produce enzymes capable of degrading and digesting cellulose.

The Liver

The liver produces bile, which contains salts that aid in the preparation of fats for absorption in the small intestine. In most animals, the bile made in the liver is stored in the gall bladder. The presence of feed in the duodenum causes the gall bladder to discharge its bile into the duodenum. Because horses lack a gall bladder, the bile is discharged directly from the liver into the duodenum.

The Small Intestine and Midgut

The pylorus is a sphincter muscle that prevents premature movement of the feed out of the stomach and into the small intestine. In fish, the structure downstream from the stomach is termed the *midgut*. In the midgut, trout have numerous out-pocketings called *pyloric ceca*. The role of pyloric ceca in fish digestion is unclear, though histologically, they are similar to the intestine. Pyloric ceca are absent in channel catfish.

The small intestine is normally considered to have two distinct parts: the uppermost part, called the *duodenum*, and the lower small intestine, which includes the jejunum and ileum. In swine nutrition, amino acid bioavailability is expressed as true ileal digestibility. Determination of true ileal digestibility for a given amino acid in a given feedstuff involves measuring the proportion of that feedstuff's amino acid that has disappeared from the gut when digesta reach the terminal ileum. This proportion is the amino acid's apparent ileal digestibility for that feedstuff. Two corrections are then applied to get true ileal digestibility. The first correction involves accounting for the amino acids that are absorbed in a form that cannot be fully utilized by the pig. The second correction involves accounting for endogenous amino acid losses. The companion application to this text uses true ileal digestibility in expressing swine amino acid requirements and in expressing feedstuff amino acid content.

Peristalsis moves food material through the small intestine and, in fish, the pyloric ceca. Enzymes secreted into the duodenum by the pancreas and by glands in the intestinal wall, along with the fish's pyloric ceca, continue the digestive process by breaking down the fragments of proteins and carbohydrates produced earlier into absorbable amino acids and monosaccharides. Bile from the liver also enters the duodenum via the bile duct to assist in fat digestion and absorption. The population of bacteria residing in the small intestine may contribute to overall carbohydrate digestion. In the dog, fermentation by these bacteria may contribute to overall starch digestion.

The tissue of the small intestine facing the lumen is lined with small, fingerlike projections called *villi*. These villi greatly increase the absorptive area of the intestine. This tremendous surface area makes possible rapid absorption of nutrients. Most of the nutrients needed by an animal must be absorbed at the small intestine.

Absorption

Absorption of most nutrients takes place across the wall of the small intestine (Figures 2–1 and 2–2). In general, there are two types of chemical actions that are applied to each of the three categories of organic compounds to make absorption possible. The first chemical action involves breaking the large organic molecule into smaller, manageable fragments. The second involves clipping off small, absorbable pieces from each fragment. For example, in the case of starch, pancreatic amylase splits starch into smaller fragments, and alpha-dextrinase from glands of the small intestine clips off individual monosaccharide molecules from these fragments. In the case of proteins, the pancreatic enzymes of trypsin and chymotrypsin continue the work of pepsin (at work in the true stomach) to break proteins down into smaller fragments of peptides. Carboxypeptidase cleaves off individual amino acids from the peptides. In the case of fats, bile made in the liver (and in most animals, secreted from the gall bladder) is used to emulsify the fat in chyme. Bile is not an enzyme, but emulsification is prerequisite for effective enzyme action on fat. Pancreatic lipase acts to cleave two of the three fatty acids on a triglyceride, leaving absorbable fatty acids and monoglyceride. For additional information regarding nutrient absorption, see Chapters 6, 7, and 8.

The Large Intestine

The large intestine is generally considered to include the cecum, the colon, and the rectum. In fish, there is little to distinguish the different portions of the intestine, and the term *hindgut* corresponds to the large intestine. As with the esophagus and small intestine, peristalsis moves food material through the large intestine. There is more variation in appearance of the large intestine from one species to another than there is of the small intestine.

The Cecum

Ceca is plural for *cecum.* While poultry have two ceca, most other livestock have just one cecum. The cecum/ceca are pouches that lie at the juncture of the lower small intestine and the remaining parts of the large intestine. Whereas the chicken's ceca are approximately 5 inches in length, the horse's cecum is 4 feet long. Because these structures are blind sacks, there is little flow through them and microbial populations are able to become established in the ceca of animals. These microbes ferment the feed material to produce high-energy volatile fatty acids and B vitamins, both potentially valuable nutrients for their host. In the horse, rabbit, and other nonruminant herbivores, the cecum plays a significant role in the digestion of fiber in feed. Although some absorption of the products of fermentation undoubtedly occurs across the wall of the large intestine, the primary site of nutrient absorption is the small intestine (Figure 2–1 and Figure 2–2). This means that much of the products of fermentation will pass in the manure. Through the practice of coprophagy, the chicken—and especially the rabbit—may ingest a portion of their fecal material and thereby pass the microbial products manufactured in the cecum through the small intestine. Without coprophagy, it is unclear how much the chicken and nonruminant herbivores benefit from the microbial activity in the cecum.

The Colon and Rectum

Although the small intestine is the primary site of nutrient absorption, some nutrient absorption undoubtedly occurs across the wall of the large intestine.

The colon of the horse starts out as a large-diameter structure (the large colon), but narrows toward the rectum (the small colon). Colic in horses refers to abdominal pain, and it has various causes. One cause is a blockage at this location where the colon diameter narrows.

The large intestine does not exist as such in chickens and fish. Chickens do not have a colon and in fish, there is no large intestine distinct from the small intestine.

The rectum is a relatively straight tube and is readily dilated for storage of feces.

The Anus, Cloaca, and Vent

The anus is the junction of the terminal part of the digestive tract and the skin. The cloaca of poultry is a chamber common to the digestive, urinary, and reproductive passages. The cloaca opens externally at the vent.

SUMMARY

Chapter 2 describes the mechanisms and organs of digestion and absorption. The comparative digestive anatomy of the livestock species is described. Whereas herbivores possess long and complex digestive tracts, carnivores possess relatively short and simple digestive tracts. The ruminant animal has a specialized compartment in its digestive system to accommodate a fermenting population of microbes that assist in the extraction of nutrients from ingested feeds. It is apparent that livestock have evolved to use different strategies to achieve the same goal: extraction and absorption of the nutrients in ingested feedstuffs.

END-OF-CHAPTER QUESTIONS

1. Compare the ratio of digestive-tract length to body length in herbivorous animals to that of carnivorous animals. Compare this ratio in ruminants and nonruminant herbivores. Explain your findings.
2. Describe the difference between digestion and absorption.
3. What nutrients are produced by the microbial inhabitants of the digestive tract in ruminant animals?
4. What are the activities and functions of the three digestive chambers upstream from the abomasum in ruminant animals? Describe the function of the ventricular groove in ruminants.
5. What are the activities and functions of the true stomach? What is chyme?
6. What are the activities and functions of the crop? What are the activities and functions of the gizzard?
7. What are the activities and functions of the small intestine? What are the activities and functions of the large intestine?
8. Give three products of the pancreas and describe their functions.
9. Where is bile made? What role does bile play in digestion?
10. Use the following terms in a description of absorption of most nonfat nutrients: *villus, blood capillary, hepatic portal system, liver,* and *blood circulation.* Use the following terms in a description of fat absorption: *fatty acid, villus, lacteal, lymphatic system,* and *blood circulation.*

REFERENCES

Calhoun, M. L. (1954). *Microscopic anatomy of the digestive system of the chicken.* Ames, IA: Iowa State University Press. Retrieved 12/1/2002 from http://www.extension.iastate.edu/publications/PM1696.pdf

Church, D. C. (Editor). (1988). *The ruminant animal, digestive physiology and nutrition, D. C. Church, p. 120.* Englewood Cliffs, NJ: Prentice-Hall.

Gillespie, J. R. (1998). *Animal science.* Clifton Park, NY: Thomson Delmar Learning.

Ishler, V., Heinrichs, J., & Varga, G. (1996). From feed to milk: Understanding rumen function. Extension Circular 422. http://www.das.psu.edu/dcn/catnut/PDF/rumen.pdf

McDougall, E. I. (1948). Studies on ruminant saliva. *Biochemical Journal. 43,* 99–109.

Merchen, N. R. (1988). Digestion, absorption and excretion in ruminants. In *The ruminant animal, digestive physiology and nutrition,* D. C. Church, editor. P. 174. Englewood Cliffs, NJ: Prentice-Hall.
Morrison, F. B. (1949). *Feeds and Feeding, 21st edition.* Ithaca, NY: The Morrison Publishing Company.
Murray, S. M., Flickinger, E. A., Patil, A. R., Merchen, N. R., Brent, J. L., Jr., & Fahey, G. C., Jr. (2001). In vitro fermentation characteristics of native and processed cereal grains and potato starch using ileal chime from dogs. *Journal of Animal Science 79,* 435–444.
Pond, W. G., Church, D. C., & Pond, K. R. (1995). *Basic animal nutrition and feeding.* NY: John Wiley & Sons.

CHAPTER 3

FEEDSTUFFS

Our business fills a great human and economic need.

LEOPOLD LOUIS-DREYFUS
GRAIN MERCHANT, LATE 1800S

INTRODUCTION

Many types of feed ingredients or feedstuffs are available to supply the nutritional needs of livestock. These feedstuffs are the raw materials that are converted into animal cells, tissues, organs, and products. A familiarity with the chemical and nutritional composition of the various classes of feedstuffs is essential in order to formulate the most economical and profitable rations. It is also important to be familiar with the various feedstuff types to plan for planting, harvesting, and storage of homegrown feedstuffs. Proper preservation of stored feedstuffs is a critical profitability factor for some types of farms and ranches.

CLASSIFICATION OF FEEDSTUFFS

A feedstuff is loosely defined as any component of a ration that serves some useful function. Feedstuffs generally are included in the ration to help meet the requirement for one or more nutrients. However, they may also be included in the ration to provide bulk, reduce oxidation, emulsify fats, provide flavor, improve animal health, or improve characteristics of the products produced by livestock.

Because there are thousands of feedstuffs used in the formulation of livestock rations, a discussion of feedstuffs must be made on the basis of groups or types of feedstuffs with common characteristics. The National Research Council (NRC) has established such groups in developing its feedstuff numbering system (National Research Council, 1972).

In the NRC system a 6-digit number, the International Feed Number (IFN), is used to identify feedstuffs. The first digit in the IFN identifies the feed type as one of the following:

1. Dry forages
2. Pasture, range plants, and feeds cut and fed green
3. Silages
4. Energy concentrates
5. Protein supplements
6. Minerals
7. Vitamins
8. Additives

The remaining five digits in the IFN are unique to the feedstuff and may be used to identify it in databases used in research (Table 3–1).

It is important to realize that the classification of feedstuffs into categories is imprecise because most feedstuffs are complex packages of multiple nutrient sources. For example, the same feedstuff may legitimately be considered either an energy feed or a protein supplement, or even a forage. For this discussion, we will use the categories identified by the first digit in the IFN and include a few subcategories.

1. Dry forages
2. Pasture, range plants, and feeds cut and fed fresh
3. Silages
   - *Hay crop silage*
   - *Small grain silage*
   - *Corn and sorghum silage*

Table 3–1
Guide to feedstuff nomenclature

Classification	1	2	3	4	5	6	7	8
Species, variety or kind	Alfalfa	Orchard-grass-Ryegrass	Corn	Oats	Soybean	Magnesium oxide	Yeast, brewers *Saccharomyces*	Lignin sulfonate
Part eaten	Hay	Aerial part	Aerial part	Grain	Seeds without hulls	—	—	—
Process(es) and treatment(s)	Sun cured	Fresh	Ensiled	Rolled	Solvent-extracted, ground	—	Dehydrated, ground	Dehydrated
Stage of maturity	Early bloom	Immature	Well eared	—	—	—	—	—
Cutting or crop	Cut 1	—	—	—	—	—	—	—
Grade, quality or guarantees	—	—	—	—	Maximum 3% crude fiber	—	—	—
IFN	1-00-108	2-03-472	3-02-823	4-03-307	5-04-612	6-02-756	7-05-527	8-02-627

From National Research Council. (1972). *Atlas of nutritional data on United States and Canadian feeds.* Washington, DC: National Academy Press.

4. Energy concentrates
   - *Cereal grains*
   - *Residues from the sugar and citrus industries*
   - *Fats and oils*
5. Protein supplements
   - *Plant protein sources*
   - *Animal protein sources*
   - *Nonprotein nitrogen sources*
6. Mineral supplements
7. Vitamin supplements
8. Additives

THE FEEDSTUFFS

Forages

In the NRC system, categories 1 through 3 include forages. These forage categories differ as to whether the crop is preserved or fed fresh, and if preserved, how. The term *roughage* is synonymous with the term *forage*, but because the term *roughage* has a somewhat negative connotation ("rough"), it has fallen out of favor with many nutritionists. Most feedstuffs classified as forage are bulky, high-fiber feedstuffs that have a low weight and low nutrient content per unit of volume. Though some forage is essential for the health of herbivorous animals, productivity on an all forage diet is usually too low to be profitable.

The protein, mineral, and vitamin content of forages varies greatly. Legumes may contain 25 percent protein. Other forages such as straws may have only 5 percent crude protein. Mineral content is also highly variable. Most forages, particularly legumes, are relatively good sources of calcium. Phosphorus content is moderate to low and potassium content is high relative to the requirement for most animals. Magnesium content of forages is usually good but under certain circumstances, animals on high-forage diets can experience magnesium deficiency. The trace minerals vary greatly depending on plant species, soil, and fertilization practices.

A number of factors may affect forage composition and nutritive value.

1. Maturity has one of the most pronounced effects. As a forage plant matures, the protein and soluble carbohydrate content decline whereas the fiber and lignin content increase. Lignin is not only indigestible itself, but also has an encrusting effect that reduces digestibility of otherwise digestible plant cell components. The presence of lignin greatly limits plant digestibility.
2. Soil fertility, fertilization, and weather are known to have a pronounced effect on quantity of forage produced. Quality of the forage may also be affected by these factors.
3. Harvesting and storage methods can have a significant effect on nutrient value of forage. For example, harvesting techniques that result in significant leaf loss will reduce the nutrient content of the feedstuff.

Grasses

The grasses include all of the wild and cultivated species used for grazing, as well as the cultivated cereal grain species. However, this section will be concerned only with grass as a forage. The discussion will be general because of the tremendous variety of grasses.

In comparison with legumes of similar maturity, the protein and calcium content of grasses is nearly always lower. The variation in other nutrients is such that none is consistently higher in either type of forage (Table 3–2).

Table 3–2
Average legume and grass nutrient analyses values

Feedstuff	Number of CP observations	CP	ADF	NDF	NEI	DE	Ca	P	Mg	K
All legume	4697	19.4	32.5	41.2	0.64	0.99	1.46	0.25	0.29	2.58
Mixed, mostly legume	3649	16.4	36.7	51.3	0.56	0.91	1.14	0.25	0.26	2.26
Mixed, mostly grass	3819	12.1	39.3	60.5	0.53	0.87	0.75	0.23	0.23	1.93
All grass	3343	10.6	38.7	64.8	0.50	0.83	0.55	0.22	0.21	1.84

From Northeast DHIA Forage Laboratory. (1995). Tables of feed composition. Ithaca, NY.

CP: Crude protein; ADF: acid detergent fiber; NDF: neutral detergent fiber; NEI: net energy for lactation; DE: digestible energy; Ca: calcium; P: phosphorous; Mg: magnesium; K: potassium.

Grasses may be classified as cool season (temperate), warm season, or tropical. Cool-season grasses grow rapidly during the cool, moist seasons of the year, and become dormant during the hot, dry seasons. Warm-season grasses grow during the hot seasons and become dormant during the cool seasons. Cool-season grasses generally mature at slower rates and deteriorate in quality less rapidly than do warm-season grasses. Tropical grasses are adapted only to tropical climates where freezing temperatures do not occur.

Some of the more commonly used cool-season grasses are ryegrass, orchardgrass, reed canarygrass, tall fescue, timothy, and smooth brome grass. Some of the more commonly used warm-season grasses are Bermudagrass, Johnsongrass, Bahiagrass, Dallisgrass, Switch grass and the Bluestem grasses.

The grasses that are cultivated for their cereal grains are also used as forage sources in pasture. Examples include barley, oats, winter wheat, and rye. These species can be used for pasture during the winter and early spring with minimal effect on grain yield.

Legumes

Alfalfa is the most common legume used for pasture, hay-crop silage, and hay. It is known as lucern in most English-speaking areas other than North America. The clovers are also legumes that are extensively used in animal diets. Common clovers include ladino, red, white, alsike, sweetclover, and subterranean clover. Lespedeza, crown vetch, kudzu, and birdsfoot trefoil are other legumes that may be fed to animals.

Through an association with bacteria in the root nodules, legumes are able to "fix" nitrogen from the atmosphere. This means that legumes can make the protein in their tissues from nitrogen in the air rather than from nitrogen in the soil. In legumes, atmospheric nitrogen is reduced to ammonia, which is used to manufacture amino acids. The amino acids are used in plant protein synthesis.

Some legumes, particularly alfalfa and white, ladino, and red clover, are prone to cause bloat in cattle. Bloat is caused by foam-producing compounds such as cytoplasmic proteins and pectins. Foam at the base of the esophagus inhibits the eructation mechanism leading to the accumulation in the rumen of normal rumen gases. As the rumen swells and pressure increases, expansion of the thoracic cavity during inhalation becomes difficult. Without relief, suffocation and/or heart failure will occur. Additional information on bloat is found in Chapter 16.

Native Pastures and Range Plants

Uncultivated native pastures and rangeland account for many millions of acres of land where the topography, soil, or environment is unsuited to intensive agricultural methods. These areas contain a wide range of grasses, sedges, forbs, and browse.

Table 3–3
Plants that may be lethal when ingested by livestock

Common name	Botanical name	Agent
Arrow grass	*Triglochin maritime*	Cyanide (HCN)
Blue flax	*Linum* spp.	Cyanide
Chokecherry	*Prunus virginiana*	Cyanide
Elderberry	*Sambuccus* spp.	Cyanide
Johnson grass (Sudan)	*Sorghum halepense*	Cyanide
Poison suckleya	*Suckleya suckleyana*	Cyanide
Serviceberry	*Amelanchier alnifolia*	Cyanide
Johnson grass	*Sorghum halepense*	Nitrate
Kochia weed	*Kochia scoparia*	Nitrate
Lamb's quarter	*Chenopodium* spp.	Nitrate
Nightshades	*Solanum* spp.	Nitrate
Pigweed	*Amaranthus* spp.	Nitrate
Russian thistle	*Salsola rali*	Nitrate
Sunflower	*Helianthus* spp.	Nitrate
Death camas	*Zigadenus* spp.	Alkaloid
Water hemlock	*Cicuta* spp.	Alkaloid
Beet tops	*Beta vulgaris*	Oxalates
Curly leafed dock	*Rumex crispus*	Oxalates
Greasewood	*Sarcobatus vermiculatus*	Oxalates
Halogeton	*Halogeton glomeratus*	Oxalates
Milkweeds	*Halogeton glomeratus*	Oxalates
Osalis, "shamrock"	*Oxalis* spp.	Oxalates
Pigweed	*Amarantus* spp.	Oxalates
Rhubarb	*Rheum rhaponticum*	Oxalates
Yew	*Taxus* spp.	Cyanide

From American Sheep Industry Association, Inc. (1996). *Sheep production handbook* (pp. 384–385). Denver, CO: C&M Press.

Miscellaneous Forage Plants

The forages used in livestock diets are chosen because they are available, they have good nutrient content, and they yield well as a crop. Nontraditional crops may become useful as forages, depending on the farm and the year.

The tops of root crops such as beets and turnips have been used successfully. Plants in the cabbage family, including kale cabbage and rape, have also been used as forages in livestock diets. Rape is sometimes planted for use by sheep as fall pasture.

Toxins in plants

Under certain conditions, some plants may accumulate toxins in their tissues. Generally, livestock will avoid such plants. However, especially during times of feed shortage, livestock may ingest toxic plants. Tables 3–3 through 3–5 list some of these plants and their toxins. The toxins that may be ingested by grazing livestock and their effects on these animals has been recently reviewed (Cheeke, 1995).

Table 3–4
Plants causing photosensitization in livestock

Common name	Botanical name	Agent
Bishop's weed	*Ammi majus*	Primary
Buckwheat	*Tagopyrum sagittatum*	Primary
Dutchman's breeches	*Thamnosma texana*	Primary
Rain lily	*Cooperia peduniculda*	Primary
Spring parsley	*Cymopterus watsoni*	Primary
St. John's wort	*Hypericum perforatum*	Primary
Agave	*Agave lechuguilla*	Secondary
Horsebrush	*Tetradymia* spp.	Secondary
Klein grass	*Panicurn coloradatum*	Secondary
Kochia	*Kochia* spp.	Secondary
Lantana	*Lantana* spp.	Secondary
Sacahuiste	*Nolina texana*	Secondary

From American Sheep Industry Association, Inc. (1996). *Sheep production handbook* (pp. 384–385). Denver, CO: C&M Press.

Table 3–5
Plants affecting the nervous system of livestock

Common name	Botanical name	Agent
Bitterweed	*Hymenoxys* spp.	Semiarid regions in U.S.
Black henbane	*Hyoscyamus niger*	Northern Rocky Mts.
Black nightshade	*Solanum migrum*	Eastern U.S.
Deadly nightshade	*Atropa belladonna*	Cultivated in gardens
Fitweed	*Corydalis caseara*	Intermountain U.S.
Horse nettle	*Solanum carolinense*	Texas/Atlantic coast
Jimson weed	*Datura stramonium*	Florida to Texas
Locoweed	*Oxytropis* spp.	Western U.S.
Locoweed	*Astragalus* spp	Western U.S.
Lupine, bluebonnet	*Lupinus* spp	North America
Paper flower	*Psilostrope* spp	Southwestern U.S.
Rayless goldenrod	*Isocoma wrightii*	Western U.S.
Silverleaf nightshade	*Solanum eleagnifolium*	Southwestern U.S.
Snakeroot	*Eupatorium rugosum*	Eastern U.S.
Twin leaf senna	*Cassia occidentalis*	U.S.
Wheat	*Triticum aestivum*	U.S.

From American Sheep Industry Association, Inc. (1996). *Sheep production handbook* (pp. 384–385). Denver, CO: C&M Press.

Dry Forages

Hay Haymaking has been practiced for many centuries as an effective method of conserving forage crops. With the development of newer methods of forage preservation, the importance of haymaking has declined in recent years. However, the form of hay is unique among feedstuffs. Its length and bulkiness are useful in maintaining the digestive health of herbivores, especially in high-grain diets.

The goal of making hay is to preserve the forage by making it dry enough through the curing process so that molds cannot grow and the enzymes of spoilage bacteria cannot function.

Both the quality and quantity of field-cured hay depends on plant maturity when cut, method of handling, moisture content, and weather conditions during harvest.

Hay should be harvested at the stage of maturity that will provide a maximum yield of nutrients per unit of land without causing damage to the next crop. The maturity stages for legumes are vegetative, bud, bloom, and mature or seed stage. The maturity stages for grasses are vegetative, boot, head (containing the blooms), and mature or seed stage. For practical nutrition purposes, the most commonly recommended stage of maturity to cut legumes and grasses is when blooms first begin to appear. For both legumes and grasses, cutting later than the recommended time results in more yield but poorer quality. Cutting earlier results in less yield but higher quality, and runs the risk of damaging the plants, particularly legumes.

During the curing process, the moisture content is reduced, thereby increasing the dry matter content. Moisture content of fresh herbage will range from 75 to 90 percent (dry matter content from 10 to 25 percent). The moisture content of the cured hay must be no more than 20 percent (dry matter no less than 80 percent) to ensure that it can be stored without marked nutritional changes. A microwave oven can be used to determine the moisture content of the curing hay crop as follows:

1. Take a sample from your hay swaths and cut into half-inch pieces.
2. Weigh out a sample of about 100 g. Record the exact weight and identify it as the wet weight.
3. Spread the sample on a microwavable plate.
4. Place the plate of sample and a glass of water in the microwave (the water is to prevent the sample from catching fire).
5. Microwave for 3 minutes.
6. Reweigh the sample and record its weight. Stir the sample.
7. Return the sample to the microwave and microwave for 1- to 2-minute intervals. Record the weight, stir, and return to the microwave.
8. Repeat step 7 until the sample looses less than 1 g between heatings. This weight is the dry weight.
9. Calculate the percent moisture content as follows:

   [(wet weight – dry weight)/wet weight] × 100 = moisture percentage

10. Calculate the percent dry matter content as 100 – (step 9).

Typical losses of hay crop dry matter from cutting to feeding are 20 to 30 percent (Undersander et al., 1994). The losses in hay making are generally associated with harvest activities and include shattering and bleaching. As hay sits in the windrow, it dries or cures unevenly. The leaves dry faster than the stems, and will tend to become brittle and may fall off. This is called shattering. Because most of the nutrition in hay is found in the leaves, harvesting procedures that result in shattering will reduce the nutritive value of hay. *Bleaching* is the term that describes hay that is overexposed to the sun. Bleaching results in loss of vitamin value. If cured hay is rained on, the hay may lose a considerable amount of its original nutritive value as the water leaches out soluble nutrients. Rain on freshly cut hay will cause little damage.

There are various principles and techniques that may be used to minimize harvest losses in hay making. Rapid drying of the cut crop ensures minimal nutritive losses. Slow drying is frequently accompanied by mold growth, which reduces palatability and nutritive value. Slow drying may also lead to a reduction in nutrient value due to the activity of plant enzymes and microorganisms or oxidation. Crimping, to crush the stems of plants, speeds up the drying process. For legumes, drying agents are available that remove the waxy cutin layer, hastening water loss. Adding preservatives such as propionic acid to the cut crop makes it possible to bale hay at a higher moisture content (25 to 30 percent). This can help with the harvest in difficult weather and will minimizing shattering.

In field-curing hay, some loss of leaves should be expected due to shattering. Dehydration is an alternative to field-curing hay that minimizes the losses due to shattering. The crop is harvested wet and moved into a dehydrating facility. Dehydration is practiced in the United States and some areas of Europe. In the United States, alfalfa is the primary crop that is dehydrated, but in Europe, grasses or grass-clover mixtures may be used. In dehydrating alfalfa, the herbage is cut, usually at prebloom, dried quickly, ground, and sometimes cubed or pelleted.

Hay is stored in cubes, bales, or stacks. Hay cubes are made by compressing long or coarsely chopped alfalfa hay. Usually, an edible glue or binder is added to make cubes approximately 1¼ to 2 inches on a side. Hay cubes have a density of 30 to 32 lb./ft^3. Because of their high density, a given storage space will be able to hold more tons of hay in the form of cubes than in any other form. Cubes may be more convenient to handle and feed than other hay packages. Though hay in cubed form is usually the most expensive way to feed hay, cubes may result in reduced wastage.

Long hay is packaged in bales or stacks. Bales come in many different sizes and shapes from the 40-lb. square bale to the round bales that weigh from 400 to 1200 lb. or more. Stacks are made by hydraulically compressing hay. Hay stacks generally weigh from 1 to 6 tons.

Once in the cube, bale, or stack, hay stored properly will maintain its nutritive value for years. However, hays that are stored too wet may lose nutritive value due to microorganism activity. Microorganism activity in wet hay reduces nutritive value in three ways:

1. Microorganisms use up nutrients.
2. Microorganisms may produce toxins.
3. Microorganisms generate heat, which may reduce nutrient availability. If enough heat is generated, spontaneous combustion could occur.

Crop Residues Straw is a poorly digested, low-nutrient content crop residue. It consists mostly of the stems that remain after the removal of the crop's seeds. The primary supply of straw comes from wheat, barley, rye, rice, and oats, but in some areas substantial amounts may be available from the grass or legume seed industry, and from other miscellaneous crops. The nutrient content and palatability of straw is low, and this limits its use in livestock diets.

Other crop residues include corn cobs, stover (corn or sorghum stalks and leaves), sugarcane bagasse, and hulls of cottonseeds, peanuts, and soybeans. Crop residues such as straw and hulls consist of the nonliving cell wall portion of the plant and, therefore, contain little of the nutrients found inside cells. In addition, the cell wall of these crop residues is usually highly lignified, meaning that the cell wall constituents are encrusted in indigestible lignin, rendering them inaccessible to microbial enzymes.

There are various chemical treatments that are capable of dissolving lignin to improve the digestibility of highly lignified crop residues. Sodium hydroxide and ammonia treatments (such as ammonium hydroxide and gaseous ammonia) are effective in dissolving lignin to increase the digestibility of cell wall constituents. Ammoniated hay toxicosis (bovine hysteria, bovine bonkers, crazy cow syndrome) is a health problem that occurs if a toxin is produced during ammoniation. This toxin is produced when reducing sugars such as glucose react with ammonia in the presence of temperatures in excess of 158° F (Perdok & Leng, 1987). Symptoms include hyperexcitability, incoordination, tremors, and convulsion.

Pasture, Range Plants, and Feeds Cut and Fed Fresh

Pasture and range, like dry forage, may consist of native and/or cultivated species, the latter used to improve productivity or versatility. The use of pasture and range allows grazing animals to harvest the forage and spread the manure. Animals on pasture or range should have access to water, shade, and a mineral mix containing salt and the mineral nutrients in which the pasture is deficient.

Range The western U.S. rangelands are vast areas used for grazing livestock. They differ from pastures mainly in size and in the fact that much of the rangeland is on public lands.

Pasture The primary incentives to use pasture are the following:

1. Less labor may be required feeding livestock pasture forage compared to green chop or preserved forages
2. The possibility that pasture is cheaper to produce than hay, silage, or green chop
3. A possible marketing advantage for pasture-fed livestock
4. Some lands may not be useful agriculturally except as pasture

However, the use of pasture does bring with it some management challenges, five of which are described below.

1. Nitrate poisoning may occur in animals grazing on grasses that have been heavily fertilized with nitrogen-containing fertilizers.
2. When ruminant animals (and on rare occasions, horses) have been pastured for some time exclusively on lush spring pasture, they begin to lose coordination and may undergo sudden convulsive seizures. This is called grass staggers or grass tetany. It is caused by a magnesium deficiency. To avoid this problem, animals grazing spring pastures should be provided with a mineral supplement containing a source of magnesium. Grass tetany is discussed in Chapter 9.
3. Early in the growing season, grasses—especially cool season grasses—have a very high water content and an excess of protein for most grazing animals. The result is that the pastured animals often have "loose" manure or diarrhea.
4. Lactating dairy cows on pasture usually lose weight and will be unable to consume the pounds of pasture needed to deliver the energy to support high production. There are four reasons for this.
   a. The high water content of pasture makes it very dilute in nutrient content.
   b. Pasture is not a high-energy feedstuff. Because pastured animals spend much of the day on pasture, it is difficult to supplement their diet with high-energy feedstuffs.
   c. Grazing requires the animal to do the harvesting that uses up feed energy.
   d. Much of the high protein of lush pasture needs to be processed and eliminated by livestock. This processing activity uses energy that exacerbates the energy shortage. Table 3–6 illustrates why concentrated energy sources are needed to create rations that can support high production.
5. The pasture represents just the forage component of the diet, and its nutrient profile is lacking in more than just magnesium and energy for high-producing animals. Animals with high production potential or the potential for rapid gains will not realize their potential when on pasture because the feeder has limited control over the animals' diet.

Table 3–6

Pasture's low dry matter content, low energy density and the increased energy requirement associated with grazing activity make it difficult for the pastured animal to physically ingest the pounds of feed needed to deliver the energy to support high production

Feedstuff	Feedstuff DM content %	Feedstuff Energy content, Mcal/lb. NEl, DM basis	Lb., as fed (DM) in a ration supporting 85-lb. milk production	Feedstuff Energy (Mcal NEl) contribution
Energy supplied by a mix of feedstuffs to a confined cow (energy required by a confined cow to make 85 lb. is assumed to be 36.9 Mcal NEl)				
Corn grain	88.1	0.90	22 (16.8)	16.8
Soybean meal	89.5	0.98	6.8 (5.8)	5.8
Corn silage	35.1	0.64	32 (11.2)	7.0
Grass silage	42.0	0.52	35 (14.7)	7.3
		Totals:	**95.8** (48.5)	36.8
Energy supplied by pasture without supplementation to a pastured cow (energy required by a pastured cow to make 85 lb. is assumed to be 47.7 Mcal NEl)				
Pasture	20.1	0.69	**346** (68.34)	47.7

DM: Dry matter; NEl: net energy for lactation (dairy).

Values taken from software provided by the Dairy NRC. The feedstuff NEl values shown here are for comparison purposes only. The Dairy NRC does not calculate ration NEl directly from feedstuff NEl values.

In spite of the challenges involved when attempting to incorporate pasture in a balanced ration, pasture can be an economical component of livestock diets.

Pasture Management All systems of pasture management are centered on the principle of controlling the frequency and severity of defoliation of individual plants. This control is exercised by management of stocking rates and the intensity and frequency of grazing. Pastures may benefit from periodic mowing. Mowing helps limit the growth of undesirable plants that have been avoided by livestock and can help maintain desirable plants in a vegetative, high-nutrient content stage of growth.

Continuous grazing is essentially unmanaged pasture. It involves stocking a pasture with animals continuously. Continuous grazing is the least costly method of pasture management, but results in the least amount of nutrient intake from pasture plants. Animals on a continuously grazed pasture avoid the plants around manure, overgraze some areas of the pasture while avoiding others, and may eat only the most palatable portions of nutritious plants. This method of pasture management makes it difficult to take maximum advantage of pasture as a source of nutrients for livestock.

Rotational grazing involves fencing off the pasture into paddocks. Animals are moved through the farm's different paddocks, allowing each paddock's forage time to recover before the animals return. When moving animals through the paddocks, it may be useful to know that animals, especially sheep, usually prefer to graze into the wind. The combination of high stocking density and short access time (usually 1 to 4 days) characteristic of rotational grazing prevents the problem of selective grazing. Rotational grazing allows the producer to effectively utilize pasture as a source of nutrients for livestock.

The daily forage allowance (DFA) is useful in predicting how much pasture will be eaten by grazing animals in an intensively managed system. Pasture consumption is a function of the pasture dry matter available, relative to the animals' potential pasture dry matter intake. The DFA is expressed as multiples of

Given:
Initial pasture mass (lb. dry matter/acre): 1250
Paddock size (acres): 25
Potential forage dry matter intake/animal (lb.): 30
Number of animals: 80
Number of days in the rotation: 6

DFA = (1250 × 25) / (30 × 80 × 6) = 2.2

Figure 3–1
Calculating the DFA

Table 3–7
Adjustment factors for pasture dry matter intake

Pasture Dry Matter Available, lb/acre	Daily Forage Allowance			
	4	3	2	1
	Pasture intake adjustment factor			
100	0.21	0.18	0.17	0.15
200	0.37	0.33	0.32	0.26
300	0.51	0.46	0.44	0.37
400	0.64	0.57	0.55	0.46
500	0.74	0.67	0.64	0.54
600	0.83	0.75	0.72	0.60
700	0.90	0.81	0.78	0.65
800	0.95	0.86	0.82	0.69
900	0.99	0.89	0.85	0.71
1,000	1	0.90	0.86	0.72
1,500	1	1	0.98	0.82
2,000	1	1	1	0.92

From Fox, D. G., Tylutki, T. P., Van Amburgh, M. E., Chase, L. E., Pell, A. N., Overton, T. R., Tedeschi, L. O., Rasmussen, C. N., & Durbal, V. M. (2000). The net carbohydrate and protein system for evaluating herd nutrition and nutrient excretion (CNCPS vol. 4.0. p. 212). Animal Science Department Mimeo 213, Cornell University, Ithaca, NY.

the potential dry matter intake of the herd. A DFA of 2 means that the herd has available twice what it could potentially eat in pasture dry matter (Figure 3–1). The companion application to this text for beef uses a value referred to as *daily forage grazed* (DFG). This differs from the DFA in that it is the forage available expressed as tons/acre, dry basis, rather than as multiples of potential dry matter intake.

The initial pasture mass (IPM) is expressed in pounds dry matter per acre, and can be estimated from hay harvesting experience, clippings, or calibrated measuring devices. In well-managed pastures, the IPM can also be estimated from the plant height. Legumes contain about 120 lb. dry matter per acre per inch of plant height. Well-managed grass pastures contain about 250 lb. dry matter per acre per inch of plant height. An orchardgrass pasture that has grown to 8 in. in height contains about 2,000 lb. of dry matter per acre (250 × 8).

Using Table 3–7, an adjustment factor can be found that may be applied to the animal's predicted pasture dry matter intake. As can be seen in Table 3–7, pasture dry matter intake declines with reduced pasture dry matter available and with reduced daily forage allowance. When pasture dry matter intake declines, performance will decline unless supplemental feed is provided or animals are moved to a new pasture.

Mixed Livestock Grazing Because herbivores differ in their grazing habits, different species grazing the same pasture may not be directly competitive. Simultaneous grazing of sheep and cattle may result in higher yields of animal products per unit of land area than single-species grazing (Nolan & Connolly, 1989). Mixed livestock grazing also may help maintain beneficial forage species on pasture (Abaye, Allen, & Fontenot, 1997).

Green Chop Green chop, sometimes called *soilage,* is herbage that has been cut and chopped in the field and fed fresh to livestock in confinement. Plants used in this manner are forage grasses, legumes, sudan grass, the corn plant, and residues of food crops used for human consumption.

A major advantage of green chop is that more usable nutrients can be salvaged per unit of land than by other methods such as pasturing, haymaking, or ensiling. A major disadvantage of green chop is the labor required; the herbage is not preserved and must be harvested and fed daily to maintain its palatability and nutritive value. When growth outruns daily need, it may be necessary to mow the crop to maintain quality.

Silages

Types of crops made into silage

Hay crop silage The primary reason for choosing to store the hay crop as silage rather than hay has to do with the labor necessary to feed the crop. Feeding the hay crop as silage requires less labor than would feeding the hay crop as hay. Weather also may be a factor involved in choosing to make silage or hay: dry hay requires more drying time than hay crop silage. There are also significant differences in the storage facilities needed.

Good silage is equally palatable to good hay, and it is well utilized. This has been verified in all species of livestock. In addition to other advantages, the fermentation process will reduce the level of some toxic substances in the fresh crop.

Silage is produced by fermenting high-moisture herbage. The goal of making silage is to turn enough carbohydrate into acid during the fermentation process so that the pH of the silage prevents the growth of spoilage bacteria. A silo is used to create the anaerobic environment necessary for fermentation to occur.

There are several different types of silos. The general categories of silos are tower or upright silos, horizontal silos, bag silos, and stack silos. Tower silos may be of the conventional type, usually made from concrete staves or poured concrete with a roof. Tower silos also may be constructed of protected metal or fiberglass. Horizontal silos may be of the bunker type that have concrete walls and (usually) a concrete floor, or they may be of the trench type (a simple excavation with a sloped floor to permit drainage). Both types of horizontal silos should be covered with heavyweight (6 mm) polyethylene (plastic) film that is securely held down to reduce surface spoilage. Polyethylene tubes, packed with silage and sealed at each end, may be used as a silo. The capacity of these tubes depends on the diameter of the face of the tube. A tube that is 12 ft. in diameter holds approximately 2 tons (as fed) per linear foot. Properly used, these tubes produce excellent quality silage although they are not reusable and bag disposal may be a problem. Stack silos are essentially a packed pile of silage covered with a polyethylene film that is held down.

In preparing the hay crop for the silo, the crop must be wilted and chopped. Recommendations for maturity, dry matter content, and chop length for crops to be ensiled are given in Table 3–8. *Wilting* is the term used to describe the process of drying the hay crop to a level suitable for ensiling, usually 30 to 40 percent dry matter (60 to 70 percent water). Moisture must be reduced if the acid produced from carbohydrates during fermentation is to reduce the overall pH effectively. The fresh cut crop will have 75 to 90 percent moisture. The proper level for ensiling depends on the crop, the chop length, and the silo type. Hay crops to be ensiled must be reduced to lower water content than corn and small-grain crops because hay crops contain less fermentable carbohydrates.

The dry matter content of the hay crop to be ensiled can be measured using a microwave oven in the same was as described for hay. The "grab test" is a method to estimate moisture content of the hay crop (Figure 3–2).

As can be seen from Figure 3–3, drying the hay crop results in significant harvest losses. As discussed earlier, these losses are due to leaf loss or shattering. Because silage is harvested at a higher water content than hay, shattering is not a problem.

Packing silage is essential to minimize oxygen for effective fermentation. Chop length and moisture content both affect ease of packing. Higher moisture content makes it easier to pack, as does smaller chop length.

A third variable that affects packing and oxygen exclusion is silo type. In all types of upright silos, packing is facilitated by the physical weight of the column of silage. The expensive sealed silos are effective in excluding oxygen, and

Table 3–8
Harvest recommendations

			Silo Type		
			Horizontal & Bag	Conventional Upright	Sealed Upright
Crop	Maturity	Length of Cut (inches)	Dry Matter Content (%)		
Corn silage	milk line ½ to ⅔	⅜ – ½	28–33	32–37	40–50
Small grain silage	milk to soft dough	¼ – ⅜	30–40	30–40	30–40
Hay crop silage	Early bloom	¼ – ⅜	30–35	35–40	40–50
High moisture grain	After physiological maturity	—	65–72	70–75	74–78

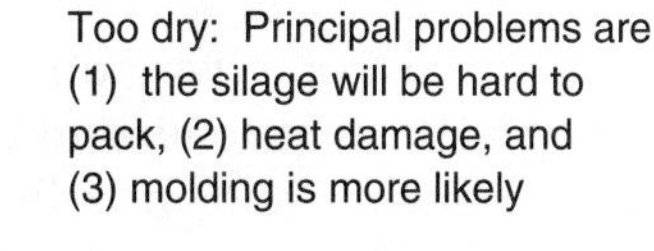
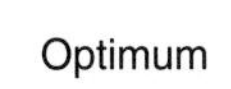
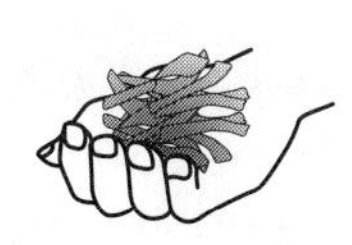

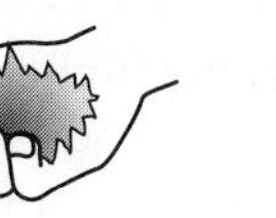

Figure 3–2
Hay crop silage (haylage) grab test results as related to whole plant dry matter content and silage production recommendations

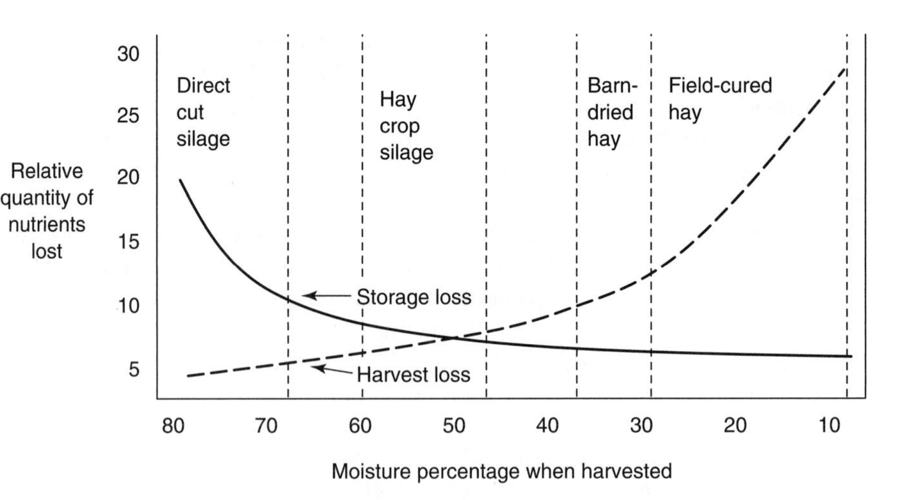

Figure 3–3
Forage losses with varying moisture contents of crop at harvest
Adapted from Hoglund, C. R., (1964). Michigan State University. *Agricultural Economics Publication # 947.*

it is possible to use larger chop length and/or lower moisture contents for silage going into these silos. Packing effectiveness in the horizontal silo depends on silo management. A little higher moisture in the crop can assist packing in this type of silo. Bag silos provide a good environment for fermentation and will produce good silage if properly managed.

Whereas the primary losses in hay making occur during harvest, the primary losses in silage making occur during storage. These losses are discussed later in this chapter under "The Fermentation Process."

Small grain silage Silage may be made from the whole plant of crops otherwise grown for their grain. Examples of small-grain crops used for silage include oats, barley, sorghum, and wheat. General recommendations for these crops are to cut when the seed is in the milk to soft-dough stage.

Corn and sorghum silage Most silage in the United States is made from the whole corn plant and from a number of sorghum varieties. Observations of the corn kernel's milk line can be useful, in conjunction with whole plant dry matter testing, to help determine when the corn crop is at the appropriate stage for ensiling. The milk line is a line that appears on the kernel at the junction between the solid and liquid phases. As the plant matures, the solid phase expands, moving down the kernel toward the cob (Figures 3–4 and 3–5). Concurrently, the plant's overall moisture content decreases. The proper dry matter level for harvest and ensiling roughly corresponds to a kernel milk line of ½ to ⅔ the way down the kernel. There will be significant variation among hybrids as to how milk-line position relates to whole plant percent dry matter content, and milk-line observations should be made in conjunction with whole plant dry matter testing.

Corn and sorghum silages are fed as forage sources. The lower the lignin content of the forage, the higher the digestibility of the forage's fiber carbohydrate, measured as neutral detergent fiber (NDF). The formula relating lignin to NDF digestibility is shown in Chapter 4 and in the dairy application's feed table as a comment behind the NDF digestibility column title. A trait found in both corn and sorghum crops called *brown-midrib* (BMR) is associated with low lignin content. Studies have shown improved growth (Colenbrander, Bauman, & Lechtenberg, 1975) and lactation (Frenchick, Johnson, Murphy, & Otterby, 1976) performance when using silages with the BMR genotype. However, BMR corn shows reduced grain yield over non-BMR corn. If it is not known at planting whether the corn crop will be used for silage or grain, BMR corn should not be planted. There appears to be no reduction in grain yield associated with the BMR train in sorghum (Cherney, 1991).

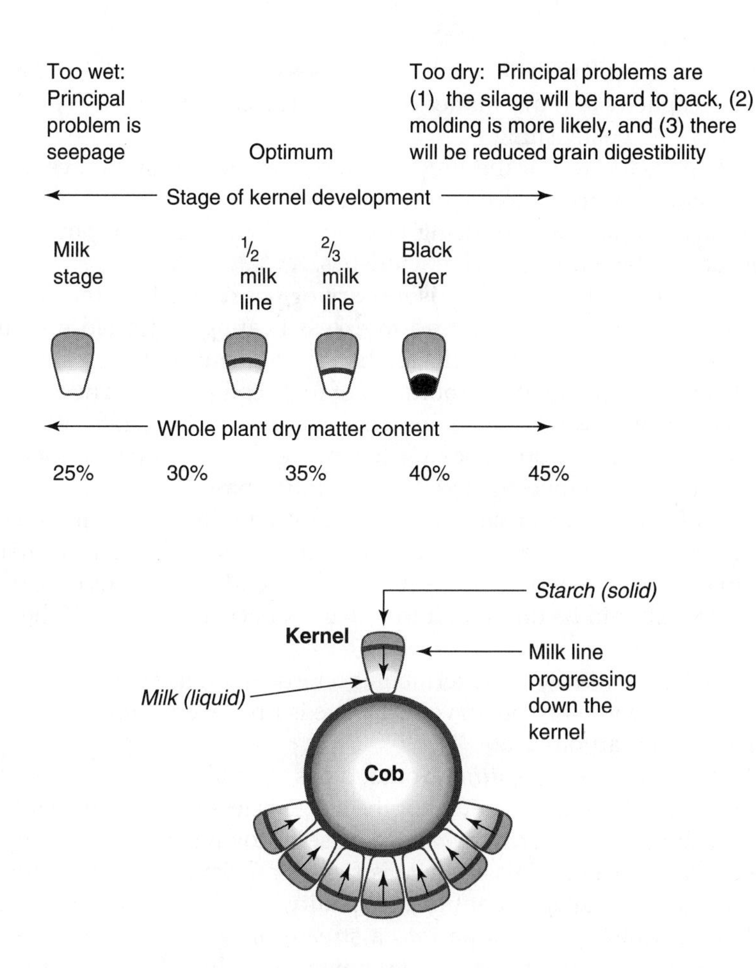

Figure 3–4
Corn plant milk line progression as related to whole plant dry matter content and silage production recommendations (note that there will be significant variation among hybrids as to how milk line position relates to whole plant % dry matter content)

Holland, C., & Kezar, W. (1995). *Pioneer forage manual—A nutritional guide.* Des Moines, IA: Pioneer Hi-Bred International, Inc.

Figure 3–5
Cross section of an ear of corn showing milk line

Miscellaneous silages Other materials have been used to make silage. Waste from food crops such as sweet corn, green beans, green peas, and potato tubers are examples. Ensiling such feedstuffs is advantageous in that it results in a more uniform feed and a known supply, and the feedstuff is preserved.

The fermentation process Once in the silo, the fresh crop will undergo changes as a result of the activity of plant enzymes and the microbes that are present on the crop in the field or that find their way into it from other routes. The plant enzymes continue to be active for the first few days after cutting, resulting in some loss of carbohydrate and protein. Plant proteins are partially broken down by cellular enzymes, resulting in an increase in nonprotein nitrogen (NPN) compounds such as amino acids. Proteins are also fermented by microbes to the gasses nitrogen tetraxide, nitrogen dioxide, and nitric acid. These gasses are highly toxic to humans. In tower silos, these gasses accumulate to dangerous levels for 3 weeks following the filling of the silo. As carbohydrate is oxidized by aerobic microorganisms and plant enzymes to CO_2 and water, heat is generated. Under normal conditions, temperatures during fermentation will peak 10° to 15° F above ambient temperature. Excessively high temperatures will result in damage to the feed protein, reducing the amount available to the animal. Heat damage can be minimized by ensuring that the crop is well packed to exclude air.

Anaerobic microorganisms become active after the oxygen in the silage has been depleted. Anaerobic microorganisms multiply rapidly, using sugars and starches, and producing lactic acid with lesser amounts of acetic, formic, propionic, and butyric acid. However, little butyric acid is present in well-preserved silage. Continued action occurs on nitrogen-containing compounds with further solubilization and production of ammonia and other nonprotein nitrogen compounds. The level of lactic acid rises in well-preserved silage, eventually reaching levels of 7 to 8 percent and the pH drops to about 3.5 to 4.0, depending on buffering capacity and dry matter content of the crop. Figure 3–6 presents a time line of events postensiling.

If the silage is too wet or the supply of fermentable carbohydrates too low, the pH will not drop to an acceptably low level. This will allow the *Clostridia* bacteria to become active, producing butyric acid and new compounds with foul odors and potentially toxic characteristics.

On the other hand, if the mass is too dry or poorly packed, the abundant oxygen present in the silage will lead to excess heating and mold growth resulting in reduced recovery of silage dry matter, nutrient losses, reduced palatability and sometimes the elaboration of toxic compounds. The amount of packing necessary depends on forage characteristics, filling rate, the tractor weight used in packing, and packing technique. The relationships between these factors have been investigated (Ruppel, Pitt, Chase, & Galton, 1995).

Unavoidable losses from ensiling Some nutrient losses are unavoidable when making silage. Respiration of the live plant cells shortly after the material is placed in the silo is unavoidable. Respiration depends on the presence of oxygen so this loss should be minimal if the silage is packed well and if the silo is reasonably air-tight.

Like respiration losses, some fermentation losses are unavoidable. The lactic acid that will eventually preserve the silage is produced from the fermentation (loss) of some carbohydrate.

Avoidable losses from ensiling Some types of silage nutrient losses are avoidable. Where dry matter content of the silage is less than 30 percent (water content is greater than 70 percent), water will seep down through the silage and drain away from the silo. This seepage is rich in nutrients and constitutes not only a loss in terms of silage quantity and quality, but also is a potentially powerful pollutant if allowed to drain into a stream or wetland. Excessively wet silage also may undergo butyric acid fermentation, which further reduces the quality and quantity of the silage.

Another avoidable loss in silage making is mold growth. Mold requires oxygen and water to grow, but mold growth is discouraged by acid. Silage will be free from mold only as long as the acid level remains high and the oxygen level remains low. The three feed-related problems associated with mold growth are (1) the mold organism uses up nutrients, (2) moldy feed is unpalatable, and (3) the mold organism may produce mycotoxins. The presence of mold does not guarantee that there are mycotoxins present, and mycotoxins may be present even if there is no visible mold. Some mold nearly always grows around the perimeter and on top of the silage because it is difficult to eliminate oxygen from these areas. The moldy feed should be discarded. This chore requires respiratory protection because of the risk of developing farmer's lung disease, an allergic reaction to the mold. Further discussion of mold growth in feedstuffs is found in Chapter 13.

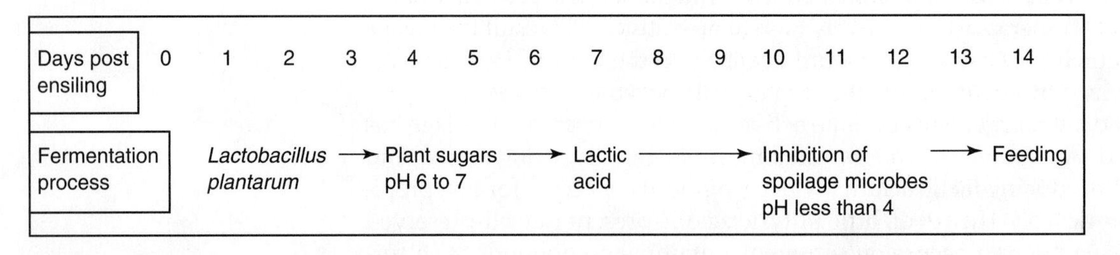

Figure 3–6
Silage fermentation timeline

Phases of silage fermentation

Phase 1

- Material placed in silo
- Plant cells use up trapped oxygen—respiration
- $C_6H_{12}O_6 + 6O_2 \rightarrow 6CO_2 + 6H_2O$ + 673 calories
- Some protein → NPN

Phase 2

- Anaerobic microorganisms begin to function
- Produce acetic acid from carbohydrate; pH drops to 4.2

Phase 3

- Acetic acid producers cease to function
- Lactic acid producers begin to produce lactic acid from carbohydrate
- Occurs 3 to 5 days after filing silo

Phase 4

- Lactic acid accumulates
- pH drops
- Temperature cools
- Final pH 3.5 to 4.0
- Microorganisms cannot attack silage

Phase 5

- If enough lactic acid has accumulated, stable storage is achieved 15 to 21 days after filling silo

Silage additives Silage additives are designed to improve the efficiency of silage fermentation. The decision as to whether to use an additive should be based on the demonstrated effectiveness of the additive, its cost, the existence of a clear need for the additive, and practicality in the use of the additive. For the most part, additives target either Phase 3 or Phase 4 of the fermentation process.

Phase 3 depends on adequate fermentable carbohydrate for the lactic acid–producing bacteria. Fermentable carbohydrate within the crop varies with the weather. Exposure to the sun increases the fermentable carbohydrate content in the living plant; ideally, the timing of harvest should be made with this in mind. Additives that have been used to increase the fermentable carbohydrate in the silage include cracked corn at 100 to 150 lb./ton of silage and molasses at 40 to 80 lb./ton of silage.

Phase 3 also depends on the presence of adequate numbers of lactic acid–producing bacteria. Microbial numbers are reduced in cool temperatures to a point at which silage fermentation may be negatively impacted. Many microbial additives or inoculants are available to help with silage preservation. These inoculants generally contain *Lactobacillus plantarum* microbial strains that have improved efficiency over naturally occurring microbes in generating lactic acid.

Phase 4 requires enough acid to lower silage pH. Organic acids such as propionic acid at 1 percent have been used successfully as silage preservation aids.

Other products are sometimes added to silage to improve its various characteristics. Limestone and urea have been added to corn silage to improve the nutrient value of this feedstuff. Certain enzymes added to silage will improve feed value, and work is underway to improve the performance and cost effectiveness of these products.

Silage troubleshooting

Problem 1: Heat damage indicated on forage analysis

- Caused by the presence of oxygen
  - poor packing
  - holes in silo

  - slow filling
  - haylage on top of corn silage or vice versa

Problem 2: Mold growth
- Caused by the presence of oxygen (see Problem 1)

Problem 3: Sour smell, slimy feel to the silage
- The acid generated was not sufficient to lower the overall pH, and clostridial organisms have begun to attack the silage in a butyric acid fermentation.
  - Problems 4 and 5 will also be evident

Problem 4: Silage feeds out hot
- Poor initial fermentation—too wet
- Oxygen exposure after feedout started
  - feedout too slow/silo too big (see *Sizing Bunker Silos*)
  - face of bunker has too much exposed surface area; as you remove silage, keep the face vertical
- Warm weather
  - this accelerates feed deterioration and may be the cause of the problem

Problem 5: Silage becomes hot after unloading
- Poor initial fermentation—too wet
- Feed bunk not clean before filling bacteria and mold become active
- Warm weather
- Other ration components may be bad (wet brewer's, other wet feeds)
- Soil, manure contamination

Problem 6: Excess surface spoilage at opening of silo
- Some spoilage is difficult to prevent along the perimeter and surface of the silo
  - minimize surface spoilage by covering the silo with a 6-mm polyethylene sheet and cover this with tires
  - use 20 to 25 tires per 100 sq. ft. This generally means tires should be touching each other over the silo surface

Differences between hay crop silage and corn silage

Density
- Corn silage density is approximately 40 to 45 as fed lb./ft^3 or about 14 dry matter lb./ft^3.
- Hay crop silage is lower in density—approximately 35 to 40 as fed lb./ft^3 or about 14 dry matter lb./ft^3.

Storage needs
- Corn silage is harvested once per year and a silo is needed that will hold a full year's inventory. Hay crop silage is harvested in multiple cuttings and, therefore, it is not necessary to have a silo available that is capable of holding a full year's inventory.

Keeping silage fresh in a bunker silo
- After feedout has begun, the silage surface or face is exposed to oxygen, which will begin to deteriorate the silage. It is necessary to remove a sufficient layer of silage each day to get to silage that has not yet begun to deteriorate. Because the corn plant, at the recommended stage for ensiling, has more fermentable carbohydrate than does the hay crop at the usual stage for ensiling, corn silage is generally more stable than hay crop silage. A removal rate of 4 to 6 in. daily is usually enough to ensure fresh corn silage at each feeding. For hay crop silage, it is recommended that 6–9 in. be removed daily, and more may be necessary in hot weather. Achieving the proper removal rate with a given number of livestock means sizing the silo properly.

Sizing bunker silos

- Silage density (lb./ft^3)
  - a typical value is 14 lb. DM/ft^3
  - for our example, we will use corn silage with a percent DM of 31 percent. The pounds per cubic foot, as fed basis is, therefore, 14/0.31 = 45.
- Side walls
  - decide how high you want the walls; as an example, use 12-ft. walls
- Daily use: for our example, we will use:
  - 250 lactating cows × 35 lb./cow = 8,750 lb./day
  - 15 dry cows × 20 lb./cow = 300 lb./day
  - 175 heifers × 10 lb./heifer = 1,750 lb./day
  - total daily use = 8,750 + 300 + 1,750 = 10,800 lb. daily silage use
- Annual use
  - 10,800 × 365 = 3,942,000 lb. annual silage use
- Spacing of walls
  - some silos have sloped sidewalls
  - to get the surface area of the silo face, you need a width measurement
  - use the average between the width across the bottom of the walls and the width across the top of the walls. We will use a value of 40 ft.
- Surface area of the face
  - 40 ft. wide × 12 ft. high = 480 sq. ft.
- If you removed 12 in. from the face, how many pounds of silage in cubic feet would you take? Our corn silage is 45 lb./cu. ft.
  - 480 sq. ft. × 45 lb./cu. ft. = 21,600 lb.
- If you removed 6 in. from the face, how many pounds of silage would you take? Daily use of 6 in. is enough to keep corn silage fresh.
  - 21,600/2 = 10,800 lb. This matches our daily use.
- With 12-ft. sidewalls spaced at 40 ft., how long would this silo have to be to hold a year's inventory? Annual use is 3,942,000 lb. One cubic foot of silage holds 21,600 lb.
  - 3,942,000 lb./21,600 lb./ft. = 182.5 ft. long.

Energy Concentrates

Images and supplemental information on energy concentrates are found in the file titled Images.ppt on the text's accompanying On-Line Companion. To view this file please go to http://www.agriculture.delmar.com and click on *Resources.* Click on *On-Line Resources* and select from the titles listed.

Feedstuffs classified as energy concentrates are those that are added to a ration primarily to increase energy density. Included are the cereal grains, some of their milling coproducts, some types of liquid feeds such as molasses, and the fats and oils. High-energy feedstuffs generally have low levels of protein. However, the high-protein oilseed meals could be included in the energy concentrates group on the basis of their energy content.

Depending on the activity of an animal, feedstuffs in this class may make up a substantial portion of an animal's total diet. As such, the energy feeds may make significant contributions of other nutrients such as amino acids, minerals, and vitamins.

Cereal Grains

Cereal grains are produced by plants in the grass family grown primarily for their seeds. The cereal grains are usually harvested and fed as the mature dry seed (approximately 15 percent moisture), but for some crops, cereal grains may be fed in a high-moisture (approximately 25 to 30 percent moisture) form that has been ensiled and fermented.

Grains are less variable in composition than forages, and it is usually acceptable to use reference values for nutrient content when formulating rations. However, soil fertility, weather, disease, and insect damage all may affect the development of seed, and under certain circumstances, its nutrient value may not be well represented by average values found in reference sources.

Because of the numerous ways of expressing energy content, and the fact that for at least some species feed energy value is determined based on feed passage rate, it is difficult to speak in generalities. In fact, historically, feed companies have kept secret the energy content of the finished feeds they sell for fear that a competitor might make improper energy comparisons between feeds to achieve a sales advantage.

The primary component responsible for the high-energy value of grains is carbohydrate. The nonfiber carbohydrate composition of most grains will range from 60 to 70 percent. The fiber carbohydrate in most grains will range from 2 to 12 percent, measured as crude fiber and 9 to 30 percent, measured as neutral detergent fiber. The fat content of most grains will vary from 1 to 6 percent, measured as ether extract. In contrast to energy content, the crude protein content of feed grains is relatively low, ranging from 8 to 12 percent. Calcium and phosphorus content of grains is low. Grains contain low to moderate levels of vitamins.

Hulls of grain are high in fiber and have a significant effect on feeding value. Most grains are processed (ground, rolled, etc.) to some extent before feeding to break the hull. This improves access for digestive enzymes to the grain's nutrients.

The United States government has established standards for assessing grain quality (USDA, 1987). The criteria for these grain standards are (1) test weights per bushel, (2) moisture content, (3) foreign material, (4) broken and damaged kernels, and (5) discoloration. Some grains have additional criteria. U.S. Grade 1 is the highest quality and U.S. Sample Grade is the lowest quality.

Corn grain (IFN 4-02-931)

Proximate analysis, percent as fed basis

- Crude protein: 8.9
- Crude fat: 3.8
- Crude fiber: 2.3

Major feed applications Corn, in areas where it grows well, will produce more digestible nutrients per unit of acreage than any other grain crop. It is very digestible and palatable for most domestic animals. Plant breeders have produced different varieties of corn to improve one or more characteristics in normal corn. High-lysine corn is a variety of corn bred to contain higher levels of the amino acids lysine, and tryptophan. Waxy corn is a variety that has a different type of starch than usual corn.

Corn grain can be preserved as high-moisture corn by ensiling. Optimal moisture levels at harvest are 25 to 30 percent. High-moisture corn should be processed before feeding for maximum feed efficiency.

Corn grain for use in livestock feed is usually ground, cracked, or rolled before being fed to animals. Many feed products are produced as coproducts of the milling of corn grain for use as food for people. The systems used for milling of corn grain fall into two categories: dry milling and wet milling (Figure 3–7).

Dry milling is used to process grain into a meal form and to extract the outer hull to produce corn flour. Corn flour is used in dry mixes such as for pancakes, doughnuts, batters, and snacks. When corn is dry milled, hominy feed may be produced as a coproduct. Hominy feed is higher than corn grain

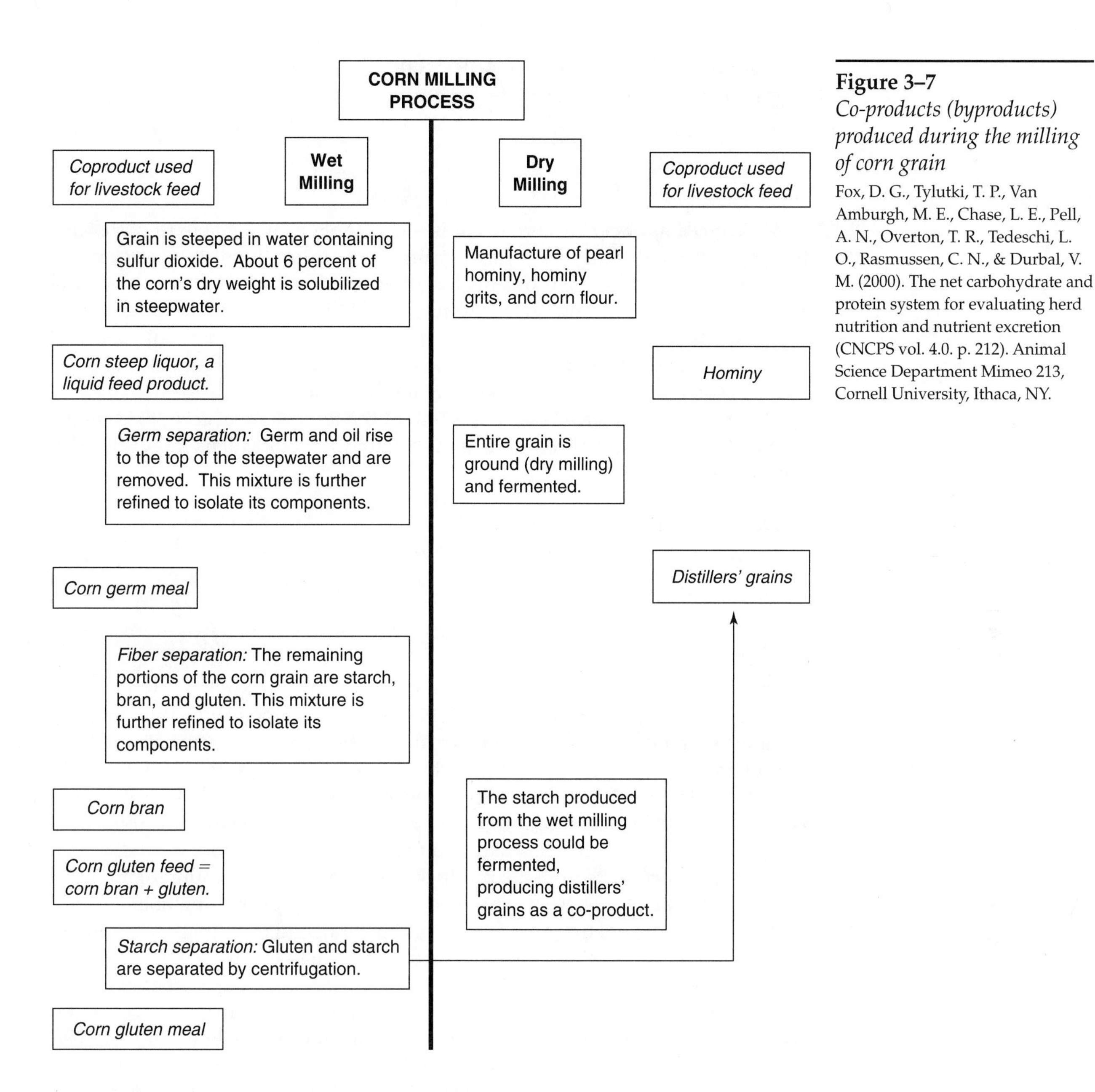

Figure 3–7
Co-products (byproducts) produced during the milling of corn grain
Fox, D. G., Tylutki, T. P., Van Amburgh, M. E., Chase, L. E., Pell, A. N., Overton, T. R., Tedeschi, L. O., Rasmussen, C. N., & Durbal, V. M. (2000). The net carbohydrate and protein system for evaluating herd nutrition and nutrient excretion (CNCPS vol. 4.0. p. 212). Animal Science Department Mimeo 213, Cornell University, Ithaca, NY.

in protein, fat, and fiber, and it is lower than corn meal in nonfiber carbohydrate.

In wet milling, corn and other grains are processed into fractions concentrated in starch, protein, fiber, and oil. Coproducts resulting from the wet milling of corn include corn gluten meal and corn gluten feed, both of which are considered to be protein supplements rather than energy concentrates.

In the distilling industries, ethyl alcohol is produced by fermenting corn and other grains. The coproduct, distillers' grains, is generally considered to be a protein supplement rather than an energy concentrate. Distillers' grains is usually produced by starting with the whole grain, grinding it to a meal as in the dry milling process, and then proceeding with the fermentation. It is also possible to produce a distillers' grains coproduct by starting with the starch produced in wet milling. A similar feedstuff, brewers' grains, is a coproduct of the brewing industry, and is also considered to be a protein supplement.

Grain sorghum or milo (IFN 4-04-383)
Proximate analysis, percent as fed basis

- Crude protein: 10.5
- Crude fat: 2.8
- Crude fiber: 2.4

Major feed applications Sorghum is a hardy plant that is able to withstand heat and drought better than most grain crops. In addition, it is resistant to some insect pests that are problems for other grains and is adaptable to a wide variety of soil types. Consequently, sorghum is grown in many areas where corn does not do well. In areas where corn grows well, sorghum will yield less than corn. The seed from all varieties is small and relatively hard, and usually requires some processing for good animal utilization.

Sorghum grains have 90 to 95 percent of the feed value of corn. Bird-resistant varieties, whose seed coats are high in tannin, are not well liked by most domestic animals.

Wheat grain (IFN 4-05-211)
Proximate analysis, percent as fed basis

- Crude protein: 14.0
- Crude fat: 1.7
- Crude fiber: 2.5

Major feed applications In the United States, wheat is rarely grown for animal feed. All commonly grown varieties were developed with flour milling qualities in mind rather than feeding values. The hard winter wheats are high in protein, averaging 13 to 15 percent, but the soft wheats have less. The amino acid distribution is better than that of most cereal grains, and wheat is a very palatable and digestible feedstuff with a relative value equal to corn. Some processing (grinding or rolling) is required for optimum utilization.

Milling coproducts from the production of wheat flour account for about 25 percent of the kernel. They are classified on the basis of decreasing fiber as bran, middlings, shorts, red dog, and feed flour. The bran and middlings are from the outer layers of the seed and contain more protein than the grain, although they are deficient in lysine and methionine as well as some other essential amino acids. These outer layers of the seed are relatively good sources of most of the water-soluble vitamins except niacin. They are low in calcium and high in phosphorus and magnesium.

Wheat is the grain used to make most breads and bread products sold in the United States. Under certain circumstances, these products may be declared unfit for human consumption but are still wholesome. Such bakery waste may be used in livestock feed.

Barley grain (IFN 4-00-549)
Proximate analysis, percent as fed basis

- Crude protein: 11.6
- Crude fat: 2.0
- Crude fiber: 6.0

Major feed applications Although a small amount goes into human food and a substantial amount is used in the brewing industry, most of the barley produced in the United States is used for animal feeding.

Barley contains more total protein and higher levels of the essential amino acids lysine, methionine, and tryptophan than corn grain. Lightweight barley tends to be higher in fiber, less digestible, and lower in energy than heavyweight barley. Barley is a very palatable feedstuff, particularly when rolled before feeding.

Oat grain (IFN 4-03-309)

Proximate analysis, percent as fed basis

- Crude protein: 12.0
- Crude fat: 4.5
- Crude fiber: 12.0

Major feed applications As a source of energy for livestock, oats is relatively unimportant. A substantial amount of the oats produced goes into human food. The protein content of oats is relatively high and the amino acid distribution is more favorable than that of corn. The hull is fibrous and poorly digested, even when ground. Oats may be milled to separate the hulls from the inner portions of the grain, called the *groats.* Oat groats have a feeding value comparable to corn, but the price is too high to compete with corn. Oat groats are not widely used except in rations where cost is a minor factor, such as early weaning diets for pigs.

Coproducts from Sugar and Citrus Industries

Molasses, cane (IFN 4-04-696)

Proximate analysis, percent as fed basis

- Crude protein: 3.0
- Crude fat: 0
- Crude fiber: 0

Major feed applications Molasses is a major coproduct of sugar production from sugar beets and sugar cane. Molasses may also be produced as a coproduct during the manufacture of dried citrus pulp. Wood molasses or hemicellulose extract is produced as a coproduct during the manufacture of some pressed wood products. Finally, starch molasses is a coproduct produced during the manufacture of dextrose from corn and sorghum.

The primary constituents of all molasses products are sugars. The sugar content of different molasses products will vary and an optical instrument called a *brix refractometer* is used to assess sugar content in molasses products. Worldwide, molasses is traded at a value of 79.5 brix.

Whereas the carbohydrate in most energy concentrates is in the form of starch, the carbohydrate in molasses is in the form of sugar. In ruminant nutrition, it is recognized that this type of carbohydrate plays a unique role in maximizing rumen productivity.

The sweet taste of molasses makes it palatable to most livestock species. In addition, molasses is of value in reducing dust, aiding pellet quality, serving as a vehicle for feeding medicants or other additives, and as a component of mixed liquid supplements. Because of its stickiness, molasses can create problems with mixing, processing, and handling equipment at the feed mill, and for this reason, the amount of molasses that can go into a mixed feed is limited.

Beet pulp (IFN 4-00-669)

Proximate analysis, percent as fed basis

- Crude protein: 9.0
- Crude fat: 0.5
- Crude fiber: 16.0

Major feed applications Dried beet pulp is the residue remaining after extraction of most of the sugar from sugar beets. Because of its high fiber content, dried beet pulp may be used as a forage alternative. It frequently has molasses added before drying, and may be sold in shredded or pelleted form. Because of the sugar content, the dried beet pulp is very palatable to most livestock.

Citrus pulp (IFN 4-01-237)

Proximate analysis, percent as fed basis

- Crude protein: 6.2
- Crude fat: 3.2
- Crude fiber: 13.0

Major feed applications Citrus pulp is the residue remaining after extraction of the juice from citrus fruit. Like beet pulp, citrus pulp is high in fiber content and so may be used as a forage alternative. Because of the residual sugar, citrus pulp is very palatable to most livestock.

Forage Substitutes

Kelp meal

Proximate analysis, percent as fed basis

- Crude protein: 7.5
- Crude fat: 3
- Crude fiber: 7

Major feed applications Kelp meal is a feed supplement made from the *Ascophyllum nodosum* seaweed (algae) growing in the North Atlantic. As a feed supplement, kelp is rich in trace minerals and it is high in fiber. Kelp is most commonly fed as part of feed that is certified organic.

Soybean hulls (IFN 1-04-560)

Proximate analysis, percent as fed basis

- Crude protein: 11.0
- Crude fat: 1.9
- Crude fiber: 36.5

Major feed applications Soybean hulls contain a high level of digestible fiber and may be heat-treated to improve feeding value. Heat-treated soybean hulls are called soybean mill run, soybean flakes, or soybran flakes. Untreated soybean hulls contain the enzyme urease, and for this reason, should not be fed with urea because the urease will rapidly convert urea to ammonia. If this ammonia is absorbed, the animal may show symptoms of breathing difficulty, incoordination, tetany, and death.

Citrus pulp and beet pulp These feedstuffs have been discussed as coproducts from the sugar and citrus industries. Because of their high fiber content, they may be used as forage substitutes.

Fats and Oils

Proximate analysis, percent as fed basis

- Crude protein: 0
- Crude fat: 82.5 to 100
- Crude fiber: 0

Major feed applications Fats fed to livestock are primarily animal fats derived from rendered beef, swine, sheep, or poultry tissues. Animal/vegetable blends are also used, but pure vegetable oils are rarely used due to their high cost.

Although most animals need a source of the essential fatty acids, these are present in most feedstuffs and supplementation is not usually required. Fats are added to rations for several reasons. As a source of energy, fats are highly digestible and supply at least 2.25 times as much energy per pound as starch, sugar, or protein. Because of their high energy value, fats can be used to increase energy density of a ration. Fats may also improve rations by reducing dustiness and increasing palatability. Feeds containing added fat usually have antioxidants added to help protect against rancidity.

High levels of fats are used in milk replacers for young ruminants. Mature ruminants, however, are less tolerant of high fat levels than are monogastrics. Concentrations in mature ruminant rations of more than 6 percent may cause reduced fiber digestibility (Palmquist & Jenkins, 1980). Rations formulated for carnivores such as cats, trout, and dogs will contain much higher fat levels than those for ruminants. Nonruminant herbivores such as horses and rabbits can tolerate large amounts of dietary fat but high fat rations are seldom beneficial for these animals. High levels of unsaturated fat in swine diets may affect pork quality for reasons explained in Chapter 8.

Protein Supplements

Images and supplemental information on protein supplements are found in the file titled Images.ppt on the text's accompanying On-Line Companion. To view this file please go to http://www.agriculture.delmar.com/ and click on *Resources*. Click on *On-Line Resources* and select from the titles listed.

The term *quality*, as it relates to protein supplements, refers to the feedstuff's amino acid content. The protein supplement is considered to be of good quality if its amino acid content resembles the amino acid requirements of the animal. Most protein supplements of animal origin are of better quality than protein supplements of plant origin. An exception is feather meal. The protein in feather meal is of only fair quality because the amino acid profile of feather meal is only a fair match to that needed by most livestock. Protein supplements are usually more expensive than energy concentrates, so optimal use is a must in any practical feeding system.

Plant Protein Sources

The most important sources of plant protein are soybeans, canola (a variety of rapeseed), and cottonseed, with lesser amounts from peanuts, flax (linseed), sunflower, sesame, and safflower. These seeds contain about 21 percent oil (dry matter basis), which is usually extracted and used for human food products. High-protein meals from these oilseeds are coproducts of the oil extraction process. When economical, the whole oilseeds themselves can be used in livestock diets. These are usually heat-treated to improve feed value. Although not as high in protein content, distillery and brewery coproducts are often economical sources of protein and other nutrients for livestock, particularly herbivores whose digestive tracts are capable of utilizing fibrous material.

Oil-bearing seeds are usually processed to extract the oil for human food. Two primary processes are used for removing the oil from these oilseed crops. This extraction is done using either the expeller method, a mechanical process, or the solvent method, a chemical process. From a feed value standpoint, the primary difference in these types is that the solvent extraction process is more efficient in removing the oil. As a result, the oilseed meal produced using solvent extraction is a bit lower in fat and energy than the oilseed meal produced using the expeller or mechanical process of oil extraction.

Protein-rich coproducts from the oil extraction process are of great value in livestock feed. These oilseed meals are high in crude protein, most being over 40 percent, as fed basis. With the exception of soybean meal, the protein in oilseed meals is of only fair quality. The essential amino acids lysine and methionine are usually the acids of most concern. Soybean meal is rich in lysine, and its protein is therefore considered to be of good quality. Calcium content in oilseed meals is usually low, but most test high in phosphorus. Usually, at least half of the phosphorus in oilseed meals is in the form of phytate, a form poorly utilized by monogastrics. Oilseed meals generally contain low to moderate levels of the vitamins.

Soybean meal 49 percent (IFN 5-04-612)
Proximate analysis, percent as fed basis

- Crude protein: 49.0
- Crude fat: 1.0
- Crude fiber: 3.0

Major feed applications In processing soybeans, the oil-extracted meal is toasted, a process that improves the value of its protein. Its protein is generally standardized at either 44 percent or 49 percent crude protein, as fed basis, depending on the amount of hull contained in the meal.

The milling of soybeans results in several coproducts that are used in livestock feeds. The value of soybean hulls as a forage substitute has been discussed. Lecithin and soapstock are produced during the refining of soybean oil. Both are liquids and both are high in energy.

When economical, whole soybeans, or "full fat" soybeans are an excellent feedstuff for ruminants. To improve feeding value, the beans are usually heat-treated, usually through roasting. The heat destroys the bean's toxins, and for ruminants, alters the protein to increase the proportion that is resistant to microbial degradation in the rumen. Excessive heat treatment, however, can render the protein totally indigestible. It appears that optimum heat treatment of soybeans intended as a protein supplement for lactating cows is heating the soybeans to a temperature of 295° F and holding them there for 30 minutes (Satter, 1994). By following these guidelines, 50 to 60 percent of the protein in roasted soybeans will be resistant to microbial degradation in the rumen.

Cottonseed meal (IFN 5-07-873)
Proximate analysis, percent as fed basis

- Crude protein: 41.5
- Crude fat: 1.5
- Crude fiber: 12.5

Major feed applications The protein of cottonseed meal is of only fair quality. Most meals are standardized at 41 percent crude protein, as fed basis. Cottonseed meal contains a yellow pigment, gossypol, which is relatively toxic to monogastric species, particularly young pigs and chicks. In addition, the gossypol in cottonseed meal results in poor egg quality. The gossypol is made in glands of the plant. These glands may be removed in processing. Also, "glandless" varieties of cotton have been developed that have very low levels of gossypol. Unfortunately, without the gossypol, high levels of insecticides must be used on these crops to achieve acceptable yields.

Cottonseed hulls are sometimes used as a forage for ruminants.

Whole cottonseed is a widely used feed for ruminants. It contains high levels of protein, fat, and fiber. Generally, whole cottonseed is fed to mature dairy cows at no higher than 6 lb. per head per day.

Other oilseed meals Other oilseeds are processed to extract the oil for use in human food, producing a high-protein coproduct that is used in livestock feed. Examples include canola meal, linseed meal, peanut meal, sunflower meal, and safflower meal.

Canola meal is made from a variety of rapeseed that has less hull and higher digestibility. Most of the toxic qualities of the earlier varieties of rapeseed have been removed in canola, but to avoid off-flavored eggs, canola meal should not be fed to brown-egg layers (NRC, 1994). Canola meal that is not properly heat processed may be goitrogenic for poultry.

Linseed meal is made from flax seed. It accounts for only a small part of the total plant protein produced in the United States. As a consequence of its reputation for enhancing hair coat appearance ("bloom"), it is widely fed to show cattle. Linseed meal is toxic to poultry (Nakaue & Arscott, 1991).

Farmers and ranchers with peanut allergies should avoid use of peanut meal in the feed to which they are exposed.

Safflower is a plant grown in limited amounts for its oil. The meal is high in fiber and low in protein unless the hulls are removed.

Sunflowers are produced for oil and seeds.

Distillers' dried grains (IFN 5-28-236)

Proximate analysis, percent as fed basis

- Crude protein: 27.0
- Crude fat: 8.0
- Crude fiber: 9.1

Major feed applications Coproducts of the distilling industries are valuable animal feedstuffs. These industries produce ethyl alcohol for fuel or in the manufacture of whiskey. The main distillery coproducts are dried distillers' grains and dried distillers' solubles, or mixtures thereof.

Brewers' dried grains (IFN 5-02-141)

Proximate analysis, percent as fed basis

- Crude protein: 25.6
- Crude fat: 6.5
- Crude fiber: 16.0

Major feed applications Coproducts of the beer-brewing industries are also used in animal feeds. The main brewers' coproducts are wet and dried brewers' grains. The wet brewers' coproduct will spoil quickly, and must be consumed within a week in cold weather and within a couple of days in warm weather. It is possible to preserve wet brewers' grains by ensiling.

Corn gluten meal (IFN 5-28-242)

Proximate analysis, percent as fed basis

- Crude protein: 60.0
- Crude fat: 2.5
- Crude fiber: 2.0

Major feed applications Corn gluten meal is the residue after removal of the kernel's starch, germ, and bran. It contains approximately 60 percent protein, the highest of any plant protein product. As with all products derived from corn grain, the protein of corn gluten meal is of only fair quality.

Corn gluten feed (IFN 5-28-243)

Proximate analysis, percent as fed basis

- Crude protein: 22.0
- Crude fat: 3.0
- Crude fiber: 8.0

Major feed applications Corn gluten feed is the residue after removal of the starch and gluten. Because corn gluten feed contains the bran portion of the corn grain, its protein content is much less than that of corn gluten meal.

Animal protein sources

The combination of feedstuffs in the balanced ration must result in the delivery of the type and quantity of amino acids required by the animal. For monogastrics, the amino acids must come directly from the combination of protein sources contained in the ration. Animal protein sources are useful because they are generally of higher quality than plant protein sources. However, palatability problems often limit the amount of animal protein products that may be included in livestock diets. The inclusion of mammalian and poultry derived proteins in livestock diets may be a factor in the transmission of disease. This practice is currently under review by the USDA and FDA.

Blood meal (IFN 5-00-380)

Proximate analysis, percent as fed basis

- Crude protein: 80.0
- Crude fat: 1.2
- Crude fiber: 1.0

Major feed applications Blood meal is a coproduct of the animal rendering industry. When dried without excessive heat as in the spray- or ring-drying process, blood meal is a high protein content, high protein quality feedstuff. If excessive heat is used in processing blood meal, however, the protein is damaged and the feedstuff's value is reduced. Most of the protein in blood meal is resistant to microbial degradation in the rumen.

Feather meal, hydrolyzed (IFN 5-03-795)

Proximate analysis, percent as fed basis

- Crude protein: 85.0
- Crude fat: 3.0
- Crude fiber: 2.0

Major feed applications There are four important issues regarding feather meal as a feedstuff.

1. Feathers test high in crude protein, but the protein in feathers is in the form of keratin, which animals cannot digest. As a feedstuff, therefore, feather meal must be hydrolyzed. Hydrolysis breaks the internal bonds of the keratin in feathers so that animal enzymes can effectively digest the protein.
2. The protein of feather meal is of only fair quality; its amino acid profile is not a good match for growing and producing livestock.
3. Feather meal is probably the least palatable of the animal protein supplements for most livestock. Its use in the ration must be minimal to avoid affecting overall ration palatability. In the dairy industry, feather meal is best kept to less than 1 lb. per animal per day. In swine and poultry diets, it should not be fed at more than 5 percent of the ration (Chiba, Ivey, Cummins, & Gamble, 1996).
4. Feather meal is inexpensive. For this reason, there is considerable incentive to try to overcome its deficiencies and use it more in animal diets.

Fish meal, Herring (IFN 5-02-000)

Proximate analysis, percent as fed basis

- Crude protein: 72.0
- Crude fat: 8.4
- Crude fiber: 0.6

Major feed applications The most common fish used to produce fish meal are menhaden, herring, anchovy, whitefish, and redfish. The definition of fish meal is such that it allows for significant variation among sources. Fish meal may be made from whole fish or fish cuttings (also described as seafood processing waste). If the origin of fish meal is fish cuttings, it may be the portion of the fish remaining after the filets have been removed. This product will test high in mineral content. Mineral content is reported as "ash" on a laboratory analysis. Fish meal with 25 percent ash or more is of low value as a feedstuff for fish. Fish meal also may or may not be produced from fish products from which the oil has been extracted.

Meat meal (IFN 5-00-385)

Proximate analysis, percent as fed basis

- Crude protein: 54.8
- Crude fat: 9.7
- Crude fiber: 2.8

Major feed applications Meat meal is a rendered product derived from the tissues of slaughtered mammals.

Meat and bone meal (IFN 5-00-388)

Proximate analysis, percent as fed basis

- Crude protein: 50.0
- Crude fat: 9.5
- Crude fiber: 2.5

Major feed applications Meat and bone meal is the rendered product from mammal tissues including bone. Because of the inclusion of bone, there is a possibility of contamination with nervous tissue (the spinal cord running inside the backbone). Nervous tissue is the source of the agent that causes bovine spongiform encephalopathy (BSE) ("mad cow disease") and therefore meat and bone meal may not be fed to ruminant animals.

Nonprotein Nitrogen

Urea

Proximate analysis, percent as fed basis

- Crude protein: 287.0
- Crude fat: 0
- Crude fiber: 0

Major feed applications Although some feedstuffs, particularly hay crop silage, contain substantial amounts of nonprotein nitrogen (NPN), from a practical point of view, NPN in formula feeds usually refers to urea. Rations containing large amounts of hay crop silage will not benefit from the addition of urea.

NPN, especially urea, is primarily used for feeding animals with a functioning rumen. The reason for this is that urea is a source of nitrogen that is converted by rumen microbes to ammonia, which is used by rumen microbes to build amino acids and protein. The microbes' protein is, in turn, used by the animal hosting the microbes when the microbes are ultimately washed out of the rumen and digested. Urea should be introduced to the diet gradually so that rumen microbes have an opportunity to adapt to its presence.

The companion application to this text predicts the amount of microbial protein produced in the rumen for dairy, beef, sheep, and goat rations. Levels of urea beyond that which results in increased microbial protein production are to be avoided as they can result in absorption of urea into the blood of the ruminant, leading to reproductive and health problems. Urea toxicity in livestock is evidenced by an ammonia odor on the animal's breath. Additional information regarding the feeding of urea and blood urea levels is found in Chapter 16.

Urea should not be fed with raw soybeans or raw soybean hulls. These feeds have urease activity that will result in the conversion of urea to ammonia, possibly leading to toxicity.

The main advantage of using urea in ruminant diets is a saving of feed dollars. In other words, amino acids can be manufactured from urea by rumen bacteria for less money than would be required to purchase intact protein.

Feed grade biuret is another source of nonprotein nitrogen. It was developed as an alternative to urea in an attempt to improve palatability and rumen-release characteristics.

Mineral Supplements

Images and supplemental information on mineral supplements are found in the file titled Images.ppt on the text's accompanying On-Line Companion. To view this file please go to http://www.agriculture.delmar.com and click on *Resources.* Click on *On-Line Resources* and select from the titles listed.

Virtually all feedstuffs have some mineral content, but supplemental mineral is usually necessary in livestock rations to meet requirements. Which minerals must be supplemented and to what level can only be determined by analyzing the mineral content of the individual feedstuffs and calculating the contribution of each mineral from each feedstuff consumed in the ration.

Laboratory analysis is the only way to accurately determine the mineral content of a feedstuff. However, it is important to realize that the digestive tract of livestock may not be able to extract and absorb all the mineral that the analysis indicates is in the feedstuff. *Bioavailability* refers to the portion of the total mineral in the feedstuff that is absorbed into the blood from the digestive tube. The most important factors affecting mineral bioavailability are the mineral source, the method and amount of processing of the mineral source, and the livestock species fed. Ideally, the mineral content of each mineral source is accompanied by the bioavailability data for the animal to be fed, but this is not usually the case.

When bioavailable mineral values are not known, the requirement values are usually adjusted upward with so-called "safety factors" to account for the unavailable mineral content of feedstuffs. Mineral excretion due to the use of safety factors has become a pollution concern.

Mineral supplementation is best achieved by mixing the sources of mineral with the rest of the ration because many mineral sources are unpalatable. This is sometimes called "force feeding" minerals. Animals do, however, have an appetite for salt and will generally eat salt to meet their requirement. Salt may be offered to livestock in loose form or in block form.

Animals that have been fed diets deficient in salt may consume excess salt when a salt source is introduced. Such animals should be offered only as much salt as they require until their appetite for salt has returned to normal.

A salt source may be fortified with other minerals so that when the animal meets its salt requirement, it has also met its requirement for these other minerals. Such a trace mineralized salt source may also be offered in loose form or in block form. An example of the formula and analysis of one type of trace mineralized salt block is shown in Table 3–9.

Vitamin Supplements

Images and supplemental information on vitamin supplements are found in the file entitled Images.ppt on the text's accompanying On-Line Companion.

Table 3–9
A formula and analysis of a trace mineralized salt block

Nutrient	Type of Guarantee	Concentration
Sodium chloride	Maximum	97.00%
Sodium chloride	Minimum	95.00%
Zinc	Minimum	0.600%
Manganese	Minimum	0.330%
Iron	Minimum	0.240%
Copper	Minimum	0.340%
Iodine	Minimum	0.130%
Cobalt	Minimum	0.120%
Ingredients		
Salt, zinc oxide, manganous carbonate, iron sulfate, copper sulfate, potassium iodide, cobalt carbonate, mineral oil, and cane molasses.		

To view this file please go to http://www.agriculture.delmar.com/ and click on *Resources.* Click on *On-Line Resources* and select from the titles listed.

Vitamin content of feedstuffs varies based on the type of feedstuff, treatment during harvest, processing method if any, and the conditions and duration of storage of the feedstuff. Given the variation, it would seem logical that analysis of feedstuff vitamin content would be routine. However, this is not the case.

Determining the vitamin content of feedstuffs is time-consuming and expensive. It involves the use of biological and/or chemical methods of evaluation.

The biological method measures a feedstuff's potency in eliminating signs of a vitamin deficiency, and will include all sources of vitamin activity rather than a single vitamin source. The feed to be tested is fed at several levels to different groups of animals, typically rats or chicks, as a supplement to a vitamin-free diet that has caused symptoms of vitamin deficiency. The response to the feed is compared to a similar group of rats or chicks receiving a standard source of the vitamin at the level required.

A chemical method involves the use of chromatographic procedures to separate the biologically active vitamin source(s).

Analyzing the content of individual vitamins in feedstuffs is usually unnecessary. It is more cost effective to supplement vitamins to their required level than it would be to test feeds and supplement to meet calculated deficiencies. As indicated in Table 10–1, not all species require dietary supplementation of all vitamins.

Liquid Feeds

Liquid feedstuffs include distillers' solubles, fish solubles, corn fermentation solubles, lecithin, soapstock, glutamic acid fermentation liquor, and propylene glycol. The most common liquid feedstuff is molasses, which is often used as a medium to carry other ingredients in liquid form. Technologies have been developed to incorporate many types of feed ingredients in liquid supplements. Intake of molasses-based liquid supplements in ruminant rations is usually limited to 1 to 3 lb. per head per day.

Additives

Images and supplemental information on additives are found in the file entitled Images.ppt on the text's accompanying On-Line Companion. To view this file please go to http://www.agriculture.delmar.com/ and click on *Resources.* Click on *On-Line Resources* and select from the titles listed.

In the feed industry, the term *additive* is used to identify ingredients in a ration that do not directly meet known nutrient requirements. Additives are usually used in small quantities and, therefore, require careful handling and mixing. The use of some additives is regulated by federal law (USDA, Center for Animal Health Monitoring). Any substance that has been implicated as a carcinogen when fed to any animal at any level may not be used as a feed additive. This is the result of the "zero tolerance" policy adopted by Congress as a result of the Delaney Clause passed by Congress in 1958.

Buffers

Buffers are included in ruminant rations to help counteract the acid generated by ruminal fermentation of grain. Examples of buffers fed to livestock include sodium bicarbonate and sodium sesquicarbonate. Sodium bicarbonate is osmotically active and may influence performance in ruminant animals by increasing liquid turnover rate and thereby dry matter intake. Though technically an alkalizing agent and not a buffer, magnesium oxide is often used in livestock rations to prevent the accumulation of acid in the digestive system.

Antioxidants

Polyunsaturated fats may react with oxygen and fall apart, leading to the creation of new compounds that are unpalatable and may be toxic. This process is called *oxidative rancidity.* Vitamins A, D, and E as well as several of the B vitamins may also be destroyed as a result of oxidative rancidity. To protect polyunsaturated fats from oxidative rancidity, antioxidants are often added (at approximately 0.25 lb./ton) to feeds. Examples include butylated hydroxytoluene (BHT), butylated hydroxyanisole (BHA), and ethoxyquin. Several vitamins themselves act as antioxidants. However, as vitamins work to protect against oxidative rancidity in the feed, they are "used up" and are no longer available to the animal consuming the feed.

Hormones

Hormones are made inside cells and are secreted by ductless glands into the blood, where they are carried to all parts of the body to exert their influence on tissues or other glands. Because fat-based compounds are absorbed by the digestive tract largely unaltered, feeding a fat-based hormone may result in absorption of that hormone. Feeding a protein-based hormone, however, will not result in absorption because proteins are digested and absorbed as amino acids. As a group, the fat-based hormones are called *steroids.* Synthetic steroids and substances that have steroid-like effects may be used as feed additives to improve feed efficiency and rate of gain. Steroids in feed may be used as a management tool to manipulate the estrous cycle so that animals all come into heat at the same time.

Antibiotics and Chemotherapeutic Agents

Most antibiotics are compounds produced by fungi, bacteria, or algae. Chemotherapeutic agents such as arsenical compounds are fed to swine and poultry for much the same reason as antibiotics.

Antibiotics and chemotherapeutic agents inhibit or destroy microorganisms in the animal that are detrimental to health or performance. In the feed industry, these substances are added at subtherapeutic levels (too low to be effective in treating disease) to reduce the incidence of subclinical bacterial infections, and to improve rate of gain and feed efficiency for approved species. Antibiotics are used in complete feeds at an inclusion rate of 2 to 10 mg/kg (parts per million); chemotherapeutic agents are usually used at somewhat higher inclusion rates. Antibiotics and chemotherapeutic agents fed at subtherapeutic levels are also helpful in controlling certain health problems such as respiratory infection, liver abscesses, foot rot, and diarrhea in livestock. These additives give the greatest response in those animals that are kept in unsanitary, stressful environments. The conditions under which these drugs may be used as feed additives are closely prescribed and care must be taken that they are used properly to avoid residues in human food. Antibiotic use as a feed additive is under review because of the risk that microorganisms that cause disease in humans may develop resistance through exposure to antibiotics in livestock feed. Regulations pertaining to the use of antibiotics and chemotherapeutic agents in the United States are found in the Feed Additive Compendium, which is published annually.

Ionophores are a unique type of antibiotic obtained from *Streptomyces* microorganisms. They are not used in treating human disease. Ionophores are fed to prevent coccidiosis in poultry and young ruminants, and to improve feed efficiency in growing and lactating ruminants.

Ionophores improve feed efficiency in ruminants by changing the microbial population in the rumen. Ionophores cause sensitive bacteria to lose potassium (a source of base) and the cell cytoplasm is, therefore, at risk of turning acid. These bacteria then use all available adenosine 5′-triphosphate (ATP) to pump acid out of the cell. With such a demand on the ATP supply, the sensitive bacteria stop growing.

Only Gram-positive bacteria are sensitive to ionophores. Gram-negative bacteria have outer membranes that keep ionophore out. In the presence of ionophore, the rumen population of Gram-negative bacteria grows along with the products they produce, and the population of Gram-positive bacteria declines along with the products they produce. The result is:

- Lower acetate-to-propionate ratio, because Gram-negative bacteria produce more propionate and Gram-positive bacteria produce more acetate.
- Decreased methane production because methane production is associated with the production of acetate (Gram-positive bacteria).
- Amino acid sparing (more efficient use of dietary protein) because some of the sensitive bacteria are guilty of converting amino acids to ammonia.
- Increased feed efficiency due to all of the above. Increased feed efficiency means that:
  - less ingested feed will be needed to meet maintenance energy needs. Therefore, a greater portion of the energy in ingested feed will be available for other functions such as growth and lactation. The companion application to this text for beef and dairy heifers applies a credit to the feed NEm value, increasing the ration NEm concentration by 12 percent when ionophore is used. In the companion application, ionophore use in dry and lactating dairy cows results in a 12 percent reduction of the net energy required for maintenance, excluding that due to stress.
  - maintenance energy needs will be more easily met with what little feed is ingested. By making it easier for the appetite-depressed cow at the start of her lactation to meet her maintenance energy needs, she will be less susceptible to all of the metabolic problems associated with an energy deficit.
- Most studies indicate that use of an ionophore results in a reduction in dry matter intake that diminishes at least some of the benefits listed above. The companion application to this text for beef animals predicts a reduction of 6 percent dry matter intake when ionophore is used. In the dairy application, the reduction in dry matter intake due to ionophore use is an inputted value from within the range of 0 to 6 percent.

Direct-Fed Microbials or Probiotics

As the name implies, probiotics are products that promote the growth of microbial populations. Probiotics, or direct-fed microbials, consist of microbes that have been selected for specific desirable characteristics. These products may be helpful in establishing a more desirable population of gastrointestinal microorganisms. The use of probiotics in livestock diets has been reviewed (Ghorbani, Morgavi, Beauchemin, & Leedle, 2002).

Nutrients as Additives

Nutrient requirements are established based on the level of nutrient needed to support health and performance in animals under usual circumstances. Animals in unusual circumstances, such as those experiencing disease or severe strain, may benefit from higher levels of some nutrients. Nutrients used to help the animal deal with unusual circumstances are appropriately described as additives. Another term that has been used to describe nutrients used in this way is *nutriceuticals.*

In the ruminant animal, feedstuffs are subject to microbial activity before reaching the site of absorption. This means that amino acids and most water-soluble vitamins that are in the ration will not be available to the ruminant animal. For this reason, various technologies have been developed to protect substances from the microbes while allowing them to be absorbed by the ruminant. This has allowed researchers to deliver nutrients directly to the animal. Feeding a rumen-protected source of the B vitamin choline improves liver performance (Piepenbrink & Overton, 2003), and may help to prevent fatty liver and ketosis in high producing cows.

Feeding niacin as a nutriceutical has reduced plasma ketone concentration (Grummer, 1993) and may help prevent ketosis in dairy animals when fed at levels well beyond the usual requirement.

Health and performance benefits have been shown when feeding unusually high levels of chromium (Lindemann et al., 1995), copper (Davis et al., 2002), and zinc (Hill et al., 2000) to swine.

Flavors

Flavoring agents are feed additives that are popular in the pet food industries. Studies show that animals given a choice will often show a preference for feed containing flavor additives. However, this preference generally does not translate to increased consumption of flavored feed over unflavored feed when no choice is offered.

A common flavor enhancer is termed *digest.* Digest is sometimes sprayed onto dry pet foods. Digest is made from poultry and beef byproducts using an enzymatic process to reduce these materials to a liquid.

Yeast and Yeast Culture

Yeast culture is used in microbiology to stimulate the growth of specific types of bacteria, particularly those that utilize lactic acid. As a feed additive, yeast culture is used to increase the vitality of bacteria inhabiting the digestive tract. It is often included in the diet to help animals weather stressful environments or conditions. Live yeast is a feed additive intended to deliver the yeast culture through the activity of the live yeast in the animal's digestive tract.

Enzymes

Animals produce enzymes in the glands of their digestive system. In addition, enzymes are produced by the microbes inhabiting animal digestive systems, and feedstuffs may contain active enzymes when eaten. Needless to say, enzymes play an important role in digestion. For this reason, there has been much research to determine if digestion would be improved by the dietary supplementation of enzymes.

To date, the only enzyme that has gained acceptance in the feed industry is phytase. Phytase is an enzyme that degrades phytin or phytic acid. Phytin is the compound containing most of the phosphorus in grains. By using phytase, producers can meet animal phosphorus requirements with less total dietary phosphorus, this leads to reduced phosphorus excretion. Because phytase is made by the microbes inhabiting the rumen, phytase is only considered for monogastric animals, particularly poultry (Perney, Cantor, Straw, & Herkelman, 1993) and swine (O'Quinn, Knabe, & Gregg, 1997).

Pigments

Xanthophylls are carotenoid pigments that are used to contribute to the coloration of livestock tissues. Astaxanthin is a xanthophyll used in salmon and trout diets to achieve an attractive flesh and/or skin pigmentation. Corn gluten meal is a source of xanthophylls that is used in poultry diets to deepen the color of egg yolk and to impart a yellow tint to the skin of broilers. Other sources of xanthophylls include marigold petal meal, dried algae, and alfalfa meal. Xanthophylls may also act as antioxidants, and as such, may have positive effects on health and reproductive performance.

Mold Inhibitors and Mycotoxin Binders

The most effective ways to keep feed from becoming moldy are to keep feed moisture content below 13.5 percent, keep feed fresh, and keep equipment clean. Mold inhibitors can be temporarily effective in stopping mold growth after it has begun or in protecting feed from the start of mold growth. Examples of mold inhibitors include organic acids (propionic, sorbic, benzoic, and acetic acids) and salts of organic acids (calcium propionate and potassium sorbate). Although it is sometimes used as a mold inhibitor, the effectiveness of copper sulfate is difficult to document.

One of the problems caused by mold growth in feed is the production of toxins called *mycotoxins.* Mycotoxin binders are products that bind mycotoxins and may thereby reduce their toxic effects on livestock consuming the unwholesome feed. Mycotoxin binders are clay compounds, an example of which is sodium bentonite. For a discussion of the problems associated with moldy feed, see Chapter 13.

Other Additives

There are many other types of substances that find their way into livestock diets to serve a variety of functions, including wormers (anthelmintic agents), fly control agents, pellet binders, bloat preventatives, and ergogenic aids. The latter are feed products designed to improve the athletic performance of horses. Products such as sodium bicarbonate, carnosine, and carnitine are sometimes fed as ergogenic aids. The feeder should be skeptical of these and other products described as ergogenic aids because in most cases, they provide no benefit and may be banned by racing authorities.

FINISHED FEEDS

Finished feeds are mixes or blends of ingredients that are handled as a single feedstuff. It is easiest to define and describe finished feeds by considering the products manufactured and sold by feed mills. Feed mills sell the following types of finished feeds.

1. Complete feeds
2. Concentrate feeds
3. Base mixes
4. Premixes

A complete feed is all necessary ration components that are inventoried by the feed mill. Since feed mills generally do not inventory forages, the complete feed contains all nonforage ration components. For nonforage eaters, the complete feed makes up the entire daily ration. It may be fed as meal, pellets, crumbles, or in the form of a liquid or paste. Complete feed may also be available in a form described as textured or coarse. A textured complete feed generally contains up to 15 percent molasses and other ingredients in a variety of forms including pellets, whole grains, and processed grains. Because of the molasses content, textured feeds may also be described as sweet feeds. For ruminant animals and any animal that is to be fed forage, a complete feed contains all ration components except the forage.

Regardless of the type of farm or ranch, livestock are always fed complete feeds. The complete feed may be fed through a total mixed ration (TMR) in which a mixer has been used to blend all ration components, including forage. Alternatively, the complete feed may be fed through a component feeding system in which at least some individual ingredients are fed unmixed. In either system, by the end of a 24-hour period, livestock have received a complete feed, and if the animal is a herbivore, forage.

Complete feeds may be purchased directly from the feed mill. However, savings may be realized by purchasing some feedstuffs that would be part of the complete feed from different sources. These purchase savings must be weighed against the additional storage, handling, and mixing expenses that would be incurred in this type of feed buying strategy.

Corn meal is the major component in most complete feeds. Often, a farm or ranch will purchase a blend of all ingredients of the complete feed except the corn meal from one source and purchase the corn meal separately. A blend of all ingredients except the corn meal is called a *concentrate* or *supplement.* The concentrate inclusion rate in a ton of complete feed is usually at least 200 lb.

Soybean meal is the second most plentiful component of most complete feeds. If the farm or ranch purchases corn meal and soybean meal separately, the remaining ingredients are purchased as a base mix or super concentrate. A base mix may still contain some of the protein source for the complete feed, but it contains all the mineral fortification, both macro and trace, and the vitamins. The base mix inclusion rate in a ton of complete feed is usually 100 to 200 lb.

The macro minerals as a group are the third most plentiful component of most complete feeds. If the farm or ranch purchases corn meal, soybean meal, and macromineral sources such as salt, limestone, dicalcium phosphate, and magnesium oxide separately, what remains to make the complete feed is one or more premixes. One premix may contain all the remaining ingredients—the trace minerals, vitamins, mold inhibitors, flavor ingredients, antibiotics, and so on. Alternatively, these ingredients may be packaged in separate premixes. The premix inclusion rate in a ton of complete feed is usually 5 to 100 lb. The smaller inclusion rate premixes will usually include a carrier to ensure that the mixing equipment used will be able to effectively mix the target ingredient. Examples of carriers include rice hulls, calcium carbonate, ground corn cobs, and wheat middlings.

SUMMARY

Chapter 3 lists and describes the general categories of feedstuffs. The categories, as identified by the NRC's IFN are:

1. Dry forages
2. Pasture, range plants, and feeds cut, and fed green
3. Silages

4. Energy concentrates
5. Protein supplements
6. Minerals
7. Vitamins
8. Additives

Some of the nutritional and agronomic characteristics of the eight feed categories are discussed. Because forages are an important part of the diet for some livestock types, and because the forages are usually homegrown, special emphasis is given to the harvesting and storage of forages. The application of feedstuffs from the various categories in developing a balanced ration for livestock is described.

END-OF-CHAPTER QUESTIONS

1. Give three examples each of grasses and legumes. Comparing grasses and legumes of similar maturity, which two nutrients are generally present at a higher concentration in legumes?
2. Why are spoilage bacteria and molds unable to grow in well-made hay?
3. Compare and contrast losses that occur when making dry hay versus hay crop silage.
4. Discuss the importance of reducing the moisture content of the hay crop prior to ensiling.
5. What factors influence the amount of packing that is necessary to effectively exclude oxygen in a hay crop put up in a horizontal silo?
6. High stocking density and short access time are characteristic of what type of pasture management?
7. Describe the following terms as they refer to mixed feeds: complete feed, concentrate, base mix, and premix.
8. What is the approximate oil content in the whole oilseeds? Among the oilseed meals, which is the source of the highest quality protein?
9. When discussing feed protein sources, to what does the term *protein quality* refer?
10. What does the following formula calculate?

    $$[(\text{wet weight} - \text{dry weight})/\text{wet weight}] \times 100$$

    What is the recommended percent dry matter range for corn silage? What is the recommended percent dry matter range for hay crop silage?

REFERENCES

Abaye, A. O., Allen, V. G., & Fontenot, J. P. (1997). Grazing sheep and cattle together or separately: Effect on soils and plants. *Agronomy Journal. 89,* 380–386.

American Sheep Industry Association, Inc. (1996). *Sheep production handbook* (pp. 384–385). Denver, CO: C&M Press.

Cheeke, P. R. (1995). Endogenous toxins and mycotoxins in forage grasses and their effects on livestock. *Journal of Animal Science. 73,* 909–918.

Cherney, D. J. R. (1991, October 10). Low-lignin, brown mid-rib genotypes and their potential for improving animal performance (pp. 13–19). *Proceedings Cornell Nutrition Conference for Feed Manufacturers,* Rochester, NY.

Chiba, L. I., Ivey, H. W., Cummins, K. A., & Gamble, B. E. (1996). Hydrolyzed feather meal as a source of amino acids for finisher pigs. *Animal Feed Science and Technology. 57,* 15–24.

Colenbrander, V. F., Bauman, L. F., & Lechtenberg, V. L. (1975). Feeding value of low lignin corn silage. *Journal of Animal Science. 41,* 332.

Davis, M. E., Maxwell, C. V., Brown, D. C., de Rodas, B. Z., Johnson, Z. B., Kegley, E. B., Hellwig, D. H., & Dvorak, R. A. (2002). Effect of dietary mannan oligosaccharides and (or) pharmacological additions of copper sulfate on growth performance and immunocompetence of weanling and growing/finishing pigs. *Journal of Animal Science. 80,* 2887–2894.

Feed Additive Compendium. (2004). Minnetonka, MN: Miller Publishing Co.

Fox, D. G., Tylutki, T. P., Van Amburgh, M. E., Chase, L. E., Pell, A. N., Overton, T. R., Tedeschi, L. O., Rasmussen, C. N., & Durbal, V. M. (2000). The net carbohydrate and protein system for evaluating herd nutrition and nutrient excretion (CNCPS vol. 4.0. p. 212). Animal Science Department Mimeo 213, Cornell University, Ithaca, NY.

Frenchick, G. E., Johnson, D. G., Murphy, J. M., & Otterby, D. E. (1976). Brown midrib corn silage in dairy cattle rations. *Journal of Dairy Science. 59,* 2126.

Ghorbani, G. R., Morgavi, D. P., Beauchemin, K. A., & Leedle, J. A. Z. (2002). Effects of bacterial direct-fed microbials on ruminal fermentation, blood variables, and the microbial populations of feedlot cattle. *Journal of Animal Science. 80,* 1977–1985.

Grummer, R. R. (1993). Etiology of lipid-related metabolic disorders in periparturient dairy cows. *Journal of Dairy Science. 76,* 3882–3896.

Hill, G. M., Cromwell, G. L., Crenshaw, T. D., Dove, C. R., Ewan, R. C., Knabe, D. A., Lewis, A. J., Libal, G. W., Mahan, D. C., Shurson, G. C., Southern, L. L., & Veum, T. L. (2000). Growth promotion effects and plasma changes from feeding high dietary concentrations of zinc and copper to weanling pigs (regional study). *Journal of Animal Science. 78*(4), 1010–1016.

Hoglund, C. R., (1964). Michigan State University. *Agricultural Economics Publication # 947.*

Holland, C., & Kezar, W. (1995). *Pioneer forage manual—A nutritional guide.* Des Moines, IA: Pioneer Hi-Bred International, Inc.

Lindemann, M. D., Wood, C. M., Harper, A. F., Kornegay, E. T., & Anderson, R. A. (1995). Dietary chromium picolinate additions improve gain:feed and carcass characteristics in growing-finishing pigs and increase litter size in reproducing sows. *Journal of Animal Science. 73,*(2) 457–465.

Nakaue, H. S., & Arscott, G. H. (1991). Feeding poultry. In D. C. Church (Editor), *Livestock feeds & feeding* (3rd edition). Englewood Cliffs, NJ: Prentice-Hall.

National Research Council. (1972). *Atlas of nutritional data on United States and Canadian feeds.* Washington, DC: National Academy Press.

National Research Council. (1994). *Nutrient requirements of poultry* (9th revised edition). Washington, DC: National Academy Press.

Nolan, T., & Connolly, J. (1989). Mixed vs. monograzing by steers and sheep. *Animal Production 48,* 519–533.

Northeast DHIA Forage Laboratory. (1995). Tables of feed composition. Ithaca, NY.

O'Quinn, P. R., Knabe, D. A., & Gregg, E. J. (1997). Efficacy of natuphos in sorghum-based diets of finishing swine. *Journal of Animal Science. 75,* 1299–1307.

Palmquist, D. L., & Jenkins, T. C. (1980). Fat in lactation rations: review. *Journal of Dairy Science. 61*(1), 1.

Perdok, H. B., & Leng, R. A. (1987). Hyperexcitability in cattle fed ammoniated roughages. *Animal Feed Science and Technology. 17,* 121–143.

Perney, K. M., Cantor, A. H., Straw, M. L., & Herkelman, K. L. (1993). *Poultry Science. 72,* 2106.

Piepenbrink, M. S. & Overton, T. R. (2003). Liver metabolism and production of cows fed increasing amounts of rumen-protected choline during the periparturient period. *Journal of Dairy Science. 86,* 1722–1733.

Ruppel, K. A., Pitt, R. E., Chase, L. E., and Galton, D. M. (1995). Bunker silo management and its relationship to forage preservation on dairy farms. *Journal of Dairy Science. 78,* 141–153.

Satter, L. D. (1994, October 18). Use of heat processed soybeans in dairy rations (pp. 19–28). *Proceedings Cornell Nutrition Conference for Feed Manufacturers,* Rochester, NY.

Undersander, D., Martin, N., Cosgrove, D., Kelling, K., Schmitt, M., Wedberg, J., Becker, R., Grau, C., Doll, J., & Rice, M. E. (1994). *Alfalfa management guide.* Published by American Society of Agronomy, Inc.; Crop Science Society of America, Inc.; Soil Science Society of America, Inc. Produced at Cooperative Extension Publications, University of Wisconsin-Extension.

USDA Center for Animal Health Monitoring. USDA: APHIS: VS. Centers for Epidemiology and Animal Health, Fort Collins, CO. Retrieved October 13, 2003 from http://www.aphis.usda.gov/vs/ceah/cahm

USDA. (1987). *The official United States standards for grain.* Washington, DC: Federal Grain Inspection Service.

CHAPTER 4

SAMPLING AND ANALYSIS

If you can measure it, you can manage it.

ANONYMOUS

INTRODUCTION

The key to balancing rations for livestock is in knowing the nutrient content of what you have to work with. Finding out the nutrient content involves using proper sampling technique and laboratory analysis. Applying this information requires an ability to interpret the analysis results reported by the laboratory.

VARIATION IN FEEDSTUFF NUTRIENT CONTENT

The nutrients required by livestock are supplied in balanced rations by feed ingredients or feedstuffs. Feedstuffs vary in the nutrients they contain and in the availability of these nutrients to livestock. This variation is quite consistent with grains, protein supplements, and mineral and vitamin supplements. In fact, for these feedstuffs, it is usually acceptable to use analysis values in reference sources. However, for forages, the variation is not consistent. The analysis of this year's alfalfa crop may be dramatically different than that for last year's crop so that average values will not be useful in developing a balanced ration.

Forages, such as alfalfa, form the foundation of diets for ruminants (beef, dairy, sheep, and goats), horses, and rabbits. However, for growing, working, and lactating animals, forages will not provide sufficient nutrients to meet requirements. Supplemental feedstuffs containing more concentrated sources of nutrients must be a part of these animals' diets. It is important to recognize that supplementation of any nutrient beyond the requirements will not benefit the animal being fed, and excessive supplementation may create toxicity problems. Therefore, supplements should be used to address specific deficiencies. Again, to know what deficiencies exist in animal diets, each lot of forage must be properly sampled and sent to a laboratory for analysis of nutrient content.

SAMPLING PROCEDURE

A knowledge of the nutrient content of a feedstuff helps determine the potential usefulness of that feedstuff in a ration. Analytical methods have been developed that make it possible to measure feedstuff nutrient content for a large number of nutrients. These values are of no practical use, however, if the sample received by the laboratory is not representative of the feedstuff as it is fed to livestock. Collecting a representative feedstuff sample is, therefore, the first step of the analytical process. Following recommended sampling procedures will help ensure that the laboratory results truly reflect the nutritive value of the feedstuff.

The following forage sampling procedures are adapted from the Dairy One Forage Laboratory mailer insert (2001). Images and supplemental information related to feed sampling are found in the file titled Images.ppt on the text's accompanying On-Line Companion. To view this file please go to http://www.agriculture.delmar.com and click on *Resources*. Click on *On-Line Resources* and select from the titles listed.

Hay

Hays of different types, cuttings, or lots should be sampled separately. Using a hay probe, bore 12 to 20 bales selected at random. Combine all core samples, and from this, submit a subsample for analysis.

Silage

Grab handfuls of silage from 12 to 20 locations in the unloaded silo pile, across the silo face, in the feed bunk, or from in front of 12 to 20 animals. All subsamples should be combined and thoroughly mixed in a clean plastic bucket to form a composite sample. Submit 1 lb. of the composite for analysis.

Pasture

Randomly select 12 to 20 sites where the animals have been grazing and clip the forage at grazing height. All subsamples should be combined and thoroughly mixed in a clean plastic bucket to form a composite sample. Take a 1-lb. composite sample, pack tightly in a plastic bag, and freeze for 12 hours prior to submitting for analysis. Freezing will help prevent marked chemical changes due to respiration or fermentation.

Grains and Ingredients

Bin storage: Randomly collect 12 to 20 samples as the grain is discharged and combine in a clean plastic bucket.

Flat storage: Grab 12 to 20 samples from various sites and combine in a clean plastic bucket. Thoroughly blend and submit a 1-lb. sample for analysis.

Note: wherever possible, a grain probe should be used to take the sample.

FEED MICROSCOPY

Feed microscopy is a complement to routine chemical quality-control analyses. It involves an evaluation by a trained microscopist to provide a rapid evaluation of ingredient or finished feed quality. Feed microscopy produces a subjective opinion that can be a useful component of a feed mill's quality-assurance program. Feed microscopy can be used to identify weeds, insects, molds, and mold spores, and otherwise contaminated, adulterated, or damaged feedstuffs arriving at the mill's receiving area.

Feed microscopy may be used to evaluate feed particle size. Particle size information may, in turn, be used to improve manufacturing processes, or adjustments may be made to improve the health and performance of livestock consuming the feed. Feed microscopy can also be used to evaluate mixing efficiency and ingredient-segregation problems.

ANALYSIS PROCEDURES AND VALUES

Proximate Analysis

One of the oldest laboratory methods for predicting the value of a feedstuff for animals is a system of chemical analysis termed *proximate analysis.* By appropriate analytical procedures, the components of a feedstuff can be divided into six categories: crude protein, crude fiber, nitrogen-free extract, crude fat (ether extract), ash, and moisture. These classifications can give some information about the usefulness of a feedstuff.

Table 4–1 shows the feedstuff carbohydrate components as partitioned in proximate analysis. Notice that the hemicellulose and lignin fractions are split between crude fiber and nitrogen-free extract. Because hemicellulose is digestible and lignin is not, feedstuffs that potentially have significant amounts of hemicellulose and/or lignin are not well described by crude fiber and nitrogen-free extract. Such feedstuffs include forages such as pastures, hay crops, corn silage and sorhgum silage, as well as some common byproducts including distillers' grains, brewers' grains, and beet pulp.

Crude Protein

The average protein molecule contains about 16 percent nitrogen. This value is used to estimate the protein content of feed samples. The Kjeldahl procedure of proximate analysis is used to measure the nitrogen content of a sample. The nitrogen value is multiplied by 6.25 to obtain the approximate protein content (Figure 4–1). Because protein is not the only feed component that contains nitrogen, this calculated value is called *crude protein.*

In livestock nutrition, crude protein is not an adequate description of feed protein value. In monogastric nutrition, the amino-acid content of protein must be known because ration balancing involves meeting amino-acid requirements using feed protein sources. In ruminant nutrition, feed protein needs to be described according to its susceptibility to microbial degradation in the rumen. Though crude protein content must be stated on the feed tags accompanying the sale of commercial feeds, it is also a legal requirement that feed tags specify

Table 4–1
Feed carbohydrates as identified in the proximate analysis procedure

Component	Relative Digestibility[1]	Crude Fiber	Nitrogen-Free Extract
Sugar	1		✓
Starch	1		✓
Hemicellulose	2	✓	✓
Cellulose	3	✓	
Lignin	4	✓	✓

[1]Using a scale from 1 (highly digestible) to 4 (indigestible).

Figure 4–1
Calculating a feedstuff's crude protein content

The average protein molecule is 16% nitrogen. *Percent* means parts per 100 parts, so this can be restated as 16 parts nitrogen/100 parts protein. The reciprocal is also true: 100 parts protein/16 parts nitrogen. If a feed sample tests at 3% nitrogen, its approximate protein content can be calculated as follows:

$$100 \text{ parts protein}/16 \text{ parts nitrogen} \times 3 \text{ parts nitrogen}/100 \text{ parts feed} = 0.1875 \text{ parts protein}/1 \text{ part feed}$$

In percentage form, this is 18.75%. The calculation can be simplified by multiplying the result of 100/16 by the feed's percentage of nitrogen:
100/16 = 6.25
In the previous example, 3% × 6.25 = 18.75%

Figure 4–2
Calculating equivalent crude protein from nonprotein nitrogen

If a 2000-lb. mix contained 30 lbs. of urea, which contains 45% nitrogen (N), the calculation of equivalent crude protein (CP) from nonprotein nitrogen, as fed basis, would be:

$$\frac{30 \text{ lb. urea}}{2000 \text{ lb. mix}} \times \frac{45 \text{ lb. N}}{100 \text{ lb. urea}} \times \frac{100 \text{ lb. CP}}{16 \text{ lb. N}} \times 100 = 4.22\%$$

the amount of this crude protein coming from nitrogen-containing compounds other than true protein. On the feed tag, this is given as percent equivalent crude protein from nonprotein nitrogen, as fed basis (Figure 4–2). In the companion application to this text, the equivalent crude protein from nonprotein nitrogen is calculated when blending feedstuffs for ruminants.

In ruminant nutrition, feed protein is described as degradable, undegradable/digestible, or undegradable/indigestible. Degradable protein is degraded in the rumen and used by the microbes. Undegradable/digestible protein is resistant to microbial activities, and arrives at the small intestine intact where it is digested and absorbed. Undegradable/indigestible protein bypasses the rumen and is indigestible in the small intestine, and emerges in the feces. The characterization of these protein fractions is described in Chapter 7.

The application of heat to a feedstuff can change the digestibility characteristics of its protein. During the fermentation process, for example, silages sometimes heat excessively, potentially changing the protein. This heat leads to a reaction between the carbohydrate and protein in the feed. Acid detergent fiber insoluble protein (ADFIP) on the forage analysis identifies unavailable protein, including that lost to heat damage.

Neutral detergent fiber insoluble protein (NDFIP) is used in protein and energy calculations in ruminant nutrition. NDFIP is an estimation of the portion of the crude protein that is true protein but slowly degraded or undegradable in the rumen. ADFIP, which is not only undegradable in the rumen but also indigestible in the small intestine, is a subset of NDFIP. NDFIP minus ADFIP identifies slowly degraded or undegradable, digestible true protein.

NDFIP is also used to calculate the digestible energy (DE) content of some feedstuffs through its use in the calculation of digestible neutral detergent fiber (NDF) and digestible nonfiber carbohydrate (NFC).

It should be noted that ADFIP and NDFIP were designed to evaluate forages and are not considered useful indicators of heat exposure with animal byproducts and nonforage plant protein sources (Satter, Dhiman, & Hsu, 1994).

Carbohydrate

For forages, the Van Soest system is used to partition forage carbohydrate components in a more useful way than is possible with the proximate analysis method. Table 4–2 shows the components of the Van Soest system.

Structural carbohydrates, measured as NDF, constitute the cell wall of a plant. The cell walls of plants can be compared to reinforced concrete in which the reinforcing steel bars are represented by the cellulose fibrils, and the concrete is represented by the matrix material that includes hemicellulose and an indigestible polymer called *lignin*. NDF includes the fiber fractions cellulose, hemicellulose, lignin, and NDFIP. Acid detergent fiber (ADF) is the NDF less the most digestible fiber component, hemicellulose. To the extent that a feedstuff's ADF content is high, it will be of low digestibility. Figures 4–3a and 4–3b show the procedures used to measure NDF and ADF content of feedstuffs.

Table 4–2
Forage carbohydrates as identified in the Van Soest procedure

Component	Relative Digestibility[1]	ADF	NDF	ADF lignin
Sugar	1			
Starch	1			
Hemicellulose	2		✓	
Cellulose	3	✓	✓	
Lignin	4	✓	✓	✓

ADF: Acid detergent fiber; NDF: Neutral detergent fiber.
[1]Using a scale from 1 (highly digestible) to 4 (indigestible).

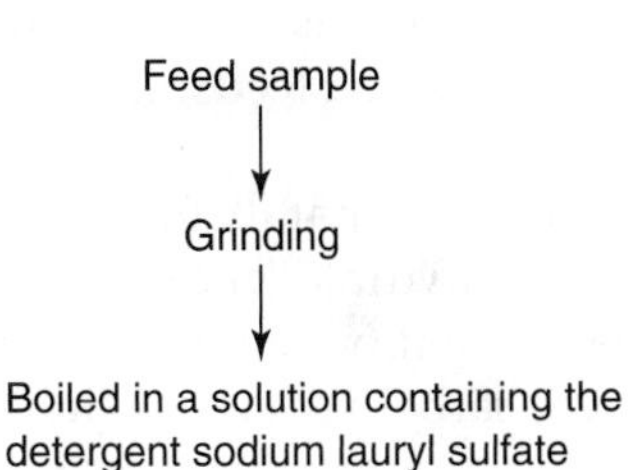

Figure 4–3a
Van Soest procedure for measuring neutral detergent fiber

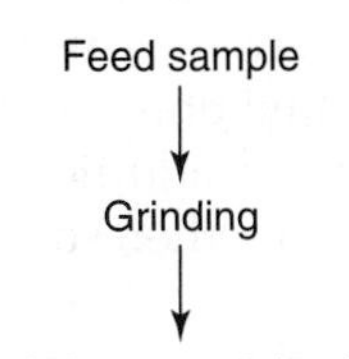

Figure 4–3b
Van Soest procedure for measuring acid detergent fiber

Table 4–3
Carbohydrate fractions in feedstuff analysis

	NSC	NFC	ADF	NDF
Starch	✓	✓		
Sugars	✓	✓		
Organic acids	✓	✓		
Pectin		✓		
Hemicellulose				✓
Cellulose			✓	✓
Lignin			✓	✓

Because NDF is a mixture of substances that vary in digestibility, it is important to identify NDF digestibility when formulating rations, especially those for ruminants. In this text's dairy application, the NDF digestibility (NDFdig) is calculated as follows (NRC, 2001):

$$NDFdig = (0.75 \times \{[NDF\% - NDFIP\%] - Lignin\%\} \times \{1 - [Lignin\% / (NDF\% - NDFIP\%)]^{0.667}\}) / NDF\%$$

Nonstructural carbohydrate (NSC) and NFC refer to the cell contents. NSC is a measured value and includes sugars and starch. NFC is a calculated value and includes starch, sugar, organic acids, and pectin. NFC percent can be calculated as follows:

$$100 - [\%\ crude\ protein + (\%\ NDF - \%\ NDFIP) + \%\ crude\ fat + \%\ ash]$$

Table 4–3 summarizes how the various feed carbohydrates are partitioned on a feed analysis. Table 4–4 gives selected feedstuffs and the analysis of their carbohydrate content.

In ruminant nutrition, a system paralleling that of the protein degradability systems is being used for carbohydrate. The carbohydrate in a given feedstuff is described, based on its degradability and digestibility characteristics. The various analysis values for these carbohydrate fractions are described in Chapter 6.

Lipids

Lipids are high-energy components of feedstuffs that contain a higher proportion of hydrogen to oxygen than do carbohydrates. The increased number of hydrogen bonds means more energy is stored in lipids than in carbohydrates. A pound of lipid will contain at least 2.25 times as much energy as a pound of carbohydrate. Lipids that exist as a liquid at room temperature are often referred to as *oils* and those that are solid at room temperature are referred to as *fats*. In feed analysis, lipid content is determined using a solvent extraction method. The solvent used is typically diethyl ether, so feedstuff lipid content is sometimes referred to as *ether extract*. In addition to lipids, compounds such as waxes and resins are also soluble in ether, so to the extent that feedstuffs contain these compounds, ether extract is not an accurate estimation of lipid content. For some applications, the ether extract percentage value is reduced by one to better approximate the true lipid content.

Ash

Ash is an approximation of the total mineral (inorganic) portion of the feed sample. It is obtained by oxidizing the sample in a furnace at 500° to 600° C.

Table 4–4
Carbohydrate fractions in various feedstuffs

	% of Dry Matter			
Feedstuff	**NSC**	**NFC**	**ADF**	**NDF**
Legume hay	27.4	27.7	32.5	41.2
Grass hay	15.4	16.0	38.7	64.8
Legume silage	24.1	19.7	35.8	45.0
Grass silage	19.1	14.0	38.7	59.4
Corn silage	37.4	38.5	25.9	46.0
Citrus pulp	59.4	59.3	25.2	22.3
Almond hulls	46.6	48.1	28.2	35.3
Corn grain ground	68.7	67.5	3.4	13.1
Hominy feed	63.0	64.6	5.6	18.8
Barley grain	60.5	59.9	7.3	21.2
Oat grain	47.5	48.7	13.4	29.5
Wheat grain	64.8	64.5	4.8	15.8
Wheat middlings	33.5	33.0	11.8	38.2
Whole cottonseed	2.0	2.6	40.3	51.0
Bakery waste	57.3	58.7	5.3	13.6
Distillers dried grains	21.4	17.4	19.1	38.7
SBM 49%	29.3	27.7	5.8	10.5
Soybeans, heated	17.2	21.8	10.7	17.6
Soybean hulls	15.9	16.0	45.3	60.6
Peanut meal	19.2	18.3	13.9	25.6
Canola meal	23.2	23.8	19.2	28.6
Blood meal	0	0	0	0
Feather meal	0	2.4	0	0
Fish meal	0	8.1	0	0

NSC: Nonstructural carbohydrate; NFC: Nonfiber carbohydrate. ADF: Acid detergent fiber. NDF: Neutral detergent fiber.

NDF and NSC values taken from Northeast DHIA, 1995. NFC values calculated from values from same source. Corn grain ground values (NSC, NFC, NDF) taken from Miller and Hoover, 1998. ADF value taken from Dairy NRC, 2001.

Ash is useful in determining the feedstuff content of organic components that cannot be measured directly, such as NFC. Recall that NFC is a calculated value and that the formula includes the feedstuff's ash content. Ash content of diets is of special concern for cats and fish. In cat nutrition, high-ash diets have been suggested as possible contributors to feline lower urinary tract disease. Fish meal is a primary component of fish diets, and a high ash content in this feedstuff may indicate that the product contains an unacceptably high bone content.

Total Digestible Nutrients

Total digestible nutrients (TDN) is a measurement of the digestible organic constituents of the feedstuff (Figure 4–4). Because any of the organic constituents could be used by the animal as a source of energy, TDN has historically been used to describe an animal's energy requirement and to assess a feedstuff's energy value for a given species of livestock. TDN is no longer used

Figure 4–4
Total digestible nutrients

Total digestible nutrients	=	
		Digestible crude protein
	+	Digestible crude fiber
	+	Digestible nitrogen-free extract
	+	Digestible ether extract or crude fat × 2.25

for these purposes, but it still may be used as an intermediate calculation to arrive at more useful values.

In the TDN formula, the digestible crude fat is multiplied by 2.25 because TDN is an assessment of a feedstuff's energy value, and a pound of fat provides 2.25 times as much energy as a pound of any of the other organic constituents of TDN.

The values used in the TDN formula are determined by multiplying a feedstuff's proximate analysis values for crude protein, crude fiber, nitrogen-free extract, and crude fat by their respective digestion coefficients. The digestion coefficients for a particular feedstuff are found by performing a digestion trial.

In a digestion trial, the percentage of each nutrient in the feedstuff is determined through proximate analysis. A weighed amount of the feedstuff is then fed to the animal during the test period. The feces are collected, weighed, and analyzed using proximate analysis. The amount of each nutrient digested is assumed to be the difference between the amount consumed and the amount excreted. The proportion of each nutrient digested is the digestion coefficient for that nutrient for that feedstuff.

The dairy NRC (2001) does not use digestion trial data to determine feedstuff TDN. Instead, TDN is calculated from feedstuff composition values obtained through *in vitro* techniques. The formulas used vary with the feedstuff type and are shown in the companion CD-ROM for dairy at **F9**.

Other Feed Analysis Issues and Values

Dietary Energy

As discussed previously, TDN was formerly used to assess a feedstuff's energy value and to describe an animal's energy requirement. For this purpose, TDN has been replaced by a system that describes the fractions of a feedstuff's energy content available to the animal and those fractions that are not available to the animal. This system and its application to animal nutrition are described in Chapter 5.

Near Infrared Reflectance Spectroscopy

Near Infrared Reflectance Spectroscopy (NIR) is a sophisticated analytical method that many feedstuff analysis laboratories have available. For many feedstuff types, NIR produces rapid results with a minimum amount of labor and will be less expensive than the traditional wet chemistry analysis.

NIR involves measuring reflected light in the near infrared region from a ground sample. The equipment uses equations that have been established for the specific type of feedstuff to be tested. An accurate description of the feedstuff sample is, therefore, critical. Because minerals do not absorb light in the near infrared region, errors in a feedstuff's mineral content will be greater in an NIR analysis than those for organic feedstuff components.

Sampling Total Mixed Rations

Total mixed rations (TMR) are rations that are mixed on the farm and include all ingredients fed. Laboratories are sometimes asked to analyze TMR samples for nutrient content to meet one or more of the following objectives:

All values are given on a dry matter basis.

$$\% \text{ DDM} = 88.9 - (\text{ADF } \% \times 0.779)$$
$$\% \text{ DMI} = 120/\% \text{ NDF}$$
$$\text{RFV} = (\% \text{ DDM} \times \% \text{ DMI})/1.29$$

Figure 4–5
Calculating relative feed value

1. To determine if the mix is meeting nutrient specifications.
2. To determine if the mixer is working properly.
3. To determine if the mixer is being used properly.

Many variables influence the results of TMR analysis, and money is better spent meeting the above goals through other means. To determine if the mix is meeting nutrient specifications, the feeder should reanalyze and confirm the nutrient composition of the individual ingredients and compare ration nutrient levels with predicted requirements. To determine if the mixer is working properly, the best testing method is as follows:

1. Add a tracer product to the mixer.
2. Take several samples of the mix.
3. Count the amount of tracer present in all samples.
4. Apply statistics to evaluate the effectiveness of the mixer.

To determine if the mixer is being used properly, employee training and supervision is the best solution.

Relative Feed Value

Another value that is often reported on a forage analysis is the relative feed value (RFV). This is calculated from the percentage digestible dry matter (DDM) and the dry matter intake (DMI) as a percentage of body weight for mature dairy cows (Figure 4–5). RFV is designed as a tool to assist in marketing forage products; it is not used in ration formulation.

Relative Forage Quality

Relative forage quality (RFQ) is intended to replace RFV. Unlike RFV, RFQ accounts for differences in NDF digestibility. The more digestible the NDF in a forage, the more the forage behaves like a grain. This means that more of this highly digestible forage can replace grain without sacrificing production, but it also means that the forage is less helpful in reducing the risk of acidosis (Chapter 20).

RFQ values will be similar to RFV values if NDF digestibility is "average." However, where NDF digestibility varies considerably, as is the case with grasses, RFQ will give a better estimate of the feedstuff's value.

Expressing Nutrient Concentration

Laboratories that perform feedstuff analyses report nutrient concentrations on two bases: as fed and dry matter. The "as fed" or "as sampled" basis is used by feed manufacturers because it is also the basis on which the feed is bought and sold. The as fed or as sampled basis is also the basis on which feed nutrient content must be reported in accordance with laws pertaining to the feed industry. However, nutritionists prefer to use nutrient analyses on a dry matter basis because a feedstuff's water content is largely irrelevant once the feedstuff is eaten. In discussing silage quality, nutrient content is expressed on the dry matter basis to ensure that accurate comparisons of nutrient content are made in feedstuffs of varying water content.

Nutrient concentration expressed on a dry matter basis will be a larger value than nutrient concentration expressed on an as fed or as sampled basis (Figure 4–6 and Figure 4–7a). The companion application to this text uses dry matter as the standard basis in expressing nutrient concentration.

Figure 4–6
Dry matter and as fed basis conversions

To convert from dry matter to as fed basis:

Nutrient content, as fed basis = Nutrient content, dry matter basis × decimal of % dry matter

To convert from as fed to dry matter basis:

Nutrient content, dry matter basis = Nutrient content, as fed basis/decimal of % dry matter

Figure 4–7a
Protein concentration expressed on as fed and dry matter basis

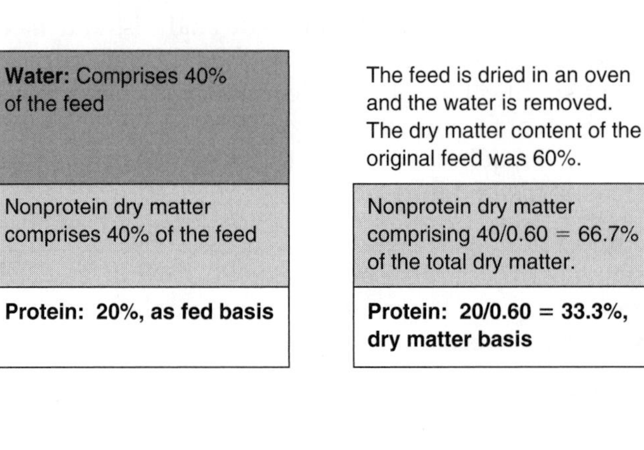

Figure 4–7b
Converting from kilocalories per pound to kilocalories per kilogram

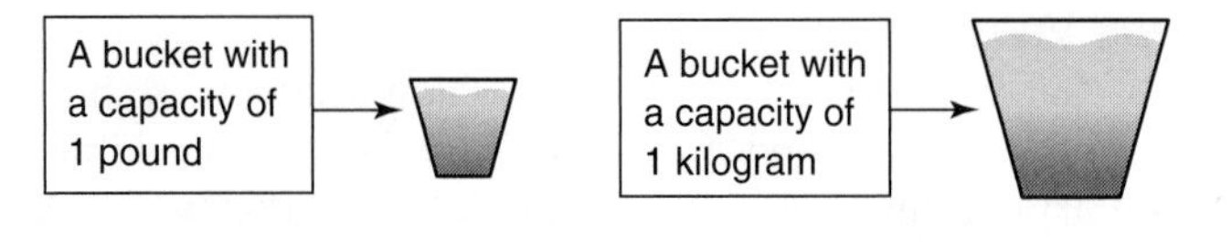

To convert from kcal/lb. to kcal/kg: kcal/lb. × 2.2046 = kcal/kg. The energy value of a given feedstuff will be a larger value when expressed as kcal/kg than it will be when expressed as kcal/lb.

Percentage is one way of expressing nutrient concentration in feedstuffs. Percentage represents parts of nutrient per 100 parts of feedstuff. The "part" may be any measured amount: pounds, grams, milligrams, kilograms, and so on. Examples of feed nutrients for which concentration is usually expressed as a percentage include dry matter, protein, amino acids, carbohydrate components, and macrominerals.

Energy, micromineral, and vitamin concentration in feedstuffs are not expressed as percentages. Energy concentration is expressed in kilocalories (Kcal) or megacalories (Mcal) of digestible energy (DE), metabolizable energy (ME), or net energy (NE) per pound of feedstuff. These energy values are described in Chapter 5. Micromineral concentration is expressed in milligrams per kilogram, which is the same as parts per million (ppm). Vitamin concentration in feedstuffs is expressed in milligrams or International Units of vitamin per kilogram or per pound of feedstuff.

Expressing Feedstuff Amounts

This text uses the pound as the standard unit in expressing feedstuff amounts. The kilogram (kg) is the standard in the scientific literature. There are 2.2046 pounds in 1 kg. Be aware that the 2.2046 value is used differently when converting from pounds to kilograms than when converting from kcal/lb. to kcal/kg. See Figure 4–7a, 4–7b, and 4–7c for conversion explanations.

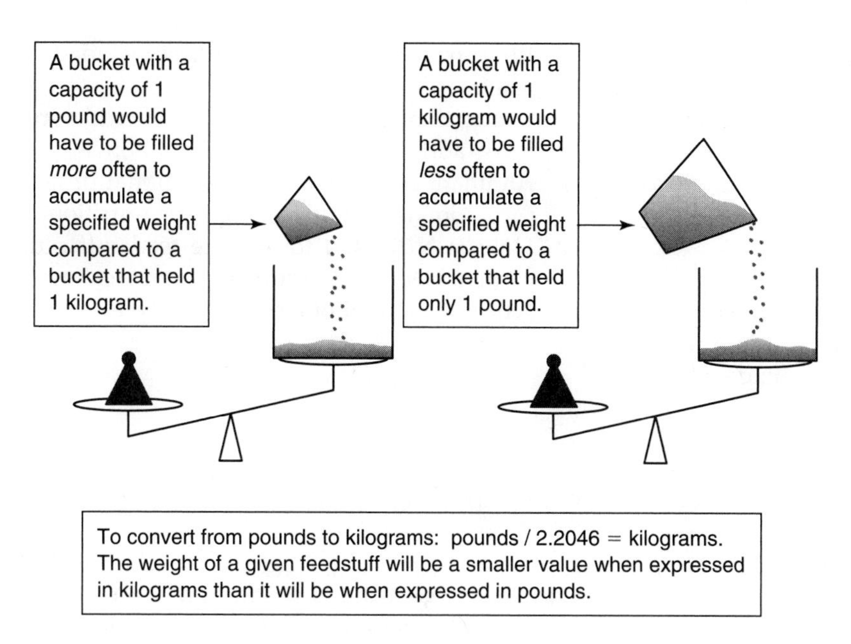

Figure 4–7c
Converting from pounds to kilograms

FEED LAWS AND LABELING

The *Official Publication* of the Association of American Feed Control Officials (2003) contains a model bill for state legislators to consider when making laws pertaining to the labeling requirements of feeds sold in their state. A primary purpose of the model bill is to promote consistency among the state laws pertaining to feed tags and thereby facilitate feed sales across state lines. The feed tag, then, is a mercantile instrument. As such, it will seldom supply sufficient information for meaningful ration formulation.

SUMMARY

Chapter 4 emphasizes the importance of laboratory analysis for forages. Proper sampling involves taking 12 to 20 samples, combining these, mixing them, and then obtaining a subsample for analysis. Feed microscopy is described as a subjective means of assessing feed quality. Analysis procedures available from feed analysis laboratories are described. Energy is perhaps the most important nutrient, but because livestock can acquire energy from any organic component of feed, assessing feed energy value is problematic. The NE system involves assessing feed energy content by systematically removing sources of wasted energy until the NE—what remains—represents the energy actually available to the animal. The feed labels as required by law are described as being of limited value in providing useful information toward the development of livestock rations.

END-OF-CHAPTER QUESTIONS

1. Describe the proper technique for acquiring a sample of a supply of feedstuff to submit for analysis.
2. What are the components of neutral detergent fiber?
3. What are the components of nonfiber carbohydrate?
4. If a given weight is expressed in units of both pounds and kilograms, which unit has the lowest numerical value?

5. If a given nutrient concentration is expressed both on an as fed and dry matter basis, which concentration value is the highest numerical value?
6. Compare the utility of the Van Soest and proximate analysis methods for expressing the carbohydrate content of forages.
7. The Kjeldahl procedure measures the nitrogen content of a feed sample. The nitrogen content is then multiplied by 6.25 to arrive at the crude protein content. What assumption is made when using the 6.25 value?
8. What fraction of the feedstuff is included in the ash fraction?
9. In addition to the concentration of crude protein, what additional protein information is used in ruminant nutrition? What additional protein information is used in monogastric nutrition?
10. How is NIR used to analyze feedstuff nutrient content? Discuss the strengths and weaknesses of NIR as a method of feedstuff analysis.

REFERENCES

Association of American Feed Control Officials. (2003). *Official Publication*. West Lafayette, IN.

Dairy One Forage Laboratory. (2001). Mailer insert. Dairy One Forage Laboratory. Ithaca, NY.

Miller, T. K., & Hoover, W. H. (1998). Nutrient analyses of feedstuffs including carbohydrates. Animal Sci. Report #1, West Virginia University.

National Research Council. (2001). *Nutrient requirements of dairy cattle* (7th revised edition.). Washington, DC: National Academy Press.

Northeast DHIA, Dairy Herd Improvement Association. (1995). Forage analysis statistics. Ithaca, NY.

Satter, L. D., Dhiman, T. R., & Hsu, J. T. (1994, October 18–20). Use of heat processed soybeans in dairy rations. In *Proceedings Cornell Nutrition Conference for Feed Manufacturers* (pp. 19–28). Rochester, NY.

CHAPTER 5

DIETARY ENERGY

The magnitude of many body processes changes in a regular fashion as the size of the organism changes. A surprising number of such processes can be described in a very simple fashion by: $M = a \times x^k$ *where* M *is the body process in question (for example metabolic rate),* x *is a measure of the size of the organism, and* a *and* k *are constants.*

JOHAN, 2000

THE CHALLENGE OF ENERGY IN NUTRITION

Energy is power used to perform work. In most instances, most of the work done by livestock involves the maintenance of life. If there is more energy available than that needed to maintain life, the extra energy will go toward the work of growth, production, and reproduction.

Livestock acquire energy from the organic compounds in feedstuffs. These compounds include carbohydrates, fats, proteins, and nucleic acids. The energy in these compounds is made available to the animal when they are oxidized during metabolism. As the electrons in these compounds are transferred to oxygen in the mitochondria of the cells, their energy is transferred to the chemical bonds of adenosine triphosphate (ATP). The energy stored in ATP is used as the immediate energy source for all the reactions of metabolism.

To address the challenge of assessing the energy value of feed, nutritionists use the net energy (NE) system, which identifies and quantifies the different energy-bearing fractions into which a feedstuff is degraded during digestion and metabolism.

THE NET ENERGY SYSTEM

Animals eat to acquire energy. In other words, appetite is driven by the animal's need for energy. This sets energy apart from the other feed nutrients required by livestock. Energy is the master nutrient. For the most part, the satisfaction of all other nutrient requirements will occur only because livestock seek energy in their feed. Feed energy value and animal energy requirement are usually described using the NE system shown in Figure 5–1.

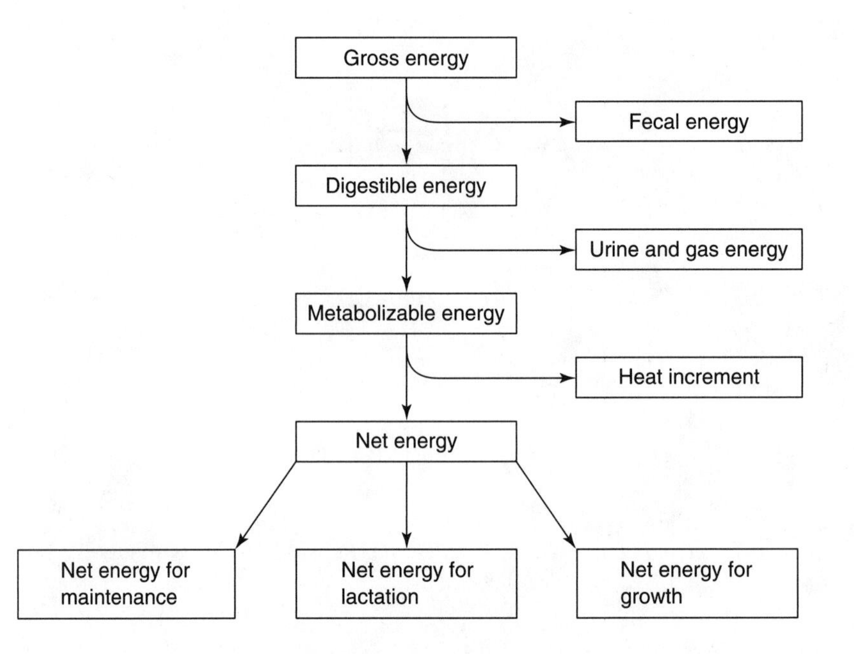

Figure 5–1
Feed energy fractions

Gross energy is the total energy contained in a feedstuff. It is determined by oxidizing (burning) the feedstuff and measuring the energy released as heat. It is not directly applicable to animal nutrition because animal digestive systems are not able to use the entire gross energy contained in a feedstuff.

Fecal energy is the energy contained in the fecal material resulting from ingestion of the feedstuff. The energy in the fecal material is determined by burning the fecal material and measuring its gross energy content.

Digestible energy (DE) is the difference between the gross energy and the fecal energy. It is the portion of the feedstuff energy assumed to be digestible. DE is approximately equivalent to total digestible nutrients (TDN) (Chapter 4). DE is the targeted energy used in horse, rabbit, and fish nutrition.

Urine and gas energy includes the energy contained in these wastes and produced during digestion of the feedstuff and subsequent metabolic activities. Energy-containing compounds in urine have been filtered out of the blood by the kidneys. These compounds may have originated in the feedstuff or may be of endogenous origin. The energy in gaseous waste (primarily methane) is important in ruminant animals, but is insignificant in other species.

Metabolizable energy (ME) is the difference between the DE and energy in urine and gas. In poultry nutrition, ME is used because the feces and waste from the urinary system are voided together. The ME for poultry is called MEn because it is corrected for nitrogen. The nitrogen correction involves accounting for feed protein (nitrogen) that was retained in the body rather than metabolized for energy. In swine and ruminant nutrition, ME is used for improved accuracy over DE. ME is also used in cat nutrition. In dog nutrition, ME is used, but because very few energy measurements have been conducted on dogs, the ME is calculated from gross energy rather than measured during a digestion trial (Figure 5–2).

Metabolic rate increases following the ingestion of feed. This results in increased energy losses as heat, which must be accounted for when determining the amount of feed energy available to the animal. The increased metabolic rate that follows the ingestion of feed has been described as the *specific dynamic effect* and the *heat increment*.

NE is calculated as the difference between the ME and the heat increment. The NE content of a feed represents the actual energy value of the feed to the

Protein has a gross energy of 2.00 Mcal/lb. The crude protein content of a feedstuff is assumed to be 80% metabolizable. 2.00 × 0.80 = 1.60. The crude protein measured in the feedstuff is multiplied by 1.60. Nitrogen-free extract has a gross energy of 1.88 Mcal/lb. The nitrogen-free extract content of a feedstuff is assumed to be 85% metabolizable. 1.88 ×.85 = 1.60. The nitrogen-free extract measured in the feedstuff is multiplied by 1.60. Fat has a gross energy of 4.26 Mcal/lb. The crude fat content of a feedstuff is assumed to be 90% metabolizable. 4.26 × 0.90 = 3.83. The crude fat measured in the feedstuff is multiplied by 3.83. Fiber is assumed to have no energy value for the dog. The ME value of a feedstuff for dogs, Mcal/lb., dry matter basis, is calculated as follows:

(crude protein × 1.60) + (nitrogen-free extract × 1.60) + (crude fat × 3.83)

Figure 5–2
Calculating feedstuff ME value for the dog

Another way to look at net energy is from the standpoint of the energy contained in the product produced. The feed net energy required to make 80 lbs. of milk is the gross energy contained in that milk; the feed net energy required to make a quantity of gain is the gross energy contained in that gain. As an example, say we have a quantity of milk that contains 40 Mcal of gross energy. Given that corn grain contains 0.78 Mcal of NEl/lb., as fed basis, it would take (40/0.78) = 51 lb. of ground corn grain to produce this milk. Say we have a quantity of meat that contains this same amount of gross energy—40 Mcal. Given that ground corn grain contains 0.57 Mcal of NEg/lb., as fed basis, it would take (40/0.57) = 70 lb. of ground corn grain to produce this gain.

Figure 5–3
Net energy in the feed is the gross energy in the product produced

animal. Animals use feed NE with varying efficiencies (Figure 5–3). The NEm (net energy for maintenance, dairy heifers) value for ground corn grain (IFN 4-02-854) is 0.98 Mcal/lb., dry matter basis. The NEl (net energy for lactation, dairy) value for ground corn grain is 0.91 Mcal/lb., dry matter basis. The NEg (net energy for gain, dairy heifers) value for the same feedstuff, however, is only 0.67 Mcal/lb., dry matter basis. NEm is only used for heifers, not for adult cows. For adult dairy cows, energy requirements and energy values of feed are expressed in NEl units when considering both maintenance and lactation functions because ME is used with similar efficiency for maintenance and lactation (Moe & Tyrrell, 1972). In the goat NRC (1981), the NE is reported without reference to function.

NE may be used for the function of reproduction. The priority that the body assigns to pregnancy and lactation is a close second behind that assigned to maintenance. For short periods of time, animals will still maintain the pregnancy and/or lactate, even if the dietary energy content is not enough to support the maintenance functions. It is important to realize that although the animal is not ingesting enough energy to meet her maintenance energy needs, these needs are still being met with reserves of energy stored as body fat (she will be losing weight).

In dairy nutrition, it is recognized that the energy value of feedstuffs must be discounted in animals in which feed is passing rapidly through the digestive tract. Such animals have high dry matter and energy intakes. Dry matter and energy intakes are often expressed as multiples of values typical of a cow at maintenance. A feedstuff consumed by a cow with an energy intake of 3× maintenance would have a higher energy value than would the same feedstuff consumed by a cow with an energy intake of 4× maintenance. In the companion CD-ROM for dairy cattle, a discount is applied to the energy value of each feedstuff in cases of high ration TDN% and/or high animal dry matter intake (DMI).

NE may be used for the functions of growth, wool production, and mohair production. Growth (or gain) will not be supported if there is not enough energy in the diet to supply the entire maintenance energy requirement. Recall (Chapter 1) that the energy expended in dealing with a stressful environment (the essence of animal strain) is part of the animal's maintenance energy requirement. Insufficient energy to support maintenance can disrupt the hair follicle cycle, resulting in reduced quality and quantity of wool and mohair production.

ENERGY AND APPETITE

For the most part, an animal's appetite for food is a reflection of its metabolic need for energy. This fact has important implications regarding feeding management. If, for example, fat is added to the diet to replace a portion of the grain, the energy density of the diet will be increased. This means that it will take fewer pounds of the fat-added ration to supply a given amount of energy. Appetite may be reduced. It may be necessary to increase the density of other nutrients when fat is added in order to prevent nutrient deficiencies and maintain performance. Conversely, if the diet energy density is reduced, increased intake may present an opportunity for cost savings by reducing the density of other nutrients. The relationship between energy density and feed intake holds over a limited range of energy densities. Feed intake will be limited by gut fill at very low energy densities. At very high energy densities, behavioral and digestive problems may result or animals may consume energy beyond their need in order to achieve satiety.

Following is a summary of how the relationship between energy density and feed intake is handled for the various species in the NRC publications and in the companion application to this text. The cat energy information comes from the AAFCO (2003).

- Poultry feed intake and nutrient requirement calculations are based on an assumed feed energy density of 1,462 kcal of MEn/lb., dry matter basis.
- Trout DMI and nutrient requirement calculations are based on an assumed feed energy density of 1,814 kcal of DE/lb., dry matter basis.
- Catfish DMI and nutrient requirement calculations are based on an assumed feed energy density of 1,512 kcal of DE/lb., dry matter basis.
- Goat DMI and nutrient requirement calculations are based on an assumed feed energy density of 1,090 kcal of ME/lb., dry matter basis.
- Dog DMI and nutrient requirement calculations are based on an assumed feed energy density of 1,665 kcal of ME/lb., dry matter basis for all dogs except those at maintenance. Requirements for dogs at maintenance are based only on body weight. To account for the effect of high dietary energy density on DMI, the author of this text has modified the prediction of DMI for dogs as follows: predicted DMI is reduced 5 percent for each increase of 45 units above 1,814 kcal of ME/lb., dry matter basis.
- Cat DMI is predicted based on the assumed energy content of the diet as associated with food type. Dry foods are assumed to have lower energy density than wet foods, and cats are predicted to eat more dry matter of a low-energy dense food than a high-energy dense food. With two exceptions, cat nutrient requirement calculations are based on an

assumed feed energy concentration of 1,814 kcal of ME/lb., dry matter basis. The exceptions are taurine and copper. Taurine in canned foods has been reported to be less available to the cat (Douglass, Fern, & Brown, 1991). Likewise, copper in extruded foods has been reported to be less available (AAFCO, 2003). To meet the cat's requirement for these two nutrients, the targeted requirement is increased when the problem food forms are identified in the input section of the companion application to this text.

- Dairy DMI and nutrient requirement calculations are based on animal, environment, and production inputs, energy density of the inputted ration, and whether or not ionophore is used. There is also a unique formula used to predict the energy value for fatty acids, fats, and animal proteins, and a single formula to predict all others. These formulas are displayed in the companion application for dairy cattle by striking **F9**.
- Beef DMI and nutrient requirement calculations are based on animal and environment inputs, ionophore use, anabolic implant use, and the energy density of the inputted ration.
- For horses, sheep, and rabbits, the NRC formulas and the companion application to this text do not use ration energy density to predict DMI.
- For swine, the NRC predicts feed intake and nutrient requirements based on the DE concentration expected from a diet based on corn grain and soybean meal diet. Such a diet would contain a DE concentration of 1,714 Mcal/lb., dry matter basis.

METABOLIC BODY SIZE

An animal's energy requirement comprises the total energy needed to perform the functions of maintenance, growth, production, and reproduction. In most cases, the requirement to perform maintenance functions will be the single largest component of the energy requirement. The maintenance energy requirement is the energy needed to carry on the activities associated with the animal's basal metabolism as well as activities (strains) associated with combating environmental stressors. Activities that are part of basal metabolism include respiration, circulation, transport of ions and metabolites, and body constituent turnover. The energy used in basal metabolic activities can be determined by measuring the heat lost from the surface of a stress-free, fasting, homeothermic animal. The basal metabolic rate is therefore related to the surface area of an animal's body.

The relationship between body weight and body surface area is body weight raised to the 0.67 power. However, Kleiber (1932) demonstrated that the relationship between body weight and basal metabolic rate is body weight raised to the 0.75 power (Figure 5-4). $BW_{kg}^{0.75}$ is referred to as *metabolic body size*. Determination of energy requirements in formulas such as this is referred to as *allometric scaling*.

In the application accompanying this text, the maintenance energy requirement is determined using $BW_{kg}^{0.75}$ for most domestic animals, but not for all. Metabolic body size predicts energy requirement as ME. Metabolic body size can only be applied in ration formulation if the ME content of the feedstuffs is available for the species considered. In the NRC publications for the horse, rabbit, and fish, feedstuff energy values are expressed as DE rather than ME. For these livestock, the energy requirement presented cannot be based on metabolic body size. The authors of the horse NRC (1989) further explain that the "values for energy requirements are not based on metabolic body size, because Pagan and Hintz (1986) found no benefit from using metabolic body weight

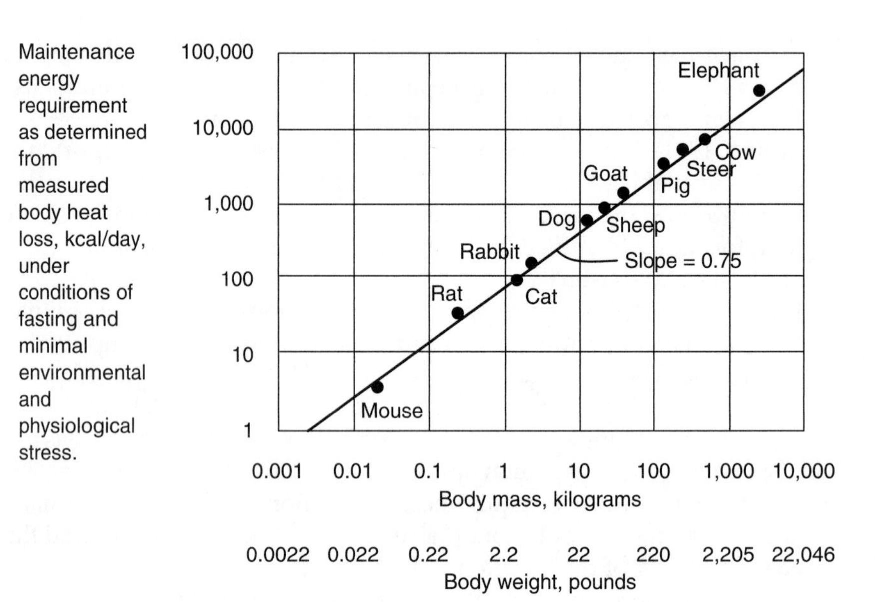

Figure 5–4
A log–log plot of maintenance energy requirement against body mass for a wide variety of animals

($kg^{0.75}$) over weight ($kg^{1.0}$) in determining the energy requirements of horses ranging in size from 125 to 856 kg." Though the cat NRC (1986) does use ME as the measurement of feedstuff energy value, the authors subscribe to the conclusion of Kendall, Blaza, and Smith (1983) who "found no extra precision when energy requirements of adult domestic cats were scaled to mass exponents of body weight (kg) of either 0.75 or 0.67, compared with unity." Payne (1965) reported dog energy requirements based on $BW_{kg}^{0.67}$. These values are reported in the dog NRC (1985) and are used in the companion application to this text. With regard to fish, one would not expect that metabolic body size as predicted by Kleiber's (1932) formula would apply because fish muscles are not working against gravity and because fish are poikilotherms. Brett (1973) reports that the maintenance energy requirement of fish is 5% to 10% that of other livestock of similar size in a thermoneutral environment.

HOW ENERGY IS MEASURED

Energy is measured in calories. A calorie is the energy needed to raise the temperature of 1 gram of water 1° C, or more specifically, from 14.5° C to 15.5° C. This value is too small for practical use in animal nutrition. A kilocalorie, or Kcal, is equal to 1,000 calories, and this is the standard energy unit for all livestock except horses and ruminants. For horses and ruminants, the megacalorie (Mcal) is the standard. One Mcal is equal to 1,000,000 calories or 1,000 kcal. A joule is an alternative to the calorie. The joule is defined in mechanical terms and can be converted to the calorie system: 1 joule = 0.239 calories. The joule has replaced the calorie in many countries and scientific journals, but is not in widespread use in the United States. Table 5–1 gives the energy requirements for selected animals.

SUMMARY

Chapter 5 defines the terminology used in describing feedstuff energy content and animal energy requirements. Metabolic body size is defined and its application in animal nutrition is discussed.

Table 5–1
Energy requirement for selected animals

Animal	Required, per Animal per Day	Required Concentration, Dry Matter Basis
Fish, channel catfish, 100 g body weight	10 kcal of DE	1,512 kcal/lb, DE
Fish, rainbow trout, 100 g body weight	5.04 kcal of DE	1,814 kcal/lb, DE
Chicken, broiler, 5 weeks of age	439 kcal of MEn[1]	1,619 kcal/lb, MEn[1]
Chicken, white egg layer, 3 lb. body weight	296.90 kcal of MEn[1]	1,496 kcal/lb, MEn[1]
Pig, growing, 45 lb. body weight	4,529 kcal of DE	1,714 kcal/lb, DE
	4,348 kcal of ME	1,645 kcal/lb, ME
Dog, growing, 30 lb. body weight	1,294 kcal of MEdog[2]	1,665 kcal/lb, MEdog[2]
Cat, growing kitten, 4.2 lb. body weight	337 kcal of ME	1826 kcal/lb, ME
Rabbit, growing, 5 weeks of age	377 kcal of DE	1,280 kcal/lb, DE
Horse, light work, 1,100 lb. body weight	25,290 kcal of DE	870 kcal /lb, DE
Goat, maintenance, 88 lb. body weight	1,970 kcal of DE	1,250 kcal /lb, DE
	1,610 kcal of ME	1,020 kcal /lb, ME
	910 kcal of NE[3]	570 kcal /lb, NE[3]
Ewe, maintenance, 110 lb. body weight	2,390 kcal of DE	1,090 kcal /lb, DE
	2,000 kcal of ME	910 kcal /lb, ME
	1,050 kcal of NEm	480 kcal /lb, NEm
Beef animal, growing, 800 lb. body weight	Based on performance desired[4]	Based on performance desired[4]
Dairy cow, lactating, 1,400 lb. body weight	Based on performance desired[4]	Based on performance desired[4]

DE: Digestible energy; ME: metabolizable energy; NE: net energy; NEm: net energy for maintenance, dairy heifers.

[1]Metabolizable energy, nitrogen-corrected. The nitrogen correction involves accounting for feed protein (nitrogen) that was retained in the body, rather than metabolized for energy. MEn is only used in poultry nutrition.

[2]Metabolizable energy, specific to dogs. Few ME values have been determined for feedstuffs fed to dogs. In the absence of actual measured values, MEdog in feedstuffs is calculated using apparent digestibilities of 85% for nitrogen-free extract, 80% for crude protein, and 90% for ether extract (fat). Apparent digestibilities are multiplied by the gross energy values of 4.15 for nitrogen-free extract, 4.4 for crude protein, and 9.4 for fat. It is assumed that fiber digestion yields no ME for the dog. The resulting values of 3.53, 3.52 and 8.46 are kcal of ME available to dogs from nitrogen-free extract, crude protein, and fat.

[3]The efficiencies of use of net energy for the different functions of maintenance, gain, pregnancy, lactation and mohair production have not been differentiated in goat nutrition. The value in this table is a sum of net energy requirements reported for maintenance, gain, pregnancy, lactation and fiber production.

[4]In beef and dairy nutrition, rather than establish an energy requirement, the animal's performance is predicted based on the ration energy level. The energy requirement, therefore, is dependent upon the level of performance desired.

END-OF-CHAPTER QUESTIONS

1. From what four types of compounds in feed can animals derive energy for metabolism?
2. Starting with the gross energy of a feedstuff, describe how one arrives at that feedstuff's NE content.
3. In canine nutrition, the ME of a feedstuff is calculated from the gross energy in that feedstuff's protein, nitrogen-free extract, and fat. Give the gross energy for each of these compounds and the percentage of each compound's gross energy that is assumed to be metabolizable.
4. Explain why metabolic processes use more pounds of corn grain to produce a quantity of meat containing 40 Mcal of gross energy than to produce a quantity of milk also containing 40 Mcal of gross energy.
5. Explain why the dairy NRC committee developed formulas that discount feed energy value in animals with high DMIs.
6. Under what conditions is animal appetite limited by dietary energy density? Under what conditions is animal appetite limited by gut fill?

Describe the relationship between appetite, energy density of the diet, and DMI.
7. For which species of domestic animals are DMI and nutrient requirements predicted by the NRC committees, based on an assumed feed energy density?
8. What is metabolic body size? Explain how metabolic body size is related to body surface area.
9. Which NRC committees do not use metabolic size in predicting maintenance energy requirement?
10. How is *calorie* defined, and how is it used in animal nutrition? How is *joule* defined, and how is it used in animal nutrition?

REFERENCES

Association of American Feed Control Officials. (2003). *Official publication*. West Lafayette, IN.

Brett, J. R. (1973). Energy expenditure of sockeye salmon during sustained performance. *J. Fish. Res. Board Can. 30*, 1799–1809.

Douglass, G. M., Fern, E. B., & Brown, R. C. (1991). Feline plasma and whole blood taurine levels as influenced by commercial dry and canned diets. *J. Nutr. 121*(suppl.), S179–S180.

Johan, (2000). Retrieved August, 2004 from http://www.anaesthetist.com/physiol/basics/scaling/Kleiber.htm

Kendall, P. T., Blaza, S. E., & Smith, P.M. (1983). Comparative digestible energy requirements of adult beagles and domestic cats for bodyweight maintenance. *J. Nutr. 113*, 1946.

Kleiber, M. (1932). Body size and metabolism. *Hilgardia. 6*, 315–353.

Moe, P. W., & Tyrrell, H. F. (1972). The net energy value of feeds for lactation. *J. Dairy Sci. 55*, 945–958.

National Research Council. (1986). *Nutrient requirements of cats*. Washington, DC: National Academy Press.

National Research Council. (1989). *Nutrient requirements of horses*. Washington, DC: National Academy Press.

Pagan, J. D., & Hintz, H. F. (1986). Composition of milk from pony mares fed various levels of digestible energy. *Cornell Vet. 76*, 139.

Payne, P. R. (1965). Assessment of the protein values of diets in relation to the requirements of the growing dog. In O. Graham-Jones (Editor), *Canine and Feline Nutrition Requirements*. London: Pergamon Press.

CHAPTER 6

CARBOHYDRATES

In the biosphere there is probably more carbohydrate than all other organic matter combined, thanks largely to the abundance in the plant world of two polymers of D-glucose, starch *and* cellulose.

A. L. LEHNINGER, 1978

INTRODUCTION

In most healthful and profitable herbivore diets, carbohydrates will be the most important source of energy. Carbohydrate nutrition has become very complex as it has become clear that there are many different compounds that make up the carbohydrates. Management of carbohydrate nutrition is one of the greatest challenges facing farmers and ranchers feeding herbivore livestock.

IMPORTANCE OF CARBOHYDRATE

Starch is the form of carbohydrate in which most plants store their reserve energy. Cellulose in a plant is structural; it gives a plant the rigidity it needs to grow against gravity. Together, starch and cellulose are the most abundant constituents of herbivore diets. When these carbohydrates are consumed, digested, and absorbed by livestock, the energy in them is released and captured during animal metabolism.

Carbohydrates are the major source of energy in herbivore diets. Because of this, and because energy is the nutrient required in greatest amounts, the physical characteristics of the carbohydrate sources in the ration will largely determine the physical characteristics of the ration.

With the possible exception of some fish, starch is easily digested by all domestic animals because domestic animals are capable of making the enzymes necessary to hydrolyze starch into absorbable monosaccharides. However, livestock do not make the enzymes necessary to hydrolyze cellulose. Livestock that are fed and utilize large amounts of cellulose-containing feedstuffs such as pasture, green chop, hay, and silage depend on a population of microbes living in their digestive tracts to produce the enzymes necessary to digest or ferment hemicellulose and cellulose.

Through the action of microbes living in the digestive tract, herbivores ferment both starch and cellulose carbohydrates into volatile fatty acids (VFA), the most important of which are acetic acid, propionic acid, and butyric acid. Ruminants, horses, and rabbits absorb these VFA and use them to meet at least a portion of their energy requirements.

STRUCTURE OF CARBOHYDRATES

Carbohydrates, or saccharides, are compounds of carbon, hydrogen, and oxygen. The hydrogen and oxygen in carbohydrates are almost always in the same mutual proportion as in water (H_2O). The chemical shorthand for monosaccharides is $(CH_2O)n$, where n is greater than or equal to three.

All carbohydrates are built from simple sugars or monosaccharides. There are two types of monosaccharides: those made with five carbons (pentoses) and those made with six carbons (hexoses). The important pentoses in feed are arabinose and xylose. Ribose is another pentose found in nucleic acids but ribose is generally not considered to be an important compound in animal nutrition. The important hexoses are glucose, fructose, galactose, and mannose. Nutritionally, the hexoses are more important than the pentoses.

Monosaccharides are built into disaccharides (two monosaccharides) and polysaccharides (many monosaccharides). The term *oligosaccharide* is used to denote a compound containing a small number of monosaccharides. A disaccharide is a type of oligosaccharide. In all of these compounds, the monosaccharides are joined together by glycosidic linkages.

Figure 6–1a shows the open chain form of the monosaccharide, glucose. In nature, the #1 carbon is usually found bonded to the hydroxyl group (—OH) at the #5 carbon to create a ring or cyclic form (Figure 6–1b). This ring formation means that the glucose molecule could have two different conformations, depending on the orientation of the bonds associated with the #1 carbon (Figures 6–2a and 6–2b).

The two conformations of glucose are called *anomers.* There is the alpha (α) anomer and the beta (β) anomer, and the difference between them has tremendous significance in animal nutrition. Whereas a glycosidic bond made between α anomers of glucose is relatively easily hydrolyzed by mammalian enzymes, a glycosidic bond made between β anomers of glucose can only be hydrolyzed by the enzymes made by microbes such as those inhabiting the digestive tract of livestock.

A disaccharide is built when two monosaccharides are bonded together in a glycosidic linkage. Other oligosaccharides and polysaccharides are com-

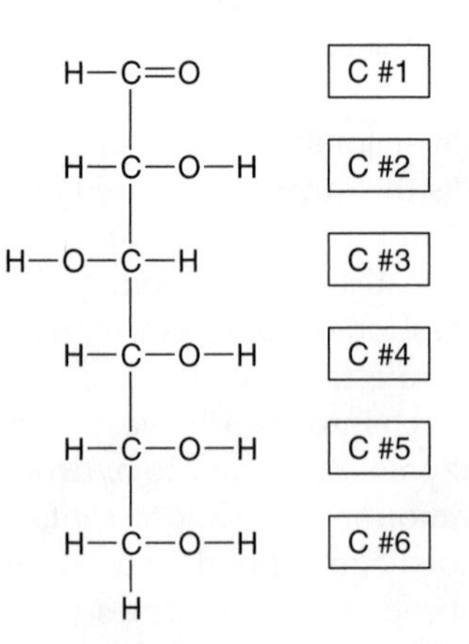

Figure 6–1a
An open-chain representation of glucose

pounds created with repeated glycosidic linkages. These linkages may be made between α glucose anomers or β glucose anomers. Though it varies with the type of carbohydrate, the carbons most likely to be involved in this type of linkage are the #1 and #4 carbons (Figure 6–3a and 6–3b).

There are also some nutritionally important mixtures in feedstuffs that contain carbohydrates. Hemicelluloses are mixtures of pentoses and hexoses. Pectins and gums are mixtures of pentoses, hexoses, and salts of complex acids. Figure 6–4 displays the variety and complexity of carbohydrates and carbohydrate mixtures.

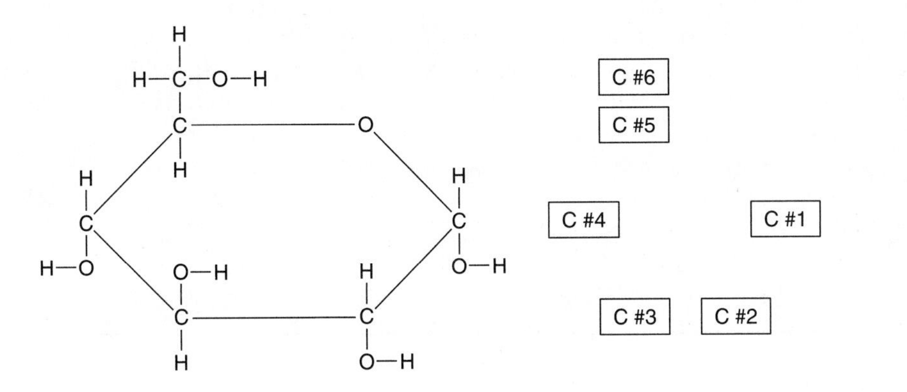

Figure 6–1b
A cyclic representation of glucose

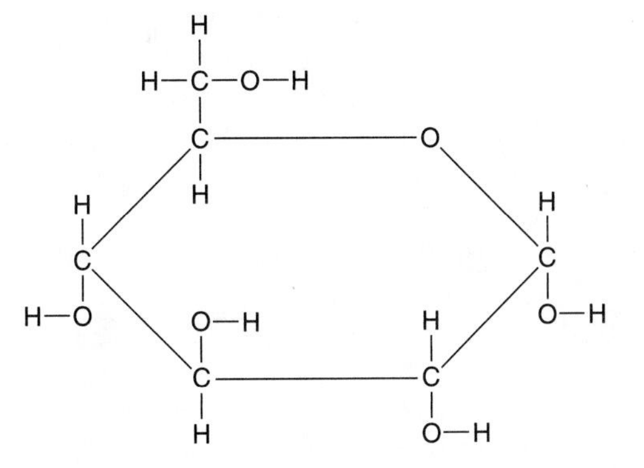

Figure 6–2a
An alpha (α) anomer of glucose

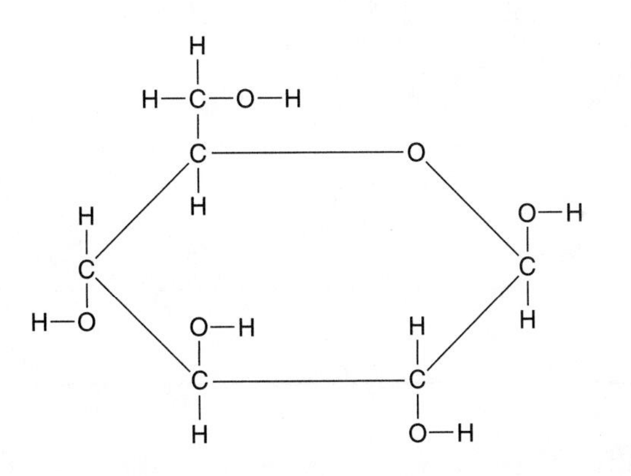

Figure 6–2b
A beta (β) anomer of glucose

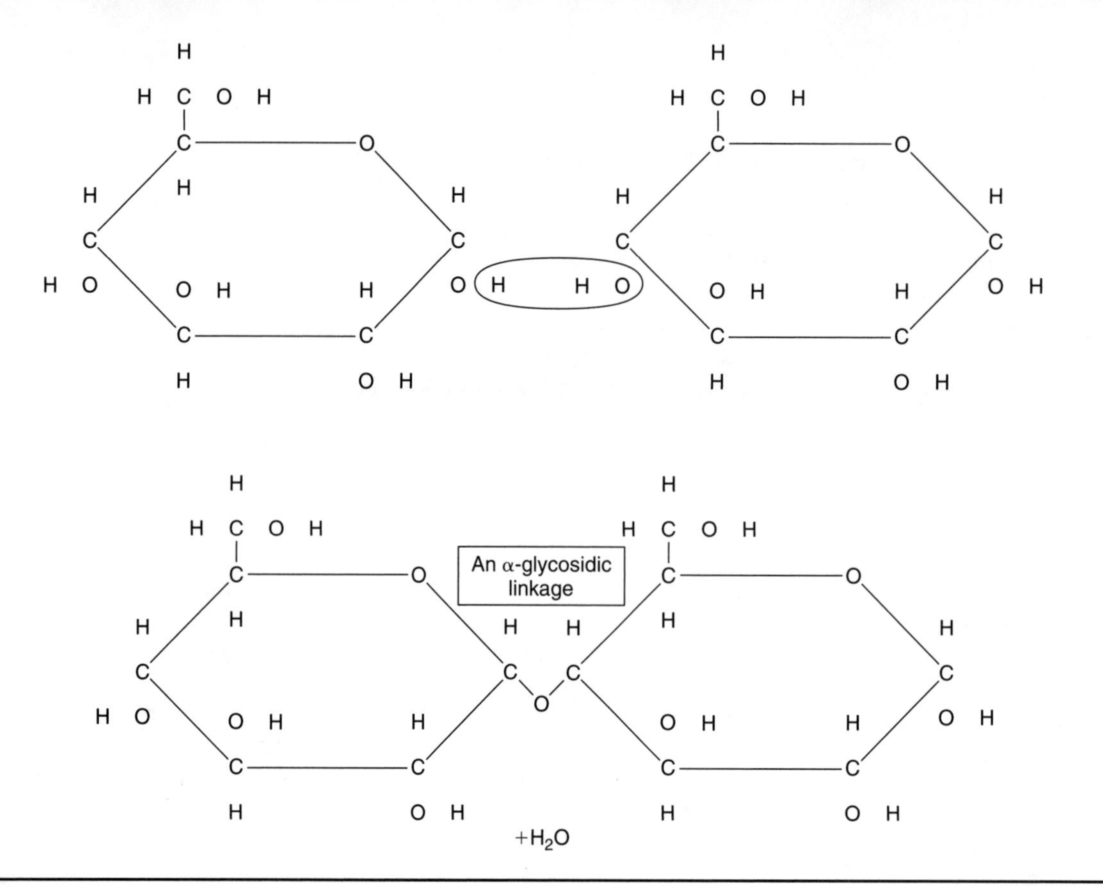

Figure 6–3a
Building a disaccharide with two α glucose anomers

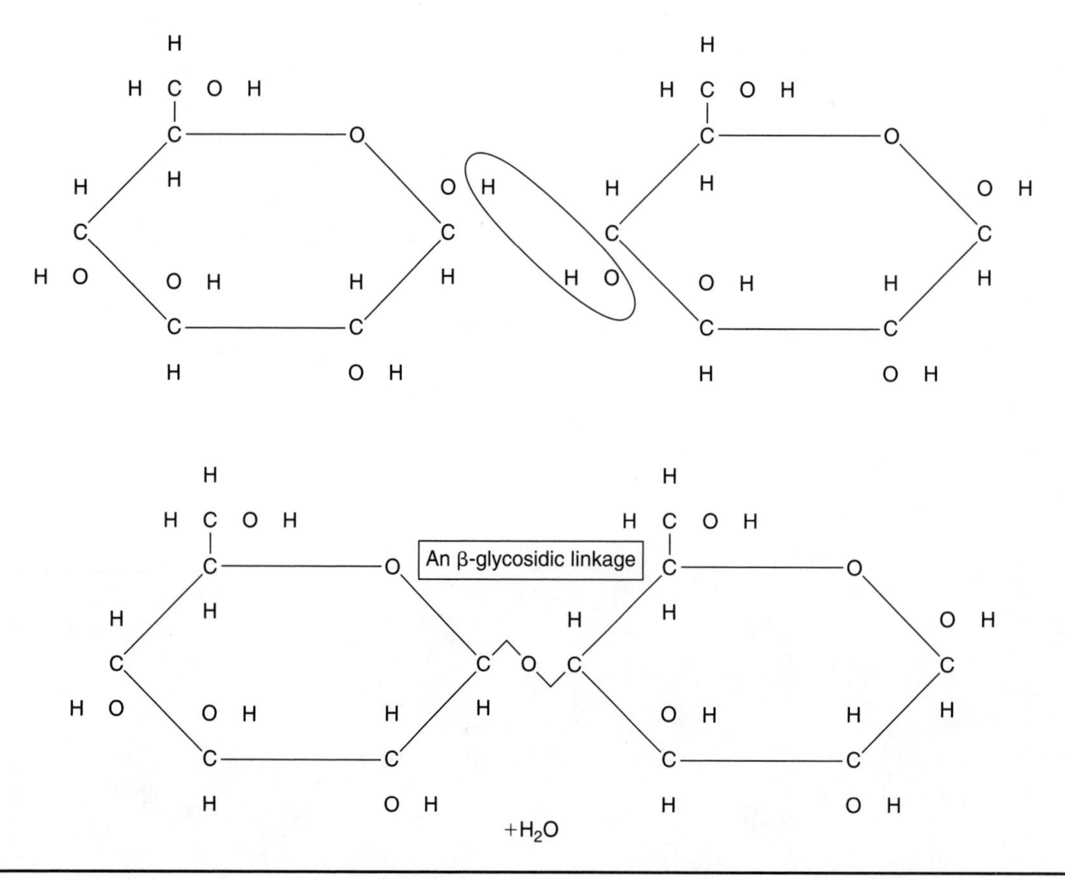

Figure 6–3b
Building a disaccharide with two β glucose anomers

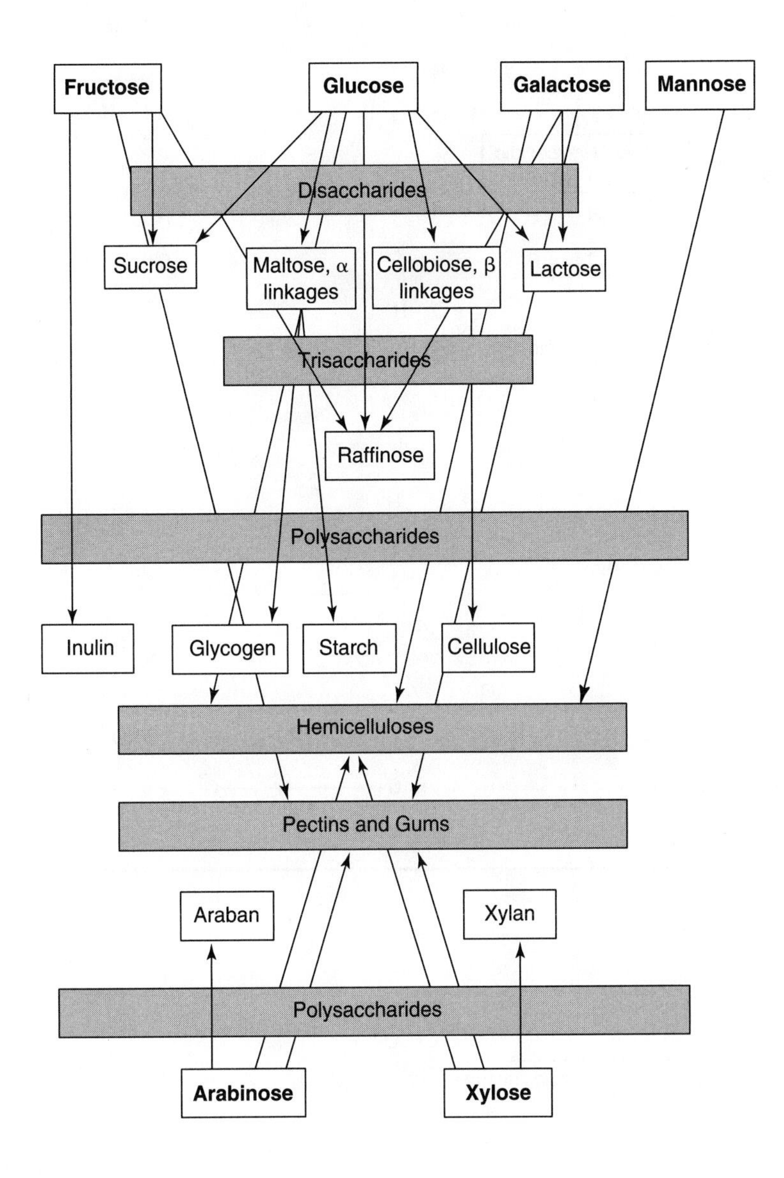

Figure 6–4
Carbohydrates built from hexoses (top row) and pentoses (bottom row). Arrows that pass over a gray rectangle point to compounds that are described by that gray rectangle.

CARBOHYDRATE ABSORPTION

Hydrolysis of the glycosidic bonds in oligosaccharides and polysaccharides is important because only the monosaccharide (α or β) may be absorbed from the intestine into the blood. This hydrolysis occurs through the action of enzymes produced either by the animal itself or by the microbes inhabiting the digestive tract.

The hydrolysis of oligosaccharides and polysaccharides with α linkages is accomplished by enzymes produced primarily in the pancreas (Figure 6–5a). Starch is the most important example of this type of carbohydrate.

The hydrolysis of oligosaccharides and polysaccharides with β linkages is accomplished by the enzymes produced by the microbes inhabiting the digestive tracts of livestock (Figure 6–5b). Cellulose is the most important example of this type of carbohydrate. Note that the enzymes working on the α and β linkages must have different structures.

The Sodium Pump

Most monosaccharides are absorbed via the sodium pump. Sodium is the main cation of extracellular fluid. Potassium is the main cation of intracellular

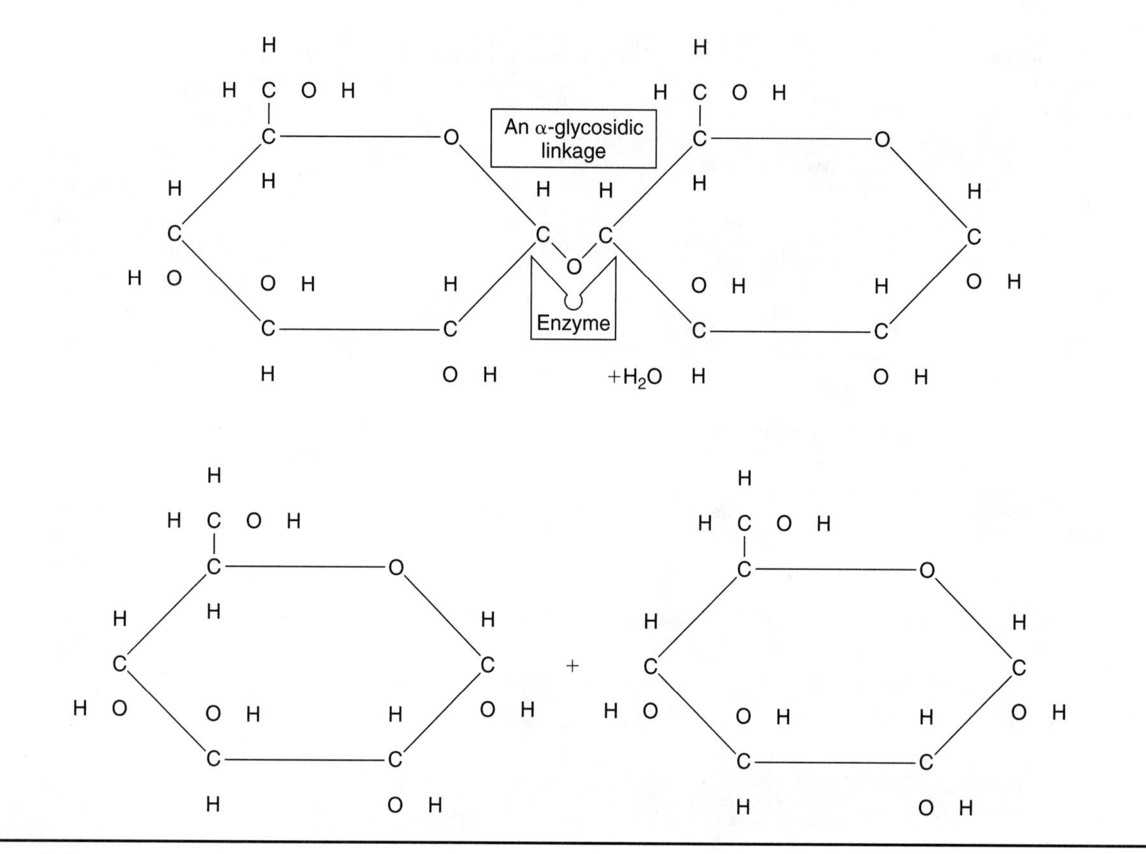

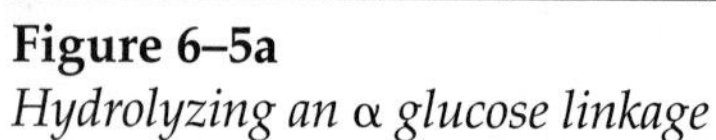

Figure 6–5a
Hydrolyzing an α glucose linkage

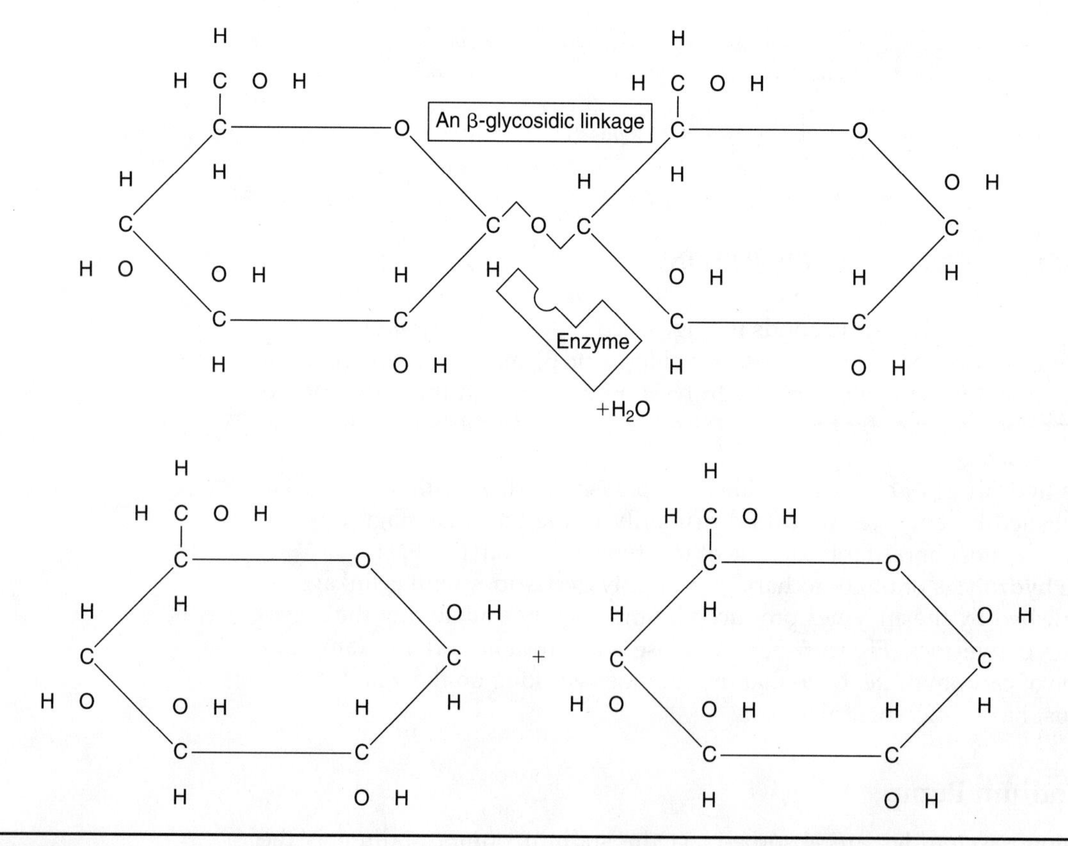

Figure 6–5b
Hydrolyzing a β glucose linkage

fluid. All cell membranes use sodium pumps (also called *sodium-potassium pumps*) to pump sodium cations out of the cell. Potassium cations are moved into the cell to maintain the cell's characteristic resting potential. But sodium cations are drawn down the concentration gradient back into the cell and the cycle continues. Most monosaccharides are absorbed into cells (from the intestinal lumen into the intestinal epithelium, and from the interstitial fluid into other cells of the body) by means of the sodium pump. The co-transport of sodium and monosaccharide by a proposed carrier molecule is illustrated in Chapter 9, Figure 9–3.

CARBOHYDRATE SOURCES

Whereas the various carbohydrates are widely distributed in plant tissues, relatively little carbohydrate is found in animal tissues. Some feed sources of the various carbohydrates are listed in Table 6–1.

MANAGEMENT OF CARBOHYDRATE FEEDING

Although carbohydrate will be the primary source of energy in most economical and healthful herbivore diets, any organic compound can be used as a source of energy. Because of this, specific requirements for carbohydrate have not been established for livestock. In spite of the fact that specific dietary requirements for carbohydrate levels do not exist, carbohydrate nutrition has become very complex. As the primary source of energy for herbivores, carbohydrates represent the primary component of the ration. The form of carbohydrate basically dictates the form of the ration, and diet form will have an impact on animal health.

As described in Chapter 4, feed carbohydrates are broadly classified as fiber and nonfiber carbohydrates. The fiber carbohydrates are included in the NDF (neutral detergent fiber) fraction of the feed and the nonfiber carbohydrates are included in the NFC fraction of the feed. The acid detergent fiber (ADF) carbohydrate fraction contains the more difficult-to-digest components of NDF. ADF is useful in evaluating the overall digestibility of feedstuffs.

The beef National Research Council (NRC) (2000) level II and the Cornell Net Carbohydrate and Protein System (CNCPS 4.0) (Fox et al., 2000) break

Table 6–1
Carbohydrate components of feed materials

Feed Source	Carbohydrate Component
Sugar cane, sugar beets	Sucrose
Starchy plants and roots	Maltose
Grains, seeds, tubers	Starch
Potatoes, tubers, artichokes	Inulin
Citrus fruits, apples	Pectins
Cottonseed, sugar beets	Raffinose
Acacia trees	Gums
Fibrous portion of plants	Cellobiose
Fibrous portion of plants	Hemicellulose
Corn cobs, wood	Xylose
Plant cell walls	Cellulose
Liver and muscle	Glycogen

Table 6–2
The carbohydrate pools used in the CNCPS and beef NRC level II

Carbohydrate Fraction	Fraction Components
A	Sugars, organic acids, oligosaccharides
B1	Starch, pectins
B2	Digestible NDF
C	Lignin

Source: Fox et al., 2000 and NRC, 2000.

Figure 6–6
Calculation of total digestible NFC

Total digestible NFC = 0.98 × (100 − [% CP + (% NDF − % NDFIP) + % fat + % ash)] × processing adjustment factor.

Figure 6–7
Calculation of the NDF digestibility coefficient

NDF digestibility coefficient = (0.75 × (% NDF − % NDFIP − % lignin) × [1 − (% lignin / (% NDF − % NDFIP))$^{0.667}$)] / % NDF

down each feedstuff's carbohydrate into four fractions (Table 6–2). These fractions differ in their rates of digestibility. Carbohydrate fraction A represents a feedstuff's most rapidly digested carbohydrate. Fraction A includes sugars and organic acids. Carbohydrate fraction B1 represents a feedstuff's content of intermediate rate of digestion carbohydrate. Fraction B1 includes starch and pectins. Carbohydrate fraction B2 represents a feedstuff's slowest digested carbohydrate. Fraction B2 includes digestible NDF. Carbohydrate fraction C includes the feedstuff's lignin content, which is indigestible. Describing a feedstuff's carbohydrate in this way gives ruminant nutritionists a great deal of information with which to evaluate carbohydrate nutrition.

In the dairy NRC (2001) and in the companion application for dairy cattle, NFC is assumed to be 98 percent digestible, but feedstuffs are assigned a processing adjustment factor that may increase or decrease NFC digestibility. Additionally, the NDFIP value is considered in the calculation of total digestible NFC for a given feedstuff (Figure 6–6).

The dairy NRC (2001) handles NDF by calculating an NDF digestibility coefficient for each feedstuff (Figure 6–7). Digestible NDF is calculated by multiplying the feedstuff's NDF contribution by its NDF digestibility coefficient.

SUMMARY

Carbohydrates are the largest component of herbivore diets. This carbohydrate may be in the form of cellulose reported in neutral detergent fiber, or it may be in the form of starch reported in NFC. Within each category, however, are dozens of compounds with more or less similar nutritional characteristics. In the dairy NRC (2001), all the NFC is assumed to be 98 percent digestible but each feedstuff is assigned an NDF digestibility coefficient. The CNCPS and beef NRC (2000) level II isolate and classify feed carbohydrate into four components, each with different digestion characteristics.

END-OF-CHAPTER QUESTIONS

1. Why is the form of dietary sources of carbohydrate important in herbivore diets?
2. Domestic herbivores do not make the enzymes necessary to digest cellulose, yet the rations formulated for these animals contain pasture, green chop, silages, and/or hay, all of which contain significant amounts of cellulose. Explain why.
3. Use the following terms in a description of the structural and nutritional differences between starch and cellulose: monosaccharide, glycosidic bond, alpha anomer, beta anomer.
4. What is hydrolysis and what role does it play in carbohydrate digestion and absorption?
5. Explain how $(CH_2O)n$ is shorthand for carbohydrate.
6. Explain the relationship between the following terms: monosaccharide, disaccharide, polysaccharide, starch, cellulose.
7. Name three components in each of the following mixtures of carbohydrate: hemicellulose and pectin.
8. Why have the NRC committees not determined specific carbohydrate requirement values for domestic animals?
9. Name and describe the four carbohydrate fractions used by the beef NRC level II and the Cornell Net Carbohydrate and Protein System (CNCPS).
10. Give a feed source for each of the following types of carbohydrate: sucrose, starch, hemicellulose, cellulose, pectin.

REFERENCES

Fox, D. G., Tylutki, T. P., Van Amburgh, M. E., Chase, L. E., Pell, A. N., Overton, T. R., Tedeschi, L. O., Rasmussen, C. N., & Durbal, V. M. (2000). The net carbohydrate and protein system for evaluating herd nutrition and nutrient excretion (CNCPS, version 4.0). Animal Science Department Mimeo 213, Cornell University, Ithaca, NY.

Lehninger, A. L. (1978). *Biochemistry* (2nd edition). New York: Worth Publishers, Inc.

National Research Council. (2000). *Nutrient requirements of beef cattle* (7th revised edition). Washington, DC: National Academy Press.

National Research Council. (2001). *Nutrient requirements of dairy cattle* (7th revised edition.). Washington, DC: National Academy Press.

CHAPTER 7

PROTEINS

It has long been known that all animals must receive in their food at least a certain minimum amount *of protein. More recently, investigations have shown that for man and for such animals as swine, poultry, dogs and rats, the* quality *or* kind *of protein is fully as important as the amount.*

F. B. MORRISON, 1949

INTRODUCTION

Protein supplements generally constitute the highest-cost ingredients of finished feeds. For this reason, it is important that the purchaser of feeds understand the nature of the requirement for protein. In fact, protein is probably not required at all by any species of livestock. Rather, it is the amino acids that constitute the feedstuff protein that are required by livestock. In ruminant nutrition, the production of microbial protein must be managed to maximize productivity and profitability.

IMPORTANCE OF PROTEIN

About 15 percent of body weight is due to protein. This protein is distributed throughout the body in tissues, enzymes, and hormones.

Protein sources for livestock rations are more expensive than carbohydrate sources. In fact, protein is usually the most costly component of a finished feed. Because of this, a purchased feed product is often identified by its percent protein content.

A major goal in protein nutrition is to create a ration that enables the animal to capture feed protein in animal flesh or animal product efficiently. The more efficiently dietary protein is captured, the less protein needs to be in the ration to achieve a given level of productivity. If the ration is not properly balanced for nutrients, a significant amount of the feed protein may not be captured. This uncaptured protein will be removed from the body, primarily by the kidneys, in the form of ammonia (fish), urea (mammals), or uric acid (poultry). The nitrogen in these excretory products is a significant source of environmental pollution.

In order to support efficient production and/or growth, rations must be balanced such that the levels of energy and protein each support similar levels of performance. A protein shortage relative to energy results in increased fat deposition because the ration energy cannot be used to synthesize protein tissue—the energy is put into storage for later use. The use of body fat to meet the body's need for energy is very inefficient and an excessive reliance on fat stores of energy will lead to health problems.

Applying a knowledge of protein nutrition to ration formulation then, has three benefits: (1) it maximizes profitability by optimizing the use of expensive dietary protein, (2) it minimizes the risk of nitrogen pollution, and (3) it helps maintain animal health.

STRUCTURE OF PROTEIN

Proteins are organic compounds containing carbon, hydrogen, oxygen, nitrogen, and sometimes sulfur. The distinguishing feature of protein molecules is the nitrogen content. The nitrogen in protein molecules is found in the amino groups of amino acids. An amino group is comprised of a nitrogen atom bonded to two hydrogens, $—NH_2$. In addition to an amino group, amino acids have a carboxylic acid group, —COOH, hence the name *amino acid.* In amino acids, both the amino group and the carboxylic acid are bonded to the same carbon. The other two entities bonded to this carbon are a hydrogen atom and an R group. The R group varies with the amino acid (Figure 7–1).

A protein molecule can be visualized as a long chain or several chains of amino acids joined together by linkages. These chains are folded in complex ways, creating the three-dimensional structure that is an important characteristic of proteins. The properties of a protein molecule are determined by the number, type, and arrangement of the amino acids that comprise the protein.

The linkages between amino acids are called *peptide bonds* (Figure 7–2). The terms *peptide* and *polypeptide* refer to chains of amino acids that are shorter than normally found in protein molecules.

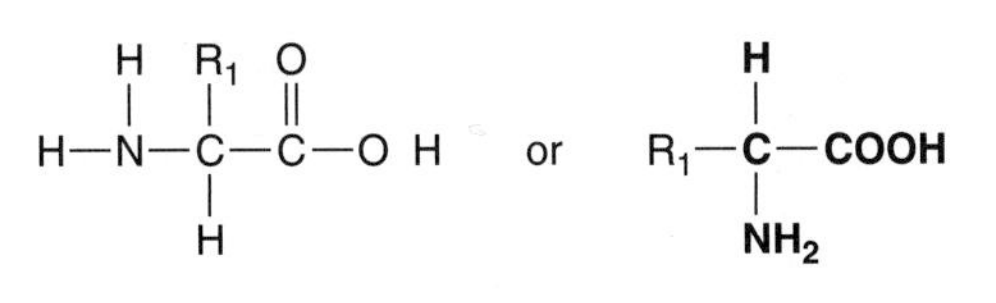

Figure 7–1
Amino acid structure

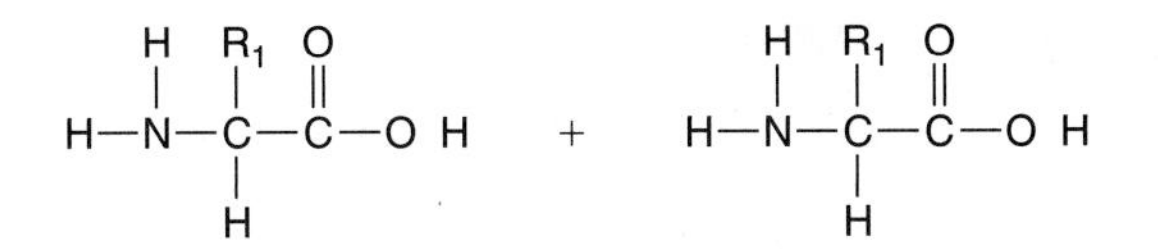

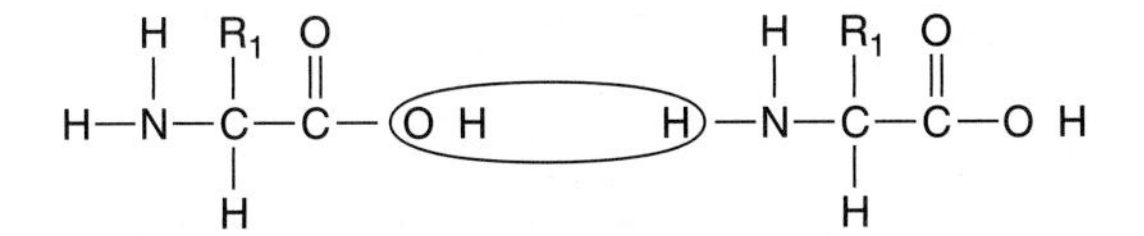

Figure 7–2
Formation of a peptide bond (in bold type) between two amino acids

AMINO ACIDS

There are 22 amino acids found in body proteins. For most species of livestock, about half of these can be manufactured by the body if supplied with extra amino groups or nonspecific nitrogen sources. These are called *nonessential* or *dispensable* amino acids. The remaining amino acids either cannot be made by the animal or cannot be made at a rate sufficient to support maximum performance. These are called *essential* or *indispensable* amino acids. They must be provided by either diet or the synthetic activities of microbes inhabiting the digestive tract. In discussing dietary protein sources, *protein quality* refers to the content of essential amino acids in the feedstuff. Figures 7–3a to 7–3j show the structure of the classic 10 essential amino acids. Memorization is helped by the phrase PVT TIM HALL (phenylalanine, valine, theonine, tryptophan, isoleucine, methionine, histidine, arginine, leucine, lysine).

Figure 7–3a
The structure of the essential amino acid lysine

Lysine (Lys)

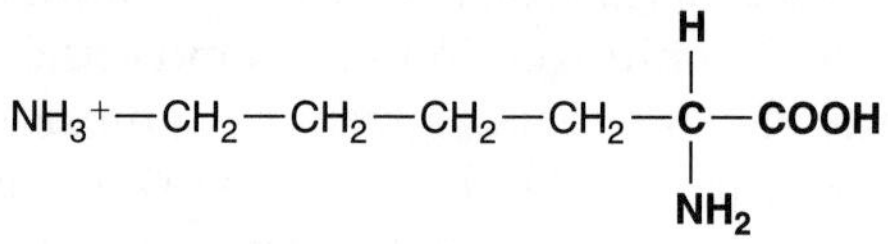

$$NH_3^+{-}CH_2{-}CH_2{-}CH_2{-}CH_2{-}\underset{NH_2}{\overset{H}{C}}{-}COOH$$

Figure 7–3b
The structure of the essential amino acid methionine

Methionine (Met)

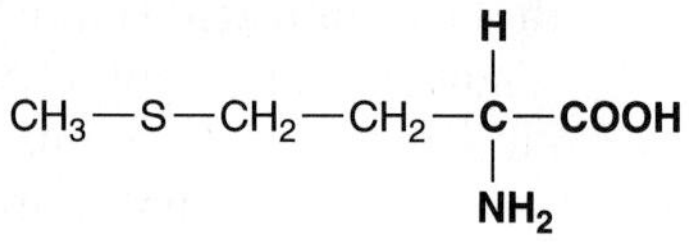

$$CH_3{-}S{-}CH_2{-}CH_2{-}\underset{NH_2}{\overset{H}{C}}{-}COOH$$

Figure 7–3c
The structure of the essential amino acid histidine

Histidine (His)

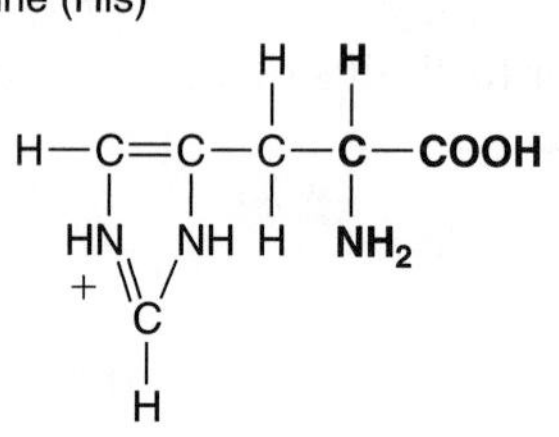

Figure 7–3d
The structure of the essential amino acid threonine

Threonine (Thr)

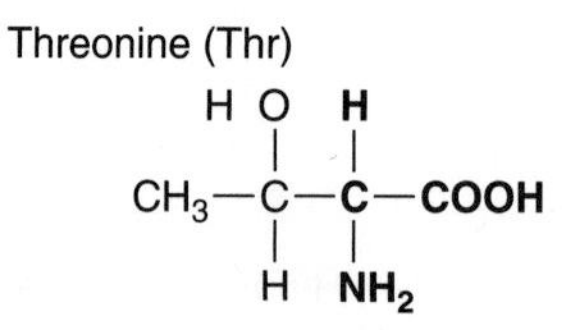

Figure 7–3e
The structure of the essential amino acid valine

Valine (Val)

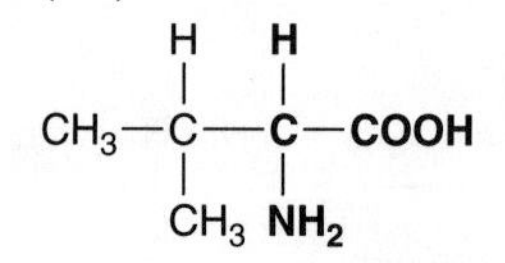

Figure 7–3f
The structure of the essential amino acid leucine

Leucine (Leu)

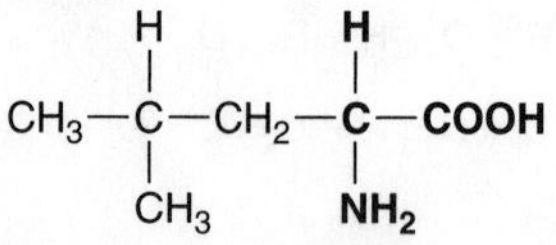

$$CH_3{-}\underset{CH_3}{\overset{H}{C}}{-}CH_2{-}\underset{NH_2}{\overset{H}{C}}{-}COOH$$

Taurine (Figure 7–3k) is another amino acid. Among domestic livestock, it is a dietary essential only for the cat. The structure of taurine is shown in Figure 7–3k.

There are some amino acids that, for some species of livestock, do not fit neatly into the essential and nonessential categories. A dietary proline requirement has been established for chicken broilers, but not for layers or other types of livestock. Methionine is an essential amino acid and cysteine is not. However, animals make cysteine from dietary methionine. Therefore, to ensure that that the animal receives adequate amounts of both methionine and cysteine, the diet either has to have enough methionine to meet both the methionine and cysteine requirements, or a lower amount of methionine and enough cysteine proper. This is of nutritional significance in the dog, pig, chicken, and fish because these animals receive minimal amino acids from microbial activity in their digestive systems, and because the requirements for methionine and cysteine for these animals are known. It should be mentioned that cysteine often occurs in proteins in its oxidized form, cystine, in which two cysteine molecules are bonded together through formation of a disulfide group.

As methionine is to cysteine, so phenylalanine is to tyrosine: given excess phenylalanine, livestock can make tyrosine. Either the diet has to have enough

Isoleucine (Ile)

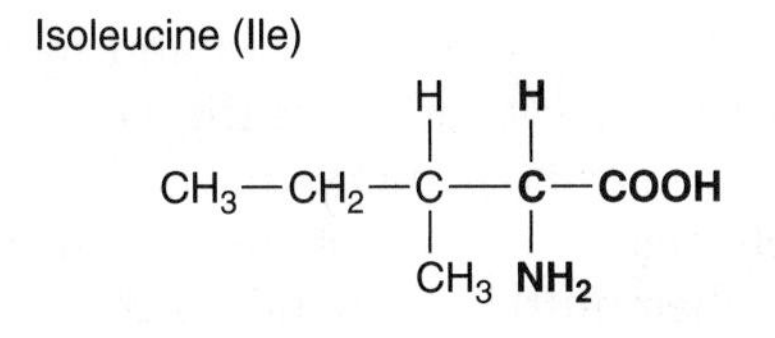

Figure 7–3g
The structure of the essential amino acid isoleucine

Tryptophan (Trp)

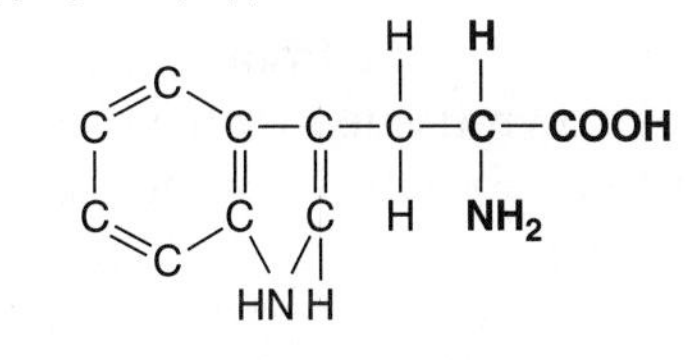

Figure 7–3h
The structure of the essential amino acid tryptophan

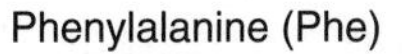

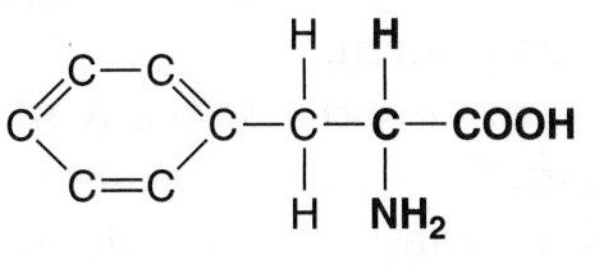

Figure 7–3i
The structure of the essential amino acid phenylalanine

Arginine (Arg)

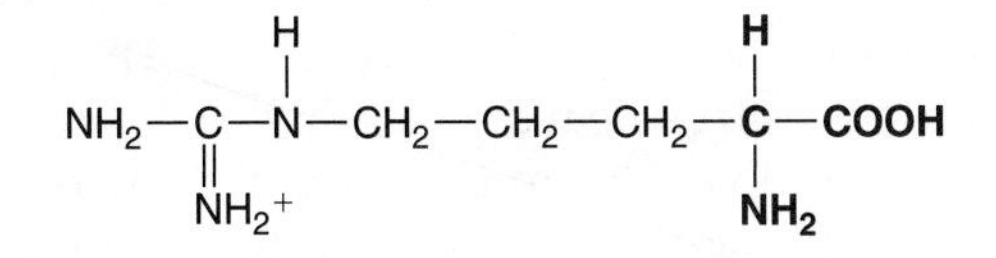

Figure 7–3j
The structure of the essential amino acid arginine

Taurine

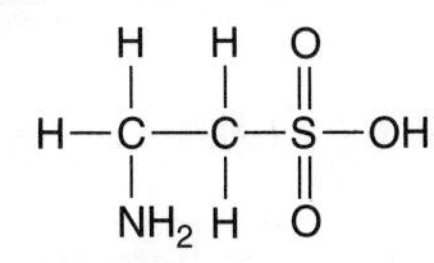

Figure 7–3k
The structure of the amino acid taurine. Among domestic livestock, taurine is an essential amino acid for the cat only

Figure 7–4
Protein hydrolysis

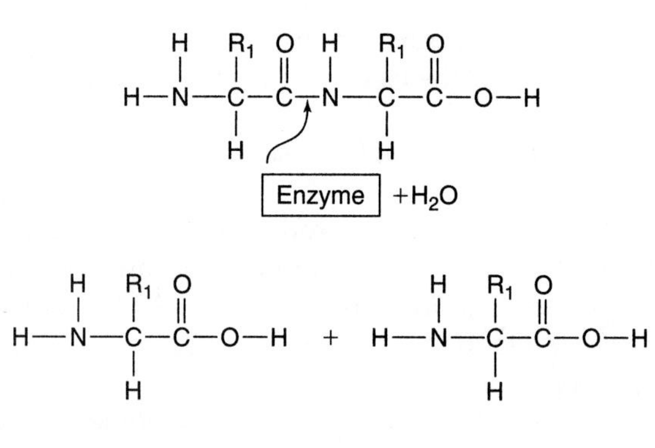

phenylalanine to meet both phenylalanine and tyrosine requirements, or a lower amount of phenylalanine and enough tyrosine proper.

In poultry, glycine can be synthesized, but the rate is not sufficient to support maximum growth. Excess serine can be converted to glycine so poultry nutritionists monitor the glycine + serine level in the diet of growing birds.

AMINO ACID ABSORPTION

Most of the protein in animal diets must be digested and hydrolyzed into the component amino acids in order to pass out of the digestive tube and into the blood (Figure 7–4). The exceptions are the peptides that may originate in the feed or may result from incompletely digested feed proteins. Absorbed peptides are responsible for food allergies but their nutritional significance is largely unknown.

As is the case with monosaccharides, some amino acids are absorbed via the sodium pump. The sodium pump is described in Chapter 6 and illustrated in Chapter 9, Figure 9–3.

Most of the amino acids absorbed come from feed protein, but some come from the proteins of the animal's own tissue that have been sloughed off from the epithelium lining the digestive tube and from the enzymes of digestive secretions. This is referred to as *endogenous protein.* Endogenous protein, in grams, is calculated as 4.72 × dry matter fed in kilograms (dairy NRC, 2001). In the companion application to this text for dairy, beef, sheep, and goats, calculations of available metabolizable protein (MP) use the sum of MP from the rumen microbes digested, MP from feed protein that was not degraded by the rumen microbes, and MP from endogenous protein.

In swine nutrition, the issue of amino acid bioavailability is addressed using ileal digestibilities. A feedstuff with known amino acid content is fed to swine that have been fitted with a cannula in the ileum. Ileal contents are collected and analyzed for amino acid content. The proportion of each amino acid that is not collected is referred to as the *apparent ileal digestibility* for that amino acid in that feedstuff. If this proportion is corrected for endogenous losses of amino acids, it is referred to as *true ileal digestibility.* The companion application to this text for swine uses true ileal digestibility.

Once absorbed, amino acids are used to build the proteins of the animal's body. Virtually every tissue in the animal body contains protein, and all cells synthesize proteins for part or all of their life cycle. In addition, some of the body's hormones and all the body's enzymes are made of protein.

MANAGEMENT OF PROTEIN FEEDING

Livestock require essential amino acids in their diets and sufficient nonspecific sources of nitrogen with which to make the nonessential amino acids. Whether

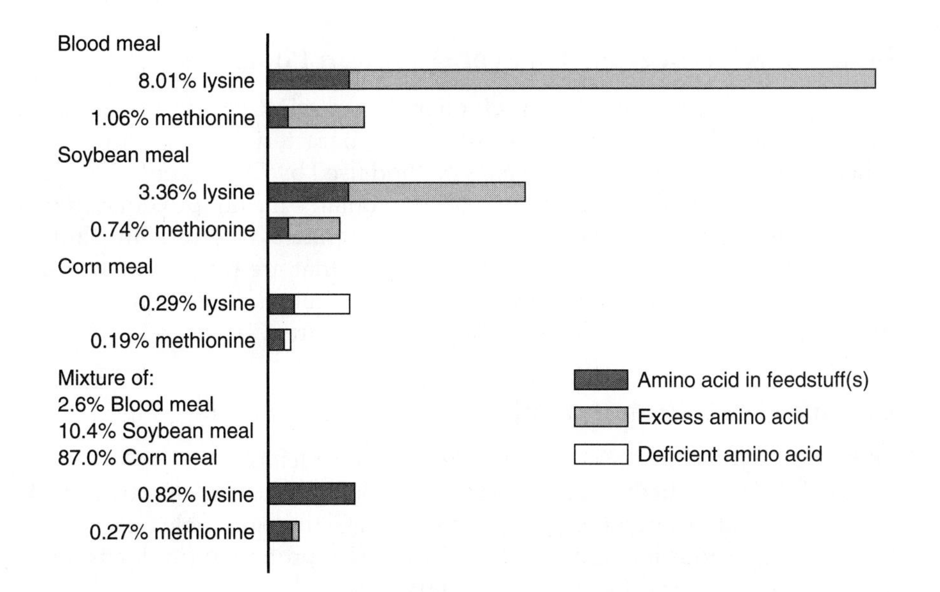

Figure 7–5
Meeting amino acid requirements and minimizing amino acid excesses in the ration of a 55-lb. pig by using a variety of protein sources. Amino acid content is expressed as percent, dry matter basis.

these come from the protein in soybean meal, alfalfa hay, or meat meal makes no difference. In fact, in the swine, poultry, fish, and pet food industries, it is routine to meet some of the essential amino acid requirements by including crystalline amino acids in the feed formula. For the most part, however, it is still most economical to meet the amino acid requirements of livestock by formulating a ration that contains a combination of natural protein sources. An example is shown in Figure 7–5. By themselves, blood meal, soybean meal, and corn meal are either excessive or deficient in lysine and methionine for a 55-lb. growing pig, but a combination of these protein sources provides the proper amino acid balance.

The microbial populations living in the digestive tracts of livestock are capable of making all amino acids needed by livestock. These amino acids are incorporated into the cells of the bacteria. If these bacterial populations exist downstream from the small intestine, the animal will probably not have access to these amino acids except through coprophagy (consumption of feces). If, as is the case with ruminants, the microbial population exists upstream from the small intestine, some of the bacteria will regularly pass through the small intestine and their component amino acids will be absorbed. Due to the synthetic activity of bacteria in the rumen, there are no amino acids that are dietary essentials for healthy, mature ruminant animals. In fact, the rumen microbes can make essential amino acids from nitrogen sources that are not even in protein form.

This is not to say, however, that protein quality is not an issue in ruminant nutrition. As described in Chapter 4, feed protein is broadly classified as degradable and undegradable. The undegradable protein is unavailable to the microbes of the rumen. This is termed *bypass protein, undegradable intake protein* (UIP) as contrasted with degradable intake protein (DIP), or *rumen undegradable protein* (RUP) as contrasted with rumen degradable protein (RDP). Because RUP in ruminant diets arrives at the small intestine unaltered by microbial activity, its initial quality determines its ultimate value. To maximize the delivery of amino acids to the mammary gland of high-producing dairy cows, 35 percent or more of the dietary protein may need to bypass the rumen.

Identifying what portion of feed protein should be RDP and what portion should be RUP is a major focus of ruminant nutrition. Nitrogen fractions or "pools" have been identified in feedstuffs. There are two systems of protein pools currently in use in dairy nutrition: the dairy NRC system and the CNCPS 4.0/beef NRC (2000) level II system (Figures 7–6 through 7–9).

The CNCPS 4.0/Beef NRC Level II Protein Pools

In the Cornell Net Carbohydrate and Protein System (CNCPS 4.0) as well as the beef NRC (2000) level II, the protein pools are characterized using an *in vitro* method (Sniffen et al., 1992). The *in vitro* method used by CNCPS gives five protein pools: A, B1, B2, B3, and C. Protein pool A contains the nonprotein nitrogen (NPN). Protein pool B1 contains the rapidly degraded true protein. Protein pool B2 contains true protein and large peptides that are potentially degradable. Protein pool B3 contains slowly degradable protein. Protein pool C contains undegradable, largely indigestible protein (Figures 7–6 and 7–7).

The Dairy NRC Protein Pools

In the dairy NRC (2001), the protein pools are characterized using an *in situ* method. The *in situ* method used by the NRC defines the protein pools as: A, rapidly degraded; B, potentially degradable; and C, undegradable (Figures 7–8 and 7–9). The various methods used to characterize protein in ruminant nutrition have been reviewed (Schwab et al., 2003).

Figure 7–6
CNCPS v.4.0/beef NRC level II protein pools (in vitro method).
Sources: Fox et al., 2000 (CNCPS 4.0) and National Research Council, 2000 (beef NRC level II).

Protein pool identification	A	B1	B2	B3	C
Protein pool characteristics	Nonprotein nitrogen, rapidly degraded	True protein, rapidly degraded	Potentially degradable protein	Mostly undegradable protein	Undegradable and indigestible protein and nitrogen sources

Figure 7–7
Components of the protein pools (determined using an in vitro method) used in the beef NRC, 2000 level II and CNCPS 4.0.
Source: National Research Council, 2000 (beef NRC level II) and Fox et al., 2000 (CNCPS 4.0).

- Protein pool A
  —Nonprotein nitrogen (NPN)
  - Ammonia
  - Peptides, amino acids
    —Rapidly converted to ammonia by and for ruminal microbes
  - Much of the degradable crude protein in preserved forages is NPN
- Protein pool B1
  —True protein, rapidly degraded to ammonia
  - Most of the degradable protein in pasture is B1
- Protein pool B2
  —True protein, potentially degradable
  - Competition between digesta passage rate (Kp) and feedstuff protein B2 pool digestion rate (Kd) determines fate of protein pool B2
- Protein pool B3
  —True protein, mostly undegradable
  - B3 protein is associated with cell wall
    —Measured as neutral detergent insoluble protein less acid detergent insoluble protein (NDFIP – ADFIP)
  - Lots of B3 in treated protein supplements
  - Lots of B3 (a large % of total, though total may be low) in some forages, fermentation co-products, animal protein co-products
- Protein pool C
  —True protein, resistant to mammalian and microbial enzymes
  —Measured as acid detergent insoluble protein (ADFIP)
  - Most of the protein associated with lignin in cell walls
  - Most of the protein associated with Maillard products
  - Most of the protein in tannin-protein complexes

In both the dairy NRC and CNCPS 4.0/beef NRC level II systems, there is a protein pool that is potentially degradable. To determine whether this protein pool becomes a part of RDP or RUP, a calculation of feed passage rate (Kp) is used. Kp is used by both the dairy NRC and CNCPS 4.0/beef NRC level II systems. It is the percent per hour disappearance rate for a given feedstuff, and it is calculated based on the characteristics of the entire ration. High-producing dairy cows fed large amounts of grain will have rapid passage rates and high Kp values. Dry cows receiving large amounts of dry hay will have slow passage rates and low Kp values.

This text and the dairy ration application use the protein pools system as defined in the dairy NRC publication. All discussion that follows pertains to the dairy NRC (2001) system.

The contribution of absorbed protein from pool B in a given feedstuff is determined using the predicted passage rate of the ration (Kp) and the inputted degradation rate, percent/hour of the feedstuff's protein pool B (Kd). Kd is a characteristic of each feedstuff, and as such, is an inputted value. Kp is calculated for each feedstuff based on feedstuff type. The formulas used to calculate feedstuff Kp in the dairy NRC (2001) and in the companion application to this text are shown in Figure 7–10. A feedstuff with a slow rate of pool B degradation, Kd, in the ration of a cow experiencing a rapid feed passage rate, Kp, will have a large RUP component in its B fraction. A feedstuff with a rapid rate of pool B degradation in the ration of a cow experiencing a slow feed passage rate will

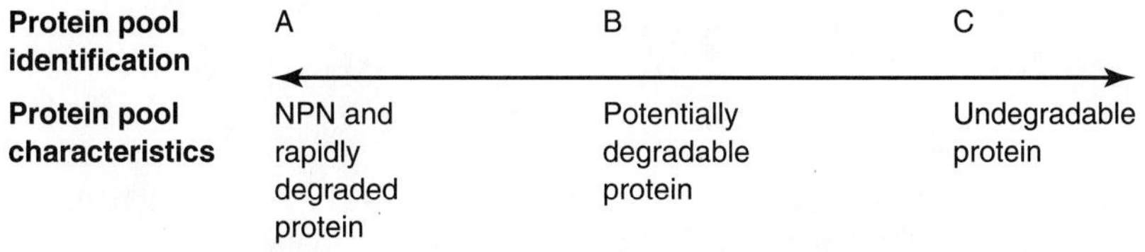

Protein pool identification	A	B	C
Protein pool characteristics	NPN and rapidly degraded protein	Potentially degradable protein	Undegradable protein

Figure 7–8
*NRC protein pools used in dairy nutrition (*in situ *method)*

- Protein pool A
  - —Non-protein nitrogen (NPN)
    - Ammonia
    - Peptides, amino acids
      - —Rapidly converted to ammonia by and for ruminal microbes
    - Much of the degradable crude protein in preserved forages is NPN
  - —True protein, rapidly degraded
    - Most of the degradable protein in pasture is true protein, rapidly degraded to ammonia
- Protein pool B
  - —True protein, potentially degradable
    - Competition between digesta passage rate (Kp) and feedstuff protein B pool digestion rate (Kd) determines fate of protein pool B
- Protein pool C
  - —True protein, undegradable
    - Digestible
      - —Lots of digestible C in treated protein supplements
      - —Lots of digestible C (a large % of total, though total may be low) in some forages, fermentation co-products, animal protein co-products
    - Indigestible
      - —Resistant to mammalian and microbial enzymes
        - Most of the protein associated with lignin in cell walls
        - Most of the protein associated with Maillard products
        - Most of the protein in tannin-protein complexes

Figure 7–9
Components of the protein pools (determined using an in situ *method) used in the dairy NRC publication*

Figure 7–10
Feedstuff passage rate formulas (Kp) from the dairy NRC 2001. DMI: Dry matter intake; BW: body weight.

For wet forages: Kp = 3.054 + (0.614 × DMI % BW)
For dry forages: Kp = 3.362 + (0.479 × DMI % BW) – (0.017 × NDF %) – (0.007 × % concentrate)
For concentrates: Kp = 2.904 + (1.375 × DMI % BW) – (0.020 × % concentrate)

Figure 7–11
How the dairy NRC calculates RUP and RDP in a feedstuff

Determining the RUP of blood meal in the following ration fed to a lactating dairy cow: legume hay 3 lb., mixed hay 5 lb., mixed silage 30 lb., corn silage 35 lb., corn meal 20 lb., soybean meal 3 lb., distillers grains 2 lb., blood meal 0.5 lb., mineral/vitamin 1.5 lb.

Given—
Table values for blood meal:
Protein pool B = 39.9% of crude protein
Protein pool C = 50.0% of crude protein
Kd (rate of degradation of the B fraction) = 1.9%/hr.

Given—
Passage rate of blood meal based on the above ration:
Kp (rate of passage from the rumen) = 7.39%/hr.

Solution:

$$\begin{aligned} RUP &= \{B \times [Kp / (Kd + Kp)]\} + C \\ &= \{39.9 \times [7.39 / (1.9 + 7.39)]\} + 50.0 \\ &= [39.9 \times (7.39 / 9.29)] + 50.0 \\ &= (39.9 \times 0.80) + 50.0 \\ &= 31.92 + 50.0 \\ &= 81.92 \end{aligned}$$

This is the % of the crude protein of blood meal in the above diet that behaves as RUP. The % RDP in the crude protein of blood meal in the above diet is 100 − % RUP = 18.08.

have a large RDP component in its B fraction. The formula used to calculate the percentage RUP in the crude protein of a given feedstuff is:

$$RUP = B\,[Kp/(Kd + Kp)] + C$$

This formula is used in an example in Figure 7–11. Because RUP and RDP are expressed as percent of crude protein, and because crude protein is either RUP or RDP, RDP may be calculated as 100 − % RUP.

In addition to passage rate, feed processing affects degradability of the B pool. Heat treatment of whole oilseeds and solvent-extracted oilseed meals shifts the potentially degradable B pool toward the undegradable end of the spectrum.

Feedstuff protein of the A pool and the degradable protein portion of the B pool constitute the rumen degradable protein. Ration RDP results in microbial growth in the rumen. A portion of the microbes are regularly washed out of the rumen and digested, and their amino acids are made available to the ruminant. Amino acid content of microbial protein, as well as that of animal tissue and bovine milk, is shown in Table 7–1.

Feedstuff proteins that fall into the undegradable portion of the B pool and the C pool make up the RUP. The digestible portion of the RUP will result in amino acids being delivered directly to the blood.

The total amino acids absorbed by ruminant animals will be the sum of that from bacteria, digestible RUP, and endogenous protein. Protein nutrition in the

Amino acid	Bovine Tissue	Bovine Milk	Rumen Bacteria[1]
Arginine	16.8	7.2	10.2
Histidine	6.3	5.5	**4.0**
Isoleucine	7.1	11.4	11.5
Leucine	17.0	19.5	**16.3**
Lysine	16.3	16.0	**15.8**
Methionine	5.1	5.5	5.2
Phenylalanine	8.9	10.0	10.2
Threonine	9.9	8.9	11.7
Tryptophan	2.5	3.0	2.7
Valine	10.1	13.0	12.5

[1]Bacteria values in bold type indicate amino acid content is below that of both tissue and milk protein.

Source: National Research Council, 2001.

Table 7–1
Amino acid contents expressed as percent of essential amino acids of tissue, milk and rumen bacteria

ruminant animal involves managing the protein pools to ensure optimum delivery of both microbial protein and digestible RUP to the small intestine.

AMINO ACID REQUIREMENTS

It is often possible, for a given ration, to identify which amino acid is falling farthest from its requirement. This amino acid is said to be first limiting. The first limiting amino acid is not necessarily the one required in the greatest quantity or the one least prevalent in the ration. By identifying the first limiting amino acid, the nutritionist has identified a factor that may be limiting performance. If other nutrients are in excess of the requirement, adding more of the first limiting amino acid to the ration may result in improved performance.

With monogastric species, the first limiting amino acid is often added to rations to improve the efficiency of protein utilization. In usual rations for horses, fish, and swine, the first limiting amino acid is lysine. For poultry, the first limiting amino acid in usual rations is methionine. Dog rations are variable to the extent that it is difficult to identify a "usual" ration. The first limiting amino acid for a given dog food formulation may be lysine, methionine, or another amino acid.

In swine nutrition, the concept of first limiting amino acid has been expanded to what is called the ideal protein. The ideal protein is the optimum dietary ratios of essential amino acids relative to lysine. These ratios are established for the functions of maintenance, protein accretion, and milk synthesis, and are shown in Table 7–2. In swine diets formulated on an ideal protein basis, amino acids are provided in the exact proportions required, and in this way, every amino acid is equally limiting. Formulating diets based on the ideal protein should reduce the amount of excess amino acids that are catabolized (Lopez et al., 1994).

In dairy nutrition, the matter has been extensively studied and the first limiting amino acid does not appear to be limiting production in usual dairy rations. In other words, the performance supported by the first limiting amino acid in ruminant animals is usually in excess of that supported by other nutrients. Supplying more of the first limiting amino acid, therefore, usually does not improve performance. The amino acids most often predicted to be first limiting in dairy rations are methionine, lysine, and, collectively, the branched

Table 7–2
The ideal protein for swine, expressed as ratios of amino acids to lysine

Amino acid	Maintenance	Protein Accretion	Milk Synthesis
Lysine	100	100	100
Arginine	−200[1]	48	66
Histidine	32	32	40
Isoleucine	75	54	55
Leucine	70	102	115
Methionine	28	27	26
Methionine + cystine	123	55	45
Phenylalanine	50	60	55
Phenylalanine + tyrosine	121	93	112
Threonine	151	60	58
Tryptophan	26	18	18
Valine	67	68	85

[1]The negative value for arginine indicates metabolic synthesis in excess of maintenance requirement.

Source: National Research Council, 1998.

chain amino acids (valine, leucine, and isoleucine). It is interesting to consider the challenges involved in developing an amino acid supplement for ruminant animals. The product must be resistant to microbial degradation, yet absorbable at the small intestine.

PROTEIN REQUIREMENTS

Table 7–3 gives crude protein requirements predicted by NRC tables and formulas for selected domestic animals. These requirements are expressed in terms of pounds and in terms of the percent of the dry diet. Remember that protein is the package in which are found the amino acids that are required by the animal. For most livestock species, the minimum protein requirement is determined from the animal's requirement for both the essential amino acids and the nitrogen sources needed for the manufacture of nonessential amino acids.

In examining Table 7–3, it is important to keep in mind the factors that determine how much protein an animal will require.

1. Growing animals accrete protein and because they have not yet reached mature body size, growing animals have relatively low dry matter intakes. Growing animals therefore require diets that contain a greater concentration of protein than do mature, idle animals.
2. Mature animals that are not pregnant or lactating use dietary protein only to maintain body tissues and replace protein secretions. The protein concentration required for mature, idle animals may be half that required for growing or producing animals.
3. Herbivores have evolved to use carbohydrate in the plants they eat as their primary source of energy. Because carnivores eat animals and few plants, their natural diet includes relatively little carbohydrate, but a relatively large amount of protein and fat. Table 7–4 gives a comparison of the two feed types. Because of the differences in diet, carnivores have evolved to use protein and fat as their primary sources of energy. Domestic carnivores,

Table 7–3

Crude protein requirements for selected animals.[1] *Percent values are given on a dry matter basis.*

	Required (pounds)	Required (%)
Fish, channel catfish, 100 g body weight	0.00198	30.00
Fish, rainbow trout, 100 g body weight	0.00114	41.11
Chicken, broiler, 5 wks. of age	0.0603	22.22
Chicken, white egg layer, 3 lb. body weight	0.0331	16.67
Pig, growing, 45 lb. body weight	0.45	16.92
Dog, growing, 30 lb. body weight	0.08	10.51
Cat, growing kitten, 4.2 lb. body weight	0.04	24.00
Rabbit, growing, 5 wks of age	0.0517	17.55
Horse, light work, 1,100 lb. body weight	2.23	7.64
Goat, maintenance, 88 lb body weight	0.14	9.04
Ewe, maintenance, 110 lb body weight	0.21	9.40
Beef animal, growing, 800 lb. body weight	See note below	See note below
Dairy cow, lactating, 1,400 lb. body weight	See note below	See note below

[1]Note: In beef and dairy nutrition, the protein "requirement" would depend on the energy level of the ration. The protein in the ration should be what is necessary to support a similar level of performance (gain or milk production) to that supported by the energy in the ration.

Table 7–4

Comparison of animal and plant composition

Nutrient Component	% in Meat and Bone Meal Tankage, IFN 5-00-387 (dry matter basis)	% in Orchardgrass Hay, Early Bloom, IFN 1-03-425 (dry matter basis)
Protein	50	12.8
Fat	12.4	2.9
Minerals	28	8.5
Other (including carbohydrate)	9.6	75.8

therefore, are fed diets containing much more protein than would be needed to build the body's protein tissue. It has become apparent, however, that domestic carnivores can use more carbohydrate than would be present in the diet of their wild counterparts. Since carbohydrate is less expensive than protein, some carbohydrate is usually included in the diets of domestic carnivores.

The Consequences of Feeding Excess Protein

All absorbable amino acids are absorbed. In other words, even if the amount of amino acids flowing through the small intestine is more that what the animal can use, all amino acids will, in most cases, still be absorbed. Absorbed amino acids are transported from the blood into the cells. Recall that the sodium pump is involved in this transport for some amino acids. Inside the cells, the breakdown of excess amino acids yields amino groups and carbon skeletons. The carbon skeletons resemble carbohydrates and are used by the animal as a source of energy. The amino groups are unstable and they are quickly converted to ammonia (NH_3). But ammonia is toxic to cells, so it is expelled back into the blood. In fish, the ammonia is excreted at the gills. In poultry, the ammonia is converted in a complex series of reactions to uric acid prior to excretion. In mammals, the bloodstream carries the ammonia to the liver. There, the enzymes of the urea cycle convert the ammonia to urea. The urea is then released into the blood, where most of it is filtered out at the kidneys and sent to the bladder for excretion. The operation of the urea cycle and the conversion of ammonia to uric acid require energy. In mammalian livestock, the energy used to process the nitrogen in unused amino acids is referred to as the *urea cost*. The urea cost becomes part of the animal's maintenance energy requirement, leaving less energy available to support other functions. In the companion application to this text for dairy and beef cattle, the urea cost is calculated and added to the maintenance energy requirement.

The elevated blood levels of urea that result from feeding excess protein have been associated with health and reproductive problems in some species. The effects of excess dietary protein are discussed in Chapter 18 and are illustrated in Figure 18–5.

The processing of extra absorbed protein is not totally efficient, and heat is generated. Feeding excess protein to a hot animal (due to high ambient temperature or high work level) can make it even more challenging for the animal to maintain normal body temperature.

If a protein excess occurs as the result of an excess amount of purchased protein supplement being fed, there is an economic consequence to feeding excess protein. Finally, animals fed protein in excess of their requirement may present an environmental threat due to the effects of excreted nitrogen in the environment.

SUMMARY

Protein nutrition in livestock rations involves providing livestock with sources of essential amino acids and nonspecific nitrogen sources with which to build the nonessential amino acids. When absorbed, amino acids provide the building blocks for the construction of protein in animal tissues and products. Ruminant nutrition involves managing the flow of protein in two sources: the microbes grown on RDP and the bypass protein or RUP.

END-OF-CHAPTER QUESTIONS

1. Explain the following statement: "Protein is probably not required by any species of livestock."
2. Name the classic 10 essential (or indispensable) amino acids.
3. Explain the relationship between the amino acid methionine (essential) and the amino acid cysteine (nonessential). Explain the relationship between the amino acid phenylalanine (essential) and the amino acid tyrosine (nonessential).
4. In protein nutrition, animals require the essential amino acids and a supply of nonspecific sources of nitrogen. Why do animals require nonspecific sources of nitrogen?
5. Explain why, in ruminant nutrition, the protein quality of the rumen undegradable protein is more important than the protein quality of the rumen degradable protein.
6. Explain the concept behind the protein pools as applied by the dairy and beef NRC committees.
7. Which of the dairy NRC protein pools is affected by feed passage rate? Which of the CNCPS protein pools is affected by feed passage rate?
8. Explain the term *ileal digestibility* as it is used in amino acid nutrition of swine.
9. Explain the ideal protein concept as applied to swine nutrition.
10. What becomes of the protein that is ingested in excess of the animal's requirement?

REFERENCES

Fox, D. G., Tylutki, T. P., Van Amburgh, M. E., Chase, L. E., Pell, A. N., Overton, T. R., Tedeschi, L. O., Rasmussen, C. N., & Durbal, V. M. (2000). The net carbohydrate and protein system for evaluating herd nutrition and nutrient excretion (CNCPS version 4.0). Animal Science Department Mimeo 213, Cornell University, Ithaca, NY.

Lopez, J., Goodband, R. D., Allee, G. L., Jessee, G. W., Nelssen, J. L., Tokach, M. D., Spiers, D., & Becker, B. A. (1994). The effects of diets formulated on an ideal protein basis on growth performance, carcass characteristics, and thermal balance of finishing gilts housed in a hot, diurnal environment. *Journal of Animal Science 72*, 367–379.

Morrison, F. B. (1949). *Feeds and feeding* (21st ed.). Ithaca, NY: Morrison Publishing Co.

National Research Council. (2000). *Nutrient requirements of beef cattle* (7th revised edition.). Washington, DC: National Academy Press.

National Research Council. (2001). *Nutrient requirements of dairy cattle* (7th revised edition). Washington, DC: National Academy Press.

National Research Council. (1998). *Nutrient requirements of swine* (10th revised edition.). Washington, DC: National Academy Press.

Schwab, C. G., Tylutki, T. P., Ordway, R. S., Sheaffer, C., & Stern, M. D. (2003). Characterization of proteins in feeds. *Journal of Dairy Science 86*(E. Supplement), E88–E103.

Sniffen, C. J., O'Connor, J. D., Van Soest, P. J., Fox, D. G., & Russell, J. B. (1992). A net carbohydrate and protein system for evaluating cattle diets. II. Carbohydrate and protein availability. *Journal of Animal Science 70*, 3562–3577.

CHAPTER 8

LIPIDS

Few livestock producers and animal production researchers or advisors are aware that the renderers must process more pounds of product than is actually sold as meat. Annually this amounts to approximately 38 billion pounds of offal, carcass bones, trimmings, fallen animals and recycled restaurant and cooking fats and oils . . . As long as there is an animal agriculture, the rendering industry and its process offers the only logical, environmentally-acceptable, biosecure procedure for recycling the inedible components of its production into quality value-added products.

G. G. PEARL, 1995

INTRODUCTION

The term *lipid* includes the fats and the oils. Lipids are important constituents of every cell of the body. Their content in feedstuffs varies considerably. As a component of livestock feed, lipids are generally in the form of triacylglycerol (formerly named *triglyceride*). Triacylglycerol in feed may deteriorate, so it is important that livestock feeders be familiar with the terms and processes that are used to assess lipid quality. In this chapter, the digestion and absorption of ingested lipid in feed is also described.

IMPORTANCE OF LIPIDS

Lipids are a diverse group of molecules that are insoluble in water but soluble in nonpolar solvents such as ether, chloroform, and benzene. Like carbohydrates, lipids are compounds of carbon, hydrogen, and oxygen, but they contain a much higher percentage of hydrogen than carbohydrates. The lipid

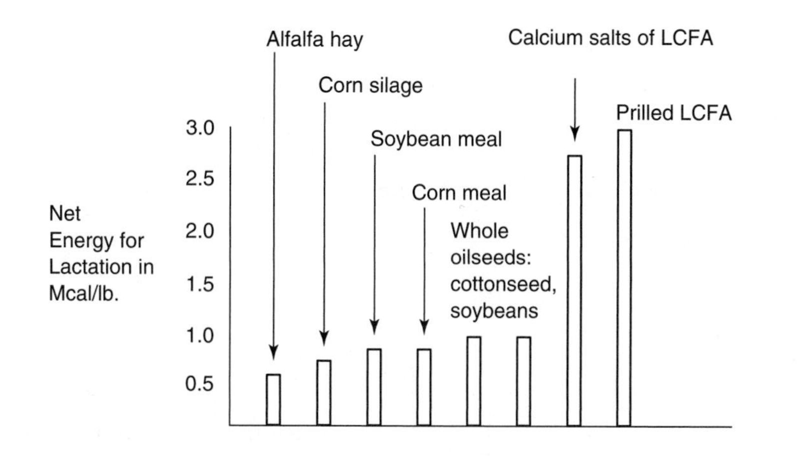

Figure 8–1
Feedstuff energy comparisons. LCFA: Long-chain fatty acids.

group includes triacylglycerol, phospholipids, glycolipids, and lipoproteins. From the standpoint of animal nutrition, the triacylglycerols are the most important lipid.

In the body, lipids serve as structural components of membranes, protective coatings on the surface of animals, cell-surface components concerned with cell recognition and tissue immunity, and storage and transport forms of metabolic fuel. In the feed, lipids increase energy density, reduce dustiness, enhance palatability, lubricate mixing and handling equipment, and act as a medium for dietary sources of the fat-soluble vitamins.

A challenge faced by everyone involved in feeding livestock is to meet the energy requirement for growing animals and high producers. The first step in boosting energy content of herbivore rations is to substitute grains for forages. But this measure alone does not achieve the performance potential of some animals, even with the highest level of grain that can be fed without negatively affecting animal health. In comparison to carbohydrate and protein, fat contains more than twice the amount of energy per pound (Figure 8–1). Adding fat will increase energy density in rations for all types of domestic animals.

STRUCTURE OF TRIACYLGLYCEROLS

Triacylglycerols are glycerol esters of fatty acids. As part of the triacylglycerol, each fatty acid is an acyl group (Figure 8–2). Glycerol is a three-carbon compound with hydroxyl (—OH) groups attached to each carbon. Each of these —OH groups is available to bond to a fatty acid. The term *ester* refers to the type of linkage between the —OH group of the glycerol and the fatty acid (Figure 8–3). A glycerol molecule whose three —OH groups have participated in bonds with long-chain fatty acids is called a *triacylglycerol.*

The glycerol is often referred to as the "backbone" of the triacylglycerol molecule. There are many possible fatty acids that may be attached to the glycerol backbone. The nature of these fatty acids gives each triacylglycerol its own unique characteristics. When triacylglycerols contain a high percentage of saturated fatty acids such as ruminant fat or pork fat (lard), they are usually solid at room temperature. Triacylglycerols that are solid at room temperature are generally referred to as *fats.* Triacylglycerols that contain a high percentage of unsaturated fatty acids such as corn oil, soybean oil, and canola oil are usually liquid at room temperature. Triacylglycerols that are liquid at room temperature are generally referred to as *oils.* Triacylglycerol as a feed ingredient is usually referred to as simply *fat* and that is the convention that will be used from this point forward in this text.

Figure 8–2
The acyl group

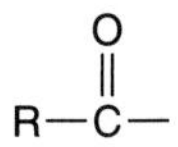

$$R-\overset{\displaystyle O}{\overset{\|}{C}}-$$

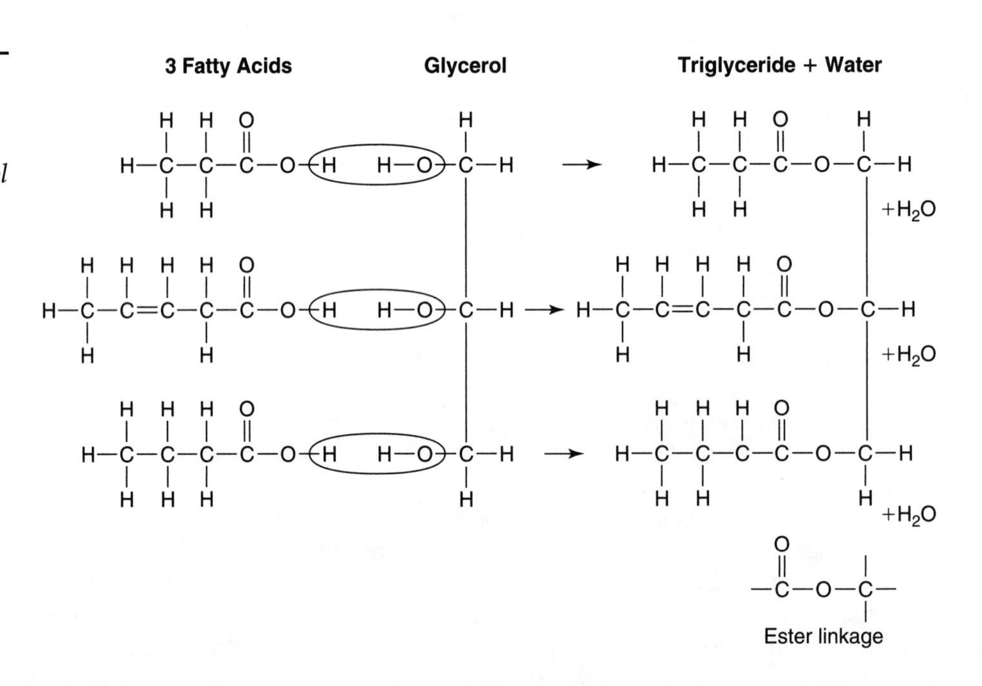

Figure 8–3
Three fatty acids form an ester bond on glycerol to make a lipid or triacylglycerol

FAT TERMINOLOGY

Biohydrogenation: Saturation of fatty acids by microbial activity.
Bypass or inert fats: A fat that when fed to a ruminant animal, has no effect on rumen microbe activity.
Chain length: Number of carbons in a fatty acid.
Ester: The type of bond used in the attachment of a fatty acid to glycerol.
Fatty acid: Fatty acids are examples of carboxylic acids, $CH_3—(CH_2)_n—COOH$.
Free fatty acid: A fatty acid that is not attached to glycerol; usually used in reference to fat quality.
Hydrogenated fat: A fat whose component fatty acids have been chemically saturated.
Iodine value: A measure of the degree of unsaturation of a fat. Each double bond takes up two atoms of iodine, so unsaturation is quantified as the number of centigrams of iodine absorbed/gram of fat.
LCFA: The longer fatty acids (roughly 14 or more carbons) are often referred to in nutrition as LCFA (long-chain fatty acids).
Lipolysis: The activity of breaking ester bonds to cleave fatty acids from glycerol.
Melting point: The temperature at which solidified fat becomes liquid.
Nonesterified fatty acid: Same as free fatty acid; usually used in reference to fat metabolism.
Prilled fat: Fat that has been processed into small spherical shapes.
Protected fat: Fat treated to prevent biohydrogenation and increase rumen inertness.
Saturated fat: A fat comprised of glycerol and of mostly saturated fatty acids. The fatty acids are "saturated" with respect to hydrogen. A saturated fatty acid has no double bonds between carbon atoms. It is usually solid at room temperature.

Fat + Sodium Hydroxide ⟶ Glycerol + Soap

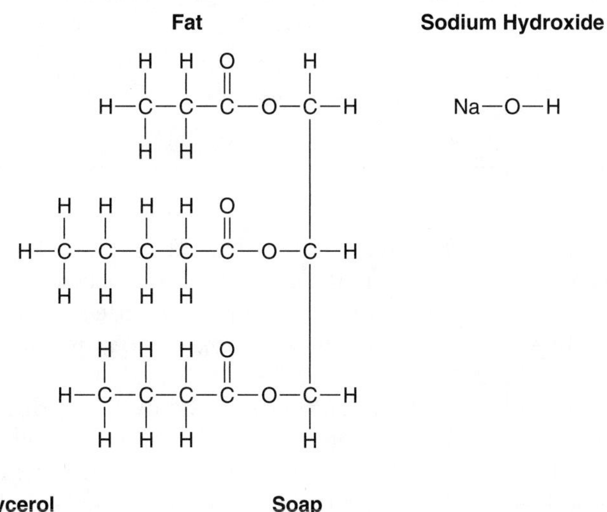

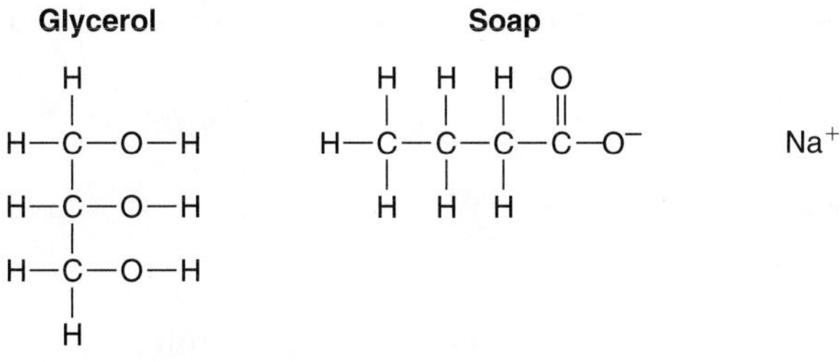

Figure 8–4
Saponifiable matter is the portion of fat that will react to form a soap

Tallow: A mixture of animal fat with a titer above 40.

Titer: The temperature at which liquefied fat becomes solid. Titer may be as much as 3° C higher than melting point.

Triacylglycerol: A molecule of glycerol plus three fatty acids. Formerly described as triglyceride.

Unsaponifiable matter: Portion of fat source that will not react to form a soap (Figure 8–4). Unsaponifiable matter is used to assess fat quality and purity.

Unsaturated fat: A fat comprised of glycerol and of mostly unsaturated fatty acids. The fatty acids are "unsaturated" with respect to hydrogen. An unsaturated fatty acid has at least one double bond between its carbon atoms. It is usually liquid at room temperature.

FAT QUALITY

The judgment of fat quality is made based on unsaponifiable matter, free fatty acid content of the fat, and degree of saturation in the fatty acids bonded to the glycerol. Fats of poor quality or fats that are kept under poor storage conditions will begin to fall apart. The process of falling apart in fats is described as *turning rancid.*

When fats turn rancid, the molecules of fat decompose, creating free fatty acids and new molecules. The presence of free fatty acids in the fat source indicates that the ester bonds in triacylglycerols have been broken. The level of free fatty acids in a feed fat gives an indication of the structural integrity of the triacylglycerols in the fat source. New molecules created as fats decompose will have reduced palatability and may even be toxic. As fats turn rancid, they will destroy fat soluble vitamins.

Table 8–1
Fat quality as identified by titer and iodine value

Fat source	Titer	Iodine Value
Beef fat	42–47	50–54
Lard (pork fat)	38	65–75
Poultry fat	32	80–85
Vegetable oils	Less than 0	105–128

Pure fat (made of 100 percent triacylglycerol) will be 100 percent saponifiable (0 percent unsaponifiable matter). That is, the fatty acids linked to the glycerol should all react in potassium or sodium hydroxide solution to form soaps. The material in a fat sample that does not form a soap is not part of a triacylglycerol and would, therefore, depress fat purity and quality.

In ruminant nutrition, fat quality is assessed by fat hardness. A hard fat is one that is solid at animal body temperature. In solid form, the fat is less likely to interfere with microbial activity in the rumen. This characteristic of noninterference is described as *rumen inertness.* Fat hardness is measured using titer and iodine values (Table 8–1).

DESCRIBING FATTY ACIDS

Fats and the fatty acids that comprise them are described as *saturated* and *unsaturated.* These terms refer to the status of hydrogen in the fatty acid: a fatty acid is saturated or unsaturated with respect to the hydrogen. Examples of saturated fatty acids are palmitic acid and stearic acid. No more hydrogens can be added to either palmitic or stearic acid. Examples of unsaturated fatty acids are linolenic acid and linoleic acid. These fatty acids can accept additional hydrogens and in doing so, will become more saturated. Fats contain a mixture of fatty acids. When a fat is described as saturated or unsaturated, the description is in reference to the nature of the fatty acids in the fat.

There are three systems used to describe the fatty acids, which along with glycerol, make up fat. All systems use carbon length and double bond information about the fatty acid.

The first system involves counting the carbons and the double bonds starting from the carboxyl end (—COOH) and giving the location of the double bonds. The first number denotes the number of carbons. The second number, preceded by a colon, gives the number of double bonds. The third number, designated as delta (Δ), indicates the number of carbon atoms counting from the carboxyl terminal to the first carbon participating in each double bond. Using this method, linolenic acid would be $18{:}3^{\Delta 9,12,15}$. Stearic acid would be 18:0 (Figures 8–5 and 8–6, respectively).

The second system involves counting the double bonds starting from the methyl or omega end (—CH_3). The carbon length and number of double bonds is designated similarly to the first method described. The third number, designated as n-, indicates the number of carbon atoms counting from the methyl terminal to the first double bond. Using this system, linolenic acid would be identified as 18:3(n-3) and could be described as an omega-3 fatty acid. Stearic acid would be 18:0.

A third system for identifying fatty acids is to simply give the number of carbons and the number of double bonds. Linolenic acid, then, would be 18:3 and stearic acid would be 18:0.

As can be seen from Tables 8–2a and 8–2b, most fatty acids contain an even number of carbons. The most abundant are those with chains between 14 and 22 carbon atoms long.

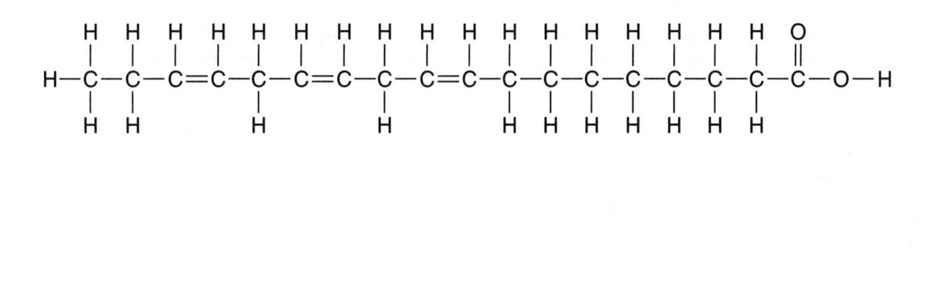

Figure 8–5
Linolenic acid, 18:3(n3) or $18:3^{\Delta 9,12,15}$. An unsaturated fatty acid, liquid at room temperature

```
    H   H   H   H   H   H   H   H   H   H   H   H   H   H   H   H   H   O
    |   |   |   |   |   |   |   |   |   |   |   |   |   |   |   |   |   ||
H—C—C—C—C—C—C—C—C—C—C—C—C—C—C—C—C—C—C—O—H
    |   |   |   |   |   |   |   |   |   |   |   |   |   |   |   |   |
    H   H   H   H   H   H   H   H   H   H   H   H   H   H   H   H   H
```

Figure 8–6
Stearic acid, 18:0. A saturated fatty acid, solid at room temperature

THE REQUIREMENT FOR FATTY ACIDS

Unlike protein nutrition, which is mostly about meeting the requirement for essential amino acids, fat nutrition is not primarily about meeting the requirement for essential fatty acids. Fat nutrition is usually about meeting energy requirements. This is because most livestock diets made with usual feedstuffs will meet the animal's requirement for essential fatty acids. The fatty acids most often described as being essential for livestock are linoleic acid, $18:2^{\Delta 9,12}$ or 18:2(n-6), arachidonic acid, $20:4^{\Delta 5,8,11,14}$ or 20:4(n-6), and linolenic acid, $18:3^{\Delta 9,12,15}$ or 18:3(n-3).

Animals are unable to synthesize n-6 or n-3 fatty acids *de novo* or to interconvert one series of fatty acids to another. However, most animals do possess the enzymes needed to convert one fatty acid to another fatty acid within the same series. Essential fatty acid requirements, therefore, may be expressed in terms of the two series rather than the individual fatty acids. Furthermore, the n-3 series is less likely to be deficient in usual diets and is not considered in balancing rations for most livestock. Fish and cats may be exceptions. Under at least some physiological conditions, for example, some types of fish may not be able to convert 18-carbon fatty acids to longer-chain, more unsaturated fatty acids of the same series (Owen, Adron, Middleton, & Cowey, 1975). In cats, there is apparently only limited ability to synthesize arachidonic acid from other n-6 fatty acids and so arachidonic acid may be a conditionally essential fatty acid for cats.

Linoleic acid is involved in maintenance of membrane function. Arachidonic acid is involved in the production of compounds called eicosanoids, which are needed for normal reproduction and platelet aggregation. A deficiency in the essential fatty acids may result in coarse hair coat, hair loss, skin lesions, or abnormalities in the reproductive system, adrenal glands, kidney, and liver (MacDonald, Anderson, Rogers, Buffington, & Morris, 1984).

FAT CONTENT OF FEED

The fat content of feedstuffs varies. The usual feedstuffs such as forages and grains contain 2 to 4 percent fat on a dry matter basis. On a dry matter basis, brewer's grains and distiller's grains contain about 10 percent fat and whole oilseeds such as soybeans and cottonseed contain about 20 percent fat. Rich sources of omega-6 fatty acids include plant oils and animal fats. Rich sources of omega-3 fatty acids include fish oils and flax.

Table 8–2a
Fats and their component fatty acids (% of total fatty acids), part 1

Carbon Atoms:	Butyric	Caproic	Caprylic	Capric	Lauric	Myristic	Myristoleic	Pentadecanoic	Palmitic	Palmitoleic	Margaric	Margaroleic	Stearic
Double Bonds	4:0	6:0	8:0	10:0	12:0	14:0	14:1	15:0	16:0	16:1	17:0	17:1	18:0
Butter fat	3.6	2.2	1.2	2.5	2.9	10.8	0.8	2.1	26.9	2.0	0.7		12.1
Tallow (mutton)				0.2	0.3	5.2	0.3	0.8	23.6	2.5	2.0	0.5	24.5
Tallow (beef)					0.1	3.2	0.9	0.5	24.3	3.7	1.5	0.8	18.6
Lard				0.1	0.1	1.5		0.1	26.0	3.3	0.4	0.2	13.5
Palm oil					0.1	1.0			44.4	0.2	0.1		4.1
Chicken fat					0.1	0.8	0.2	0.1	25.3	7.2	0.1	.01	6.5
Peanut oil						0.1			11.1	0.2	0.1	0.1	2.4
Cottonseed oil					0.1	0.7			21.6	0.6	0.1	0.1	2.6
Rapeseed oil						0.1			3.8	0.3			1.2
Canola oil						0.1			4.1	0.3	0.1		1.8
Oat oil						0.2			17.1	0.5			1.4
Corn oil						0.1			10.9	0.2	0.1		2.0
Soybean oil						0.1			10.6	0.1	0.1		4.0
Sunflower oil						0.1			7.0	0.1	0.1		4.5
Safflower oil						0.1			6.8	0.1			2.3
Cod liver oil						3.2			13.5	9.8			
Linseed oil									5.3				4.1
Menhaden oil						7.3			19.0	9.0			4.2

Table 8–2b
Fats and their component fatty acids (% of total fatty acids), part 2

	Oleic	Linoleic	Linolenic	Arachidic	Gadoleic	Arachidonic	Eicosa-pentaenoic	Behenic	Erucic	Docosa-hexaenoic	Lignoceric	Iodine Value
Carbon Atoms: Double Bonds	**18:1**	**18:2 n-6**	**18:3 n-3**	**20:0**	**20:1**	**20:4 n-6**	**20:5 n-3**	**22:0**	**22:1**	**22:6 n-3**	**24:0**	**—**
Butter fat	28.5	3.2	0.4		0.1							25–42
Tallow (mutton)	33.3	4.0	1.3									35–46
Tallow (beef)	42.6	2.6	0.7	0.2	0.3							40–55
Lard	43.9	9.5	0.4	0.2	0.7							48–65
Palm oil	39.3	10.0	0.4	0.3				0.1				50–55
Chicken fat	37.7	20.6	0.8	0.2	0.3							74–80
Peanut oil	46.7	32.0		1.3	1.6			2.9			1.5	84–100
Cottonseed oil	18.6	54.4	0.7	0.3				0.2				98–118
Rapeseed oil	18.5	14.5	11.0	0.7	6.6			0.5	41.1		1.0	100–115
Canola oil	60.9	21.0	8.8	0.7	1.0			0.3	0.7		0.2	100–115
Oat oil	33.4	44.8		0.2	2.4							105–110
Corn oil	25.4	59.6	1.2	0.4				0.1				118–128
Soybean oil	23.2	53.7	7.6	0.3				0.3				123–139
Sunflower oil	18.7	67.5	0.8	0.4	0.1			0.7				125–140
Safflower oil	12.0	77.7	0.4	0.3	0.1			0.2				140–150
Cod liver oil	23.7	1.4	0.6		7.4	1.6	11.2		5.1	12.6		155–173
Linseed oil	20.2	12.7	53.3									169–192
Menhaden oil	13.2	1.3	0.3		2.0	0.2	11.0		0.6	9.1		170–200

FAT ABSORPTION

Ingested fats and oils do not mix in the water moving through the digestive tube. For absorption to take place, the system uses emulsifiers to suspend small globules of fat in the water. The emulsifiers used are bile and phospholipids. Bile is made in the liver and secreted into the duodenum. Phospholipids look like triacylglycerols, except at least one fatty acid has been replaced by a compound containing phosphorus. Lecithin is an important type of phospholipid. Fish diets generally contain high levels of fat, and lecithin is often added to ensure that the dietary fat is emulsified adequately.

The small globules of fat are more efficiently attacked by the pancreas's fat-digesting lipase enzyme than is nonemulsified fat. During digestion, lipase breaks down triacylglycerols into free fatty acids and monoacylglycerols or glycerol. Glycerol and short-chain fatty acids are absorbed directly into the bloodstream. The LCFA and the monoacylglycerols are coated in bile salts. These emulsified fat packets are called *micelles.* Micelles are transported through the watery intestinal environment to the site of absorption at the intestinal wall. The LCFA and monoacylglycerols diffuse into the cells, leaving the rest of the micelle behind. Inside the intestinal cells, the monoacylglycerols and LCFA are then rebuilt into new triacylglycerols.

In intestinal cells, triacylglycerols are packaged into lipoproteins. Lipoproteins contain both lipid and protein.

Short-chain fatty acids and glycerol, as well as carbohydrates, proteins, minerals, water soluble vitamins, and water, will pass from the intestinal cell into the blood vessels of the mesentery during absorption. The mesenteric blood vessels carry absorbed materials to the portal vein and on to the liver. However, the lipoproteins made from monoacylglycerols and LCFA are passed from the intestinal cells into the lymphatic system during absorption in most species of domestic animals. In all livestock except poultry, absorbed lipoproteins move into the lymphatics, thereby bypassing the liver. The lymphatic system merges with the circulatory system at veins in the neck. At this location, the lipoproteins enter the blood. In poultry, absorbed fat is carried directly to the portal circulation and to the liver.

CHYLOMICRON METABOLISM

Outside the intestinal cell, the lipoproteins made from absorbed monoacylglycerols and LCFA are often called *chylomicrons.* Binding structures called *apolipoproteins* are embedded on the surface of a chylomicron. An enzyme called *lipoprotein lipase* is responsible for hydrolysis of circulating triacylglycerols in chylomicrons. This enzyme is located on the inside wall of blood vessels and it binds the apolipoprotein. The triacylglycerols inside the chylomicrons are then hydrolyzed into fatty acids and glycerol.

Cells in the vicinity of the fat hydrolysis absorb most of the fatty acids and glycerol. If the animal needs energy to meet its energy requirement, the cells may prepare the fatty acids for entrance into the tricarboxylic acid (TCA) cycle where they are oxidized for fuel. If the intake of energy is in excess of the animal's requirement, the cells may use the absorbed fatty acids to reform triacylglycerols and put them into storage. Adipose tissue is made of cells that specialize in fat storage. Concentrations of adipose tissue are called *fat depots.*

Absorption of monoacylglycerols and LCFA is summarized in Figure 8–7.

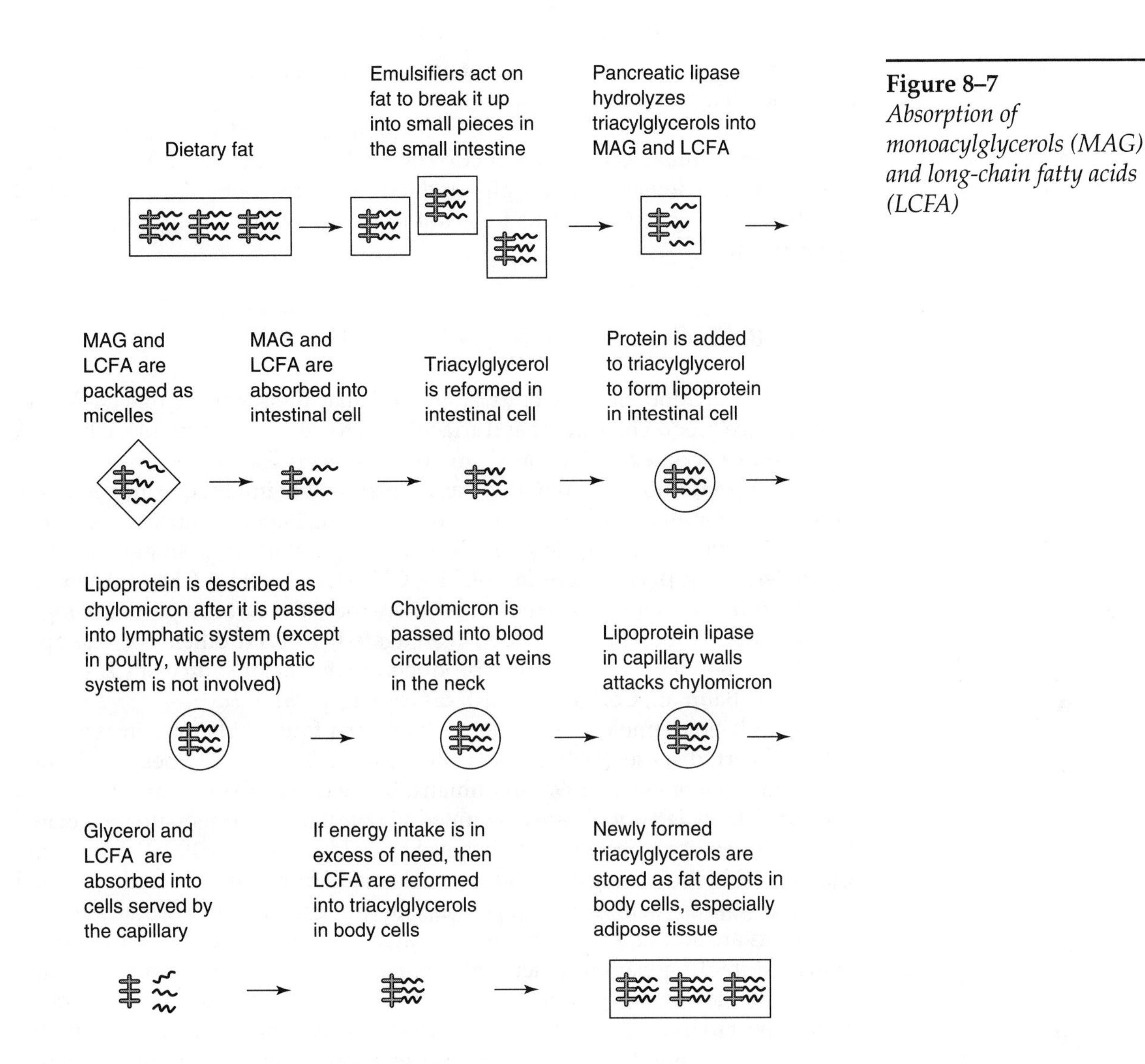

Figure 8–7
Absorption of monoacylglycerols (MAG) and long-chain fatty acids (LCFA)

FAT STORAGE IN MONOGASTRICS AND RUMINANTS

In monogastrics, fatty acids are usually not changed from the point of ingestion to the point of storage in fat depots. This means that the same fatty acids in the feed fat are present in the micelles in the small intestine, the chylomicrons in the blood, and in the reformed triacylglycerols in fat depots. The triacylglycerols of fat depots in monogastrics, therefore, tend to have a similar fatty-acid composition to that of dietary fat. This explains the cause for "soft pork," a problem in the swine industry occurring as a result of feeding a diet rich in unsaturated fat (Averette Gatlin, See, Hansen, & Odle, 2003).

In ruminant animals, however, fatty acids are changed by the action of the microbes in the rumen. The microbes in the rumen obtain energy by oxidizing organic compounds, primarily carbohydrates. The oxidation of one substance requires the presence of an oxidizing agent, something that can be reduced. In most biological situations, oxygen serves as the ultimate oxidizing agent, accepting hydrogen to become water (H_2O). However, there is little oxygen in the rumen. Microbes in the rumen must reduce all potential oxidizing agents including unsaturated fats. When an unsaturated fat is reduced, it becomes a more saturated fat. The triacylglycerols of fat depots in ruminants, therefore, tend to contain saturated fat regardless of the type of fat in the diet.

It should be noted, however, that for both monogastrics and ruminants, where energy intake is in excess of current needs, triacylglycerols may be manufactured from any absorbed compound that contains energy and stored in fat depots. The composition of this fat is not related to either dietary fat or microbial activity.

DIETARY FAT AND HUMAN HEALTH

In human nutrition, studies have suggested that diets rich in omega-3 fatty acids reduce blood cholesterol and triacylglycerol levels (Simopoulos, 2002). A major source of omega-3 fatty acids in human diets is fish oil.

Conjugated linoleic acid (CLA) has been shown to inhibit carcinogenesis in experimental animals (National Research Council, 1996). In most types of unsaturated fatty acids, the double bonds are separated by a single —CH_2— (methylene) group. For example, —CH = CH—CH_2—CH = CH—. An exception exists in the conjugated fatty acids where there the carbons participating in adjacent double bonds are bonded (conjugated) to each other. For example, —CH = CH—CH = CH—. Dairy products are the major source of CLA in human diets (Bauman, Corl, Baumgard, & Griinari, 1998).

There is epidemiological evidence that trans fatty acids may increase the risk of heart disease (Willett, Stampfer, Manson, Colditz, Speizer, Rosner, Sampson, & Hennekens, 1993) in humans. In the diet of the U.S. adult, 80 to 90 percent of trans fatty acids are consumed via partially hydrogenated vegetable oils (Feldman, Kris-Etherton, Kritchevsky, & Lichtenstein, 1996). Margarine and vegetable shortening are food products made from partially hydrogenated vegetable oils. In manufacturing partially hydrogenated vegetable oils, hydrogen atoms are added to the unsaturated fatty acids in vegetable oil through an industrial hydrogenation (chemical reduction) process. As hydrogens are added, double bonds are eliminated and the fatty acid becomes more saturated with respect to hydrogen. This results in a more stable fatty acid and one with a higher melting point, which is the goal in making margarine and vegetable shortening from vegetable oils. Another change to the unsaturated fat takes place during the industrial hydrogenation process. Some of the unsaturated bonds are transformed to a trans configuration.

As vegetable oils, the hydrogens are usually both on the same side of the double bond. This is called the *cis configuration* (Figure 8–8). During the process of making margarine and vegetable shortening, some of the double bonds are changed so that the hydrogens are on opposite sides of the double bond. This is called the *trans configuration* (Figure 8–9). Trans fatty acids are unsaturated fatty

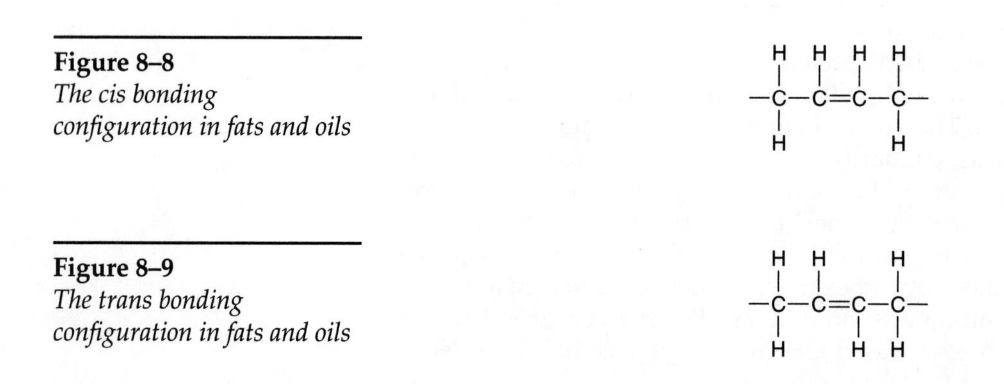

Figure 8–8
The cis bonding configuration in fats and oils

Figure 8–9
The trans bonding configuration in fats and oils

acids that contain a trans double bond. In contrast to the more typical cis configuration, the trans double bond results in a more rigid, linear molecule with a higher melting point.

Trans fatty acids also occur in nature. The fat in milk produced by cows fed diets deficient in forage has a higher content of trans fatty acids (Chapter 18). It is not clear if trans fatty acids that occur naturally have the same effect on heart health as those produced by hydrogenated vegetable oils.

SUMMARY

In the feed industry, both fats and oils are generally referred to as *fat*. Most feedstuffs contain 2 to 4 percent fat. Though livestock do require essential fatty acids, this requirement is generally met with usual feedstuffs. The triacylglycerols that make up a molecule of fat are composed of a single glycerol molecule bonded to three fatty acids. There is great variety in the types of fatty acids that may be found in fat, and the fatty acid composition of fat determines its feeding value. Measurements of fat quality include titer and iodine value. Supplemental fat can improve a diet by increasing the energy density, reducing dustiness, enhancing palatability, and lubricating mixing and handling equipment. Most ingested fat is absorbed into the lymphatic system, thereby bypassing the liver.

END-OF-CHAPTER QUESTIONS

1. Relate the following terms to one another in a single sentence: lipid, fat, oil, saturation, unsaturation.
2. Why is the fat in dairy products and meat products from beef, lamb, and goat highly saturated?
3. Explain why the iodine value of lard (pork fat) is more dependent on the type of dietary fat than is the iodine value of ruminant fat.
4. What is the relationship between rancidity and free fatty acid content of fat?
5. What is the relationship between the purity of a fat sample and its saponification value?
6. What is a triacylglycerol? What is a triglyceride?
7. A fat sample that has a high titer value would have a low iodine value. Explain why.
8. Describe the structure of a fatty acid identified as 18:2(n-6).
9. The requirement for fatty acids is sometimes described as a requirement for a series of fatty acids. One such series would include all those fatty acids in which the first carbon participating in a double bond is the 6th carbon counting from the methyl terminal (omega-6). Explain why the requirement for fatty acids is sometimes expressed as a requirement for a series of fatty acids.
10. Explain the difference between poultry and other domestic animals in how the LCFA derived from triacylglycerol is absorbed into the bloodstream.

REFERENCES

Averette Gatlin, L., See, M. T., Hansen, J. A., & Odle, J. (2003). Hydrogenated dietary fat improves pork quality of pigs from two lean genotypes. *Journal of Animal Science 81*, 1989–1997.

Bauman, D. E., Corl, B. A., Baumgard, L. H., & Griinari, J. M. (1998, October 20–22). Trans fatty acids, conjugated linoleic acid and milk fat synthesis. In *Proceedings of the Cornell nutrition conference for feed manufacturers* (pp. 95–103). Rochester, NY.

Feldman, E. B., Kris-Etherton, P. M., Kritchevsky, D., & Lichtenstein, A. H. (1996). Position paper on trans fatty acids. Special Task Force Report. *American Journal of Clinical Nutrition 63*, 663–670.

MacDonald, M. L., Anderson, B. C., Rogers, Q. R., Buffington, C. A., & Morris, J. G. (1984). Essential fatty acid requirements of cats: Pathology of essential fatty acid deficiency. *American Journal of Veterinary Research 45*, 1310–1317.

National Research Council. (1996). *Carcinogens and anticarcinogens in the human diet.* Washington, DC: National Academy Press.

Owen, J. M., Adron, J. W., Middleton, C., & Cowey, C. B. (1975). Elongation and desaturation of dietary fatty acids in turbot (*Scophthalmus maximus*) and rainbow trout (*Salmo gairdneri*). *Lipids 10*, 528–531.

Pearle, G. G. (1995, October 24–26). The fats and protein research foundation. In *Proceedings Cornell nutrition conference for feed manufacturers* (pp. 1–2). Rochester, NY.

Simopoulos, A. P., (2002). The importance of the ratio of omega-6/omega-3 essential fatty acids. *Biomedicine & Pharmacology* (Uncorrected Proof).

Willett, W. C., Stampfer, M. J., Manson, J. E., Colditz, G. A., Speizer, F. E., Rosner, B. A., Sampson, L. A., & and Hennekens, C. H. (1993). Intake of trans fatty acids and risk of coronary heart disease among women. *Lancet 341*, 581–585.

CHAPTER 9

MINERALS

The use of mineral supplements when they are not needed is not only a waste of money, but also may in some cases be actually injurious.

F. B. MORRISON, 1949

INTRODUCTION

Mineral nutrients are inorganic compounds that play roles in the metabolism of livestock. Some are required in relative large quantities, others in relatively small quantities, but the quantity required has no bearing on the severity of a deficiency. Mineral content and bioavailability in feedstuffs varies. Feeding excess minerals may result in interactions that reduce the bioavailability of other minerals.

IMPORTANCE OF MINERALS

The animal body is constructed of water and both organic and inorganic components. Protein and fat are the primary organic constituents of the animal body. The elements making up these organic molecules are carbon, hydrogen, oxygen, and nitrogen. Sulfur is also found in amino acids, but is usually not considered to be an organic element.

The inorganic constituents of the body make up the minerals. At least 22 minerals are used in livestock metabolism, although the number of minerals that require dietary supplementation for most livestock is fewer than 15. The minerals are classified into two categories based on the amount required by livestock: major or macrominerals, and trace or microminerals (Table 9–1). This classification is unrelated to the mineral's biological importance.

Some minerals participate in the catalytic activity of enzymes. Others serve a structural function. Some minerals participate in the body's acid–base and electrolyte balances.

Table 9–1
Animal mineral nutrients

Major or Macrominerals	Trace or Microminerals
Calcium	Iodine
Phosphorus	Iron
Sodium	Manganese
Potassium	Copper
Magnesium	Molybdenum
Chloride	Zinc
Sulfur	Selenium
	Chromium
	Cobalt

ABSORPTION OF MINERALS

Bioavailability refers to the proportion of the ingested nutrient that the digestive and absorptive mechanisms are actually able to bring into the bloodstream. Mineral nutrition is made difficult due to the varying bioavailabilities of different mineral sources. To address this problem, the dairy NRC (2001) publication calculates requirements for minerals at the tissue level and gives bioavailable mineral values in feedstuffs. Each mineral in each feedstuff is assigned an absorption coefficient that is used to calculate the amount of mineral in consumed feeds that is actually absorbed. The companion application to this text uses the dairy NRC (2001) mineral bioavailabilities in the dairy and dog ration files. The advantage to using absorbed values is that it improves on the accuracy of ration formulation and reduces the need for large safety factors. A safety factor is a nutrient excess deliberately included in the ration to account for variable nutrient bioavailabilities.

A problem with mineral excesses is the potential for interaction of these excesses with other minerals. A mineral excess can impair or improve the absorption of another mineral. Likewise, a mineral deficiency can impair or improve absorption of another mineral. Figure 9–1 illustrates these types of interactions.

The previous edition of the dairy NRC publication *Nutrient Requirements of Dairy Cattle* (1989) used the value of 0.38 for bioavailability of calcium from all sources. However, data in Table 9–2 demonstrate that mineral bioavailability is not consistent among the various feedstuffs. Therefore, the use of a single value for mineral bioavailability is imprecise and may result in considerable ration deficiencies or excesses of minerals.

The concern over phosphorus pollution resulting from the widespread use of safety factors for phosphorus has prompted swine and poultry nutritionists to develop bioavailable requirement values for this nutrient and to measure phosphorus bioavailability in feedstuffs. Phosphorus bioavailability in swine and poultry nutrition is taken as phosphorus in compounds other than phytic acid. Phytic acid is the principle form of phosphorus in cereal grains and oilseed meals. Phytic acid is also identified as phytate or phytin.

Efforts to improve the bioavailability of minerals have included both physical and chemical techniques. Inorganic mineral supplements that are finely ground are going to be dusty, but more bioavailable than supplements that are more granular. The sulfate, carbonate, chloride, and oxide forms are the most common types of inorganic trace minerals used in animal diets. Oxide forms are usually the least bioavailable of the four.

Trace mineral sources complexed with organic molecules, though more expensive, are often more bioavailable than inorganic forms of the trace mineral. The AAFCO (2003) defines six categories of organic trace mineral products:

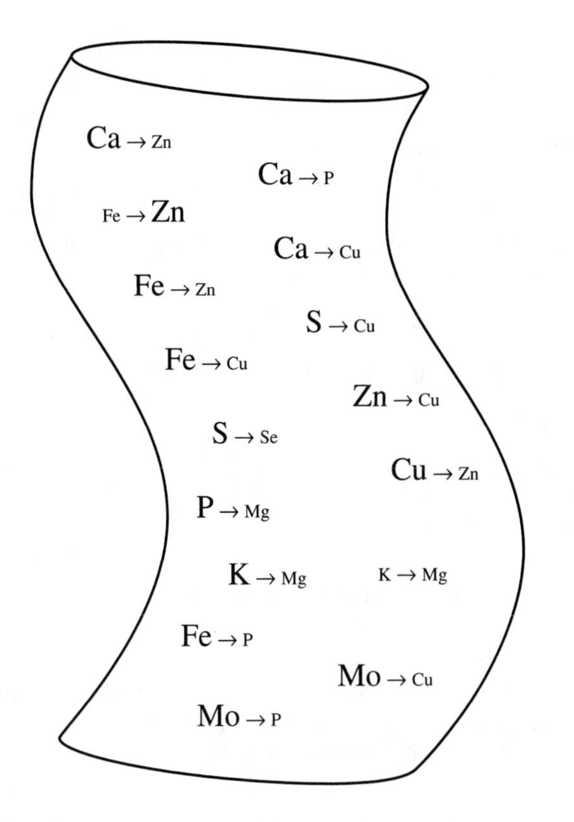

Figure 9–1
Mineral interactions affecting absorption from the digestive tract. Large-size font indicates mineral excess or increased absorption. Small-size font indicates mineral deficiency or decreased absorption.

Table 9–2
Mineral bioavailability data for selected feedstuffs as indicated in the dairy NRC, 2001

Feedstuff	Ca	P	Mg	Cl	K	Na	S	Co	Cu	I	Fe	Mn	Se	Zn
Grass and legume hay	0.30	0.64	0.16	0.90	0.90	0.90	1.00	1.00	0.04	0.85	0.10	0.0075	1.00	0.15
Corn grain, ground IFN 4-02-854	0.60	0.70	0.16	0.90	0.90	0.90	1.00	1.00	0.04	0.85	0.10	0.0075	1.00	0.15
Soybean meal 48% IFN 5-20-638	0.60	0.70	0.16	0.90	0.90	0.90	1.00	1.00	0.04	0.85	0.10	0.0075	1.00	0.15
Blood meal IFN 5-26-006	0.60	0.70	0.16	0.90	0.90	0.90	1.00	1.00	0.04	0.85	0.10	0.0075	1.00	0.15
The most commonly used inorganic supplement	0.75	0.75	0.70	0.90	0.90	0.90	1.00	1.00	0.05	0.85	0.20	0.012	1.00	0.20

From National Research Council, 2001.

1. Metal amino acid complex
2. Metal (specific amino acid) complex
3. Metal amino acid chelate
4. Metal polysaccharide complex
5. Metal propionate
6. Metal proteinate

When discussing minerals in animal nutrition, caution must be exercised to avoid confounding mineral requirements expressed on a bioavailable basis with laboratory analyses of feedstuff mineral content. Bioavailable livestock mineral requirements will usually be lower values than requirements for total dietary mineral intake. In discussing mineral requirements, it must be made clear which system is being used.

In addition to bioavailablity characteristics of the mineral source, conditions inside the animal can affect mineral absorption. For example, in ruminant animals, the rumen wall is one of the primary sites of absorption for magnesium. However, absorption is dependent on magnesium solubility, which is affected by rumen pH. Diets that tend to increase rumen pH, such as high forage diets, diets high in potassium, or diets containing rumen active buffer, tend to reduce magnesium absorption.

Free-choice, ad-lib, or *ad libitum* mineral feeding involves making a supply of mineral available to animals separate from the rest of the feed. Most minerals are unpalatable to livestock, but free-choice mineral feeding can work because animals do have an appetite for salt. With a prediction of salt intake, it is possible to fortify a salt source with the needed mineral nutrients and meet animal mineral requirements through free-choice feeding.

MAJOR OR MACROMINERALS

Calcium

Table 9–3 gives calcium requirements for selected livestock.

Functions

Calcium is the most abundant mineral in the body, and is deposited in the teeth and skeleton. Except in fish, moderate amounts of calcium may be withdrawn from bone when calcium intake is not sufficient to meet requirements (such as late pregnancy and lactation). In fish, the scales are an important site of calcium metabolism and deposition. Calcium plays a role in many functions in the body including muscle contraction, blood clotting, nerve function, and acid–base balance. Adequate calcium nutrition is dependent on the nutritional status of phosphorus and vitamin D.

Calcium Deficiency

Deficiency of calcium, phosphorus, and/or vitamin D results in poor bone development, easily fractured bones, reduced milk yield, nervous symptoms, and reduced growth. Calcium needs of laying hens are higher than for any other

Table 9–3
Calcium requirements for selected livestock[1]

	Required (g)	Required (%), DM basis
Fish, channel catfish, 100-g body weight	Unknown	Unknown
Fish, rainbow trout, 100-g body weight	.014	1.11
Chicken, broiler, 5 wks of age	1.23	1.0
Chicken, white egg layer, 3-lb. body weight	3.2498	3.61
Pig, growing, 45-lb. body weight	7.99	0.67
Dog, growing, 30-lb. body weight	2.07	0.59
Cat, growing kitten, 4.2-lb. body weight	0.84	1.00
Rabbit, growing, 5 wks of age	0.79	0.59
Horse, light work, 1,100-lb. body weight	30.85	0.23
Goat, maintenance, 88-lb. body weight	2.25	0.31
Ewe, maintenance, 110-lb. body weight	1.996	0.20
Beef animal, growing, 800-lb. body weight	64.46	0.74
Dairy cow, lactating, 1,400-lb. body weight	66.55	0.27

[1]With the exception of dairy, fish, and dog, all requirements are given as total dietary calcium. Dairy, fish and dog values are given as bioavailable calcium.

animal species. Lack of adequate calcium in the diet of laying hens results in soft-shelled eggs.

Feed Sources

Supplemental feed sources include limestone (34.0% Ca), calcium carbonate $CaCO_3$ (39.39% Ca), and ground oyster shells (38.0% Ca). Fish can absorb calcium from their environment, primarily through the gills. In some fish, the digestive system is not a major site of calcium absorption, making dietary supplementation of calcium unnecessary.

Toxicity and Special Issues

Excess calcium interferes with phosphorus utilization, and increases the requirement for zinc and vitamin K. Table 9–4 gives the safe upper limits for calcium in livestock diets as presented in the NRC (1980).

Phosphorus

Table 9–5 gives phosphorus requirements for selected livestock.

Functions

Phosphorus is the second most abundant mineral in the body. Phosphorus is a component of bones, teeth, cell membranes, and many enzymes. Phosphorus plays an important role in energy production and utilization (Figure 9–2). Phosphorus also functions in cell growth, maintenance of acid–base balance, and osmotic balance.

Phosphorus Deficiency

Adequate phosphorus nutrition is dependent on the nutritional status of calcium and vitamin D. Phosphorus deficiency is likely in grazing animals that do

1. ATP. Cells use the energy contained in the compound adenosine triphosphate (ATP). One molecule of ATP contains three phosphate groups, each of which contains the mineral **phosphorus**. After the energy is used, the ATP loses one of its phosphate groups and becomes adenosine diphosphate (ADP). Feed energy can be used to attach another phosphate group to ADP to regenerate ATP.
2. Glycolysis. Oxidation of glucose during cellular respiration begins with glycolysis taking place in the cell cytosol. During glycolysis, a 6-carbon glucose molecule is oxidized to two 3-carbon pyruvate molecules. As a component of the activated pyruvate kinase enzyme, **potassium**, **magnesium** and **manganese** are used to break glucose down to pyruvate, capturing the energy released in ATP.
3. Tricarboxylic Acid (TCA) Cycle. Most of the pyruvate made during glycolysis, as well as some amino acids not needed to build protein, are converted to acetyl coenzyme A (acetyl CoA). This molecule links glycolysis, which occurs in the cytosol, with the TCA cycle, which occurs in the matrix of the cell's mitochondria. In this cycle, a series of oxidations and reductions transfers chemical energy, in the form of electrons, from pyruvate derivatives to several **sulfur**- and **phosphorus**-containing coenzymes.
4. Electron Transport Chain. The electron transport chain occurs on the inner mitochondrial membrane. It uses the coenzymes reduced during the TCA cycle to generate more ATP. Compounds containing **iron**, **sulfur**, and **copper** are used as carriers in a series of oxidation-reduction reactions in which electrons are passed from higher energy carriers to lower energy carriers until they are finally passed to oxygen, creating water. The energy released is captured in ATP.

Figure 9–2
A summary of mineral roles in energy production

Table 9–4
Safe upper limits of minerals in livestock rations, expressed as total mineral in cat diets, bioavailable mineral in others

Mineral	Fish[1]	Chickens	Swine	Dogs	Cats	Rabbits	Horses	Goats[2]	Sheep	Beef animals	Dairy cows
Calcium	3%[3]	6%/1.2%[4]	2%[3]	2.05%[5]	10×[6]	4.5%[7]	2%	2%	2%	2%	2%
Phosphorus	4.5%[3]	1%[3]	1.5%	1.44%[5]	10×[6]	1%	1%	0.6%	0.6%	1%	1%
Magnesium	0.32%[8]	0.3%	0.3%	0.3%[9]	8.75×[10]	0.3%	0.3%	0.5%	0.5%	0.5%	0.5%
Potassium	2.1%[3]	2%	2%	2%[9]	10×[6]	3%	3%	3%	3%	3%	3%
Sodium	1.8%[3]	0.79%	3.1%	3.1%[9]	3×[6]	1.2%	1.18%	3.5%	10×[6]	3.5%/1.6%[11]	3.5%/1.6%[11]
Chloride	2.7%[3]	1.33%	4.9%	4.9%[9]	10×[6]	2%	1.82%	See sodium chloride	See sodium chloride	See sodium chloride	5.5%/2.5%[11]
Sodium chloride (salt)	—	See sodium and chloride	See sodium and chloride	See sodium and chloride	—	See sodium and chloride	3%	9%	9%	9%/4%[11]	see sodium and chloride
Sulfur	—	—	—	—	—	—	0.4%[3]	0.4%	0.4%	0.4%	0.4%
Cobalt	—	—	—	—	—	10 mg/kg	10 mg/kg	10 mg/kg	10 mg/kg	10 mg/kg	10 mg/kg
Copper	730 mg/kg[8]	300 mg/kg	250 mg/kg	250 mg/kg[9]	10([6]	300 mg/kg[3]	800 mg/kg	40 mg/kg[3]	25 mg/kg	100 mg/kg	100 mg/kg
Iodine	13.5 mg/kg[3]	300 mg/kg	400 mg/kg	400 mg/kg[9]	10([6]	300 mg/kg[3]	5 mg/kg	50 mg/kg	50 mg/kg	50 mg/kg	50 mg/kg
Iron	1,380 mg/kg[8]	1,000 mg/kg	3,000 mg/kg	3,000 mg/kg[9]	100×[6]	500 mg/kg	500 mg/kg	500 mg/kg	500 mg/kg	1,000 mg/kg	1,000 mg/kg
Manganese	RT: 100 mg/kg[3], CC: 40 mg/kg[3]	2,000 mg/kg	400 mg/kg	400 mg/kg[9]	10×[6]	400 mg/kg	400 mg/kg	1,000 mg/kg	1,000 mg/kg	1,000 mg/kg	1,000 mg/kg
Selenium	RT: 13 mg/kg[8], CC: 15 mg/kg[8]	2 mg/kg	2 mg/kg	2 mg/kg[9]	50×[10]	2 mg/kg	2 mg/kg	2 mg/kg	2 mg/kg	2 mg/kg	2 mg/kg
Zinc	1,000 mg/kg[8]	1,000 mg/kg	1,000 mg/kg	1,000 mg/kg[9]	500 mg/kg[3]	500 mg/kg	500 mg/kg	300 mg/kg	300 mg/kg	500 mg/kg	500 mg/kg

[1]RT: Rainbow trout; CC: channel catfish.
[2]Except for copper, the goat values are taken from the values for sheep.
[3]Safe upper limit predicted by author.
[4]6% for white egg layers, white-egg breeders (Author); 1.2% for others (NRC, 1980).
[5]Safe upper limit taken from dog NRC, 1985.
[6]Safe upper limit predicted by author as a multiple of the requirement.
[7]Safe upper limit taken from rabbit NRC, 1977.
[8]Safe upper limit taken from fish NRC, 1993.
[9]Safe upper limit taken from swine NRC, 1998.
[10]Safe upper limit taken from cat NRC, 1986.
[11]First number: nonlactating animal; second number: lactating animal.
Unless noted otherwise, data from National Research Council, 1980.

Table 9–5
Phosphorus requirements for selected livestock[1]

	Required (g)	Required (%), DM basis
Fish, channel catfish, 100-g body weight	0.015	0.50
Fish, rainbow trout, 100-g body weight	0.0084	0.67
Chicken, broiler, 5 wks of age	0.4897	0.39
Chicken, white egg layer, 3-lb. body weight	0.2489	0.28
Pig, growing, 45-lb. body weight	3.06	0.26
Dog, growing, 30-lb. body weight	1.55	0.44
Cat, growing kitten, 4.2-lb. body weight	0.67	0.80
Rabbit, growing, 5 wks of age	0.464	0.35
Horse, light work, 1,100-lb. body weight	22.00	0.17
Goat, maintenance, 88-lb. body weight	1.51	0.21
Ewe, maintenance, 110-lb. body weight	1.796	0.18
Beef animal, growing, 800-lb. body weight	33.37	0.38
Dairy cow, lactating, 1,400-lb. body weight	61.54	0.25

[1]With the exception of dairy, fish, dog, swine and poultry, all requirements are given as total dietary phosphorus. Dairy, fish, dog, swine and poultry values are given as bioavailable phosphorus.

DM: dry matter.

not receive supplemental phosphorus. At higher-than-recommended levels of supplemental calcium, dietary phosphorus absorption is decreased. Phosphorus is a component of bone, but bones do not function as a reserve of phosphorus to be tapped in times of dietary deficiency to the extent that they do for calcium. Phosphorus deficiency results in reduced growth and feed efficiency, reduced milk production and fragile bones.

Feed Sources

Most feedstuffs that contain phosphorus also contain calcium, but the most popular calcium supplements contain very little phosphorus. For this reason, when formulating rations, it is most practical to add the phosphorus supplement first because doing so may eliminate the need to add any calcium supplement. Phosphorus bioavailability varies in both organic and inorganic feedstuffs. Phytate phosphorus in feedstuffs of plant origin is largely unavailable to monogastrics. Most of the phosphorus in cereal grains and oilseed meals is in the form of phytate. In ruminant animals, microbes make the enzyme phytase that breaks down phytate, making the phosphorus available for absorption. Inclusion of phytase in the diets of monogastric animals increases the bioavailability of phosphorus and creates an opportunity to reduce ration phosphorus content. This reduces ration cost and phosphorus excretion. Phosphorus is a more important feed nutrient in fish than is calcium because phosphorus is absorbed primarily from the digestive tract whereas calcium is absorbed primarily at the gills. In fish, warm-water species appear to be less able to extract the phosphorus in fishmeal feedstuffs than do cold-water species (Watanabe et al.,1980). The use of phytase in fish diets is complicated by the fact that the enzyme product must be stable in water.

Supplemental feed sources for phosphorus include dicalcium phosphate $CaHPO_4 \cdot 2(H_2O)$, 19.3% P, bone meal 12.86% P, calcium phosphate, monobasic $Ca(H_2PO_4)_2 \cdot H_2O$, 21.6% P, sodium phosphate, monobasic $NaH_2PO_4 \cdot H_2O$, 22.5% P.

Toxicity and Special Issues

Excessive phosphorus intake may cause bone resorption, changes in blood composition, and urinary calculi.

Phosphorus is now recognized as one of the most important agricultural pollutants. This has stimulated interest in the use of phytase in monogastric species and research in mineral bioavailability generally. Table 9–4 gives the safe upper limits for phosphorus in livestock diets as presented in the NRC (1980).

The Calcium-to-Phosphorus Ratio Efficiency of absorption of phosphorus declines with increasing calcium intake. The calcium-to-phosphorus ratio has been used historically to ensure adequate phosphorus absorption by establishing the relationship between these two important mineral nutrients in diets of domestic animals. The target has usually been a ratio of 1.5 to 2 parts calcium to 1 part phosphorus. The more recent NRC publications, however, have tended to diminish the importance of the ratio and to focus instead on meeting requirements and minimizing excesses. The dairy NRC (2001) publication states that the ratio is of little significance if requirements for phosphorus and calcium are met. No differences in milk yield, persistency of milk production, milk composition, or reproductive performance were found in cows during early lactation when they were fed diets with calcium-to-phosphorus ratios ranging from 1:1 to 8:1.

The swine NRC (1998) and the companion application to this text for swine give two recommended ratios, depending on whether the ratio is calculated using total dietary phosphorus or available phosphorus. When the Ca:P ratio is based on total phosphorus, the recommendation is to have 1.1 to 1.25 times as much calcium as phosphorus. When the Ca:P ratio is based on available phosphorus, the recommendation is to have 2 to 3 times as much calcium as phosphorus. If the latter ratio is within acceptable limits, then the ratio calculated using total phosphorus can be ignored.

Sodium

Table 9–6 gives sodium requirements for selected livestock.

Table 9–6
Sodium requirements for selected livestock[1]

	Required (g)	Required (%), DM basis
Fish, channel catfish, 100-g body weight	unknown	unknown
Fish, rainbow trout, 100-g body weight	0.0084	0.67
Chicken, broiler, 5 wks of age	0.2050	0.17
Chicken, white-egg layer, 3-lb. body weight	0.1500	0.17
Pig, growing, 45-lb. body weight	1.33	0.11
Dog, growing, 30-lb. body weight	0.19	0.06
Cat, growing kitten, 4.2-lb. body weight	0.17	0.20
Rabbit, growing, 5 wks of age	0.619	0.46
Horse, light work, 1,100-lb. body weight	52.94	0.40
Goat, maintenance, 88-lb. body weight	0.65	0.09
Ewe, maintenance, 110-lb. body weight	0.898	0.09
Beef animal, growing, 800-lb. body weight	6.13	0.07
Dairy cow, lactating, 1,400-lb. body weight	48.07	0.19

[1]With the exception of dairy, fish, and dog, all requirements are given as total dietary sodium. Dairy, fish and dog values are given as bioavailable sodium.

DM: dry matter.

Functions

About two thirds of body water is located inside the body's cells in the intracellular fluid, and the remaining third is outside the body's cells in the extracellular fluid. Sodium is the principal inorganic cation of extracellular fluids. As such, it, along with potassium and chloride, determines to a large extent whether water enters or leaves the cells through the process of osmosis. Sodium, as a cation, is a source of base, which is used in the body to maintain the acid–base equilibrium. Sodium also plays a role in heart and nerve function. The kidneys of livestock have the ability to conserve sodium in times of dietary shortage and to excrete sodium in times of dietary excess. In ruminants, sodium in saliva is a component of compounds that buffer the acid generated during fermentation.

The sodium pump (Figure 9–3) plays a role in the transport of sugars and amino acids from the interstitial fluid and intestinal lumen into cell cytoplasm. All cells of the body have sodium pumps and a significant portion of the adenosine triphosphate (ATP) generated by the cell fuels these pumps. Sodium is constantly leaking into the cell cytoplasm by diffusion, and the cell must constantly pump sodium out of the cell to maintain function. Some types of sugars and amino acids are transported into the cell by coupling with the diffusion of sodium.

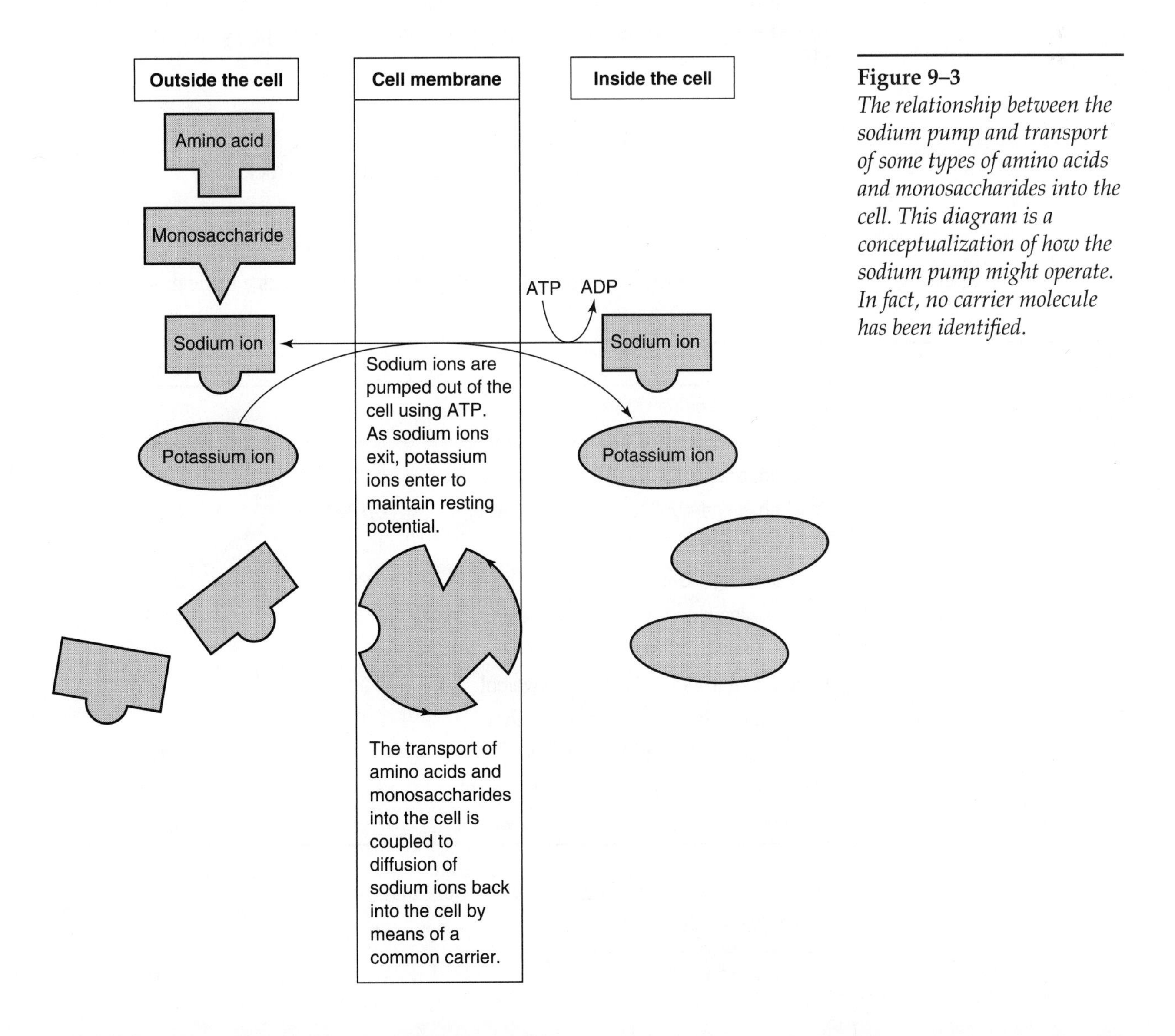

Figure 9–3
The relationship between the sodium pump and transport of some types of amino acids and monosaccharides into the cell. This diagram is a conceptualization of how the sodium pump might operate. In fact, no carrier molecule has been identified.

Sodium Deficiency

A sodium deficiency may not manifest itself for several days. A deficiency of sodium results in pica, the term for abnormal appetite. Sodium-deficient animals have been observed drinking urine, and licking and chewing soil and objects. These behaviors, however, should not be taken as definitive evidence of sodium deficiency. Other symptoms of deficiency include loss of body weight, rough hair coat, and incoordination.

Feed Sources

The bioavailability of sodium in most feeds and in water is very high. Supplemental feed sources include sodium chloride NaCl, 39.34% Na, sodium bicarbonate $NaHCO_3$, 27.0% Na, sodium phosphate monobasic $NaH_2PO_4 \cdot H_2O$, 16.68% Na, sodium sulfate decahydrate $Na_2SO_4 \cdot 10H_2O$, 14.27% Na.

Toxicity and Special Issues

Excess sodium intake could occur as a result of excess salt (sodium chloride) consumed in feed or water. Signs of sodium toxicity include nervousness, staggering, seizures, paralysis, and death. With an adequate supply of good quality drinking water, livestock can tolerate large excesses of sodium chloride intake. In the case of freshwater fish, high dietary intakes of salt depress growth and feed efficiency (Zaugg & McLain, 1969; Salman & Eddy, 1988). Table 9–4 gives the safe upper limits for sodium in livestock diets as presented in the NRC (1980).

Chloride

Table 9–7 gives chloride requirements for selected livestock.

Functions

The nutritionally important chloride ion should not be confused with molecular chlorine, which exists under biological conditions as a toxic gas. Chloride

Table 9–7
Chloride requirements for selected livestock[1]

	Required (g)	Required (%), DM basis
Fish, channel catfish, 100-g body weight	unknown	unknown
Fish, rainbow trout, 100-g body weight	0.0126	1.00
Chicken, broiler, 5 wks of age	0.2050	0.17
Chicken, white-egg layer, 3-lb. body weight	0.1300	0.14
Pig, growing 45-lb. body weight	1.07	0.09
Dog, growing, 30-lb. body weight	0.30	0.08
Cat, growing kitten, 4.2-lb. body weight	0.25	0.30
Rabbit, growing, 5 wks	0.619	0.46
Horse, light work, 1,100-lb. body weight	93.96	0.71
Goat, maintenance, 88-lb. body weight	2.18	0.30
Ewe, maintenance, 110-lb. body weight	1.21	0.12
Beef animal, growing, 800-lb. body weight	13.28[2]	0.15[2]
Dairy cow, lactating, 1,400-lb. body weight	57.99	0.23

[1]With the exception of dairy, fish and dog, all requirements are given as total dietary chloride. Dairy, fish and dog values are given as bioavailable chloride.

[2]Calculated from the requirement for salt.

DM: Dry matter.

is the principal inorganic anion of extracellular fluids. Chloride, as an anion, is a source of acid that is used to maintain acid–base equilibrium within the body. It is also used to manufacture hydrochloric acid in the stomach, abomasum, and proventriculus. Chloride, along with ions of sodium and potassium, is important in the process of osmosis that determines fluid balance of body cells.

Chloride Deficiency

Chloride deficiency signs include anorexia and alkalosis. A deficiency may be created by inadequate dietary supply or to a loss of gastric juices (vomiting).

Feed Sources

The bioavailability of chloride in most feeds and in water is very high. Supplemental feed sources include sodium chloride NaCl, 60.66% Cl, potassium chloride KCl, 47.3% Cl, ammonium chloride NH_4Cl, 66.28% Cl.

Toxicity and Special Issues

Excess chloride in the absence of a neutralizing cation can result in a disturbance of acid–base balance. In the case of freshwater fish, high dietary intakes of sodium chloride depress growth and feed efficiency (Zaugg & McLain, 1969; Salman & Eddy, 1988). Table 9–4 gives the safe upper limits for chloride in livestock diets as presented in the NRC (1980).

Potassium

Table 9–8 gives potassium requirements for selected livestock.

Functions

Potassium is the third most abundant mineral in the body. In contrast to sodium and chloride, most of the body's potassium is found inside body cells in the intracellular fluid. Potassium is involved in neuromuscular function.

Table 9–8

Potassium requirements for selected livestock[1]

	Required (g)	Required (%), DM basis
Fish, channel catfish, 100-g body weight	unknown	unknown
Fish, rainbow trout, 100-g body weight	0.0098	0.78
Chicken, broiler, 5 wks of age	0.41	0.33
Chicken, white egg layer, 3-lb. body weight	0.1500	0.17
Pig, growing, 45-lb. body weight	3.06	0.26
Dog, growing, 30-lb. body weight	1.55	0.44
Cat, growing kitten, 4.2-lb. body weight	0.50	0.60
Rabbit, growing, 5 wks	1.251	0.94
Horse, light work, 1,100-lb. body weight	34.93	0.26
Goat, maintenance, 88-lb. body weight	3.59	0.50
Ewe, maintenance, 110-lb. body weight	4.990	0.50
Beef animal, growing, 800-lb. body weight	52.55	0.60
Dairy cow, lactating, 1,400-lb. body weight	240.16	0.97

[1]With the exception of dairy, fish and dog, all requirements are given as total dietary potassium. Dairy, fish and dog values are given as bioavailable potassium.

DM: Dry matter.

Potassium is involved in numerous enzyme systems, including those involved in energy production (Figure 9–2). Potassium is also involved in acid–base regulation and water balance. It is essential for normal function of nervous, muscle, kidney, and cardiac tissues.

Potassium Deficiency

Unlike the case with calcium, the body has minimal storage capability for potassium. Potassium deficiency symptoms included anorexia, inactivity, and impaired heart function. In fish, potassium deficiency resulted in convulsions and tetany (Shearer, 1988).

Feed Sources

Supplemental feed sources include potassium chloride KCl, 50.0% K, potassium sulfate K_2SO_4, 41.8% K, potassium carbonate K_2CO_3, 56.6% K.

Toxicity and Special Issues

Excesses are tolerated if good quality drinking water is available for use in excretion. In the dairy industry, excess potassium has been associated with metabolic problems. These are discussed in the Dietary Cation-Anion Difference section later in this chapter. Table 9–4 gives the safe upper limits for potassium in livestock diets as presented in the NRC (1980).

Magnesium

Table 9–9 gives magnesium requirements for selected livestock.

Functions

Magnesium is necessary for normal bone development. Magnesium is a cofactor in many enzyme systems including those involved in energy production (Figure 9–2). Adequate magnesium is necessary for proper function of the parathyroid hormone that acts to maintain proper blood calcium level. Magnesium is also vital for normal nerve and muscle function.

Table 9–9
Magnesium requirements for selected livestock[1]

	Required (g)	Required (%), DM basis
Fish, channel catfish, 100-g body weight	0.00132	0.04
Fish, rainbow trout, 100-g body weight	0.0007	0.06
Chicken, broiler, 5 wks of age	0.0820	0.067
Chicken, white egg layer, 3-lb. body weight	0.0500	0.056
Pig, growing, 45-lb. body weight	0.53	0.04
Dog, growing, 30-lb. body weight	0.14	0.04
Cat, growing kitten, 4.2-lb. body weight	0.07	0.08
Rabbit, growing, 5 wks	0.0464	0.035
Horse, light work, 1,100-lb. body weight	11.63	0.09
Goat, maintenance, 88-lb. body weight	0.86	0.12
Ewe, maintenance, 110-lb. body weight	1.197	0.12
Beef animal, growing, 800-lb. body weight	8.76	0.10
Dairy cow, lactating, 1,400-lb. body weight	7.61	0.03

[1]With the exception of dairy, fish and dog, all requirements are given as total dietary magnesium. Dairy, fish and dog values are given as bioavailable magnesium.

DM: Dry matter.

Magnesium Deficiency

Signs of magnesium deficiency include reduced growth, nervous and muscular symptoms, weakness, loss of equilibrium, and tetany followed by death. Tetany describes a state of painful, sustained muscle contraction.

Grass tetany is a poorly understood syndrome that appears to be triggered by low blood magnesium (hypomagnesemia). This situation results from either a deficiency of magnesium in the ration or low bioavailability of magnesium in the ration. Typically, grass tetany occurs when pastures first turn green. This corresponds to a stage when magnesium bioavailability in the plant is particularly low. Animals grazing such pastures without access to supplemental magnesium may have symptoms ranging from nervousness and hypersensitivity to frenzied running, falling, and convulsing.

Feed Sources

The magnesium in grains is generally more available than the magnesium in forages. Magnesium solubility and the potential for absorption declines as pH rises above 6.5. This has implications where pH at the absorption site can change. In adult ruminants, the primary absorption site for magnesium is the rumen. Fish can obtain magnesium from either their diet or their environment. Supplemental feed sources include magnesium oxide MgO, 56.2% Mg, magnesium carbonate or magnesite $MgCO_3$, 30.81% Mg and magnesium hydroxide $Mg(OH)_2$, 41.7% Mg.

Toxicity and Special Issues

Livestock generally have the ability to excrete excess absorbed magnesium via the kidneys. The body apparently does not have a magnesium storage depot from which to draw magnesium during times of dietary inadequacy. Table 9–4 gives the safe upper limits for magnesium in livestock diets as presented in the NRC (1980).

Sulfur

Table 9–10 gives sulfur requirements for selected livestock.

Table 9–10
Sulfur requirements for selected livestock[1]

	Required (g)	Required (%), DM basis
Fish, channel catfish, 100-g body weight	0	0
Fish, rainbow trout, 100-g body weight	0	0
Chicken, broiler, 5 wks of age	0	0
Chicken, white egg layer, 3-lb. body weight	0	0
Pig, growing, 45-lb. body weight	0	0
Dog, growing, 30-lb. body weight	0	0
Cat, growing kitten, 4.2-lb. body weight	0	0
Rabbit, growing, 5 wks	0	0
Horse, light work, 1,100-lb. body weight	22.50	0.17
Goat, maintenance, 88-lb. body weight	1.15	0.16
Ewe, maintenance, 110-lb. body weight	1.397	0.14
Beef animal, growing, 800-lb. body weight	13.14	0.15
Dairy cow, lactating, 1,400-lb. body weight	52.14	0.21

[1]With the exception of dairy, all requirements are given as total dietary sulfur. Dairy values are given as bioavailable sulfur. Animals with a requirement of 0 for sulfur do not require sulfur except as a component of the sulfur-containing amino acids.

DM: Dry matter.

Functions

The B vitamins thiamin and biotin contain sulfur. Sulfur is also contained in the amino acids methionine and cysteine, which are found in proteins throughout the body. Sulfur is involved in the production of energy from feed nutrients (Figure 9–2).

Sulfur Deficiency

Elemental sulfur is not required by monogastric animals. The dietary sulfur in rations for these animals must be in the form of biotin, thiamin, methionine, or cysteine. Ruminant animals can make use of elemental sulfur through the synthetic activities of microbes living in the rumen. A sulfur deficiency in ruminant animals would initially manifest itself as reduced rumen productivity.

Feed Sources

For monogastrics, the sulfur needs are met through the intake of sulfur-containing compounds. The essential amino acid that contains sulfur is methionine. The nonessential amino acid that contains sulfur is cysteine (cystine is the oxidized form of cysteine). In ruminant animals, inorganic sulfur can be added to the diet to provide the rumen microbes with available sulfur with which to make methionine, thiamin, and biotin. Supplemental feed sources include ammonium sulfate $(NH_4)_2SO_4$, 24.1% S, manganese sulfate monohydrate $MnSO_4 \cdot H_2O$, 19.0% S, magnesium sulfate heptahydrate $MgSO_4 \cdot 7H_2O$, 13.3% S.

Toxicity and Special Issues

Excess inorganic sulfur (sulfates and sulfites) will reduce dry matter intake and can interfere with absorption of copper and selenium. Signs of toxicity involve nervous and muscle symptoms.

In ruminants, consumption of excess sulfur can lead to a disorder called polioencephalomalacia, abbreviated PEM, and also referred to as *cerebrocortical necrosis.* In the beef industry, PEM is sometimes called "blind staggers." The consumption of excess sulfur in the affected animal, often accompanied by a low rumen pH, leads to the production of excessive amounts of hydrogen sulfide in the rumen that then may be absorbed and carried to the body's cells, where it interferes with normal metabolism (Gould, 1998). Thiamin deficiency has also been associated with outbreaks of PEM.

Sulfites (SO_3^{-2}) are more toxic than sulfates (SO_4^{-2}), though sulfates may cause severe diarrhea. Table 9–4 gives the safe upper limits for sulfur in livestock diets as presented in the NRC (1980).

TRACE OR MICROMINERALS

The fact that trace minerals are needed in small quantities is not an indication of their importance. A deficiency of any one of them can make an animal just as sick or produce abnormalities just as severe as would a deficiency of one of the major minerals.

Cobalt

Table 9–11 gives cobalt requirements for selected livestock.

Functions

Cobalt is a component of vitamin B_{12}, and it appears that this is the only use for cobalt in livestock. Gluconeogenesis uses vitamin B_{12}, made by microbes from cobalt, to turn some glucogenic compounds into glucose.

Table 9–11 *Cobalt requirements for selected livestock*[1]

	Required (mg)	Required (mg/kg or ppm, DM basis)
Fish, channel catfish, 100-g body weight	0	0
Fish, rainbow trout, 100-g body weight	0	0
Chicken, broiler, 5 wks of age	0	0
Chicken, white egg layer, 3-lb. body weight	0	0
Pig, growing, 45-lb. body weight	0	0
Dog, growing, 30-lb. body weight	0	0
Cat, growing kitten, 4.2-lb. body weight	0	0
Rabbit, growing, 5 wks	0.0046	0.035
Horse, light work, 1,100-lb. body weight	1.47	0.11
Goat, maintenance, 88-lb. body weight	0.07	0.10
Ewe, maintenance, 110-lb. body weight	0.100	0.10
Beef animal, growing, 800-lb. body weight	0.88	0.10
Dairy cow, lactating, 1,400-lb. body weight	2.87	0.12

[1]With the exception of dairy, all requirements are given as total dietary cobalt. Dairy values are given as bioavailable cobalt. Animals with a requirement of 0 for cobalt do not require cobalt except as a component of vitamin B_{12}.

DM: Dry matter.

Cobalt Deficiency

Signs of cobalt deficiency in ruminants are the same as signs of vitamin B_{12} deficiency in monogastrics. Deficiency signs include nervous symptoms, liver problems, impaired immune function, and blood changes.

Feed Sources

Vitamin B_{12} is synthesized by the microbes inhabiting the digestive tracts of livestock, and some monogastrics may have access to some vitamin B_{12} from this microbial activity. Nevertheless, it is routine practice to add a source of vitamin B_{12} to monogastric diets. It is assumed that the microbial activity in the rumen satisfies the ruminant animal's need for vitamin B_{12}, provided dietary cobalt is adequate. Sources of supplemental cobalt include cobalt carbonate $CoCO_3$, 460,000 mg/kg Co, Cobalt carbonate hexahydrate $CoCo_3{\cdot}6H_20$, 259,000 mg/kg Co and cobalt dichloride hexahydrate $CoCl_2{\cdot}6H_2O$, 247,000 mg/kg Co. The cobalt in the oxide, chloride, and sulfate forms is less bioavailable.

Toxicity and Special Issues

Excess cobalt results in reduced growth and anemia. Table 9–4 gives the safe upper limits for cobalt in livestock diets as presented in the NRC (1980).

Copper

Table 9–12 gives copper requirements for selected livestock.

Functions

Copper is needed in hemoglobin synthesis. It is used in energy production (Figure 9–2). Copper is a component of many enzyme systems including those involved in the production of the hair and the skin pigment melanin, the formation of connective tissues, and the function of the immune system.

Table 9–12
Copper requirements for selected livestock[1]

	Required (mg)	Required (mg/kg or ppm, DM basis)
Fish, channel catfish, 100-g body weight	0.0167	5.56
Fish, rainbow trout, 100-g body weight	0.0042	3.33
Chicken, broiler, 5 wks of age	1.0935	8.89
Chicken, white egg layer, 3-lb. body weight	unknown	unknown
Pig, growing, 45-lb. body weight	5.33	4.44
Dog, growing, 30-lb. body weight	1.03	2.94
Cat, growing kitten, 4.2-lb. body weight	0.42	5.00
Rabbit, growing, 5 wks	0.782	5.85
Horse, light work, 1,100-lb. body weight	147.0	11.0
Goat, maintenance, 88-lb. body weight	5.03	7.00
Ewe, maintenance, 110-lb. body weight	6.99	7.00
Beef animal, growing, 800-lb. body weight	88	10.0
Dairy cow, lactating, 1,400-lb. body weight	10.71	0.43

[1]With the exception of dairy, fish and dog, all requirements are given as total dietary copper. Dairy, fish and dog values are given as bioavailable copper.

DM: Dry matter.

Copper Deficiency

A copper deficiency may produce symptoms of retarded growth, impaired pigmentation, anemia, impaired reproductive function, fragile bones, and impaired immune function. High intakes of sulfur and/or molybdenum interfere with copper utilization.

Feed Sources

Supplemental feed sources include copper sulfate pentahydrate $CuSO_4 \cdot 5H_2O$, 254,500 mg/kg Cu and copper chloride dihydrate $CuCl_2 \cdot 2H_2O$, 372,000 mg/kg Cu. A source of higher bioavailable copper is the organic trace mineral copper lysine.

Toxicity and Special Issues

Copper fed at a level in excess of the nutritional requirement stimulates growth in pigs. The mechanism for this effect is unknown. Copper toxicity in fish results in reduced feed efficiency and elevated liver copper levels. Sheep are susceptible to both copper deficiency and to copper toxicity. Deficiency in young lambs results in a condition called *neonatal ataxia* or "swayback." Affected lambs are born weak and may die because of an inability to nurse. Genetic and dietary factors (molybdenum and sulfur) influence copper requirements. In cat nutrition, the process of extrusion apparently reduces the availability of copper. Since dry and semi-moist cat foods are usually extruded products, the cat's requirement for copper is adjusted upward if the source is a dry or semi-moist food. Table 9–4 gives the safe upper limits for copper in domestic animal diets as presented in the NRC (1980).

Iodine

Table 9–13 gives iodine requirements for selected livestock.

Functions

Iodine is a component of the thyroid hormones that regulate metabolic rate.

Table 9–13
Iodine requirements for selected livestock[1]

	Required (mg)	Required (mg/kg or ppm, DM basis)
Fish, channel catfish, 100-g body weight	0.0037	1.22
Fish, rainbow trout, 100-g body weight	0.00154	1.22
Chicken, broiler, 5 wks of age	0.0478	0.39
Chicken, white egg layer, 3-lb. body weight	0.0035	0.039
Pig, growing, 45-lb. body weight	0.19	0.16
Dog, growing, 30-lb. body weight	0.21	0.59
Cat, growing kitten, 4.2-lb. body weight	0.03	0.35
Rabbit, growing, 5 wks	0.0309	0.232
Horse, light work, 1,100-lb. body weight	2.91	0.22
Goat, maintenance, 88-lb. body weight	0.07	0.10
Ewe, maintenance, 110-lb. body weight	0.10	0.10
Beef animal, growing, 800-lb. body weight	4.38	0.50
Dairy cow, lactating, 1,400-lb. body weight	9.53	0.38

[1]With the exception of dairy, fish and dog, all requirements are given as total dietary iodine. Dairy, fish and dog values are given as bioavailable iodine.

DM: Dry matter.

Iodine Deficiency

Iodine deficiency results in impaired thyroid function. This results in reduced thyroid hormone production and reduced metabolic rate. The most obvious symptom of iodine deficiency is goiter, in which an enlarged thyroid gland is evident.

Feed Sources

Iodine in feed is in the forms of iodide and iodate. Supplemental feed sources include ethylenediamine dihydroiodide or EDDI $C_2H_8N_22HI$, 803,400 mg/kg I, calcium iodate $Ca(IO_3)_2$, 635,000 mg/kg I and potassium iodide KI, 681,700 mg/kg I. Iodine is often added to sodium chloride to produce iodized salt. Fish can absorb iodine from feed or their environment.

Toxicity and Special Issues

Dairy cows consuming excessive levels of iodine produce milk that contains high iodine levels. Because humans are more sensitive to iodine toxicity than cattle, this is a public health issue. Table 9–4 gives the safe upper limits for iodine in livestock diets as presented in the NRC (1980).

Iron

Table 9–14 gives iron requirements for selected livestock.

Functions

Iron is a component of hemoglobin in red blood cells. It also functions in energy production (Figure 9–2). Iron is a component of many enzyme systems.

Iron Deficiency

Iron deficiency results in anemia manifested as a deficiency in production of red blood cells. Iron deficiency also results in impaired immune function.

Table 9–14
Iron requirements for selected livestock[1]

	Required (mg)	Required (mg/kg or ppm, DM basis)
Fish, channel catfish, 100-g body weight	0.1000	33.3
Fish, rainbow trout, 100-g body weight	0.084	66.7
Chicken, broiler, 5 wks of age	10.9347	89
Chicken, white egg layer, 3-lb. body weight	4.4997	50.00
Pig, growing 45-lb. body weight	80	67
Dog, growing, 30-lb. body weight	11.3	31.9
Cat, growing kitten, 4.2-lb. body weight	6.71	80.00
Rabbit, growing, 5 wks	7.81	58
Horse, light work, 1,100-lb. body weight	588.2	44
Goat, maintenance, 88-lb. body weight	21.6	30.0
Ewe, maintenance, 110-lb. body weight	29.9	30.0
Beef animal, growing, 800-lb. body weight	438	50.0
Dairy cow, lactating, 1,400-lb. body weight	38	1.53

[1]With the exception of dairy, fish and dog, all requirements are given as total dietary iron. Dairy, fish and dog values are given as bioavailable iron.

DM: Dry matter.

Feed Sources

Supplemental feed sources include ferrous sulfate heptahydrate $FeSO_4 \cdot 7H_2O$, 218,400 mg/kg Fe and ferric chloride hexahydrate $FeCl_3 \cdot 6H_2O$, 207,000 mg/kg Fe. Ferric oxide and ferrous carbonate are iron sources with lower bioavailabilities. A source of higher bioavailable iron is the organic trace mineral iron methionine. Feed is considered the major iron source for fish. Fish do have the ability, however, to absorb iron from the environment through the gills.

Toxicity and Special Issues

Baby pigs must consume more iron than that provided in the sow's milk or they will become anemic. The solution to this problem has been to give pigs an injection containing iron within the first 3 days of life. Excess iron fed to fish results in reduced growth, increased mortality, and liver damage.

Iron in water is more available than iron in feed, though dietary adjustments based on iron water content have not been established. Excess iron can interfere with absorption of dietary copper and zinc. Table 9–4 gives the safe upper limits for iron in livestock diets as presented in the NRC (1980).

Manganese

Table 9–15 gives manganese requirements for selected livestock.

Functions

Manganese is a component of enzyme systems including those involved with energy production (Figure 9–2). Manganese is also needed for proper bone growth.

Manganese Deficiency

Manganese deficiency results in reduced growth, increased fat deposition, skeletal abnormalities, and impaired reproductive function.

Table 9–15
Manganese requirements for selected livestock[1]

	Required (mg)	Required (mg/kg or ppm, DM basis)
Fish, channel catfish, 100-g body weight	0.0080	2.67
Fish, rainbow trout, 100-g body weight	0.0182	14.44
Chicken, broiler, 5 wks of age	8.201	66.67
Chicken, white egg layer, 3-lb. body weight	2.000	22.22
Pig, growing, 45-lb. body weight	2.7	2.2
Dog, growing, 30-lb. body weight	1.81	5.14
Cat, growing kitten, 4.2-lb. body weight	0.63	7.5
Rabbit, growing, 5 wks	1.33	9.94
Horse, light work, 1,100-lb. body weight	588.2	44
Goat, maintenance, 88-lb. body weight	14.4	20.0
Ewe, maintenance, 110-lb. body weight	20.0	20.0
Beef animal, growing, 800-lb. body weight	175	20.0
Dairy cow, lactating, 1,400-lb. body weight	2.41	0.10

[1]With the exception of dairy, fish and dog, all requirements are given as total dietary manganese. Dairy, fish and dog values are given as bioavailable manganese.

DM: Dry matter.

Feed Sources

Supplemental feed sources include manganous sulfate monohydrate $MnSO_4{\cdot}H_2O$, 325,069 mg/kg Mn and, manganous chloride tetrahydrate $MnCl_2{\cdot}4H_2O$, 277,000 mg/kg Mn. A source of higher bioavailable manganese is the organic trace mineral manganese methionine. Fish may absorb manganese from either feed or the environment.

Toxicity and Special Issues

Excess manganese causes reduced growth rate. Table 9–4 gives the safe upper limits for manganese in livestock diets as presented in the NRC (1980).

Selenium

Table 9–16 gives selenium requirements for selected livestock.

Functions

Selenium is a component of the enzyme glutathione peroxidase that protects the cell membranes from peroxide damage. Selenium also plays a role in thyroid metabolism and reproductive function.

Selenium shares its antioxidant function with vitamin E, and these nutrients appear to have a mutual sparing effect on one another. Animals receiving rations that are marginal in either selenium or vitamin E will respond to additional amounts of the other nutrient. The nature of the relationship, however, has not been quantified.

Selenium Deficiency

Selenium deficiency causes diarrhea, necrosis of the liver, a paleness of the skeletal muscles (called white muscle disease in swine and stiff lamb disease), heart problems, reproductive problems, reduced milk production, and impaired immune function.

Table 9–16
Selenium requirements for selected livestock[1]

	Required (mg)	Required (mg/kg or ppm, DM basis)
Fish, channel catfish, 100-g body weight	0.0008	0.28
Fish, rainbow trout, 100-g body weight	0.00042	0.33
Chicken, broiler, 5 wks of age	0.0205	0.167
Chicken, white egg layer, 3-lb. body weight	0.0060	0.067
Pig, growing, 45-lb. body weight	0.20	0.17
Dog, growing, 30-lb. body weight	0.039	0.11
Cat, growing kitten, 4.2-lb. body weight	0.01	0.10
Rabbit, growing, 5 wks	0	0
Horse, light work, 1,100-lb. body weight	1.47	0.11
Goat, maintenance, 88-lb. body weight	0.07	0.10
Ewe, maintenance, 110-lb. body weight	0.10	0.10
Beef animal, growing, 800-lb. body weight	0.88	0.10
Dairy cow, lactating, 1,400-lb. body weight	7.82	0.30

[1]With the exception of dairy, fish and dog, all requirements are given as total dietary selenium. Dairy, fish and dog values are given as bioavailable selenium.

DM: Dry matter.

Feed Sources

Plant selenium content is related to soil selenium content. In areas where the soils are low in selenium, selenium supplementation is necessary in order to avoid deficiencies. In areas where the soils are rich in selenium, selenium toxicity may be a concern. Selenium supplementation is regulated by the Food and Drug Administration because of the potential toxicity of this mineral. In some areas of the United States (northeast, northwest, southeast), crops will contain little selenium and supplementation will be necessary. In other areas, selenium content in some feedstuffs may be high, causing toxicity in animals that consume these feedstuffs over prolonged periods. Supplemental feed sources include sodium selenite Na_2SeO_3, 456,000 mg/kg Se and sodium selenate tetrahydrate $Na_2SeO_4 \cdot 10H_2O$, 213,920 mg/kg Se. Other sources of selenium include selenium enriched yeast and selenomethionine. Fish can absorb selenium through the gills.

Toxicity and Special Issues

Selenium is toxic at relatively low levels of excess. Rations containing 2 ppm are considered toxic. Systems, organs, and tissues affected by toxicity include the liver, kidney, and skin. Also seen are reduced growth rate, poor feed efficiency, and high mortality.

Concerns over potential environmental pollution from selenium use in agriculture have resulted in pressure on lawmakers to reduce the legal limit for selenium in livestock diets. Table 9–4 gives the safe upper limits for selenium in livestock diets as presented in the NRC (1980).

Zinc

Table 9–17 gives zinc requirements for selected livestock.

Functions

Zinc functions in many enzyme systems, including those involved in the production of the hormone *insulin*. Zinc is associated with carbohydrate, protein,

Table 9–17
Zinc requirements for selected livestock[1]

	Required (mg)	Required (mg/kg or ppm, DM basis)
Fish, channel catfish, 100-g body weight	0.0667	22.22
Fish, rainbow trout, 100-g body weight	0.042	33.33
Chicken, broiler, 5 wks of age	5.4674	44.44
Chicken, white egg layer, 3-lb. body weight	3.4998	38.89
Pig, growing, 45-lb. body weight	79.9	66.7
Dog, growing, 30-lb. body weight	12.5	35.6
Cat, growing kitten, 4.2-lb. body weight	6.29	75.00
Rabbit, growing, 5 wks	7.81	58.48
Horse, light work, 1,100-lb. body weight	588.2	44
Goat, maintenance, 88-lb. body weight	7.18	10.00
Ewe, maintenance, 110-lb. body weight	20.0	20.0
Beef animal, growing, 800-lb. body weight	263	30.0
Dairy cow, lactating, 1,400-lb. body weight	180.58	7.27

[1]With the exception of dairy, fish and dog, all requirements are given as total dietary zinc. Dairy, fish and dog values are given as bioavailable zinc.

DM: Dry matter.

and lipid metabolism. Zinc is also involved in milk synthesis, tissue repair, sperm production, and immune function.

Zinc Deficiency

Skin and/or hair problems, identified as parakeratosis, have been associated with zinc deficiency. These problems are characterized by thick, rough, scaly skin, and sometimes hair loss. Other problems include reduced growth rate, and impaired reproductive development and function in males and females. In fish, a zinc deficiency results in lens cataracts, erosion of fins and skin, and reduced egg production and hatchability.

Feed Sources

Supplemental feed sources include zinc sulfate monohydrate $ZnSO_4 \cdot H_2O$, 363,600 mg/kg Zn, zinc chloride $ZnCl_2$, 479,700 mg/kg Zn and zinc carbonate $ZnCO_3$, 521,400 mg/kg Zn. Sources of lower bioavailable zinc are zinc oxide and zinc sulfide. A source of higher bioavailable zinc is the organic trace mineral zinc methionine. Fish can absorb zinc from both feed and their environment.

Toxicity and Special Issues

The presence of phytate in plants has been demonstrated to reduce zinc absorption. The use of phytase enzymes has improved the bioavailability of zinc from plant sources. The zinc requirement may be increased when excessive levels of calcium are fed. Zinc fed at a level in excess of the nutritional requirement has resulted in increased weight gain in pigs. The mechanism for this effect is unknown. Excessive levels of zinc interfere with absorption and metabolism of copper. Excessive levels have also been associated with anemia, arthritis, and digestive problems. Excessive zinc levels may interfere with copper absorption. Table 9–4 gives the safe upper limits for zinc in livestock diets as presented in the NRC (1980).

Chromium

Functions

Chromium is believed to work as a cofactor with insulin, and as such, appears to be involved in carbohydrate metabolism. Chromium may also be involved in lipid, protein, and nucleic acid metabolism. The specific function of chromium is not known, but supplemental chromium in swine diets has sometimes resulted in improved feed utilization, carcass characteristics, and reproductive performance (Lindemann, Wood, Harper, Kornegay, & Anderson, 1995). In cattle, studies have shown that supplemental chromium has the potential to improve liver function (Besong, Jackson, Trammell, & Akay, 2001) and increase dry matter intake and milk production (Hayirli, Bremmer, Bertics, Socha, & Grummer, 2001).

In species other than swine and cattle, chromium has been less well studied and nutritional information is scanty. Even with swine and cattle, there is not yet enough information to establish a quantitative requirement for chromium.

Chromium Deficiency

The requirement for dietary chromium has not been established, so symptoms of deficiency are unknown.

Feed Sources

The variable responses to supplemental chromium may be due to inconsistent bioavailability of the chromium source used (NRC, 1998). In the inorganic form such as chromium chloride and chromium oxide, the chromium is poorly absorbed. When complexed with an organic compound as in chromium picolinate and chromium nicotinate, absorption is greatly increased. Chromium in yeast products is also efficiently absorbed.

Toxicity and Special Issues

Chromium toxicity can result in pathologic changes in the DNA within the nucleus.

BUFFERS

Buffers are compounds that resist changes in potential hydrogen ion concentrations (pH) in either direction. In nutrition, it is common to apply the term *buffer* to chemical compounds that are only capable of resisting a reduction in pH; that is, that are only capable of neutralizing acids.

Activities in the rumen generate volatile fatty acids (VFA) and sometimes lactic acid. The most important VFA produced are acetic, propionic, and butyric acids. If the acid production is excessive, rumen pH may drop below the normal range. When this happens, rumen microbe efficiency, dry matter intake, and productivity decline. Buffers in ruminant diets have been shown to improve rumen fermentation by resisting a pH shift due to the production of excess acid by rumen microbes.

THE DIETARY CATION–ANION DIFFERENCE AND ELECTROLYTE BALANCE

As is the case with buffers, the dietary cation–anion difference (DCAD) and electrolyte balance (EB) address the acid–base status of the animal. However, the primary use of buffers in animal nutrition is in diets for ruminants where the application is usually to improve the acid–base status in the rumen.

Cations	Anions
Sodium	Chloride
Potassium	Sulfur
Calcium	Phosphorus
Magnesium	

Table 9–18
Important cations and anions in animal nutrition

The DCAD and EB address the acid–base status of the blood and, therefore, apply to all livestock.

In swine and poultry nutrition, this relationship is referred to as the *electrolyte balance*. In ruminant nutrition, it is referred to as the *dietary cation–anion difference*. The important cations and anions are shown in Table 9–18.

Both the EB and the DCAD are attempts to address the fact that minerals have certain characteristics in common, and for some bodily functions, it is these characteristics rather than the minerals themselves that are important.

Minerals as nutrients function as ions. Cations are positively charged ions and anions are negatively charged ions. According to Stewart's theory (Stewart, 1983):

- The electrical charge of a solution must always be neutral.
- If cations > anions, the cations replace H^+. As H^+ concentration declines, OH^- concentration increases. Increased OH^- concentration is measured as increased pH.
- If anions > cations, the anions replace OH^-. As OH^- concentration declines, H^+ concentration increases. Increased H^+ concentration is measured as reduced pH.

Therefore, what the DCAD and EB are addressing is the effect that the dietary minerals are having on blood pH. Just as the body has mechanisms that maintain blood oxygen content within very narrow limits, it also has mechanisms that maintain blood pH within very narrow limits.

Not all of the cations and anions are considered in most formulas. In swine and poultry nutrition, the usual EB formula is Na + K – Cl, where the ions are usually expressed in milliequivalents per kilogram of diet. The companion application to this text for swine and poultry gives the EB in milliequivalents per kilogram on a dry matter basis. In the dairy industry, numerous formulas have been developed for the DCAD. These values use ion concentration in milliequivalents per kilogram on a dry matter basis. A "simple" DCAD formula is $(Na^+ + K^+) - (Cl^- + S^{-2})$. This formula is used in the companion application to this text for sheep and goats. A "complex" DCAD formula is $[Na^+ + K^+ + (0.15 \times Ca^{+2}) + (0.15 \times Mg^{+2})] - [Cl^- + (0.25 \times S^{-2}) + (0.5 \times P^{-3})]$. This last formula (Fox, Tylutki, Van Amburgh, Chase, Pell, Overton, Tedeschi, Rasmussen, & Durbal, 2000) assigns coefficients to the major dietary cations and anions based on their acidifying or alkalizing potential. It is used in the companion application for dairy cattle.

Recommendations for DCAD and EB are imprecise. A discussion of these recommendations is given in chapters on the species to which DCAD or EB is applied.

SUMMARY

The mineral content of feedstuff varies. The bioavailability of the minerals in feedstuff also varies. To account for this, most mineral requirements are established with "safety factors" built in. These safety factors usually result in excess nutrients excreted, and this has become an environmental problem. The dairy NRC (2001) gives mineral requirements based on tissue requirements and presents bioavailability data for minerals in the usual feedstuffs fed to dairy animals. Mineral interactions are complex, and excesses of one mineral may result in deficiencies of others. The acid–base status of livestock may be manipulated nutritionally through the feeding of buffers, and through adjustment of the EB and the DCAD.

END-OF-CHAPTER QUESTIONS

1. Define *bioavailability* as it is applied to mineral nutrition.
2. What is a proteinated (or chelated) mineral and how is it used in mineral nutrition?
3. To avoid symptoms of deficiency, what mineral nutrient is usually administered to the suckling pig via injection?
4. What mineral nutrients are involved in ATP production through glycolysis? What mineral nutrients are involved in the tricarboxylic acid cycle? What mineral nutrients are involved in ATP production via the electron transport chain?
5. What is the significance of the DCAD and EB in animal nutrition?
6. Which livestock have a dietary requirement for cobalt? How are these animals different from those that do not have a cobalt requirement? What is cobalt in these animals used for?
7. Name seven major or macrominerals and give a supplemental feed source for each. Name nine trace or microminerals and give a supplemental feed source for each.
8. Give 10 examples of mineral interactions in which a dietary excess of one mineral may interfere with absorption and thereby lead to a deficiency of another.
9. For which domestic animal is the calcium requirement, expressed as a percent of total diet, the highest?
10. What is the role of the enzyme *phytase* in the mineral nutrition of some domestic animals?

REFERENCES

Association of American Feed Control Officials. (2003). *Official Publication*. West Lafayette, IN.

Besong, S., Jackson, J. A., Trammell, D. S., & Akay, V. (2001). Influence of supplemental chromium on concentrations of liver triglyceride, blood metabolites and rumen VFA profile in steers fed a moderately high fat diet. *Journal of Dairy Science. 84*, 1679–1685.

Fox, D. G., Tylutki, T. P., Van Amburgh, M. E., Chase, L. E., Pell, A. N., Overton, T. R., Tedeschi, L. O., Rasmussen, C. N., & Durbal, V. M. (2000). The net carbohydrate and protein system for evaluating herd nutrition and nutrient excretion (CNCPS, version 4.0). Animal Science Department Mimeo 213, Cornell University, Ithaca, NY.

Gould, D. H. (1998). Polioencephalomalacia. *Journal of Animal Science. 76*, 309–314.

Hayirli, A, Bremmer, D. R., Bertics, S. J., Socha, M. T., & Grummer, R. R. (2001). Effect of chromium supplementation on production and metabolic parameters in periparturient dairy cows. *Journal of Dairy Science. 84*:1218–1230.

Lindemann, M. D., Wood, C. M., Harper, A. F., Kornegay, E. T., & Anderson, R. A. (1995). Dietary chromium picolinate additions improve gain:feed and carcass characteristics in growing-finishing pigs and increase litter size in reproducing sows. *Journal of Animal Science. 73*, 457–465.

Morrison, F. B. (1949). *Feeds and feeding* (21st ed.). Ithaca, NY: Morrison Publishing Co.

National Research Council. (1977). *Nutrient requirements of rabbits* (2nd revised edition). Washington, DC: National Academy Press.

National Research Council. (1980). *Mineral tolerance of domestic animals*. Washington, DC: National Academy Press.

National Research Council. (1985). *Nutrient requirements of dogs* (revised). Washington, DC: National Academy Press.

National Research Council. (1986). *Nutrient requirements of cats* (revised edition). Washington, DC: National Academy Press.

National Researach Council. (1989). *Nutrient requirements of dairy cattle* (6th revised edition). Washington, DC: National Academy Press.

National Research Council. (1993). *Nutrient requirements of fish*. Washington, DC: National Academy Press.

National Research Council. (1998). *Nutrient requirements of swine* (10th revised edition). Washington, DC: National Academy Press.

National Research Council. (2001). *Nutrient requirements of dairy cattle* (7th revised edition). Washington, DC: National Academy Press.

Salman, N. A., & Eddy, F. B. (1988). Effect of dietary sodium chloride on growth, food intake and conversion efficiency in rainbow trout (*Salmo gairdneri* Richardson). *Aquaculture 70*, 131–144.

Shearer, K. D. (1988). Dietary potassium requirements of juvenile chinook salmon. *Aquaculture 73*, 119–130.

Stewart, P. A. (1983). Modern quantitative acid-base chemistry. *Canadian Journal of Physiology and Pharmacology. 61*, 1444–1461.

Watanabe, T., Murakami, A., Takeuchi, L., Nose, T., & Ogino, C. (1980). Requirement of chum salmon held in freshwater for dietary phosphorus. *Bulletin of the Japanese Society of Scientific Fisheries 46*, 361–367.

Zaugg, W. S., & McLain, L. R. (1969). Inorganic salt effects on growth, salt water adaptation, and gill ATPase of Pacific salmon. In W. W. Neuhaus & J. E. Halver (Editors.), *Fish in Research* (pp. 293–306). New York: Academic Press.

CHAPTER 10

VITAMINS

The story of the vitamins, one of the most important episodes in the history of biochemistry, has touched profoundly on man's health and well-being and on our understanding of the catalytic processes taking place in the metabolism of living organisms.

A. L. LEHNINGER, 1978

INTRODUCTION

Vitamin nutrients are organic compounds that play roles in the metabolism of livestock. As a group, the various vitamins have very little in common structurally and are generally classified as either fat soluble or water soluble. For some vitamins, the symptoms of deficiency can be eliminated with any one of several compounds. For these vitamins, the vitamin content of feedstuffs is expressed in units of potency rather than in weighed amounts of a specific compound, and animal requirements are expressed in International Units (IU). Not all vitamins are dietary essentials for all livestock.

IMPORTANCE OF VITAMINS AND VITAMIN SUPPLEMENTATION

Vitamins are essential organic substances needed in small amounts in the diet for normal function, growth, and maintenance of body tissues. Being organic, they are composed primarily of carbon and hydrogen, but may also contain oxygen, phosphorus, sulfur, or other elements. Most vitamins function as cofactors in enzyme reactions that synthesize important chemicals, including the nonessential amino acids. Vitamins also play vital roles in calcium balance and in the extraction of energy from carbohydrate, fat, and protein. Figure 10–1 summarizes the role of the B vitamins in energy production. Because the functions of vitamins are so diverse, a lack of a particular vitamin can have widespread effects.

Livestock require 14 vitamins, but not all of these are dietary essentials for all species. Some are synthesized within animal tissues, and some are synthesized

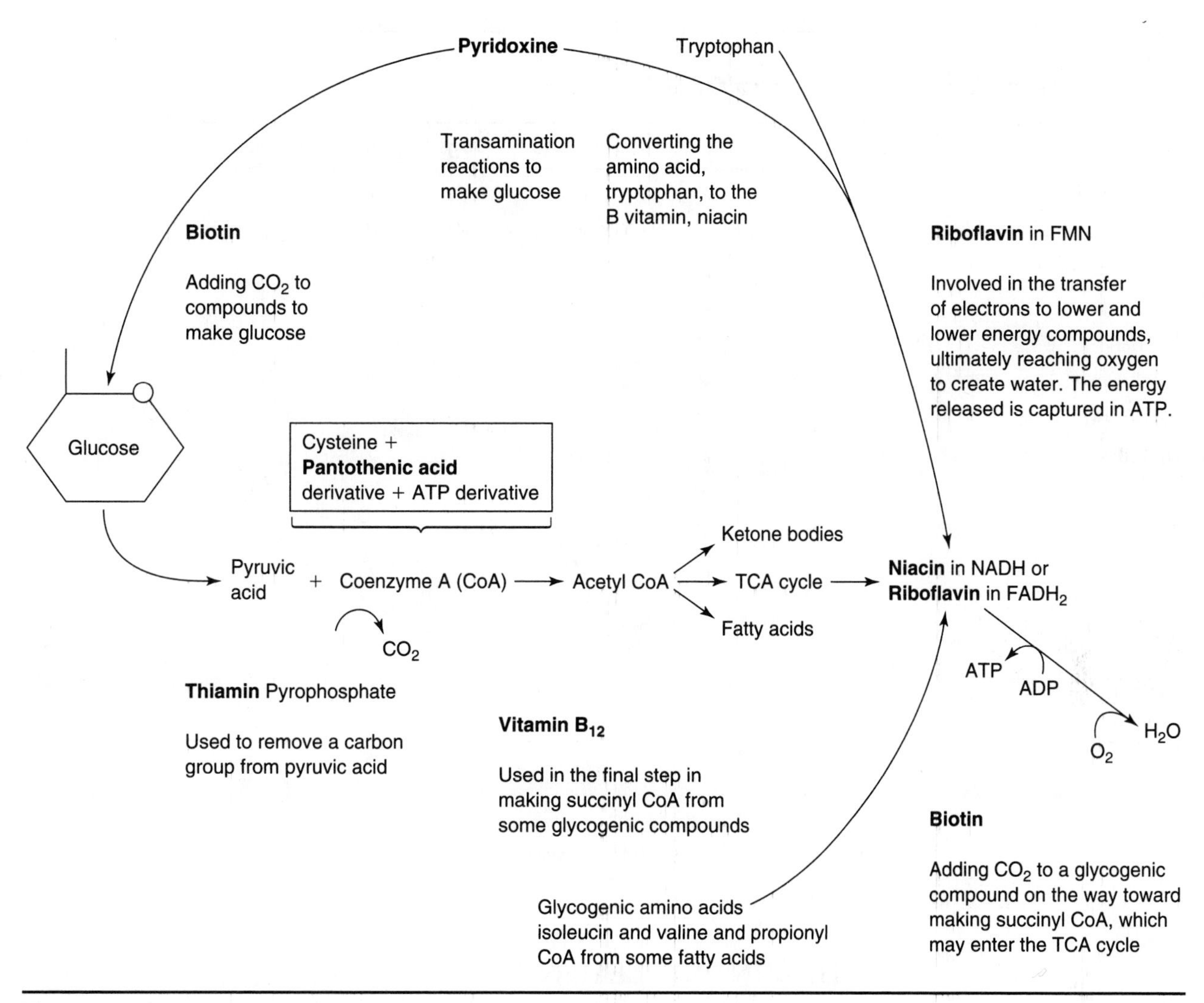

Figure 10–1
A summary of B-vitamin involvement in the production of ATP

and made available to livestock by the microbes inhabiting the digestive tract. Those vitamins that are dietary essentials may or may not be found in sufficient quantities in the usual feedstuffs fed to that animal. Table 10–1 lists the usual vitamins supplemented in livestock diets.

CLASSIFICATION OF VITAMINS

Vitamins are classified as fat soluble and water soluble. Solubility characteristics of vitamins have consequences relating to the mechanism of absorption and body storage. Vitamins A, D, E, and K are fat soluble. The B complex and vitamin C are water soluble.

VITAMIN POTENCY

The daily requirements for all vitamins except A, D, and E are expressed in milligrams (mg) or micrograms (μg). Daily requirements for vitamins A, D, and E are expressed in units of potency. There are multiple sources of these vitamins with varying potency. Potency refers to the ability of the source to remove signs

Table 10–1
Usual vitamins supplemented in rations of healthy adult livestock

Species	Ruminants	Swine	Chickens	Dogs	Cats	Horses	Rabbits	Fish
Vitamin A	✓	✓	✓	✓	✓	✓	✓	✓
Vitamin D	✓	✓	✓	✓	✓	✓	✓	✓
Vitamin E	✓	✓	✓	✓	✓	✓	✓	✓
Vitamin K		✓	✓	✓	✓		✓	✓
Thiamin		✓	✓	✓	✓		✓	✓
Riboflavin		✓	✓	✓	✓		✓	✓
Pyridoxine		✓	✓	✓	✓		✓	✓
Pantothenic acid		✓	✓	✓	✓		✓	✓
Niacin		✓	✓	✓	✓		✓	✓
B_{12}		✓	✓	✓	✓		✓	✓
Choline		✓	✓	✓	✓		✓	✓
Biotin		✓	✓	✓	✓			✓
Folacin (folate or folic acid)		✓	✓	✓	✓		✓	✓
Myoinositol					✓			✓
Vitamin C								✓

of vitamin deficiency. The potency unit that is used in the feed industry is the International Unit (IU). It is sometimes referred to as the U.S. Pharmacoepeia unit (USP).

VITAMIN SUPPLEMENTATION

The vitamins that livestock require can be supplied through consumption of fresh feedstuffs rich in vitamin content, addition of vitamins to the ration, or injection of vitamins into animals. Given the fact that during at least some seasons of the year, livestock are fed feedstuffs that have been in storage for several months, it is not possible to meet livestock requirements through the first option during all seasons of the year. The third option—injection of vitamins—will be the most expensive method of vitamin supplementation. For most livestock during most seasons of the year, vitamin supplementation will be accomplished through the addition of vitamin supplements to the diet. In fact, vitamin premixes containing the required types and amounts of vitamins are usually added to livestock rations without regard for the vitamins that may be present in the ration from other ingredients.

VITAMIN STABILITY

Vitamins are biologically active chemicals that are sensitive to their environment. Many vitamins contain functional groups and double bonds that are liable to react with oxygen. Such a reaction would destroy the vitamin. When vitamin-fortified formulations are mixed and transported, their exposure to oxygen is increased, and unless the vitamin is protected, losses may be significant.

Processes such as pelleting and extruding will result in increased destruction of vitamins (Table 10–2). To ensure that the desired level of vitamin is present in

Table 10–2
Vitamin retention percentage under three different conditions

Vitamin	Pelleting: 178°–185° F	Expanding/ Extruding: 241°–248° F	Storage with Trace Minerals and Vitamins (no choline chloride) after 3 Months	Storage with Trace Minerals and Vitamins (incl. choline chloride) after 3 Months
A beadlet cross-linked	95	65	90	84
A beadlet non-spray cong. non–cross-linked	78	46	80	50
D_3 beadlet (A/D_3)X-Link	94	87	93	86
D_3 drum dried	88	70	84	75
E acetate 50%	94	83	93	88
E alcohol	60	10	14	7
K-Menadione dimethylpyrimidinol bisulfite (MPB)	75	34	79	62
K-Menadione sodium bisulfite (MSB)	59	12	73	34
Thiamin mononitrate	88	71	90	70
Thiamin HCl	84	50	72	57
Riboflavin	91	91	92	85
Pyridoxine	90	78	91	83
B_{12}	97	87	98	95
Calcium pantothenate	91	79	96	79
Folic acid	90	67	90	77
Biotin	90	66	92	83
Niacin	91	68	90	85
Choline chloride	98	95	—	97
C-ascorbyl phosphate	94	84	90	86
C-ascorbic acid	50	10	59	58

From Coelho, 2002.

the finished feed, manufacturers of pelleted and extruded feed products must consider the cost and stability of various vitamin sources and overfortify their formulations accordingly.

Vitamin-fortified feed products stored for prolonged periods in the presence of moisture, rancid fat, acid or alkaline pH, trace minerals, choline chloride or heat will have less vitamin content than the original product. Vitamin-fortified feed products should be protected from oxygen and stored in a cool, dry location. They should be used within 3 months.

THE FAT-SOLUBLE VITAMINS

Because body fat is not turned over as often as water, the storage capability for fat-soluble vitamins is significant. Deficiencies of fat-soluble vitamins, therefore, may not immediately occur, even if the ration is devoid of vitamin. Nevertheless, in developing rations, it is best to meet daily required levels of the fat-soluble vitamins.

The mechanics of absorption of fat-soluble vitamins has much in common with the absorption of fat or triacylglycerol. In the intestine, the fat-soluble vitamins are transported to the site of absorption via the micelles. In the intestinal cell, the fat-soluble vitamin is packaged into a lipoprotein particle. The lipoproteins containing the fat-soluble vitamins are passed into the lymphatic system, and at that point are referred to as *chylomicrons.* The chylomicrons are passed into the bloodstream, where they are attacked by enzymes in the walls of the capillaries. What remains of the chylomicron after the triacylglycerols are dispensed with is called the *remnant.* It is the remnant that contains the fat-soluble vitamin. The remnant remains in the bloodstream and is picked up by the cells for their use or by the liver for storage and redistribution.

Vitamin A

Table 10–3 gives vitamin A requirements for selected livestock.

Sources

The potency of 1 IU of vitamin A is delivered by 0.3 micrograms (μg) crystalline vitamin A alcohol (retinol), or 0.344 μg of vitamin A acetate, or 0.55 μg of vitamin A palmitate. Pure forms of vitamin A may be found in animal tissues and animal products. Vitamin A alcohol, vitamin A acetate, and vitamin A palmitate may also be made synthetically.

None of these three forms of vitamin A occurs in plant material. However, fresh plant material does contain compounds that are vitamin A precursors that livestock are capable of converting to vitamin A. This conversion takes place in the intestinal cells during the absorption process. The precursors are called *carotenes* or *carotenoids* and they include alpha carotene, beta carotene, gamma carotene, and cryptoxanthin. As indicated in Table 10–4, carotenoids are converted to vitamin A with varying efficiencies, depending on the species. It is important therefore, that feedstuffs assigned an IU number based on carotenoid content be reevaluated before use with a different species of livestock.

Function and Signs of Deficiency

Vitamin A has been shown to be essential for maintenance of epithelial tissue. Because epithelial tissue is found throughout the body, many and diverse

Table 10–3
Vitamin A requirement for selected livestock

	Required (IU)	Required (IU/lb.)
Fish, channel catfish, 100-g body weight	6.667	1,008
Fish, rainbow trout, 100-g body weight	3.500	1,260
Chicken, broiler, 5 wks of age	452.00	1,667
Chicken, white egg layer, 3-lb. body weight	300.02	1,512
Pig, growing, 45-lb. body weight	1,732	655
Dog, growing, 30-lb. body weight	1,308	1,683
Cat, growing kitten, 4.2-lb. body weight	754	4,082
Rabbit, growing, 5 wks of age	937	3,183
Horse, light work, 1,100-lb. body weight	29,409	1,008
Goat, maintenance, 88-lb. body weight	1,175	742
Ewe, maintenance, 110-lb. body weight	2,736	1,244
Beef animal, growing, 800-lb. body weight	19,267	998
Dairy cow, lactating, 1,400-lb. body weight	69,854	1,276

symptoms will accompany a vitamin A deficiency. Vision, growth, reproduction, and resistance to infection will be impaired with a vitamin A deficiency. Other symptoms include lacrimation, tissue keratinization, urinary calculi, hind-legs paralysis, bone-shape abnormalities, and elevated cerebrospinal fluid pressure. As with all the fat-soluble vitamins, the body does have some storage ability and dietary deficiency will not lead immediately to deficiency symptoms.

Hypervitaminosis A

Prolonged intake of excess vitamin A has been reported to cause various problems including rough hair coat, skin cracks, tremors, hyperirritability, and bone fragility. The safe upper limit for vitamin A is 4 to 10 times the requirement for nonruminants and about 30 times the requirement for ruminants (NRC, 1987). Excessive dietary levels of carotenoid compounds are not converted to vitamin A and so do not create toxicity. Table 10–5 gives the safe upper limits for vitamin A in livestock diets as presented by the NRC (1987).

Vitamin D

Table 10–6 gives vitamin D requirements for selected livestock.

Sources

As a nutrient, vitamin D exists in two forms. Vitamin D_2 comes from plant sources and vitamin D_3 comes from animal sources. The chemical name of vitamin D_2 is *ergocalciferol.* The chemical name of vitamin D_3 is *cholecalciferol.* For poultry and fish, vitamin D_2 has very low activity, so diets requiring supplementation should include the D_3 form of the vitamin. Vitamin D_3 is found in fish oils and is made by irradiation of animal sterol.

Whether it is vitamin D_2 or vitamin D_3, vitamin D content of feedstuffs is measured in terms of potency in alleviating symptoms of rickets. The unit of potency for vitamin D_2 is the international unit (IU). The unit of potency for vitamin D_3 is the international chick unit (ICU). Both the IU and ICU are defined as the antirachitic activity of 0.025 µg of cholecalciferol. When discussing vitamin D, the IU may be applied to either vitamin D_2 or vitamin D_3 but the ICU may be applied only to vitamin D_3.

Function and Signs of Deficiency

The function of vitamin D is to facilitate mobilization, transport, absorption, and use of calcium and phosphorus in concert with the actions of the hormones of the thyroid and parathyroid glands. A deficiency causes a disturbance in calcium and phosphorus absorption and metabolism. Deficiency symptoms in young animals result in insufficient bone mineralization leading to rickets.

Table 10–4
Conversion efficiencies from beta carotene to International Units of vitamin A

Species	Number of IU converted from 1 mg Beta Carotene
Poultry	1667
Swine	267
Ruminants	400
Horses	400
Rabbits	400 (Not confirmed in controlled studies)
Dogs	Capable of conversion but efficiency unknown
Cats	Lack the ability to convert beta carotene to vitamin A
Fish	Capable of conversion but efficiency unknown

Table 10–5
Safe upper limits of vitamins in livestock rations for long term feeding[1]

	Fish	Chickens	Swine	Dogs	Cats	Rabbits	Horses	Goats	Sheep	Beef Animals	Dairy Cows
Vitamin A	4×	4×	4×	4×	4×	4×	4×	30×	30×	30×	30×
Vitamin D (D_3 for fish & chickens, D_2 for others)	20×	4×	4×	4×	4×	4×	4×	4×	4×	4×	4×
Vitamin E	20×	20×	20×	20×	20×	20×	20×	20×	20×	20×	20×
Vitamin K	1000×	1000×	1000×	1000×	1000×	1000×	—	—	—	—	—
Vitamin C	1.11%, DM basis	—	—	—	—	—	—	—	—	—	—
Thiamin (B_1)	1000×	1000×	1000×	1000×	1000×	1000×	—	—	—	—	—
Niacin	350 mg/kg BW	350 mg/kg BW	350 mg/kg BW	350 mg/kg BW	350 mg/kg BW	350 mg/kg BW	—	—	—	—	—
Riboflavin (B_2)	20×	20×	20×	20×	20×	20×	—	—	—	—	—
Pyridoxine (B_6)	50×	50×	50×	50×	50×	50×	—	—	—	—	—
Folacin	1000×	1000×	1000×	1000×	1000×	1000×	—	—	—	—	—
Pantothenic acid	100×	100×	100×	100×	100×	100×	—	—	—	—	—
Biotin	10×	10×	10×	10×	10×	—	—	—	—	—	—
Vitamin B_{12}	300×	300×	300×	300×	300×	300×	—	—	—	—	—
Choline	10×	2×	10×	3×	10×	10×	—	—	—	—	—
Myoinositol	None	—	—	—	None	—	—	—	—	—	—

[1]Unless given otherwise, the safe upper limit is expressed as a multiple (×) of the requirement. Entries of "—" indicate that the vitamin is not supplemented in usual rations. Entries of "None" indicate that the vitamin is supplemented but toxicity has not been studied. Save upper limits for all except choline in fish, swine, cat and rabbit rations are taken from National Research Council, 1987. The safe upper limit for choline in fish, swine, cat and rabbit rations was estimated by the author.

Table 10–6
Vitamin D requirement for selected livestock

	Required (IU)	Required (IU/lb.)
Fish, channel catfish, 100-g body weight (vit D_3)	1.667	252
Fish, rainbow trout, 100-g body weight (vit D_3)	3.36	1,210
Chicken, broiler, 5 wks of age (vit D_3)	60.27	222
Chicken, white egg layer, 3-lb. body weight (vit D_3)	29.98	151
Pig, growing, 45-lb. body weight	200	76
Dog, growing, 30-lb. body weight	142	183
Cat, growing kitten, 4.2-lb. body weight	62.8	340
Rabbit, growing, 5 wks of age	140	477
Horse, light work, 1,100-lb. body weight	4,407	151
Goat, maintenance, 88-lb. body weight	241	152
Ewe, maintenance, 110-lb. body weight	277	126
Beef animal, growing, 800-lb. body weight	2,409	125
Dairy cow, lactating, 1,400-lb. body weight	19,051	348

Deficiency symptoms in adult animals result in diminished mineral content of bone. This is called *osteoporosis.* Parturient paresis (milk fever) may result from a vitamin D deficiency, primarily in dairy cows, but the disorder has also been reported in ewes and doe goats. All livestock with the possible exception of fish can make vitamin D if exposed to sunlight. As with all the fat-soluble vitamins, the body does have some storage ability and dietary deficiency will not lead immediately to deficiency symptoms.

Hypervitaminosis D

Prolonged intake of excess vitamin D can lead to calcification of the soft tissues of the body including the heart muscle, kidney, and lung. The safe upper limit for vitamin D_3 is 4 to 10 times the requirement (NRC, 1987). Vitamin D_3 is 10 to 20 times more toxic than is vitamin D_2 so the upper limit for vitamin D_2 would be 40 to 100 times the requirement. Table 10–5 gives the safe upper limits for vitamin D in livestock diets as presented by the NRC (1987).

Vitamin E

Table 10–7 gives vitamin E requirements for selected livestock.

Sources

The potency of 1 IU of vitamin E is delivered by 1 mg DL-α-tocopheryl acetate, 0.735 mg D-α-tocopheryl acetate, 0.671 mg D-α-tocopherol, or 0.909 mg DL-α-tocopherol. Natural sources of vitamin E are vegetable oils, but vitamin E is generally supplemented to the level required using one of the tocopheryl or tocopherol compounds.

Functions and Signs of Deficiency

Adequate vitamin E is necessary for normal reproductive function in male and female animals. Vitamin E functions as an antioxidant to protect the lipids in cell membranes from oxidation. A deficiency of this antioxidant results in numerous and varied symptoms including muscle degeneration, anemia, digestive disorders, impaired immune function, liver problems, and sudden death. Vitamin E and other antioxidant vitamins are sometimes added to feed to function as antioxidants and prevent the degradation of unsaturated fats. It is important to

Table 10–7
Vitamin E requirement for selected livestock

	Required (IU)	Required (IU/lb.)
Fish, channel catfish, 100-g body weight	0.1667	25.2
Fish, rainbow trout, 100-g body weight	0.07	25.2
Chicken, broiler, 5 wks of age	3.0133	11.11
Chicken, white egg layer, 3-lb. body weight	0.5004	2.52
Pig, growing, 45-lb. body weight	14.7	5.5
Dog, growing, 30-lb. body weight	7.89	10.16
Cat, growing kitten, 4.2-lb. body weight	2.51	13.6
Rabbit, growing, 5 wks of age	7.82	26.55
Horse, light work, 1,100-lb. body weight	1,178	40
Goat, maintenance, 88-lb. body weight	10.8	6.8
Ewe, maintenance, 110-lb. body weight	15.0	6.8
Beef animal, growing, 800-lb. body weight	438	22.7
Dairy cow, lactating, 1,400-lb. body weight	508	9.3

Table 10–8
Vitamin K requirement for selected livestock[1]

	Required (mg)	Required (mg/kg)
Fish, channel catfish, 100-g body weight	unknown	unknown
Fish, rainbow trout, 100-g body weight	unknown	unknown
Chicken, broiler, 5 wks of age	0.0683	0.56
Chicken, white egg layer, 3-lb. body weight	0.0500	0.56
Pig, growing, 45-lb. body weight	0.67	0.56
Dog, growing, 30-lb. body weight	unknown	unknown
Cat, growing kitten, 4.2-lb. body weight	0.01	0.10
Rabbit, growing, 5 wks of age	0.3122	2.34
Horse, light work, 1,100-lb. body weight	0	0
Goat, maintenance, 88-lb. body weight	0	0
Ewe, maintenance, 110-lb. body weight	0	0
Beef animal, growing, 800-lb. body weight	0	0
Dairy cow, lactating, 1,400-lb. body weight	0	0

[1]Animals with a requirement of 0 do not require a dietary source of vitamin K because microbial synthesis of vitamin K in the digestive tract is adequate to meet the requirement. This assumes the animal is healthy.

know that as they protect unsaturated fats, these vitamins are consumed and will not be present at the levels in the original formulation.

Vitamin E shares its antioxidant function with selenium, and these nutrients appear to have a mutual sparing effect on one another. Animals receiving diets that are marginal in either vitamin E or selenium will respond to additional amounts of the other nutrient. The nature of the relationship, however, has not been quantified.

Hypervitaminosis E

Excess dietary vitamin E is rare in the feed and nutrition industries because the vitamin is so expensive. Table 10–5 gives the safe upper limits for vitamin E in livestock diets as presented by the NRC (1987).

Vitamin K

Table 10–8 gives vitamin K requirements for selected livestock.

Sources

Unlike the other fat-soluble vitamins, vitamin K requirement is not expressed in IU but rather in milligrams. Sources of vitamin K are found naturally in both animal tissues and fresh plant material. Synthetic vitamin K is identified by the chemical name *menadione.* Menadione is extremely unstable, so pure vitamin K is not utilized in the feed industry. There are numerous forms of vitamin K that have improved stability, all of which are formed by reaction with sodium bisulfite or derivatives of sodium bisulfite. The vitamin K requirement is given as menadione equivalent, so it is important to use the menadione content of the vitamin K source to meet the requirement. Three examples of menadione sources are menadione sodium bisulfite (MSB) (50 percent menadione); menadione sodium bisulfite complex (MSBC) (33 percent menadione); and menadione dimethylpyrimidinol bisulfite (MPB) (45 percent menadione).

Functions and Signs of Deficiency

Vitamin K is used in the body's blood-clotting mechanism. A deficiency results in reduced ability to form clots, and may result in unchecked internal hemor-

Table 10–9
Thiamin requirement for selected livestock[1]

	Required (mg)	Required (mg/kg)
Fish, channel catfish, 100-g body weight	0.0033	1.11
Fish, rainbow trout, 100-g body weight	0.0014	1.11
Chicken, broiler, 5 wks of age	0.2460	2.00
Chicken, white egg layer, 3-lb. body weight	0.0700	0.78
Pig, growing, 45-lb. body weight	1.33	1.11
Dog, growing, 30-lb. body weight	0.35	0.99
Cat, growing kitten, 4.2-lb. body weight	0.42	5.00
Rabbit, growing, 5 wks of age	0.3122	2.34
Horse, light work, 1,100-lb. body weight	0	0
Goat, maintenance, 88-lb. body weight	0	0
Ewe, maintenance, 110-lb. body weight	0	0
Beef animal, growing, 800-lb. body weight	0	0
Dairy cow, lactating, 1,400-lb. body weight	0	0

[1]Animals with a requirement of 0 do not require a dietary source of thiamin because microbial synthesis of thiamin in the digestive tract is adequate to meet the requirement. This assumes the animal is healthy.

rhages and death. The microbes living in animals' digestive systems make vitamin K, and without a source of dietary vitamin K, antibiotic therapy may result in vitamin K deficiency. Dicumarol, the active ingredient in some rat poisons, kills rats by virtue of its anti–vitamin K activity. Dicumarol is also produced by a mold that sometimes grows in hay made from sweet clover.

Hypervitaminosis K

Animals tolerate large excesses of this vitamin and toxicity is not a problem. Table 10–5 gives the safe upper limits for vitamin K in livestock diets as presented by the NRC (1987).

THE WATER-SOLUBLE VITAMINS

Because these vitamins are soluble in water, the body has the capability of disposing of any excess through activities at the kidneys. This means that problems due to excesses are unlikely, but it also means that daily intake is essential.

The water-soluble vitamins include vitamin C and the B vitamins. A numbering system for the B vitamins has been established but this system has not been well accepted in its entirety, so some of the B vitamins are referred to by their number in addition to their name, but others are not.

The original "vitamin B" is now known to include a number of specific chemical compounds now referred to as the *B complex.* These include thiamin (B_1), riboflavin (B_2), nicotinic acid or niacin (B_3), pantothenic acid (B_5), pyridoxine (B_6), biotin (B_7), folic acid, folacin, or folate (B_9), cyanocobalamin (B_{12}), choline, and myoinositol.

Thiamin (B_1)

Table 10–9 gives thiamin requirements for selected livestock.

Sources

Thiamin is usually supplemented in the form of thiamin mononitrate, which is 91.9 percent thiamin. Cereal grains contain significant amounts of thiamin and diets containing cereal grains will contain adequate thiamin for most animals. Microorganisms inhabiting the digestive systems of livestock synthesize thiamin.

Functions and Signs of Deficiency

Thiamin is involved in nervous system function and it participates in the production of adenosine triphosphate (ATP) from organic molecules (Figure 10–1). Deficiency symptoms include appetite suppression and digestive problems, reduced weight gain, nervous disorders, and heart-muscle degeneration. Polioencephalomalacia (PEM) or "blind staggers" is a disorder in ruminants that may be caused by a thiamin deficiency. PEM may also be caused by excessive sulfur intake and is discussed in Chapter 9 under Sulfur. In the chicken, reproductive function is impaired by thiamin deficiency.

Raw fishmeal contains thiaminase, an enzyme that destroys thiamin after contact for a prolonged period of time. A thiamin deficiency could, therefore, be created in animals fed diets containing fishmeal. Since fish diets often contain more than 50 percent fishmeal, fishmeal treatment, feed freshness, and adequate thiamin fortification are especially critical.

Hypervitaminosis Thiamin

Table 10–5 gives the safe upper limits for thiamin in livestock diets as presented by the NRC (1987).

Riboflavin (B_2)

Table 10–10 gives riboflavin requirements for selected livestock.

Sources

Riboflavin is usually supplemented in the pure form. This water-soluble vitamin, along with vitamin B_{12}, are the ones most likely to be deficient in unsupplemented diets made up of ingredients normally fed to herbivores. Much of the riboflavin that exists in plant material has low bioavailability for most species. Riboflavin is made by microbes inhabiting the digestive tracts of livestock.

Table 10–10
Riboflavin requirements for selected livestock[1]

	Required (mg)	Required (mg/kg)
Fish, channel catfish, 100-g body weight	0.0300	10.00
Fish, rainbow trout, 100-g body weight	0.0056	4.44
Chicken, broiler, 5 wks of age	0.4921	4.00
Chicken, white egg layer, 3-lb. body weight	0.2500	2.78
Pig, growing, 45-lb. body weight	3.33	2.78
Dog, growing, 30-lb. body weight	0.88	2.50
Cat, growing kitten, 4.2-lb. body weight	0.34	4.00
Rabbit, growing, 5 wks of age	0.938	7.02
Horse, light work, 1,100-lb. body weight	0	0
Goat, maintenance, 88-lb. body weight	0	0
Ewe, maintenance, 110-lb. body weight	0	0
Beef animal, growing, 800-lb. body weight	0	0
Dairy cow, lactating, 1,400-lb. body weight	0	0

[1]Animals with a requirement of 0 do not require a dietary source of riboflavin because microbial synthesis of riboflavin in the digestive tract is adequate to meet the requirement. This assumes the animal is healthy.

Function and Signs of Deficiency

Riboflavin is a cofactor for many enzyme systems. Riboflavin participates in the production of ATP from organic molecules (Figure 10–1). Riboflavin also functions as an antioxidant. Deficiency symptoms include poor hair coat, neurological problems, and changes in blood composition. In the chicken, reproductive function is impaired by riboflavin deficiency.

Hypervitaminosis Riboflavin

Table 10–5 gives the safe upper limits for riboflavin in livestock diets as presented by the NRC (1987).

Pyridoxine (B_6)

Table 10–11 gives pyridoxine requirements for selected livestock.

Sources

The term *pyridoxine* is used to include the compounds pyridoxine, pyridoxal, and pyridoxamine. Pyridoxine is found naturally in most fresh feedstuffs. However, pyridoxine is destroyed by heat and some of the pyridoxine that exists in plant material is not bioavailable. Microorganisms inhabiting the digestive systems of livestock synthesize pyridoxine.

Functions and Signs of Deficiency

Pyridoxine is involved in the metabolism of proteins. Pyridoxine is a cofactor for many enzyme systems, including those involved in the synthesis of neurotransmitters. Pyridoxine participates in the production of ATP from organic molecules (Figure 10–1). A deficiency results in reduced growth rate, impaired immune function, fatty liver, nervous symptoms, and blood composition changes. In the chicken, reproductive function is impaired by pyridoxine deficiency.

Hypervitaminosis Pyridoxine

Table 10–5 gives the safe upper limits for pyridoxine in livestock diets as presented by the NRC (1987).

Table 10–11
Pyridoxine requirements for selected livestock[1]

	Required (mg)	Required (mg/kg)
Fish, channel catfish, 100-g body weight	0.0100	3.33
Fish, rainbow trout, 100-g body weight	0.0042	3.33
Chicken, broiler, 5 wks of age	0.4784	3.89
Chicken, white egg layer, 3-lb. body weight	0.2500	2.78
Pig, growing, 45-lb. body weight	1.33	1.11
Dog, growing, 30-lb. body weight	0.39	1.10
Cat, growing kitten, 4.2-lb. body weight	0.34	4.00
Rabbit, growing, 5 wks of age	6.249	46.78
Horse, light work, 1,100-lb. body weight	0	0
Goat, maintenance, 88-lb. body weight	0	0
Ewe, maintenance, 110-lb. body weight	0	0
Beef animal, growing, 800-lb. body weight	0	0
Dairy cow, lactating, 1,400-lb. body weight	0	0

[1]Animals with a requirement of 0 do not require a dietary source of pyridoxine because microbial synthesis of pyridoxine in the digestive tract is adequate to meet the requirement. This assumes the animal is healthy.

Table 10–12
Cyanocobalamin (vitamin B_{12}) requirements for selected livestock[1]

	Required (μg)	Required (μg/kg)
Fish, channel catfish, 100-g body weight	Unknown	Unknown
Fish, rainbow trout, 100-g body weight	0.014	11.11
Chicken, broiler, 5 wks of age	1.3668	11.11
Chicken, white egg layer, 3-lb. body weight	0.4000	4.44
Pig, growing, 45-lb. body weight	13.32	11.11
Dog, growing, 30-lb. body weight	9.1	26
Cat, growing kitten, 4.2-lb. body weight	1.68	0.02
Rabbit, growing, 5 wks of age	0	0
Horse, light work, 1,100-lb. body weight	0	0
Goat, maintenance, 88-lb. body weight	0	0
Ewe, maintenance, 110-lb. body weight	0	0
Beef animal, growing, 800-lb. body weight	0	0
Dairy cow, lactating, 1,400-lb. body weight	0	0

[1]Animals with a requirement of 0 do not require a dietary source of vitamin B_{12} because microbial synthesis of vitamin B_{12} in the digestive tract is adequate to meet the requirement. This assumes the animal is healthy and that the diet contains adequate cobalt which microbes use to make vitamin B_{12}.

Cyanocobalamin (B_{12})

Table 10–12 gives vitamin B_{12} requirements for selected livestock.

Sources

Cyanocobalamin (more common referred to as B_{12}) contains cobalt. Vitamin B_{12} does not occur in plant materials, so herbivores depend on the synthetic activities of the microbes inhabiting their digestive tracts for their supply of vitamin B_{12}. Because vitamin B_{12} contains cobalt, this micromineral must be supplied in the diet if the animal's requirement is to be met through microbial synthesis. Monogastric animals that have access to their feces can meet their B_{12} requirement through coprophagy.

Functions and Signs of Deficiency

Vitamin B_{12} is a cofactor for many enzyme systems, including those involved in protein metabolism and DNA synthesis. Vitamin B_{12} participates in the production of ATP from organic molecules (Figure 10–1). A deficiency of B_{12} results in poor hair coat, kidney damage, reduced growth, nervous disorders, and blood composition changes. In the chicken, reproductive function is impaired by B_{12} deficiency.

Vitamin B_{12} is used in the production of red blood cells. Pernicious anemia is caused by a lack of functional red blood cells due to inadequate availability of vitamin B_{12}. In humans, this form of anemia can also be caused by infection with the tapeworm *Diphyllobothrium latum.* Though pernicious anemia caused by parasites has not been documented in livestock, it is possible that nutritional deficiencies in livestock may also be caused or aggravated by worm infections. Parasite prevention programs play an essential role in maximizing feed efficiency in livestock.

Hypervitaminosis B_{12}

Table 10–5 gives the safe upper limits for vitamin B_{12} in livestock diets as presented by the NRC (1987).

Table 10–13
Pantothenic acid (pan. acid) requirements for selected livestock[1]

	Required (mg)	Required (mg/kg)
Fish, channel catfish, 100-g body weight	0.0500	16.67
Fish, rainbow trout, 100-g body weight	0.028	22.22
Chicken, broiler, 5 wks of age	1.3668	11.11
Chicken, white egg layer, 3-lb. body weight	0.2000	2.22
Pig, growing, 45-lb. body weight	10.66	8.89
Dog, growing, 30-lb. body weight	3.49	9.91
Cat, growing kitten, 4.2-lb. body weight	0.42	5.00
Rabbit, growing, 5 wks of age	3.124	23.39
Horse, light work, 1,100-lb. body weight	0	0
Goat, maintenance, 88-lb. body weight	0	0
Ewe, maintenance, 110-lb. body weight	0	0
Beef animal, growing, 800-lb. body weight	0	0
Dairy cow, lactating, 1,400-lb. body weight	0	0

[1]Animals with a requirement of 0 do not require a dietary source of pantothenic acid because microbial synthesis of pantothenic acid in the digestive tract is adequate to meet the requirement. This assumes the animal is healthy.

Myoinositol

Sources

Inositol in feedstuffs exists in several forms, but the biologically active form is myoinositol. Although there is an established myoinositol requirement for rainbow trout, catfish can apparently synthesize inositol in their livers and intestines (Burtle & Lovell, 1989), and no dietary requirement has been demonstrated for catfish. Ruminants can utilize the myoinositol found in phytic acid and do not respond to supplemental dietary myoinositol (Gerloff, Herdt, Emery, & Wells, 1984; Grummer, Armentano, & Marcus, 1987). Cats may have a requirement for myoinositol. For the remaining species of livestock, inositol does not appear to be deficient in healthy animals fed usual rations.

Functions and Signs of Deficiency

Inositol is a structural component of phospholipids that are found in cell membranes. It is involved in the metabolism and transport of lipids, and many cellular processes, including liver glycogenolysis (the conversion of glycogen stores to glucose) and insulin release from the pancreas. Of all domestic animals considered in this text, only rainbow trout have an established myoinositol requirement. Symptoms of myoinositol deficiency in rainbow trout include anorexia, poor growth, anemia, fin erosion, fatty liver, dark skin coloration, delayed gastric emptying, and decreased cholinesterase and aminotransferase activities.

Hypervitaminosis Myoinositol

There are insufficient data available to support estimates of the maximum tolerable levels of dietary myoinositol.

Pantothenic Acid (B_5)

Table 10–13 gives pantothenic acid requirements for selected livestock.

Sources

Pantothenic acid is found naturally in most fresh feedstuffs. It is usually supplemented in diets for monogastrics, however, because the level present in feedstuffs and the bioavailability is usually unknown. The most commonly used supplement for pantothenic acid is calcium pantothenate. It is a salt that exists as a D and L isomer. The D isomer of calcium pantothenate has 92 percent activity and the racemic mixture of the calcium salt has 46 percent activity. Microorganisms inhabiting the digestive systems of livestock synthesize pantothenic acid.

Functions and Signs of Deficiency

Pantothenic acid participates in the production of ATP from organic molecules (Figure 10–1). Pantothenic acid is also necessary to maintain normal skin health and is involved in fat metabolism and ketone synthesis. A deficiency results in skin problems, nervous disorders, digestive problems, and impaired immune function. In swine, a deficiency of pantothenic acid results in an abnormal gait called "goose-stepping." In the chicken, reproductive function is impaired by pantothenic acid deficiency.

Hypervitaminosis Pantothenic Acid

Table 10–5 gives the safe upper limits for pantothenic acid in livestock diets as presented by the NRC (1987).

Nicotinic Acid or Niacin (B_3)

Table 10–14 gives niacin requirements for selected livestock.

Sources

Niacin supplements include nicotinic acid and nicotinamide (niacinamide). Relative to nicotinic acid, nicatinamide is 124 percent as bioavailable for chicks (NRC, 1998). The niacin in corn, oats, wheat, and grain sorghum is in a bound form that is unavailable to monogastric animals. Niacin from these sources must not be considered in balancing rations for these animals. Except in cats,

Table 10–14
Niacin requirements for selected livestock. Values for fish, dogs and swine represent bioavailable niacin. Values for other species are requirements based on total dietary niacin.[1]

	Required (mg)	Required (mg/kg)
Fish, channel catfish, 100-g body weight	0.0467	15.56
Fish, rainbow trout, 100-g body weight	0.014	11.11
Chicken, broiler, 5 wks of age	4.1005	33.33
Chicken, white egg layer, 3-lb. body weight	0.9999	11.11
Pig, growing, 45-lb. body weight	13.32	11.11
Dog, growing, 30-lb. body weight	3.88	11.01
Cat, growing kitten, 4.2-lb. body weight	5.03	60.00
Rabbit, growing, 5 wks of age	28.12	210.5
Horse, light work, 1,100-lb. body weight	0	0
Goat, maintenance, 88-lb. body weight	0	0
Ewe, maintenance, 110-lb. body weight	0	0
Beef animal, growing, 800-lb. body weight	0	0
Dairy cow, lactating, 1,400-lb. body weight	0	0

[1]Animals with a requirement of 0 do not require a dietary source of niacin because microbial synthesis of niacin in the digestive tract is adequate to meet the requirement. This assumes the animal is healthy.

niacin can be made metabolically from a dietary excess of the amino acid tryptophan. Microorganisms inhabiting the digestive systems of livestock synthesize niacin.

Functions and Signs of Deficiency

Niacin participates in the production of ATP from organic molecules (Figure 10–1). Niacin is also necessary to maintain normal skin health. A deficiency of niacin results in reduced growth, digestive disturbances, and skin problems. In lactating dairy cattle, a high level of dietary niacin has shown to reduce plasma ketone concentration (Grummer, 1993) and may help reduce the incidence of ketosis.

Hypervitaminosis Niacin

High doses of niacin have been shown to reduce fat mobilization in transition cows and may thereby play a role in ketosis prevention (Overton, 2001). Table 10–5 gives the safe upper limits for niacin in livestock diets as presented by the NRC (1987).

Folacin, Folate, or Folic Acid (B_9)

Table 10–15 gives folacin requirements for selected livestock.

Sources

Folacin and *folate* are terms for compounds that provide folic acid activity. Most fresh feedstuffs provide sufficient folacin for most animals. Microorganisms inhabiting the digestive systems of livestock synthesize folacin. Monogastric animals that have access to their feces can meet their folacin requirement through coprophagy.

Functions and Signs of Deficiency

Folacin is involved in the manufacture of red and white blood cells. Folacin is also involved in the conversion of serine to glycine, which is of special significance in poultry because for these species, glycine is an amino acid that cannot

Table 10–15
Folacin (folate or folic acid) requirements for selected livestock[1]

	Required (mg)	Required (mg/kg)
Fish, channel catfish, 100-g body weight	0.0050	1.67
Fish, rainbow trout, 100-g body weight	0.0014	1.11
Chicken, broiler, 5 wks of age	0.0752	0.61
Chickens, white egg layer, 3-lb. body weight	0.0250	0.28
Pig, growing, 45-lb. body weight	0.40	0.33
Dog, growing, 30-lb. body weight	0.07	0.20
Cat, growing kitten, 4.2-lb. body weight	0.07	0.80
Rabbit, growing, 5 wks of age	0.1561	1.17
Horse, light work, 1,100-lb. body weight	0	0
Goat, maintenance, 88-lb. body weight	0	0
Ewe, maintenance, 110-lb. body weight	0	0
Beef animal, growing, 800-lb. body weight	0	0
Dairy cow, lactating, 1,400-lb. body weight	0	0

[1]Animals with a requirement of 0 do not require a dietary source of folacin because microbial synthesis of folacin in the digestive tract is adequate to meet the requirement. This assumes the animal is healthy.

Table 10–16
Biotin requirements for selected livestock[1]

	Required (mg)	Required (mg/kg)
Fish, channel catfish, 100-g body weight	Unknown	Unknown
Fish, rainbow trout, 100-g body weight	0.00021	0.167
Chicken, broiler, 5 wks of age	0.0205	0.17
Chicken, white egg layer, 3-lb. body weight	0.0100	0.11
Pig, growing, 45-lb. body weight	0.07	0.06
Dog, growing, 30-lb. body weight	Unknown	Unknown
Cat, growing kitten, 4.2-lb. body weight	0.01	0.07
Rabbit, growing, 5 wks of age	0	0
Horse, light work, 1,100-lb. body weight	0	0
Goat, maintenance, 88-lb. body weight	0	0
Ewe, maintenance, 110-lb. body weight	0	0
Beef animal, growing, 800-lb. body weight	0	0
Dairy cow, lactating, 1,400-lb. body weight	0	0

[1]Animals with a requirement of 0 do not require a dietary source of biotin because microbial synthesis of biotin in the digestive tract is adequate to meet the requirement. This assumes the animal is healthy.

be synthesized in amounts needed to support maximum growth. Symptoms of folacin deficiency include changes in blood composition, poor skin condition, and problems with bone and cartilage development. In the chicken, reproductive function is impaired by folacin deficiency.

Hypervitaminosis Folic Acid

Table 10–5 gives the safe upper limits for folic acid in livestock diets as presented by the NRC (1987).

Biotin (B_7)

Table 10–16 gives biotin requirements for selected livestock.

Sources

Much of the biotin in feedstuffs is bound to lysine. Its availability to the animal depends on whether the lysine is in a protein that the animal can digest. Microorganisms inhabiting the digestive systems of livestock synthesize biotin.

Functions and Signs of Deficiency

Biotin participates in the production of ATP from organic molecules (Figure 10–1). Biotin plays an important role in gluconeogenesis and fatty acid synthesis. Biotin is involved in maintenance of collagen. Symptoms of biotin deficiency include skin, eye, and foot problems. In the chicken, reproductive function is impaired by biotin deficiency. Raw egg white contains a compound —avidin—that forms an indigestible complex with biotin, and animals fed raw egg white may show signs of biotin deficiency.

Hypervitaminosis Biotin

Table 10–5 gives the safe upper limits for biotin in livestock diets as presented by the NRC (1987).

Choline

Table 10–17 gives choline requirements for selected livestock.

Table 10–17
Choline requirements for selected livestock[1]

	Required (mg)	Required (mg/kg)
Fish, channel catfish, 100-g body weight	1.332	444
Fish, rainbow trout, 100-g body weight	1.4	1,111
Chicken, broiler, 5 wks of age	136.68	1,111
Chicken, white egg layer, 3-lb. body weight	104.99	1,167
Pig, growing, 45-lb. body weight	400	333
Dog, growing, 30-lb. body weight	440	1,248
Cat, growing kitten, 4.2-lb. body weight	201	2,400
Rabbit, growing, 5 wks of age	187	1,403
Horse, light work, 1,100-lb. body weight	0	0
Goat, maintenance, 88-lb. body weight	0	0
Ewe, maintenance, 110-lb. body weight	0	0
Beef animal, growing, 800-lb. body weight	0	0
Dairy cow, lactating, 1,400-lb. body weight	0	0

[1]Animals with a requirement of 0 do not require a dietary source of choline because microbial synthesis of choline in the digestive tract is adequate to meet the requirement. This assumes the animal is healthy.

Sources

Soybean meal is rich in bioavailable choline, and livestock whose diets include soybean meal may not require supplemental choline. Lecithin is a phospholipid with varying nitrogenous components, one of which may be choline. Choline is usually supplemented as choline chloride, which has 74.6 percent choline activity. There is also a choline chloride product available to the feed industry that has a 60 percent choline activity. Choline chloride is extremely hygroscopic, meaning it absorbs moisture from the atmosphere. It should be stored in a sealed container. It is also liable to react with other vitamins in the presence of trace minerals; if prolonged storage is anticipated, choline chloride should be stored unmixed (Table 10–2). With the help of vitamin B_{12} and folacin, the liver can synthesize choline from the amino acids methionine and serine.

Functions and Signs of Deficiency

Choline is a nitrogen-containing compound that is classified as a B vitamin even though the level required is more in line with amino acids than vitamins. Choline functions in metabolism as a methylating agent and plays important roles in the formation of cell membranes and fatty acid metabolism. Choline is a component of the neurotransmitter acetylcholine. Deficiency symptoms include reduced growth, blood composition changes, and fat infiltration into the liver and kidney. In the chicken, a choline deficiency leads to perosis and impaired reproductive function. Betaine is a widely distributed compound that can be used interchangeably with choline to meet the growing animal's need for methylating agents and is, therefore, important in sparing choline. Vitamin B_{12} may also have a choline-sparing effect.

Hypervitaminosis Choline

In lactating dairy cattle, studies in which high doses of rumen-protected choline chloride were fed have suggested that choline may be a limiting nutrient for milk production (Erdman & Sharma, 1991). High doses of rumen-protected choline have also resulted in improved liver performance in early-lactation dairy cows (Piepenbrink & Overton, 2003). Table 10–5 gives the safe upper limits for choline in livestock diets as presented by the NRC (1987).

Table 10–18
Vitamin C (ascorbic acid) requirements for selected livestock[1]

	Required (mg)	Required (mg/kg)
Fish, channel catfish, 100-g body weight	0.1667	55.6
Fish, rainbow trout, 100-g body weight	0.07	55.6
Chicken, broiler, 5 wks of age	0	0
Chicken, white egg layer, 3-lb. body weight	0	0
Pig, growing, 45-lb. body weight	0	0
Dog, growing, 30-lb. body weight	0	0
Cat, growing kitten, 4.2-lb. body weight	0	0
Rabbit, growing, 5 wks of age	0	0
Horse, light work, 1,100-lb. body weight	0	0
Goat, maintenance, 88-lb. body weight	0	0
Ewe, maintenance, 110-lb. body weight	0	0
Beef animal, growing, 800-lb. body weight	0	0
Dairy cow, lactating, 1,400-lb. body weight	0	0

[1]Animals with a requirement of 0 do not require a dietary source of vitamin C because synthesis of vitamin C in animal tissues is adequate to meet the requirement. This assumes the animal is healthy.

Vitamin C (Ascorbic Acid)

Table 10–18 gives vitamin C requirements for selected livestock.

Sources

Most animals can synthesize vitamin C in their tissues from glucose and related compounds. Exceptions include fish, guinea pigs, and primates.

Vitamin C is extremely labile and difficult to maintain in feed products. It is destroyed by many environmental factors. Fish nutrition presents a special challenge regarding vitamin C because for some species of fish, the feed is extruded to produce a floating feed. The extrusion process is stressful even on the most stable vitamins, so mixed feeds containing unprotected forms of vitamin C that are extruded will suffer high losses of the vitamin (Table 10–2). This situation has been addressed in four ways:

1. The feed is overfortified with vitamin C to ensure that the final product still contains enough vitamin C to meet the fish's requirement.
2. An expensive protected form of vitamin C is used that can better survive the extrusion process.
3. Supplemental vitamin C is added at the tank, raceway, or pond.
4. The feeder settles for a sinking pellet to avoid the extrusion process.

Functions and Signs of Deficiency

Vitamin C is an antioxidant. As such, it helps protect lipids in cell membranes from destruction through reaction with oxygen. It also facilitates iron absorption. It is involved in amino acid metabolism, and in the growth and maintenance of bone and collagen. Collagen is a protein that helps reinforce connective tissues throughout the body. Signs of vitamin C deficiency include structural deformities and abnormalities of supportive cartilage. Anemia and small hemorrhages have also been reported, presumably because the body's connective tissue begins to break down. Vitamin C deficiency also results in impaired wound healing, immune response, and reproductive function. In fish, it has been reported that dietary and environmental contaminants such as heavy metals (Yamamoto & Inoue, 1985) and pesticides (Mayer, Mehrle, & Crutcher, 1978) increase vitamin C requirements.

Hypervitaminosis C (ascorbic acid)

Table 10–5 gives the safe upper limits for vitamin C in livestock diets as presented by the NRC (1987).

FACTORS INFLUENCING THE VITAMIN NEEDS OF LIVESTOCK

Factors that may influence the vitamin needs of livestock under commercial production include:

- Environmental stressors leading to animal strain, infectious disease, internal and external parasites, and other conditions that lower feed intake and/or reduce intestinal absorption of vitamins
- Disturbance of intestinal microflora
- Bioavailability and/or stability of various vitamin sources in certain feedstuffs
- Interrelationship of certain vitamins with other nutrients

SUMMARY

Vitamins play essential and varied roles in animal metabolism. Livestock generally have some ability to store the fat-soluble vitamins during periods of excess intake, but the water-soluble vitamins must be ingested at required levels daily. Because vitamins are biologically active compounds, vitamins in mixed feeds will deteriorate over time. Processing of vitamin-supplemented feed may result in the destruction of the vitamins.

END-OF-CHAPTER QUESTIONS

1. Antioxidants help protect lipids in cell membranes and elsewhere in the body from destruction through reaction with oxygen. Name a fat-soluble vitamin that functions as an antioxidant. Name a B vitamin that functions as an antioxidant. Name a third vitamin that functions as an antioxidant.
2. Unlike the case with most other nutrients, it is not routine to quantify the vitamin content in feedstuffs prior to deciding on the level that must be supplemented. Explain why.
3. Fish require vitamin C. Fish feed is always processed into a pelleted form. These two facts present the manufacturer of fish feed with a special challenge. Explain why.
4. What vitamin is destroyed by prolonged contact with a compound in raw fishmeal? What term describes the deficiency of this vitamin seen in ruminants that are fed large amounts of raw fishmeal?
5. What characteristic of choline chloride leads to the destruction of other vitamins with which it might be stored?
6. What domestic animal lacks the ability to convert beta carotene into vitamin A?
7. Name the fat-soluble vitamins. Of what nutritional consequence is the fact that these vitamins are soluble in fat? Name the water-soluble vitamins. Of what nutritional consequence is the fact that these vitamins are soluble in water?
8. Name four species of domestic animals that, after maturity, do not require a dietary source of B vitamins.
9. Name two vitamins for which the safe upper limit is less than 10 times the requirement for at least some livestock species. Name two vitamins for which the safe upper limit is more than 100 times the requirement.
10. Carotenoid compounds are sources of vitamin A for most animals. Can an animal experience vitamin A toxicity (hypervitaminosis A) from the ingestion of an excess of these carotenoid compounds? Explain why or why not.

REFERENCES

Burtle, G. J., & Lovell, R. T. (1989). Lack of response of channel catfish *(Ictalurus punctatus)* to dietary myo-inositol. *Canadian Journal of Fisheries and Aquatic Sciences. 46,* 218–222.

Coelho, M. (2002). Vitamin stability in premixes and feeds, A practical approach in ruminant diets. In *Proceedings 13th annual Florida ruminant nutrition symposium.* Retrieved 12/1/2003 from: http://www.animal.ufl.edu/dairy/2002ruminantconference/coelho.pdf

Erdman, R. A., & Sharma, R. D. (1991). Effect of dietary Rumen-protected choline in lactating dairy cows. *Journal of Dairy Science. 74,* 1641–1647.

Gerloff, B. J., Herdt, T. H., Emery, R. S., & Wells, W. W. (1984). Inositol as a lipotropic agent in dairy cattle diets. *Journal of Animal Science. 59,* 806–812.

Grummer, R. R. (1993). Etiology of lipid-related metabolic disorders in periparturient dairy cows. *Journal of Dairy Science. 76,* 3882–3896.

Grummer, R. R., Armentano, L. E., & Marcus, M. S. (1987). Lactation response to short-term abomasal infusion of choline, inositol and soy lecithin. *Journal of Dairy Science. 70,* 2518–2524.

Lehninger, A. L. (1978). *Biochemistry* (2nd ed.). New York: Worth Publishers, Inc.

Mayer, F. L., Mehrle, P. M., & Crutcher, P. L. (1978). Interactions of toxaphene and vitamin C in channel catfish. *Transactions of the American Fisheries Society. 107,* 326–333.

National Research Council. (1987). *Vitamin tolerance in animals.* Washington, DC: National Academy Press.

National Research Council. (1998). *Nutrient requirements of swine* (10th revised edition). Washington, DC: National Academy Press.

Overton, T. R. (2001, November 16–17). Managing the metabolism of transition cows. In *Proceedings: Transition Cows Conference* (pp. 107–116). Canandaigua, NY.

Piepenbrink, M. S., & Overton, T. R. (2003). Liver metabolism and production of cows fed increasing amounts of rumen-protected choline during the periparturient period. *Journal of Dairy Science. 86,* 1722–1733.

Yamamoto, Y., & Inoue, M. (1985). Effects of dietary ascorbic acid and dehydroascorbic acid on the acute cadmium toxicity in rainbow trout. *Bulletin of the Japanese Society of Scientific Fisheries. 51,* 1299–1303.

CHAPTER 11

WATER

Because it is familiar and ubiquitous, water is often regarded as a bland, inert liquid; a mere space filler in living organisms. Actually, however, it is a highly reactive substance with unusual properties that distinguish it strikingly from most other common liquids.

A. L. LEHNINGER, 1978

INTRODUCTION

Livestock require a greater quantity of water than feed daily. If animals are unable to consume adequate water to meet their needs, or if the water consumed is of poor quality, production and health will decline. Given the importance of water, water quality and water consumption by livestock should be monitored.

THE REQUIREMENT FOR WATER

Water is a nutrient and an important constituent of the animal's body and of animal products. The animal's requirement for water is influenced by type of animal, environmental temperature and humidity, rate and composition of gain, pregnancy, lactation, type of diet, feed intake, and activity level.

FUNCTIONS OF WATER

The functions of water include:

- Softening feed and preparing it for maceration and passage through the digestive tube
- Playing a role in hydrolysis, whereby feed components are chemically degraded into usable compounds in the digestive tube, as well as in the cells of the body
- Serving as a carrier by transporting nutrients and hormones and allowing animals to void the waste products of metabolism
- Cooling the body through evaporation from the body's surfaces and the respiratory system

FEED, METABOLIC WATER, AND WATER CONSERVATION

Water content of feedstuffs varies. Most grains and hays contain 8 to 15 percent water, high-moisture grain contains 25 to 35 percent water, pasture and green chop contain 80 to 90 percent water, and silages contain 55 to 75 percent water.

The animal also has available metabolic water created within the body during metabolism. The water in feed and metabolic water are usually not enough to meet the water requirements of livestock, and a source of drinking water should be provided.

The chicken is very efficient at water conservation. Unlike other livestock, the chicken has no bladder for storage of urine. Instead, the urine made at the kidneys flows through ureters into the terminal part of the digestive tract to mix with feces from the digestive tract. Fluids in this area are transported back into the colon and cecum where water and salts are reabsorbed into the blood.

EVALUATING WATER INTAKE

Several factors affect the water requirement of livestock, including environmental temperature and humidity; whether or not the animal is exercising, lactating or producing other products; the age and body size of the animal; and the nature of the diet. Water content of animals and a few animal products are given in Table 11–1.

General guidelines to use in evaluating water intake are given in Table 11–2. Ideally, the water intake values in Tables 11–2a through 11–2j are met by providing free-choice or *ad libitum* water to livestock. Guidelines for waterer space for group-fed livestock are given in Table 11–3.

Table 11–1
Water composition of animals and animal products

Newborn animals	70–80%
Adult animals, thin	70%
Adult animals, fat	50%
Meat	70%
Chicken egg	65%
Whole milk (bovine)	87%

Table 11–2a
Predicted water intake by sheep

kg/day. DMI is dry matter intake daily.	$3.86 \times \text{DMI kg} - 0.99$
kg/day. T is temperature in degrees Celsius.	$[(0.18 \times T) + 1.25] \times \text{DMI kg}$

From Forbes, 1968.

Table 11–2b
Predicted water intake by goats

g/day (maintenance)	$145.6 \times \text{kg body weight}^{0.75}$
kg/day (lactation)	1.43 × kg milk produced daily in addition to maintenance
g/day (meat goats)	680 g/day

Goats at maintenance and lactating: Morand-Fehr and Sauvant, 1978. Meat goats: Devendra, 1967.

Table 11–2c
Predicted water intake by beef animals (amounts are in gallons/day unless otherwise indicated)

Feedlot steers (L/day). $-19.76 + (0.4202 \times MT) + (0.1329 \times DMI) - (6.5966 \times PP) - (1.1739 \times DS)$

Growing heifers, steers, bulls

Weight (lb.)	At 40° F	At 50° F	At 60° F	At 70° F	At 80° F	At 90° F
400	4.0	4.3	5.0	5.8	6.7	9.5
600	5.3	5.8	6.6	7.8	8.9	12.7
800	6.3	6.8	7.9	9.2	10.6	15.0

Finishing cattle

Weight (lb.)	At 40° F	At 50° F	At 60° F	At 70° F	At 80° F	At 90° F
600	6.0	6.5	7.4	8.7	10.2	14.3
800	7.3	7.9	9.1	10.7	12.3	17.4
1,000	8.7	9.4	10.8	12.6	14.5	20.6

Wintering pregnant beef cows

Weight (lb.)	At 40° F	At 50° F	At 60° F	At 70° F
900	6.7	7.2	8.3	9.7
1,100	6.0	6.5	7.4	8.7

Lactating beef cows

Weight (lb.)	At 40° F	At 50° F	At 60° F	At 70° F	At 80° F	At 90° F
900	11.4	12.6	14.5	16.9	17.9	16.2

Mature beef bulls

Weight (lb.)	At 40° F	At 50° F	At 60° F	At 70° F	At 80° F	At 90° F
1,400	8.0	8.6	9.9	11.7	13.4	19.0
1,600+	8.7	9.4	10.8	12.6	14.5	20.6

MT: Maximum temperature (in degrees Fahrenheit); DMI: dry matter intake (in kg); PP: precipitation (in cm/day); DS: percent of dietary salt.

Winchester and Morris, 1956.

Table 11–2d
Predicted water intake by dairy animals

Lactating dairy cows (kg/day).	14.3 + (1.28 × milk kg daily) + (0.32 × DM% of diet)
Dry cows (kg/day).	−10.34 + (0.2296 × DM% of diet) + (2.212 × DMI kg) daily + (0.03944 × CP% of diet)
Calves and heifers	
1st week of life, (kg/day)	1 kg
4th week of life, (kg/day)	2.5 kg

DM: Dry matter; CP: crude protein.

Lactating cows: Dahlborn, et al., 1998; dry cows: Holter and Urban, 1992; calves and heifers: Winchester and Morris, 1956.

Table 11–2e
Predicted water intake by horses

Horses, water to feed dry matter ratio	2.9-3.6 lb. water to 1 lb. feed dry matter
Lactation	Increases need by 50–70% above maintenance
Work	Increases need by 20–300% above maintenance

Horses, water to feed ratio: Fonnesbeck, 1968; horses, lactation: National Research Council, 1989; horses, working: Carlson and Mansmann, 1974.

Table 11–2f
Predicted water intake by swine

Growing/finishing pigs	2.5 kg water per kg feed
Adult swine, not lactating	80 ml/day per kg of body weight
Lactating sows	12–40 L/day
Weanling pigs 3–6 wks	0.149+(3.053 × kg dry feed intake)/day
Non-pregnant gilts	11.5 L/day
Pregnant gilts	20 L/day
Boars	10–15 L/day

Growing/finishing pigs: Cumby, 1986; adult swine, not lactating: Yang, Howard, & McFarlane, 1981; lactating swine: Lightfoot and Armsby, 1984; weanling pigs: Gill, Brooks, & Carpenter, 1986; non-pregnant gilts and pregnant gilts: Bauer, 1982; boars: Straub, Weniger, Tawfik, & Steinhauf, 1976.

Table 11–2g
Predicted water intake by chickens

Age (weeks)	Broiler Chickens (ml/day)	White Leghorn Chickens (ml/day)
1	32	29
2	69	43
3	104	—
4	143	71
5	179	—
6	214	100
7	250	—
8	286	114
9	—	—
10	—	129
11	—	—
12	—	143
13	—	—
14	—	157
15	—	—
16	—	171
17	—	—
18	—	186
19	—	—
20	—	229
	Laying hens	150–300 ml/day

National Research Council, 1994.

Table 11–2h
Predicted water intake by dogs

Growing puppy and idle adult	2–3 × weight of dry matter intake
Lactating, hot weather, exercise	4 × weight of dry matter intake

National Research Council, 1985.

Table 11–2i
Predicted water intake by cats

30 ml/lb. body weight or

Water consumption = 3 × weight of dry matter intake

Anderson, R.S. 1983. Fluid balance and diet. In *Proceedings of the 7th Kal Kan Symposium*, pp. 19–25. Ohio State University, Columbus, OH.

3 times as much water as feed dry matter	
Doe and litter	1 gallon daily

Ensminger, and Olentine, 1978.

Table 11–2j
Predicted water intake by rabbits

Table 11–3
Waterer space per head for group fed livestock

Livestock Type (reference)	Waterer Space per Head
Weaned calves—dairy or beef (Heldt)	12 in. or 1 waterer per 12–20 head
Lactating dairy cows—free stall (Grant & Keown, 1996; Jones, 1998)	24 in. per 10–15 cows or 1 waterer per 20–25 head
Finishing beef—feedlot (Taylor, 1994)	1 waterer per 20–40 head
Feeder lamb (North Dakota State U, 1996; Brevik)	1 automatic waterer per 25 head
Lamb—water bowl (North Dakota State U, 1996)	1 bowl per 50–75 lambs
Lamb—open tank (North Dakota State U, 1996)	12 inches per 25–40 lambs
Ewe—water bowl (North Dakota State U, 1996)	1 bowl per 40–50 ewes
Ewe—open tank (North Dakota State U, 1996)	12 inches per 15–25 ewes
Ewe with lambs—water bowl (North Dakota State U, 1996)	1 bowl per 40–50 ewes
Ewe with lambs—open tank (North Dakota State U, 1996)	12 in. per 15–25 ewes
Swine (Bundy, Diggins, & Christensen, 1984)	1 nipple waterer per 20 pigs
Chickens, layers (Clauer)	1 in. (waterer lip should be level with the bird's back)
Chickens, broilers (Jansen, 2001)	0.25 in.

One reason for inadequate water consumption by livestock is limited accessibility. In addition to the obvious causes for limited access to water (frozen water sources, empty containers), water access may be limited by inadequate flow rate and by a failure to adjust waterers properly. In poultry, water is taken into the beak by gravity. For birds to be able to use automatic waterers, the water source must be adjusted to a height that is level with the bird's back. Likewise, in swine, nipple waterers should be adjusted to about shoulder level in order to maximize access and reduce waste. If water quality is poor, the voluntary intake may be reduced. Stray voltage will discourage water consumption.

Stray voltage is a voltage difference between two surfaces. It could affect water consumption if the voltage difference is between the waterer and the floor. When the animal is standing on the floor and attempting to drink, it completes the electrical circuit and a current flows through its body, shocking the animal.

Inadequate water intake leads to reduced dry matter intake, reduced growth and performance, and ultimately serious health problems. On the other hand, excessive water intake may be a sign of high mineral content in the water or in the diet, or excessive dietary protein or nonprotein nitrogen.

The water intake guidelines in Table 11–2 must be compared to a measured intake. A water meter should be used. To evaluate water intake, the manager should keep track of the number and type of animals consuming water, and monitor the consumption for 10 days. Add the metered value to the calculated water from the ration(s). If the water use is 15 percent off from the predicted intake, altered water intake is likely responsible for performance and/or health problems.

What to Do If a Water-Quality Problem Is Suspected

Table 11–4 gives some characteristics of poor-quality water, the causes, and possible actions that may improve the water quality. If a water problem is suspected, the first thing to do is to check to see that waterers are functioning properly. Next, check for stray voltage. If mechanical and electrical causes are ruled out, give animals an alternative source. The fire department may be able to help. While on the alternative source, look for a consumption or performance change. Table 11–5 summarizes performance problems due to poor-quality water. Note that some contaminants may not affect palatability.

Table 11–4
Characteristics of poor-quality water

Characteristic	Cause	Solution
High bacteria count	Contaminated water; faulty well construction	Disinfect water system and correct faults or develop another source of supply
Reddish-brown color on sink	Fe, Mn, S	Install water conditioner
Black specks in water	Fe, Mn	Filter water
Water feels greasy	Fe, Mn, S	Install water conditioner or filter
Rotten egg odor	Hydrogen sulfide or bacteria	Disinfect, install water conditioner or filter
Rotten egg odor in hot water only	Defective hot water heater	Service or replace heater
Corroded valves, rusty colored water	Low pH	Treat water with soda ash
Suspended reddish slime	Iron bacteria	Disinfect well and pipes
Cloudy water following rain	Surface water in well or sediment	Reconstruct dug well or raise submersible pump

Table 11–5
Performance problems due to poor-quality water

Water Quality Item	Performance Problem
High sulfate	Loose manure, increased need for Se, Vitamin E, Cu
High magnesium	Loose manure
Hydrogen sulfide	Anemia, increased need for Se, Vitamin E, Cu
High iron	Increased need for P, Cu
Acidic water (below pH 6.5)	Acidosis, reduced milk production and milk fat, reduced growth, dry matter intake, increased incidence of disease, metabolic problems, infertility
Alkaline water (above pH 8.5)	Similar problems to acid water, problems with rumen synthesis
High copper	Liver damage
High nitrate	Reproductive problems, poor gains
High nitrite	Respiratory distress, death
High lead	Death
Bacterial contamination	Diarrhea, infections
Blue-green algae	Disease, death

From National Research Council, 1974.

The six criteria most often considered in assessing water quality are as follows.

1. Odor and taste: These are referred to as *organoleptic* properties, and are useful in assessing water quality for both domestic animals and humans.
2. pH, salinity, total dissolved solids (TDS), total dissolved oxygen, and hardness: These are referred to as *physiochemical* properties. Salinity refers to the sodium chloride content of water. TDS and total soluble salts (TSS) measure the sodium chloride, bicarbonate, sulfate, calcium, magnesium, silica, iron, nitrate, strontium, potassium, carbonate, phosphorus, boron, and fluoride content of water (National Research Council, 2001) (Table 11–6). Water hardness measures principally calcium and magnesium, but also zinc, iron, strontium, aluminum, and manganese. The primary problem associated with very hard water (Table 11–7) is that it results in the accumulation of scale in water delivery systems that may impair water availability. In dairy cattle, the hardness of the water had no effect on water intake or animal performance (Graf & Holdaway, 1952; Blosser & Soni, 1957).
3. Presence of organic chemical contaminants, heavy metals, toxic minerals: Pesticides and herbicides may come from a variety of sources, including agriculture, urban storm water runoff, and residential uses. Other organic chemical contaminants may come from poorly operated gasoline stations. Heavy metals and toxic minerals are byproducts of industrial processes and petroleum production. The presence of these substances in water makes it unsuitable for livestock. Testing is done to identify the specific substance, then that information is used to identify the source.
4. Presence of excess mineral or compounds such as nitrates, nitrites, and sulfates: Nitrate (NO_3^-) and nitrite (NO_2^-) are forms of nitrogen that may exist in water. Their source is primarily nitrogenous fertilizers leaching from septic tanks, sewage, and contamination of runoff water by animal wastes. The nitrogenous fertilizers contain nitrogen as nitrate, and the safe upper limit for livestock of nitrate-N is 100 ppm. Under some conditions, the nitrate may be reduced to nitrite. Nitrite-N is extremely dangerous, and water containing 10 ppm is the safe upper limit for livestock (Task Force on Water Quality Guidelines, 1987). Sulfates are the primary cause of water-quality problems in well water in many regions of North America (McLeese, Patience, Wolynetz, & Christison, 1991). The form of sulfate determines its toxicity; sodium sulfate is the most toxic. Other forms of sulfate in water are calcium, iron, and magnesium. Excess sulfate consumption results in diarrhea. The safe upper limit is reported as 1,000 ppm (Task Force on Water Quality Guidelines, 1987).
5. Presence of bacteria: Coliform bacteria are found in human and animal feces. A common water test is for total coliforms. Results of the total coliform test are reported as a most probable number (MPN). An MPN value of 0 is re-

Table 11–6
Evaluation of water quality livestock and poultry based on total dissolved solids

TDS (ppm)	Suitability
<1,000	No risk to livestock
1,000–2,999	Mild diarrhea (watery droppings in poultry) possible
3,000–4,999	May cause temporary refusal of water. May cause diarrhea (watery droppings in poultry) and reduced performance. For poultry, increased mortality and decreased growth is likely.
5,000–6,999	Avoid using for breeding stock. May be used where maximum performance is not required. Not acceptable for poultry.
>7,000	Consumption of this water will result in health problems and/or poor performance. Not acceptable for poultry.

From National Research Council, 1974.

Table 11–7
Water hardness

Category	Hardness (mg of calcium carbonate equivalent per liter)
Soft	0–60
Moderately hard	61–120
Hard	121–180
Very hard	>180

ported as satisfactory, 1–8 is reported as unsatisfactory, over 8 is reported as unsafe. These classifications are based on water for human consumption. The effect of coliform bacteria in water on the health of ruminants is unknown. While the presence of coliform bacteria in water indicates surface contamination, the presence of other types of bacteria in water may suggest other problems.

6. Presence of radioactive contaminants: Radiation can be the result of the decay of natural deposits or the result of oil and gas production and mining activities. If detected, the level of radiation in the water should be quantified to determine if it poses a threat.

What to Do If a Water-Quality Problem Is Confirmed

If a water-quality problem is confirmed, the manager must then begin the time-consuming and expensive process of identifying the source of the contamination. At this point, it is best to consult with a specialist.

Routine Water Tests

Routine water testing is essential to monitor quality before it becomes a problem. Testing should be done a minimum of once per year and the routine tests should be done for pH, nitrate, coliform bacteria, and total bacteria.

TWO FINAL NOTES REGARDING WATER FOR LIVESTOCK

1. Even if you do not have and have never had a water-quality problem, it would be useful to have water quality documented in order to identify and rectify future water-quality problems that may occur.
2. Always test the water quality before buying land on which to raise livestock.

SUMMARY

Although metabolic water and the water in feedstuffs is available to help meet the water needs of livestock, supplemental drinking water is necessary. In most cases, livestock should have free access to water at all times. Water consumption guidelines, as well as some of the methods to assess water quality, are presented.

END-OF-CHAPTER QUESTIONS

1. List five factors that affect an animal's requirement for water.
2. Besides drinking water, to what other sources of water does the animal normally have access?
3. What does the presence of coliform bacteria in well water suggest?
4. Nitrate and nitrite are common water contaminants. What is the difference between nitrate and nitrite? Which is more toxic?
5. Give four situations that would limit water accessibility to livestock.

6. Define stray voltage and explain how it could affect water consumption by livestock.
7. List five characteristics of poor-quality water. For each characteristic of poor-quality water listed, give a cause and suggest a solution.
8. List five water-quality problems and the performance problems that would be expected in livestock that consume affected water.
9. Assume that a livestock operation has an animal performance problem. The operators have eliminated nutrition and other aspects of management as possible causes. Describe the procedure one would follow to determine if poor water quality is responsible for the animal performance problem.
10. How is the chicken unique among domestic animals in its ability to conserve water?

REFERENCES

Bauer, W. (1982). Der Trankwasserverbrauch guster, hochtragender und laktierender Jungsauen. (Consumption of drinking water by nonpregnant, pregnant and lactating gilts). *Archiv fur experimentelle veterinarmedizin. 36,* 823–827.

Blosser, T. H., & Soni, B. K. (1957). Comparative influence of hard and soft water on milk production of dairy cows. *Journal of Dairy Science. 40,* 1519–1524.

Brevik, T. J. (undated). Housing your flock. *Sheep production & management.* Retrieved 6/17/2005 from: http://cecommerce.uwex.edu/pdfs/A2830.pdf

Bundy, C. E., Diggins, R. V., & Christensen, V. W. (1984). *Swine production,* 5th edition. Englewood Cliffs, N.J.: Prentice-Hall, Inc.

Carlson, G. P. & Mansmann, R .A.. (1974). Serum electrolyte plasma protein alterations in horses used in endurance rides. *Journal of the American Veterinary Medicine Association. 165,* 262.

Clauer, P. J. (undated). Management Requirements for Laying Flocks. *Virginia Polytechnic Institute and State University, Virginia Cooperative* Extension. Retrieved 10/30/2003 from: http://www.ext.vt.edu/pubs/poultry/factsheets/3.html

Cumby, T. R. (1986). Design requirements of liquid feeding systems for pigs: A review. *Journal of Agricultural Engineering Research. 34,* 153–172.

Dahlborn, K., Akerlind, M., & Gustafson, G. (1998). Water intake by dairy cows selected for high or low milk-fat percentage when fed two forage to concentrate ratios with hay or silage. *Swedish Journal of Agricultural Research. 28,* 167–176.

Devendra, C. (1967). Studies in the nutrition of the indigenous goat of Malaya. 4. The free-water intake of pen-fed goats. 5. Food conversion efficiency, economic efficiency and feeding standards for goats. *Malaysian Agricultural Journal. 46,* 191.

Ensminger, M. E. & Olentine, C. G. , Jr. 1978. *Feeds and nutrition complete.* Clovis, CA: The Ensminger Publishing Company.

Fonnesbeck, P. V. (1968). Consumption and excretion of water by horses receiving all hay and hay-grain diets. *Journal of Animal Science. 27,* 1350–1356.

Forbes, J. M. (1968). The water intake of ewes. *British Journal of Nutrition. 22,* 33.

Gill, B. P., Brooks P. H. & Carpenter, J. L.. 1986. The water intake of weaned pigs from 3 to 6 weeks of age. *Animal Production.* 42:470 (Abstract).

Graf, G. C., & Holdaway, C. W. (1952). A comparison of hard and commercially softened water in the ration of lactating dairy cows. *Journal of Dairy Science. 35,* 998–1000.

Grant, R., & Keown, J. (1996). Managing dairy cattle for cow comfort and maximum intake. *Cooperative Extension, Institute of Agriculture and Natural Resources, University of Nebraska-Lincoln, NebGuide.* Retrieved 10/30/2003 from: http://www.ianr.unl.edu/pubs/dairy/g1256.htm

Heldt, J. (undated). Beef Tips-Backgrounding Opportunities. *Land O Lakes/Harvest States, Beef Feeds.* Retrieved 10/30/2003 from: http://www.beeflinks.com/backgrounding.htm

Holter, J. B., & Urban, W. E., Jr. 1992. Water partitioning and intake in dry and lactating Holstein cows. *Journal of Dairy Science. 75,* 1472–1479.

Jansen, H. (2001). Broiler lighting programs, poultry fact sheet. *Nova Scotia Department of Agriculture and Fisheries.* Retrieved 12/1/2003 from: http://www.gov.ns.ca/nsaf/elibrary/archive/lives/poultry/broilers/brolight.htm

Jones, G. A. (1998). Bottlenecks in the milk factory, the ABCs of cow comfort. *Feed Facts, Dairy.* Retrieved 10/30/2003 from: http://www.moormans.com/dairy/DairyFF/dairyjun98/botnecks.htm

Lehninger, A. L. (1978). *Biochemistry,* 2nd ed. New York: Worth Publishers, Inc.

Lightfoot, A. L. & Armsby, A. W. (1984). Water consumption and slurry production of dry and lactating sows. *Animal Production. 38,* 541 (Abstr.)

McLeese, J. M., Patience, J. F., Wolynetz, M. S., & Christison, G. I. (1991). Evaluation of the quality of ground water supplies used on Saskatchewan swine farms. *Canadian Journal of Animal Science. 71,* 191–203.

Morand-Fehr, P. & Sauvant, D. (1978). Caprins, chap. 15. In: *Alimentation des ruminant.* Institut National de la Recherche Agronomique. Paris, France.

National Research Council. (1974). *Nutrients and toxic substances in water for livestock and poultry.* Washington, D.C.: National Academy Press.

National Research Council. (1985). *Nutrient requirements of dogs,* (revised). Washington, D.C.: National Academy Press.

National Research Council. (1989). *Nutrient requirements of horses* (5th revised edition). Washington, D.C.: National Academy Press.

National Research Council. (1994). *Nutrient requirements of poultry* (9th revised edition). Washington, D.C.: National Academy Press

National Research Council. (2001). *Nutrient requirements of dairy cattle* (7th revised edition). Washington, DC: National Academy Press.

North Dakota State University Extension Service. (1996). Space allotments. In: *Sheep pocket guide.* Retrieved 10/30/2003 from: http://www.ext.nodak.edu/extpubs/ansci/sheep/as989-6.htm

Straub, G., Weniger, J. H., Tawfik E. S. & Steinhauf, D. (1976). The effects of high environmental temperatures on fattening performance and growth of boars. *Livestock Production Science. 3,* 65–74

Task Force on Water Quality Guidelines. (1987). Livestock watering. Canadian water quality guidelines. Ottawa, Ontario: Inland Waters Directorate.

Taylor, R.E. (1994). *Beef production and management decisions,* 2nd edition. New York: MacMillan Publishing Co.

Winchester, C. F. & Morris, M. J. 1956. Water intake rates of cattle. *Journal of Animal Science. 15,* 722–740.

Yang, T. S., Howard, B. & McFarlane, W. V. (1981). Effects of food on drinking behaviour of growing pigs. *Applied Animal Ethology. 7,* 259–270.

CHAPTER 12

RATION EVALUATION AND FORMULATION

Perhaps, science will have replaced the art when the addition of totally defined nutrients, the removal of metabolic wastes, monitoring of physical and chemical conditions of growth, and harvesting of the crop have become automated.

L. S. DIAMOND, 1983

INTRODUCTION

Balanced livestock rations are mixtures of feedstuffs that, when consumed within 24 hours, will supply the animal with the proper levels of all required nutrients. Balancing or formulating profitable livestock rations requires knowing the levels of nutrients required, the nutrient content and price of available feedstuffs, and how much the animal will eat. Because the prices of many feedstuffs change daily, the use of published recipe rations is not recommended.

NUTRITIONAL PHASES IN THE PRODUCTION CYCLE

At the start of each chapter in this text dealing with feeding livestock (even-numbered chapters, 14 through 34) is a figure describing the nutritional phases in the production cycle of that species. These figures illustrate where in the production cycle the diets typically change. The reason for the change may be because the animal's nutritional needs change, but it also may be because the usual production situation involves changing access to feeds (for example, when animals are put out on pasture). It is important to note that the diagrams illustrate the animal's progression through the nutritional phases over time. There are many animal, management, and environmental factors that can increase or reduce the number of days between phases.

THE CONCEPT OF A "BALANCED NUTRIENT"

As a ration is developed, it should be kept in mind what it means for a nutrient to be balanced. To the feeder, balance means that the diet contains amounts of nutrients such that the animal's health and performance will not be constrained by the nutrient. There may also be an environmental component to the concept of balance. In some cases, a diet nutrient that is out of balance may not constrain animal health and performance, but may be excreted in large amounts and have detrimental effects on the environment.

The definition of nutrient balance implies a range of nutrient level. Usually, a balanced nutrient contains no less than the predicted requirement. The upper limit of balance, however, is problematic. Some excess is acceptable. Animal systems are capable of processing and excreting excesses of many nutrients. The upper limit of balance varies both by nutrient and by consequence. The upper limit for some minerals is established to prevent a toxic reaction in the animal. For other minerals, the upper limit may be established based on the consequences to the environment of mineral in the manure. For protein and some vitamins, the upper limit of balance may be based on the consequence to ration cost. For energy, the upper limit is established based on a consideration of body condition and future health.

In the companion files to this text, a ration nutrient outside the range of balance is highlighted in red. Keeping in mind the factors that are considered when defining balance, there may be situations where the formulator can justify levels outside the range of balance, particularly for nutrient excesses.

VOLUNTARY FEED INTAKE

In order to develop a balanced ration, one must know the animal's daily voluntary feed intake. The factors affecting voluntary feed intake are many and complex. Consider the animal variable. There may be considerable variation in daily feed intake among individual animals, particularly for companion animals. Consider the feedstuff variable. There are some feedstuffs that may impact overall intake of the ration for all individuals. Other variables that will affect voluntary feed intake are those associated with the environment and the management of the farm, ranch, or kennel.

DRY MATTER INTAKE VERSUS AS FED INTAKE

For some species of livestock—fish, poultry and swine—the usual feedstuffs considered by the National Research Council (NRC) committees are all about the same in water content—10 percent water and 90 percent dry matter. NRC committees working on these species have decided not to bother with the issue of as fed values versus dry matter values. These committees report feedstuff nutrient content, required nutrient concentrations, and feed intake values on an as fed basis. In discussions in this text that apply to all the NRC committees, reference will be made to the *dry matter/feed* intake requirement. The files associated with this text uses dry matter basis for feedstuff nutrient content, targeted nutrient concentrations, and predicted intake targets for all species.

DRY MATTER/FEED INTAKE AND THE ENERGY DENSITY OF THE RATION

Appetite is driven by the need for energy in all species of livestock, so the dry matter/feed intake of animals is dependent on the energy density of the ration. Animals will eat fewer pounds of dry matter/feed from rations of high energy density than they will from rations of low energy density because it will take less dry matter/feed from high-energy rations to meet their energy require-

Table 12–1
Dry matter/feed intake as presented in eleven NRC publications

NRC Nutrient Requirement Publication	As fed intake requirement implied in nutrient concentration, as fed basis. Energy density specified.	As fed intake requirement reported. Energy density specified.	Dry matter intake requirement implied in nutrient concentration, dry matter basis. Energy density specified.	Dry matter intake requirement reported. Energy density specified.	Dry matter intake requirement reported. Value is independent of energy density.	Dry matter intake requirement reported. Value is dynamic; calculated from prediction equation in which energy density is a variable.
Rabbit, 1977	✓					
Goats, 1981				✓		
Sheep, 1985					✓	
Cats, 1986		✓				
Dogs, 1985			✓			
Horses, 1989	✓		✓			
Fish, 1993		✓				
Poultry, 1994		✓				
Beef, 1996						✓
Swine, 1998		✓				
Dairy, 2001						✓

ments. The NRC committees have accounted for the influence of energy density on intake in different ways. The dairy and beef committees have developed formulas that predict dry matter intake (DMI) based on many variables. The energy density of the ration is one example. The goat, dog, cat, horse, fish, swine, poultry, and rabbit committees specify the ration energy density that goes with reported nutrient requirements. The sheep committee reports DMI with no consideration of ration energy density.

Table 12–1 summarizes the manner in which the NRC committees have reported predicted dry matter/feed intake requirements.

RATION EVALUATION

A ration is a combination of feedstuffs. There are at least three rations that may be discussed when evaluating rations and animal performance.

1. The ration developed by the nutritionist.
2. The ration fed by the feeder.
3. The ration consumed by the animal.

Ideally, the three are the same, but this is not always the case.

Ration evaluation usually refers to an evaluation of balance. Each of the three rations may be evaluated for balance in two ways.

1. *Evaluating the supply of nutrients.* One could compare the amounts of nutrients supplied by the ration to the amount of nutrients required by the animal. For

example, one could compare the pounds of protein, the megacalories of energy, the milligrams of trace mineral, and the IU of vitamin in the ration to the pounds, megacalories, milligrams, and IU required by the animal.

2. *Evaluating the concentration of nutrients.* One could also determine the concentration of the nutrients in the ration and compare this to the concentration required by the animal. Concentration is a calculation that involves not only the supply of nutrient but also the pounds of feed that contains the supply. Concentrations are expressed in percent, units per pound, and milligrams per kilogram (parts per million). One could compare the percent protein, the megacalories per pound of energy, the parts per million of trace mineral, and the IU per pound of vitamin in the ration to the percent, megacalories per pound, parts per million, and IU per pound required by the animal.

These two evaluations will lead to the same conclusions if the ration is indeed balanced; that is, if the ration contains the required level of nutrients and it is consumed (nothing is left uneaten), and the animal is satiated (the animal is no longer hungry).

The status of any nutrient in the ration, relative to its requirement, has three possible positions: it can be equal to the requirement, it can be in excess of the requirement, or it can be short of the requirement. This is also the case for the dry matter/feed intake requirement. The pounds of dry matter/feed in the ration may match the animal's requirement, the ration may contain more pounds of dry matter/feed than the animal can consume, the ration may be short of the dry matter/feed intake requirement with free-choice feeding, or the ration may be short of the dry matter/feed intake requirement with limit feeding. Since a calculation of nutrient concentration is made using both the nutrient supply in the ration and the dry matter/feed pounds of ration, an inaccurate prediction of dry matter/feed intake requirement will affect the evaluation of ration nutrient concentration. Table 12–2 summarizes the impact on the animal if an inaccurate prediction of dry matter/feed intake is used.

RATION FORMULATION

While ration evaluation involves examining existing rations for balance, ration formulation involves building balanced rations from available feedstuffs. Ration formulation is also called *ration balancing.*

Predicting the Dry Matter/Feed Intake Requirement to Formulate a Balanced Ration

The appetite of livestock is driven by the need for energy and it is measured as dry matter/feed intake. Dry matter/feed is not a nutrient required by the animal's tissues as are vitamins, minerals, and amino acids. Nevertheless, dry matter/feed intake is treated as a requirement in ration formulation in the same way as these nutrients. Because it is treated similarly to the nutrient requirements, the animal's predicted dry matter/feed intake is referred to as a requirement. Recall from Chapter 1 that the companion files to this text use the term *target* instead of requirement for nutrients and DMI because ranges rather than single values are used in ration balancing.

As with energy, livestock know their requirement for salt (NaCl). While it is usually best to include the salt source with the energy source, salt may be offered to livestock separately and fed on a free-choice basis. Free-choice salt products are often fortified with other nutrients at levels that ensure the

Table 12–2
The consequence of an unbalanced ration with respect to animal dry matter/feed intake. In all cases, the ration formulated contains the correct number of pounds of protein required by the animal.

Cases: Dry Matter/Feed Supplied by Formulated Ration vs. Animal Dry Matter/Feed Intake Requirement	Impact on the Animal					
	Animal Finishes Ration	Animal Is Satiated	Protein Intake = Protein Requirement	Protein Intake > Protein Requirement	Protein Intake < Protein Requirement	Animal Experiences Behavioral and/or Digestive Health Problems
Case 1: Formulated ration matches animal dry matter/feed intake requirement	✓	✓	✓			
Case 2: Formulated ration is in excess of animal dry matter/feed intake requirement		✓			✓	
Case 3: Formulated ration falls short of animal dry matter/feed intake requirement—ration offered free choice	✓	✓		✓		
Case 4: Formulated ration falls short of animal dry matter/feed intake requirement—ration is limit formulated	✓		✓			✓

animal will meet requirements for these other nutrients when it has met its salt requirement.

The importance of working with an accurate prediction of dry matter/feed intake when balancing rations cannot be overstated. This is not an easy prediction to make due to the number of variables that will influence dry matter/feed intake. There are formulas and tables that are useful in predicting DMI of livestock, but a measured value is always preferable.

Defining Performance Criteria

When formulating a ration, performance criteria must be defined because ration formulation involves constructing a ration that will supply the nutrients needed to support the performance criteria.

Companion Animals

A challenge with companion animal nutrition, like that of human nutrition, is defining appropriate performance criteria. Longevity is certainly a legitimate performance criterion, but there is little data available for any species on the level of dry matter/feed intake, protein, carbohydrate, fat, minerals, and vitamins needed for maximum longevity. The characteristics of speed, strength,

Table 12–3
Strategies used to choose performance criteria for a group of livestock

Strategy	Advantage	Disadvantage
Balance ration for the average performance potential of the group	Of all the options, this usually results in the fewest number of animals with nutrient excesses or deficiencies	Nutrient excesses for lowest-potential animals, nutrient deficiencies for highest-potential animals
Balance ration for the highest performance potential within the group	Most-productive animals will realize their potential	Nutrient excesses for all but the most productive
Balance ration for the lowest performance potential within the group	No nutrient excesses	Most-productive animals will not realize their potential

and various personality traits are other performance criteria that may be popular for companion animals, but non-nutritional factors have much more of an impact on these characteristics than does nutrition. For some types of companion animals, an important performance criterion is work. Working horses and sled dogs need to receive a ration formulated to meet the nutrient requirements that will support this performance criterion.

Aside from the performance criterion of work, companion animals are fed rations formulated to support life. This includes the physiological activities involved in maintenance, gestation, lactation, and growth. Because the performance criteria for companion animals are essentially the same today as they were when the species were first domesticated, the application of science and technology to companion animal nutrition is somewhat different than it is for livestock whose genetics have been systematically altered over time to better serve mankind. In companion animal nutrition, the focus is on preventing malnutrition rather than improving performance. This has implications on the approach to ration formulation and on the conclusions drawn from ration evaluation.

Animals Raised for Food and Fiber Production

The performance criteria for animals raised for food and fiber are targets in ration formulation. This is relatively straightforward when balancing a ration for a single animal. On farms and ranches, however, rations are formulated for groups of livestock, so performance criteria must be established for each group.

Table 12–3 gives the strategies used to choose performance criteria when balancing a ration for a group, and the advantages and disadvantages of each strategy. Farms and ranches with the ability to feed animals in multiple groups should have the ability to sort animals to create relative uniformity in performance potential within each group. In a uniform group, there will be little difference in the rations formulated based on any of the strategies described in Table 12–3. Conversely, if a single ration is to be fed to a group with widely divergent performance potential, there will be significant differences in the rations formulated based on the strategies in Table 12–3. In this case, the feeder will have to evaluate the strategies in Table 12–3 for their impact on profitability, the environment, and the animal.

LINEAR PROGRAMMING IN RATION OPTIMIZATION

When constructing a ration for livestock, it is important to think in terms of nutrients, not feedstuffs. Using different sources of nutrients (different feedstuffs), it is possible to develop multiple balanced rations that will support a given set of performance criteria. These balanced rations will have different costs. Linear programming is a method in which the least cost ration is chosen from among all possible balanced rations.

Linear programming (LP) involves using matrix algebra to find the combination of feedstuffs that will produce a balanced ration at the least cost (an optimized ration). Using LP software, the nutritionist enters constraints into the computer to establish the range of feedstuff amounts and nutrient content deemed acceptable. LP software then directs the computer to find a solution. There are many software products available to solve LP problems.

Setting Feedstuff Constraints or Restrictions

There may be some feedstuffs which, for one reason or another, need to be included in the ration at a "no less than a specified amount." A large inventory and a special feedstuff quality would be two possible reasons to ensure that a ration contains a minimum amount of a given feedstuff. Conversely, some feedstuffs need to be limited in the ration and a "maximum allowable" needs to be established. Low-palatability feedstuffs or feedstuffs with other undesirable qualities fall into this category.

Setting Nutrient Constraints or Restrictions

It usually is not possible to develop a ration in which all nutrients are balanced at 100 percent of the requirement. Neither is this necessarily a goal in developing rations for livestock. The goal is to meet requirements with a minimum amount of excesses.

Nutrient constraints are usually expressed as a percentage of the animal's requirement for the nutrient. The more the ration is constrained, the more difficult it will be to find a solution. Generally, it is best to initially constrain only dry matter/feed intake, energy, and protein, and for the herbivores, neutral detergent fiber (NDF). After getting a reasonable ration balanced for these nutrients, the other nutrients may be tackled one at a time using constraints based on the level of these nutrients in the last ration made.

Limitations of Linear Programming in Developing Least-Cost Rations

The computer will develop the cheapest possible balanced ration given the constraints placed on feedstuffs and nutrients. Unfortunately, the science of nutrition has not elucidated all the dietary factors that affect animal performance. As a result of the deficits in our knowledge, important inputs are undoubtedly lacking. The least-cost ration, therefore, may not be the ration that produces the most profitable production or the most efficient gains in the animal. Nevertheless, least-cost ration formulation can be useful in helping the nutritionist discover new alternatives when developing rations.

RATION FORMULATION BY ITERATION

Although Microsoft Excel offers an LP algorithm named *Solver*, it is not incorporated in the Excel Workbooks associated with this text. Instead, ration formulation is accomplished by iteration. In ration formulation by iteration, an initial ration is developed and the nutritionist looks for weaknesses in that ration identified by the software. The nutritionist then develops strategies for rectifying those weaknesses.

A livestock ration is a mixture of feedstuffs that is consumed by the animal within a 24-hour period. The ration components added together equal the ration. Put another way, the sum of the ration components is equal to one, or 100 percent. Nutritionists and feeders must recognize this fact because it is wrong to simply add more of a particular feedstuff to a balanced ration without removing an equal amount of dry matter from some other feedstuff(s). Doing so raises the ration component total to over one (more than 100 percent). Likewise, it is wrong to reduce the amount of a particular feedstuff without adding an equal amount of dry matter from something else because the ration total is then reduced to less than 100 percent. In either case, one is no longer working with the ration consumed by the animal, because fed free choice, the animal will consume feed at 100 percent of the dry matter/feed requirement.

When formulating a ration by iteration, it is important to follow a procedure that ensures the ration improves with each iteration. Integral to that procedure is ensuring that every iteration results in a ration that can be consumed by the animal—whenever a feedstuff is added to the ration to meet a nutrient requirement—a corresponding amount of dry matter should be removed from the ration in other feedstuffs so that one is always working with 100 percent of the ration consumed.

In general, the order in which nutrients should be balanced is:

1. Dry matter. Begin by including the feedstuff you expect to comprise the bulk of the balanced ration to the animals' dry matter requirement. For ruminants, horses, and rabbits this feedstuff will be the forage available to the feeder. For swine and poultry, it will be an energy concentrate. For cats, fish, and dogs it will usually be a protein supplement.
2. Energy. Animals eat to satisfy their need for energy. Productive ruminants, horses, and rabbits will not meet their energy requirement on forage alone.
3. Protein/first limiting amino acid. Forages and energy concentrates may not contain enough protein and essential amino acids to meet the requirements of domestic animals.
4. Phosphorus. Forages, energy concentrates, and protein supplements may not contain enough phosphorus to meet the requirements of domestic animals.

This order is useful for two reasons. First, it ensures that the least expensive feedstuffs will be included at the highest acceptable rate, thus keeping ration cost down. Second, forages, energy concentrates, protein supplements, and phosphorus supplements all contain many nutrients, so it will not be necessary to include a special feedstuff to meet each nutrient requirement. Many of the required nutrients not listed here will be balanced after balancing phosphorus. For those that are not balanced, it will be necessary to find a specific source to include in the ration. Usually, the addition of these feedstuffs will not impact balances already achieved. Use of the Goal Seek function in Microsoft Excel expedites this final phase of ration formulation by iteration.

RATION FORMULATION BY HAND CALCULATION: THE SQUARE METHOD

The square method is a technique that has been used to determine the mix proportions of two feedstuffs that meet specifications for two nutrients. For example, the two feedstuffs may be corn meal and soybean meal, and the nutrients may be energy and protein. The following thirteen steps show how the square method could be applied in development of a ration for a ewe.

Step 1—Animal information summary: 154-lb. ewe, 7th week lactation, suckling singles.

According to the sheep NRC (1985), the nutrient requirements for this ewe are:

DM (lb.)	DE (Mcal)	CP (lb.)
5.5	7.2	0.73

Step 2—Feedstuff information summary: We will assume we have feedstuffs available with the following as fed analyses (NRC, 1985):

	DM (%)	DE (Mcal/lb.)	CP (%)
Ryegrass hay, sun-cured	86	1.02	6.5
Corn grain, 54 lb./bushel	88	1.54	8.9
Soybean meal, solvent extracted	90	1.58	44.8

Step 3—Finding the amount of nutrients contributed by the forage: Based on hay inventory and experience feeding the sheep, it is decided that hay will be fed at a rate of 3.2 lb./head/day. This amount of hay provides an amount of DE and CP as calculated below:

	DM (lb.)	DE (Mcal)	CP (lb.)
3.2 lb. ryegrass hay	2.75	3.26	0.21

Step 4—Finding the amount of nutrients short of the requirement after the forage contribution: In comparing the amounts of nutrients from Step 3 to the ewe's nutrient requirements, it is apparent that hay alone will not supply nutrients at the required level. The difference is calculated below:

	DM (lbs.)	DE (Mcal)	CP (lbs.)
Nutrients still required after hay: Ewe requirements—level supplied by hay	2.75	3.94	0.52

Step 5—Choosing an energy concentration for the grain mix: A grain mix will be used to make up the shortage of nutrients. Because the level of energy is similar in corn grain and soybean meal, and because our grain mix will be comprised of some mixture of these two, we will assume our grain mix energy concentration will be a value equal to the average DE for these two feedstuffs: 1.56 Mcal/lb.

Step 6—Finding the number of pounds of grain mix needed to make up the energy shortfall:

$$\frac{3.94 \text{ Mcal}}{1.56 \text{ Mcal/lb.}} = 2.53 \text{ lb. grain mix}$$

Note that this is reasonable, given that the 3.2 lb. of hay falls short of the ewe's DMI by 2.75 lb.

Step 7—Determining the Protein Concentration That Must Be in the Grain Mix

$$\frac{0.52 \text{ lb. protein}}{2.53 \text{ lb. grain}} \times 100 = 20.55\%$$

Step 8—Finding the mixture of concentrate feedstuffs (corn grain and soybean meal) that will contain the protein concentration decided in Step 7:

Step 8a. Draw a square.

Step 8b. Enter the target percentage of protein decided on in Step 7 in the center of the square.

Step 8c. Label the top half of the square the energy half. Label the bottom half of the square the protein half. Everything that is subsequently written in the top half pertains to energy and the energy concentrate. Everything that is subsequently written in the bottom half pertains to protein and the protein supplement.

Step 8d. Write the energy concentrate's protein content outside the square at the top left corner. Write the protein supplement's protein content outside the square at the bottom left corner.

Step 9—Subtracting diagonally: Use absolute values. Write the difference between the protein in the energy concentrate and the target percent protein outside the square at the lower right. Write the difference between the protein in the protein supplement and the target percent protein outside the square at the upper right.

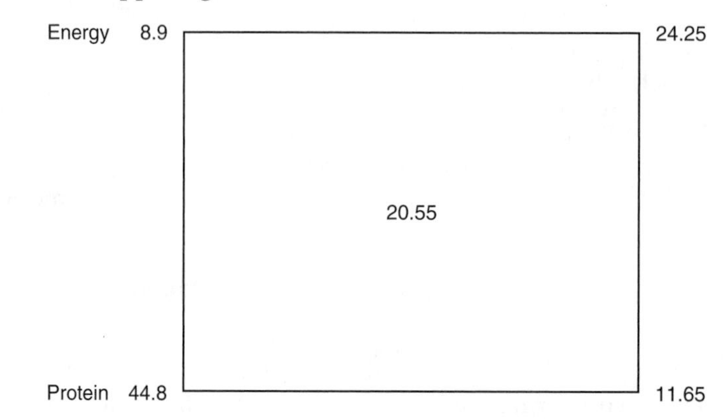

Step 10—Finding the percentage of each concentrate in the mix: Sum the values at the right side of the square. Divide each of the values at the right of the square by this sum and multiply by 100 to find the percentage of the sum represented by each value. These are the percentages corn grain and soybean meal in a mixture that contains the target percent protein.

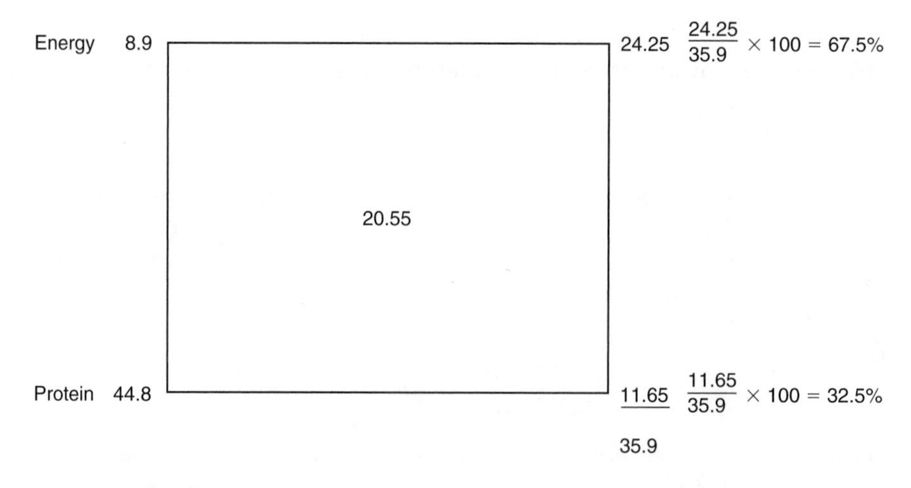

Step 11—Finding the pounds of energy concentrate and protein supplement to feed: Multiply the percent values found in Step 10 by the amount of concentrate needed to meet the energy requirement (Step 6).

67.5% of 2.53 = 1.71 lb. corn grain
32.5% of 2.53 = 0.82 lb. soybean meal

Step 12—Checking the status of the nutrients in the ration relative to the animal's requirement:

	DM (lb.)	DE (Mcal)	CP (lb.)
3.2 lb. ryegrass hay	2.75	3.26	0.21
1.71 lb. corn grain	1.50	2.63	0.15
0.82 lb. soybean meal	0.74	1.30	0.37
Total	4.99	7.19	0.73
Requirement	5.5	7.2	0.73

The total DM in this ration is within the ewe's limit for DMI. Energy and protein requirements are met.

Step 13—Finding an appropriate formulation of minerals and vitamins: For animals whose nutritional program is not closely managed, it may be acceptable to use a generic sheep mineral and vitamin supplement.

PROPRIETARY INFORMATION: THE FEED ANALYSIS VERSUS THE FEED FORMULA

Purchasers of commercial feeds should acquire the nutrient analysis of the products they buy and use this information to evaluate the ration status of all nutrients using computer software. The feed's nutrient analysis probably will not be found on the feed tag but it should be available from the feed mill that made the feed. The nutrient analysis of a commercial feed product should not be treated as proprietary information by the feed mill. Feed products are purchased for the purpose of meeting animal nutrient requirements. Feeders and/or their nutritionists need to know the nutrient content of commercial feeds in the ration in order to use them for their intended purpose.

What is appropriately considered to be proprietary information by the feed mill are the formulas of the feeds they make. Unlike the feed analysis, there is no reason to insist that the feed mill make available information about the percentage of soybean meal, linseed meal, corn, or oats in the purchased feed product.

PRESENTING THE BALANCED RATION

There are two ways to present the ration to livestock: the ration may be limit fed or livestock may receive the ration free choice. Limit feeding implies that the feeder, not the animal, has decided the amount of dry matter/feed that will be consumed. Limit feeding is sometimes called *controlled feeding* or *hand feeding.* With this method, a feeder weighs the amount of feed that the animal is to consume and the animal is given access to that amount. Portions of the ration may be presented at different times as meals, but the amount offered will be restricted to the amount weighed and fed. It is also possible to use limit feeding with groups of animals. The total feed presented is the number of animals multiplied by the prescribed feed intake per animal. A modification of limit feeding called *time-restricted meal feeding* is sometimes used with gestating sows and with companion animals. In this method, the animal is able to consume an amount of feed limited by the duration of access to the feeder.

Free-choice feeding is also called *ad lib (ad libitum)* feeding, and animals fed this way may be said to be on full feed. Free-choice feeding implies that the animal, not the feeder, will decide the amount of dry matter/feed that will be consumed. With free-choice feeding, the ration is offered to the animal(s) in a quantity greater than that amount predicted to be consumed (usually 5 to 15 percent more). The strategy here is to ensure that there is always plenty of feed available for all animals to consume whenever they take a meal. This strategy ensures that even though a given operation's livestock groups may not be perfectly uniform in performance potential, all animals in each group will meet their dry matter/feed intake requirement.

Whether limit feeding or feeding free choice, it is important not to allow feed refusals to accumulate. Always add feed to an empty, cleaned feeder.

Table 12–4 gives the advantages and disadvantages of the two feeding methods. Ultimately, farmers and ranchers will need to choose which has the greatest impact on profitability: the cost of wasted feed (free-choice feeding) or

Table 12–4
Advantages and disadvantages of two different feeding methods

Feeding Method	Advantages	Disadvantages
Limit feeding	No feed is wasted.	The dregs of the ration may not be consumed. Less aggressive animals in a group may not get their share. Animals may be limited in their performance as a result of not eating to their dry matter/feed requirement.
Free choice feeding	Every animal in the group will meet its dry matter/feed requirement and performance potential will not be limited by energy intake.	There may be a significant amount of the ration that is left uneaten.

Table 12–5
Feeder space per head for group fed livestock

Livestock Type (reference)	Feeder Space per Head (inches, unless specified otherwise)
Dairy heifers less than 6 months (Leadley & Sojda, 2000)	12
Dairy heifers greater than 6 months (Leadley & Sojda, 2000)	18
Creep feeding beef calves (Minish & Fox, 1982)	4
Weaned beef calves (Heldt)	18–26 (once/day feeding)
Weaned beef calves (Heldt)	8–12 (twice/day feeding)
Weaned beef calves (Heldt)	3–5 (fed free choice)
Lactating dairy cows (Grant & Keown, 1996; Jones)	18–30 (freestall, fed free choice)
Finishing beef, 600 lb-market (Taylor, 1994)	22–26 (feedlot, limit fed)
Finishing beef, 600 lb-market (Taylor, 1994)	6 (feedlot, fed free choice)
Feeder lamb (NDSU, 1996; Brevik)	3–4 (fed free choice)
Feeder lamb (NDSU, 1996; Brevik)	9–12 (limit fed)
Ewe (NDSU, 1996; Brevik)	8–12 (fed free choice)
Ewe (NDSU, 1996; Brevik)	16–20 (limit fed)
Ewe and lambs (NDSU, 1996)	8–12 (fed free choice)
Ewe and lambs (NDSU, 1996)	16–20 (limit fed)
Pastured horses, mature (Lawrence, 2002)	Space feeders 16–26 feet apart
Pastured horses, young (Lawrence, 2002)	Space feeders 10–16 feet apart
Piglet, 11 pounds (Iowa State University)	3 (fed free choice)
Nursery pig (Iowa State Univeristy)	1.3-2 (fed free choice)
Grower/finisher pig (Iowa State University)	2.4–3 (fed free choice)
Grower/finisher pig (Iowa State University)	1.1 × shoulder width (limit fed)
Sow (Iowa State University)	16
Chickens, layers (Clauer)	3 (feeder lip should be level with the bird's back)
Chickens, broilers (Jansen, 2001)	0:5-0.7

the missed income from animals that would have consumed more nutrients if given the opportunity (limit feeding).

Table 12–5 gives feeder-space recommendations for group-fed animals. Group-fed animals offered feed free choice do not compete for feed as intensely as they do when they are limit fed. As a result, less feeder space is required for group-fed animals fed free choice.

SUMMARY

Ration formulation involves the development of rations containing a balance of nutrients. The concept of balance is discussed. The amount of dry matter/feed the animal consumes each day is essential information in the formulation of a balanced ration. Many factors affect dry matter/feed intake by livestock. Issues that must be considered when choosing a feeding method are discussed. Nutrient requirements are established based on animal performance criteria, and the challenge of defining performance criteria for companion animals is discussed.

END-OF-CHAPTER QUESTIONS

1. Define the term *balanced ration.*
2. Give three factors that affect the animal's voluntary feed intake. Discuss the relationship between energy density of the ration and expected DMI.
3. A ration for a growing chicken (pullet), as formulated, meets the metabolizable energy requirement but includes more dry matter than the bird can eat. What can be expected from the pullet(s) fed this ration?
4. A nutrient requirement may be expressed in weight or mass units (e.g., pounds or milligrams) and in concentration units (e.g., percent or milligrams per kilogram). What two values are used to calculate the concentration requirement value?
5. The energy requirement may be expressed both in kilocalories and in kilocalories per pound. Under what conditions will a ration meet the kilocalories required but be in excess of the kilocalories per pound required? The vitamin E requirement is expressed both in international units (IU) and in IU/lb. Under what conditions will a ration meet the IU required but be short on the IU/lb. required?
6. Why is it desirable to sort animals that are to be fed into groups so that the performance potentials of all members of a given group are similar?
7. Discuss the issues involved in establishment of nutrient requirements with regard to performance criteria.
8. When balancing a ration by iteration, why should mineral nutrients be balanced after energy and protein?
9. Describe the value of the feed tag as a source of information for ration balancing.
10. Define the following feed presentation techniques: limit feeding, hand feeding, free-choice feeding, *ad libitum* feeding, time-restricted meal feeding.

REFERENCES

Brevik, T. J. (undated). Housing your flock. *Sheep production & management.* Retrieved 6/17/2005 from: http://cecommerce.uwex.edu/pdfs/A2830.pdf

Clauer, P. J. (undated). Management requirements for laying flocks. *Virginia Polytechnic Institute and State University, Virginia Cooperative Extension.* Retrieved 10/30/2003 from: http://www.ext.vt.edu/pubs/poultry/factsheets/3.html

Diamond, L. S. (1983). Lumen dwelling protozoa: *Entamoeba,* trichomonads, and *Giardia.* In *In vitro cultivation of protozoan parasites.* pp. 65–109, J.B. Jensen (editor). Boca Raton, Fl: CRC Press.

Grant, R. & Keown, J. (1996). Managing dairy cattle for cow comfort and maximum intake. *Cooperative Extension, Institute of Agriculture and Natural Resources, University of Nebraska-Lincoln, NebGuide.* Retrieved 10/30/2003 from: http://www.ianr.unl.edu/pubs/dairy/g1256.htm

Heldt, J. (undated). Beef Tips-Backgrounding Opportunities. *Land O Lakes/Harvest States Beef Feeds.* Retrieved 10/30/2003 from: http://www.beeflinks.com/backgrounding.htm

Iowa State University, College of Veterinary Medicine. (undated). Swine health management—feed. Retrieved 12/1/2003 from: http://www.vetmed.iastate.edu/departments/vdpam/swine/healthmgt/feed/feederspace.asp

Jansen, H. (2001). *Broiler lighting programs, poultry fact sheet.* Nova Scotia Department of Agriculture and Fisheries. Retrieved 12/1/2003 from: http://www.gov.ns.ca/nsaf/elibrary/archive/lives/poultry/broilers/brolight.htm

Jones, G. A. (1998). Bottlenecks in the milk factory, the ABCs of cow comfort. *Feed Facts, Dairy.* Retrieved 10/30/2003 from: http://www.moormans.com/dairy/DairyFF/dairyjun98/botnecks.htm

Lawrence, L. M. (2002). Feeding Horses. In *Livestock feeds & feeding,* 5th ed. Kellems & Church. Upper Saddle River, NJ: Prentice Hall.

Leadley, S. & Sojda, P. (2000). Feed bunk space for heifers. *Calving ease.* Retrieved 12/1/2002 from: http://www.calfnotes.com/pdffiles/CNCE1100.pdf

Minish, G. L. & Fox, D. G. (1982). *Beef production and management,* 2nd edition. Reston, VA: Reston Publishing Company, Inc.

National Research Council. (1977). *Nutrient requirements of rabbits.* Washington, D.C.: National Academy Press.

National Research Council. (1981). *Nutrient requirements of goats: Angora, Dairy, and Meat Goats in Temperate and Tropical Countries.* Washington, D.C.: National Academy Press.

National Research Council. (1985). *Nutrient requirements of dogs.* Washington, D.C.: National Academy Press.

National Research Council. (1985). *Nutrient requirements of sheep,* 6th revised edition. Washington, DC: National Academy Press.

National Research Council. (1986). *Nutrient requirements of cats.* Washington, D.C.: National Academy Press.

National Research Council. (1989). *Nutrient requirements of horses,* 5th revised edition. Washington, D.C.: National Academy Press.

National Research Council. (1993). *Nutrient requirements of fish.* Washington, D.C.: National Academy Press.

National Research Council. (1994). *Nutrient requirements of poultry,* 9th revised edition. Washington, D.C.: National Academy Press.

National Research Council. (1998). *Nutrient requirements of swine,* 10th revised edition. Washington, D.C.: National Academy Press.

National Research Council. (2000). *Nutrient requirements of beef cattle* 7th revised edition. Update. Washington, D.C.: National Academy Press.

National Research Council. (2001). *Nutrient requirements of dairy cattle,* 7th revised edition. Washington, D.C.: National Academy Press.

North Dakota State University Extension service. (1996). *Sheep pocket guide.* Retrieved 12/1/2003 from: http://www.ext.nodak.edu/extpubs/ansci/sheep/as989-6.htm

Taylor, R. E. (1994). *Beef production and management decisions,* 2nd Ed. New York: MacMillan Publishing Co.

CHAPTER 13

FEED MILLING

Numerous factors (e.g., filling the mixer beyond rated capacity; worn or altered equipment; poor mixer design; improper sequencing of ingredient additions; build-up of ingredient residues; leaking discharge gates and liquid addition systems; variations in composition and quality of ingredients; weighing errors; and postmixing segregation) are associated with variation in complete feeds. Nonetheless, inadequate mixing is implicated (probably justifiably) as the primary cause of the inadequate diet uniformity that can result in reduced animal performance and failure to comply with feed regulations . . .

TRAYLOR, BEHNKE, HANCOCK, & HINES, 1996

INTRODUCTION

The activities that take place at the feed mill include receiving ingredients, processing them individually, mixing them, processing the mix, packaging, and shipping finished feeds. Knowing the activities that take place at the feed mill gives the purchaser of feed mill products a greater understanding of the quality issues involved. In this chapter, the various feed processing methods and their effect on feed value are described.

PURCHASING

At the feed mill, ingredient cost represents 75 to 80 percent of the cost of producing a finished feed. It would be easy to take the position that a feed manufacturer should use nothing but the highest-quality ingredients, regardless of the price. This position, however, would likely lead to financial ruin for the feed mill. Successful feed manufacturers produce quality feeds using wholesome feed ingredients, but they maintain a flexibility in their purchase policies. This flexibility allows them to take advantage of opportunities to make ingredient purchases that will result in finished feeds with the greatest spread between their production cost and their value in the marketplace.

It is critical to the success of the mill that the ingredient buyer know the economic characteristics of the feedstuffs and make purchasing decisions accordingly. These economic characteristics fall into three categories, as defined in Table 13–1.

Price of Commodities

The primary economic characteristic to consider when buying commodities is price. The price of a commodity is often volatile, making the timing of purchases critical. The purchaser must know something about the factors affecting the supply and demand of the commodity to be purchased. For example, the price of whole cottonseed is lowest when supply is greatest (during the last 3 months of the year). National and international weather and politics will also have an impact on the price of commodities.

Quasi-Commodities

Although the seller of a quasi-commodity may work hard trying to differentiate his or her product from other quasi-commodities, the ingredient buyer must recognize that a primary economic characteristic of quasi-commodities will always be price. There may, however, be subtle quality differences and the purchaser must decide which of the available quasi-commodities represents the greatest value. Distillers' dried grains, for example, is produced from the distillation of grain for whiskey and fuel alcohol. There are differences in these two coproducts and the individual buyer must decide if the higher price of whiskey distillers is justified by increased value.

Table 13–1
Factors considered when making ingredient purchases at the feed mill

Ingredient Characterization	Quality	Supply	Number of Sellers	Primary Consideration(s) in Purchasing	Examples
Commodity	All sources will be of similar quality	At some price, there is always a supply available	Large	Price	Most grains and oilseed meals, whole cottonseed, wheat middlings, urea, molasses, most buffers, minerals, and vitamins
Quasi-commodity	May be some product differences	May be limited	Limited	Price, quality	Oats, roasted soybeans, amino acids, most byproducts including distillers' grains, blood meal, and tallow
Branded product	Usually are significant product differences	Limited	Few	Quality	Organic trace minerals, most additives including microbial products and rumen-protected products

Branded Products

Branded products are those few feedstuffs where price is not the primary concern in choosing a source of supply. There will be few sellers of a given branded product and the primary concern of the ingredient buyer may be to maintain a good relationship with the chosen supplier. An example of a branded product is rumen-protected methionine. The companies that manufacture this type of product use different technologies to protect the methionine from degradation in the rumen. Some of the protection technologies work better than others. Some of the technologies work too well and the methionine is protected from digestion entirely. An informed buyer will buy from a company offering the best technology.

PROCESS FLOW

At first glance, the feed mill looks like an impossibly complex manufacturing facility. A closer examination, however, reveals that all manufacturing activity at the feed mill follows a simple pattern. Feed is repetitively elevated and dropped with a processing activity in between each series. After the last series, the feed is dropped into a truck or bag. A process flow diagram illustrates the movement of feedstuffs and feed through the feed mill. It also illustrates where and how feed is conveyed through the feed mill and at what point in the flow each processing activity takes place. The cost centers that are usually considered in a process flow diagram are shown in Figure 13–1a. An example of a process flow diagram is shown in Figure 13–1b.

FEED CONVEYING EQUIPMENT

Screw Conveyors

Description: The screw of the screw conveyor runs through a trough or pipe. Material placed in the trough or pipe is moved along its length in a smooth, spiral motion by the rotation of the screw.

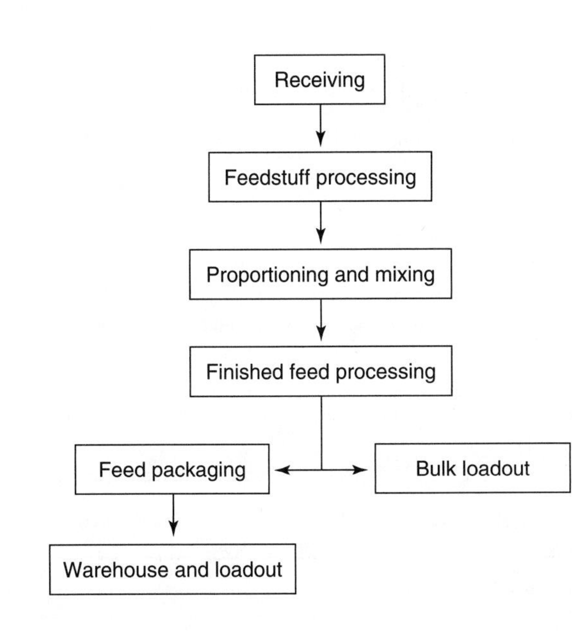

Figure 13–1a
The usual cost centers considered in the process flow diagram of a feed mill

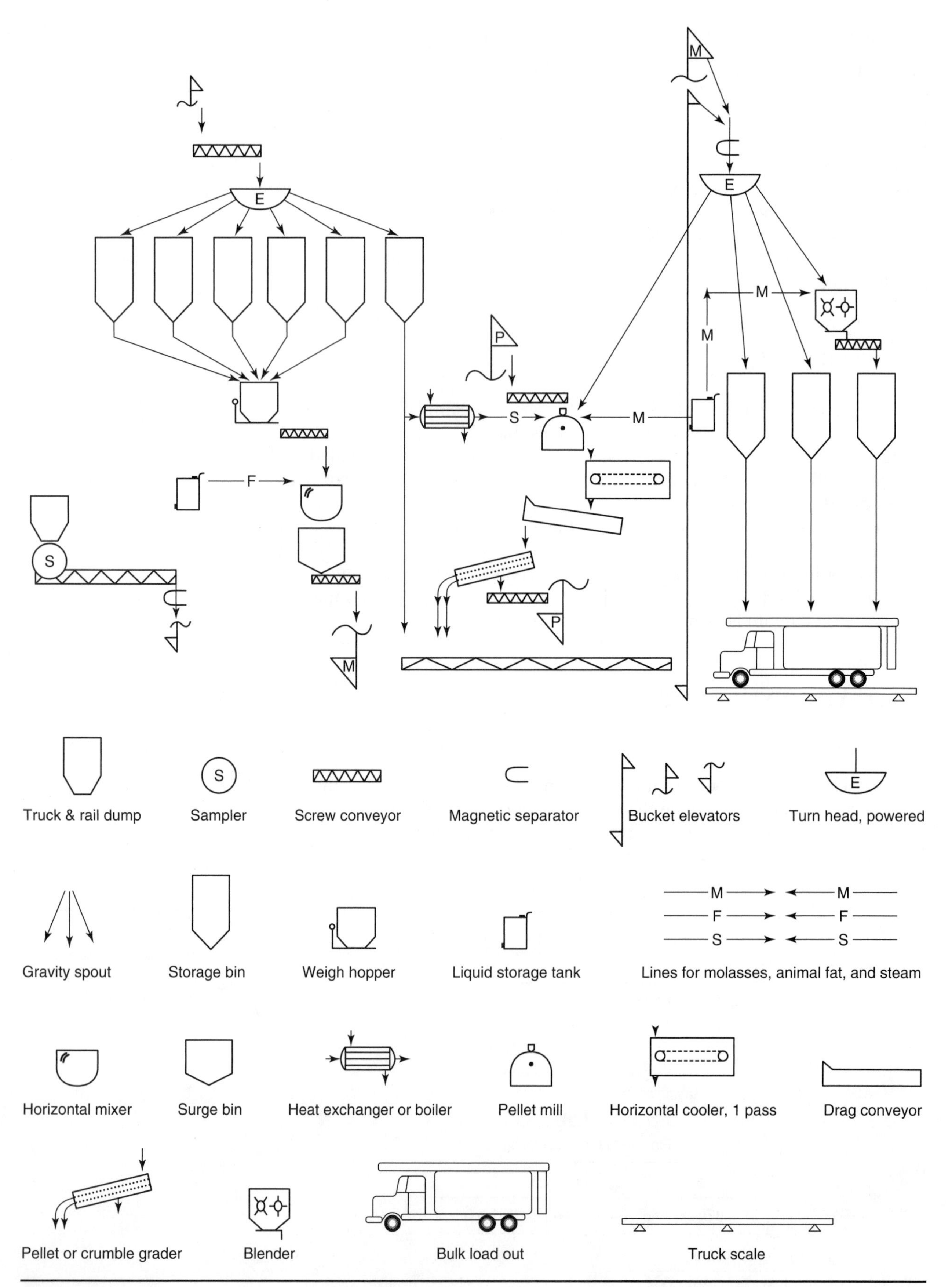

Figure 13–1b
Process flow diagram for a simple feed mill

Application: Screw conveyors perform horizontal, inclined, and sometimes vertical transit of most types of dry ingredients. Screw conveyors may be used as feeders to control the flow rate of materials. Screw conveyors are also used in in-line mixing systems where the feed is mixed as it is conveyed. There will usually be some residue left in the bottom of the trough, making a screw conveyor undesirable for medicated feeds.

Belt Conveyors

Description: Belts may be troughed or flat.

Application: Belt conveyors perform horizontal, inclined, or declined transit of dry feedstuffs. For these applications, belt conveyors are the most economical to use in terms of the cost per ton moved. Belt conveyors are gentler on pelleted feed than screw conveyors, producing fewer broken pellets. They are also used for moving bagged feed.

Drag Conveyors

Description: The drag conveyor uses a moving paddle or bar to push feed over a stationary surface. The stationary surface may be a trough or it may be flat.

Application: The drag conveyor performs horizontal and incline transit of dry feedstuffs. This type of conveyor is nearly self-cleaning and is, therefore, useful with medicated feeds.

Pneumatic Conveyors

Description: Pneumatic systems convey feedstuffs by using air flow to pick up and carry particles through a conveyor tube. Conveying air may be blown (positive pressure) or sucked (negative pressure). Pneumatic systems use more power than mechanical systems.

Application: Pneumatic conveyors perform horizontal and vertical transit of most dry feedstuffs provided that the proper air velocity is used. Blown pellets will result in a large amount of fines. Minerals are the most common feedstuff conveyed pneumatically. Dust containment will be necessary.

Bucket Elevators

Description: Bucket elevators are also called *legs.* These are the most efficient means of elevating ingredients. Bucket elevators consist of a belt or chain running in a vertical direction to which buckets are attached.

Application: Bucket elevators perform upward transit of grain, pelleted ingredients, soft ingredients, and finished feeds.

Pumps

Description: Mechanical pumps are the second most common machines in the world (after electric motors). They act by creating negative and/or positive pressure, which, coupled with valves, move liquid through a tube.

Application: Pumps perform horizontal and inclined transit of liquids.

Gravity Flow

Description: Gravity is the least expensive way to convey feedstuffs in declined transit.
Application: Gravity flow is used in declined transit. It may be used for all types of feedstuffs.

RECEIVING

Most mills are designed so that feedstuff ingredients can be received by truck or rail. The receiving shed is designed to minimize shrinkage; that is, losses of ingredients due to wind, rain, snow, and wildlife (Figure 13–2). Ingredients are dumped into a pit and elevated to storage facilities at the top of the mill. Prior to elevation, feed mills perform several operations on feedstuffs to ensure that the products that go into the mill's ingredient bins are suitable for making livestock feed (Table 13–2).

MYCOTOXINS

There are 1.6 million species of fungi or mold, and more than half of all the microorganisms living in soil are fungi (Vilgalys, 2002). Many of these fungi live inside plant seeds, roots, stems, and leaves. Mold organisms growing on feedstuffs and feeds reduce feed quantity and nutrient content. Under circumstances that are not well understood, some of these fungi produce toxins called *mycotoxins.* Although over 300 mycotoxins are known, those currently of most concern in the feed industry are:

- Aflatoxin from the genus *Aspergillus*
- Ochratoxins from the genus *Aspergillus*
- Ochratoxins from the genus *Penicillium*

Table 13–2
Quality assurance operations on incoming ingredients

Operation	Purpose
Magnets	Remove magnetic metals
Scalpers	Remove oversized chunks of product and trash
Screeners	Remove fines
Moisture tests	Check for excess moisture, which would reduce value and could lead to mold growth
Visual tests including microscopy	Verify the product; check for mold, foreign material, insects, other contaminants
Olfactory tests	Check for wholesomeness and absence of mold growth
Sample taken and kept for 90 days	With proper record-keeping, a future feed problem can be traced back to its ingredients and ingredient suppliers
Mycotoxin tests	Mycotoxins are toxins produced by molds. There are rapid tests available that can determine if mycotoxins are present

Figure 13–2
Evaluating the cost of shrinkage

Feed mills usually do a better job minimizing shrinkage than do individual feeding operations. The cost due to shrinkage must be considered along with the facilities and labor involved in handling commodities when evaluating the economics of purchasing finished feeds from feed mills vs. purchasing commodities to mix finished feeds on the farm or ranch.

- Fumonisin from the genus *Fusarium*
- Zearalenone from the genus *Fusarium*
- Trichothecenes (Deoxynivalenol [DON] or vomitoxin) from the genus *Fusarium*
- Trichothecenes (T-2 toxin) from the genus *Fusarium*

The term *ochratoxin* describes a category of chemically related mycotoxins, specific examples of which are produced by the last two genera listed above. Likewise, the term *trichothecenes* describes a category of chemically related mycotoxins, some of which are produced by the genus *Fusarium.*

The data presented in Table 13–3 give a sense of how prevalent mycotoxins are in livestock feed.

Mycotoxins produce a wide range of health and performance problems in livestock that consume them. Mycotoxin-contaminated feed may cause permanent damage to vital organs so that even after the mycotoxin contaminated feed is removed, affected animals may never return to previous health and performance levels.

Consumption by horses of grain contaminated with *Fusarium* can lead to leukoencephalomalacia, resulting in central nervous system damage and death. Zearalenone has estrogen-like characteristics and can interfere with reproductive development and performance. Aflatoxin has been shown to suppress immune function and thereby increase susceptibility to mastitis in dairy cattle (Brown, Pier, Richard, & Krogstad, 1981). Aflatoxin consumed by lactating dairy cattle is secreted into the milk, where it can become a serious public health problem. Aflatoxin is a known carcinogen and can cause birth defects in humans. Table 13–4 presents suggested limits for mycotoxins in livestock feed.

Mold growth can occur in any feed component. The focus in this section will be on the components of feed produced by the feed mill. For a discussion of mold growth in forages, see Chapter 3.

Aside from visible mold growth, symptoms of mold in feedstuffs and finished feeds include feed that needs to be pounded down from bins, occasional feed refusal by animals, slight "off" aroma of feeds, and darkening of grain ingredients.

Fundamental in preventing mold growth in feedstuffs and finished feeds are moisture control, preventing "seeding," and maximizing turnover.

Moisture Control

Moisture is the most important factor in determining if and how rapidly mold will grow in grains and grain mixtures. The moisture in grains can come from three different sources: (1) the grain mixture's component feedstuffs, (2) the feed manufacturing processes, and (3) the environment in which feed is stored.

To discourage mold growth, the average moisture content of feedstuffs should be reduced to 13.5 percent or less. It should be realized that this value is

Table 13–3
Summary of mycotoxin incidence in 586 samples of livestock feed collected in North Carolina from 1989–1993

Genus of mold responsible:	*Aspergillus*	*Fusarium*	*Fusarium*	*Fusarium*	*Fusarium*
Mycotoxin	Aflatoxin	Fumonisin	Zearalenone	Trichothecenes-Deoxynivalenol (DON or vomitoxin)	Trichothecenes-T-2 toxin
Incidence in feed	11.4%	28.4%	18%	67.5%	4.2%

From Jones, Genter, Hagler, Hansen, Mowrey, Poore, & Whitlow, (1994). Understanding and coping with effects of mycotoxins in livestock feed and forage. North Carolina State University, Department of Agricultural Research Publication number ARS-83.

Table 13–4
Suggested limits for mycotoxins in livestock feed (parts per billion)

Mycotoxin	Class of Animal	Lowest Maximum Level in Any Contaminated Feedstuff or Total Ration	Source of Recommendation
Aflatoxin	Immature animals	20 (any feedstuff)	USFDA, 2000
	Dairy animals	20 (any feedstuff)	USFDA, 2000
	Breeding beef cattle	100 (any feedstuff)	USFDA, 2000
	Finishing beef cattle	300 (any feedstuff)	USFDA, 2000
	Laying hens	100 (any feedstuff)	USFDA, 2000
	Broiler poultry	20 (total ration)	Jones, Genter, Hagler, Hansen, Mowrey, Poore, & Whitlow, 1994
	Laying hens	50 (total ration)	Jones, Genter, Hagler, Hansen, Mowrey, Poore, & Whitlow, 1994
	Breeding swine	100 (any feedstuff)	USFDA, 2000
	Finishing swine >100 lb.	200 (any feedstuff)	USFDA, 2000
	Swine, <75 lb.	20 (total ration)	Jones, Genter, Hagler, Hansen, Mowrey, Poore, & Whitlow, 1994
	Swine, 75–125 lb.	50 (total ration)	Jones, Genter, Hagler, Hansen, Mowrey, Poore, & Whitlow, 1994
	Breeding swine	50 (total ration)	Jones, Genter, Hagler, Hansen, Mowrey, Poore, & Whitlow, 1994
Fumonisin	Horses	1,000 (total ration)	USFDA, 2001
	Swine	10,000 (total ration)	USFDA, 2001
	Broiler poultry	50,000 (total ration)	USFDA, 2001
	Laying hens	15,000 (total ration)	USFDA, 2001
	Beef cattle	30,000 (total ration)	USFDA, 2001
	Sheep, goats	30,000 (total ration)	USFDA, 2001
	Dairy cattle	15,000 (total ration)	USFDA, 2001
	Catfish	10,000 (total ration)	USFDA, 2001
	Dogs and cats	5,000 (total ration)	USFDA, 2001
	Rabbits	1,000 (total ration)	USFDA, 2001
Vomitoxin (DON)	Finishing beef cattle over 4 months old	10,000 (any feedstuff)	Herrman & Trigo-Stockli, 2002
	Chickens	10,000 (any feedstuff)	Herrman & Trigo-Stockli, 2002
	Swine	300 (total ration)	Jones, Genter, Hagler, Hansen, Mowrey, Poore, & Whitlow, 1994
	Swine	5,000 (any feedstuff)	Herrman & Trigo-Stockli, 2002
	All other animals	5,000 (any feedstuff)	Herrman & Trigo-Stockli, 2002
Zearalenone	Breeding swine	100 (total ration)	Jones, Genter, Hagler, Hansen, Mowrey, Poore, & Whitlow, 1994
	Growing swine	200 (total ration)	Jones, Genter, Hagler, Hansen, Mowrey, Poore, & Whitlow, 1994
	Dairy cattle	250 (total ration)	Jones, Genter, Hagler, Hansen, Mowrey, Poore, & Whitlow, 1994
	All other animals	500 (any feedstuff)	Indiana State Chemist, 1996
T-2 toxin	Swine	200 (total ration)	Jones, Genter, Hagler, Hansen, Mowrey, Poore, & Whitlow, 1994
	All other animals	500 (any feedstuff)	Indiana State Chemist, 1996
Ochratoxin		5 (any feedstuff)	Hawk, 2005

a calculated *average*, which means that in a bin full of a given feed or feedstuff, there will be pockets where the moisture content is above the average and pockets where the moisture content is below the average. As a result, even with an acceptable average for feedstuff moisture content, it is possible that mold growth may occur in isolated areas of the storage bin. Grain bins are often aerated to dry out pockets of high moisture content.

When grains are ground, the disruption of the hull barrier makes it easier for mold to grow on the grain. Grinding followed by processing methods that use steam may encourage mold growth if the processed products are not dried properly before storage or shipment. If these processed feed products are properly dried, the feeds produced may have reduced mold counts compared to the feeds before processing. However, in contrast to the mold organisms, at least some mycotoxins are heat resistant. In any case, any improvement due to processing is temporary, and if the moisture content is excessive, pelleted feeds are actually more susceptible to mold growth than are nonpelleted feeds.

Warm air can hold more water vapor than cool air. At an air temperature of 86° F, for example, a cubic meter of air can hold 26 g of water in vapor form. If a cubic meter of 86° air actually held 26 g of water, it would be saturated in terms of water vapor. The temperature at which air is saturated with regard to water vapor is the dew point. If air is cooled to a temperature that is below the dew point, it will be forced to drop some of its water vapor. This water vapor condenses on the nearest surface as a liquid. During the normal heating and cooling of the day, the air temperature inside a feed bin may fall below the dew point (Figure 13–3). If this happens, moisture will condense on the top of grain and on the inside walls of the grain bin. These are the locations in grain bins where moldy feed is most often found. Aeration and regular bin emptying and cleaning are the primary means of preventing problems due to mold growth.

Moisture from the environment can also contaminate feed through leaks in roofs and covers of storage facilities, augers, and truck compartments.

Preventing Seeding

During its manufacture and shipment, fresh feed may come in contact with old feed that has lodged or caked in various areas of the feed conveying, processing, storage, and delivery systems. This old feed is often very moldy and may contaminate the fresher feed with mold spores. Feed mills can prevent this problem through practices and procedures that tend to minimizing lodging and caking, and through frequent cleaning and disinfesting of bulk bins, trucks, and equipment.

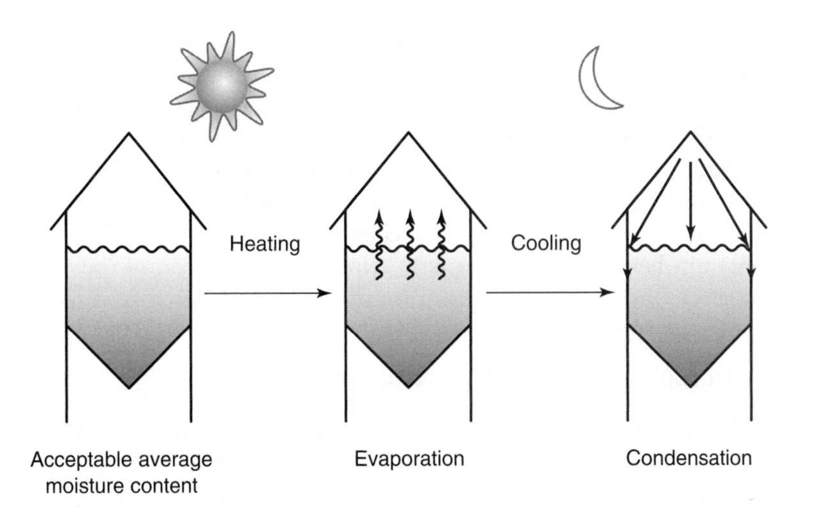

Figure 13–3
Moisture migration due to daily temperature fluctuations

Maximizing Turnover

Under perfect conditions, grain can be stored indefinitely without spoiling. However, perfect storage conditions are difficult to achieve in the feed industry. Given this fact, the longer the duration of storage, the more likely feedstuffs and feeds will be moldy. By maintaining high turnover rates, feed mills can prevent mold from gaining a foothold in their feed products. High turnover rates are achieved by closely matching purchase quantities with sales volume.

Mold Inhibitors

It is a common practice in the feed industry to use mold inhibitors, usually organic acids, to control the growth of molds in the products they sell. While the inhibitors may be effective in reducing the growth of molds, they cannot destroy mycotoxins that have already been produced. Mold inhibitor products will prevent the growth of molds for a period of time, but the concentration of these products begins to decline as soon as they are applied. Below a certain concentration, they are no longer effective in preventing mold growth. Therefore, feeds must be fed up on the farm or ranch fairly quickly if they are to be fed wholesome. How quickly depends on the status of mold growth when the feed left the feed mill, whether or not mold inhibitors were used, and the conditions under which the feed is stored and fed on the farm.

Mycotoxin Binders

Moldy feeds or treated feeds that have been moldy may contain mycotoxins. Farmers and ranchers may use these feeds by diluting them with wholesome feeds, thereby diluting mycotoxins to harmless levels. Adding binding agents may also be helpful. Most binding agents are clay compounds that reportedly act by binding the mycotoxins and carrying them harmlessly through the gut.

PROPORTIONING AND MIXING

Batch Mixing Systems

An image of a typical horizontal mixer used in most batch mixing systems is found in the file titled Images.ppt on the text's accompanying On-Line Companion. To view this file please go to http://www.agriculture.delmar.com and click on *Resources.* Click on *On-Line Resources* and select from the titles listed.

Batch mixing is the most commonly used mixing system. In batch mixing, the operator enters a customer's formula into a computer and the computer directs the proper bins to discharge the ingredients into the scale hopper. The scale hopper is the holding vessel for the scale. The computer shuts down bin discharge when the proper amount has entered the scale hopper. When the amount of all ingredients described by the formula is in the scale hopper, the product is conveyed to the mixer. Mixers usually require 3 minutes of mixing time. After the feed is mixed, the bottom of the mixer opens and the contents are discharged into a surge bin below the mixer. The surge bin is designed to catch the entire mixer contents when the mixer bottom opens. To do this, the surge bin must have 150 percent of the capacity of the mixer. The reason for using a surge bin beneath the mixer is to expedite production of the next batch. From the surge bin, the mixed feed or mash is elevated into mixed feed bins.

In-Line or Continuous Proportioning Systems

Some mills do not make feed by the batch as described above. Instead, these mills have a screw conveyor that runs below the ingredient bins. There is no scale for measuring ingredient weight in most in-line systems. Instead, calibration is done on each ingredient to determine the weight of feedstuff moved with each revolution of the drive shaft of a volumetric feeder. In the in-line system, the volumetric feeders discharge into the screw conveyor in sequence, starting with the bin containing the formula ingredient that is closest to the beginning of the screw conveyor. As the pile of feed passes beneath the next bin containing a formula ingredient, the bin is directed to discharge the appropriate amount into the screw conveyor. In this way, the screw conveyor not only conveys the feedstuffs, but also mixes them together, and after the last ingredient is added, only a small amount of final mixing is needed.

Comparison of Batch Mixing and In-Line Mixing Systems

The advantage of the in-line or continuous mixing system over the batch system is that once the formula is entered into the computer, the bins will discharge feedstuffs and the screw conveyor will mix feed nonstop. This means that it will be more efficient and therefore cheaper to make a large quantity of a given formula using the in-line mixing system as compared to the batch mixing system.

The in-line mixing system was the predominant system used in the feed industry when feed mixes were simple and few in number. However, today, with the increasing number of ingredients available and the demand for customized mixes, newer mills usually employ the batch mixing system. This is because with the batch system (compared to an in-line system), there is less time needed to transition from one formula to another. Also, to add a new ingredient to an in-line mill usually involves manually weighing out the feedstuff and adding it to the end of the screw conveyor. In a batch mixing system, it is a relatively simple project to install another bin above the mixer.

Liquid Feeds

The most common liquid feeds used in feed milling are molasses and fat. These feedstuffs present special issues when conveying and mixing.

Flowability of Liquid Feedstuffs

Many fats handled by the feed mill are solid at room temperature and must be heated for effective conveying and mixing. Fats are, therefore, usually maintained in storage at between 120° and 140° F. Molasses is traded in world markets at a standard sugar content measured as 79.5 Brix. At this sugar content, molasses is too viscous for efficient conveying and mixing, so feed mills will usually dilute molasses with water. The viscosity of molasses is also reduced with heat, and feed mills will sometimes heat the molasses to between 70° and 100° F to improve flowability.

Mixing Liquid Feedstuffs

Liquid feedstuffs may be added to the mixer along with the other ingredients, but there are other options for the feed mill. In fact, many mills avoid adding molasses to the mixer because of concern that a sticky residue of molasses may carry over from one batch to the next. The risk associated with adding fat to the mixer is that if it is not sufficiently heated, it may not mix well and will form unsightly fat balls.

One alternative method to fat and molasses addition is to spray them on cooled or warm pellets. For pelleted feed, liquids may also be added to the conditioner of the pellet mill. For mash or textured feed, liquids may be added and blended into the feed after it has been discharged from the mixer.

FEEDSTUFF PROCESSING AND FINISHED FEED PROCESSING

Images of feed processing equipment are found in the file titled Images.ppt on the text's accompanying On-Line Companion. To view this file please go to http://www.agriculture.delmar.com and click on *Resources*. Click on *On-Line Resources* and select from the titles listed.

Feedstuffs are often subjected to some type of processing before feeding to assist the digestive system in extracting the nutrients contained in them. This processing may occur on individual feedstuffs or on mixtures of feedstuffs. The methods of feedstuff processing fall into one of two categories: nonthermal and thermal.

Nonthermal Processing of Feedstuffs

The primary nonthermal method of processing feedstuffs involves particle size reduction or "grinding." The operation of grinding may be performed on raw ingredients ("pregrinding") or on mixtures of raw ingredients ("postgrinding"). Pregrinding is more commonly used in the feed industry. There are three advantages to grinding.

1. Ground feedstuffs have an increased exposed surface area for attachment by microbes and enzymes of the digestive system.
2. Ground feedstuffs have improved mixing characteristics, and mixes of ground grains are less likely to separate during handling.
3. Grinding of the grain mixture is essential for successful pelleting.

Grinding is effected through compression, impact, attrition, and shear (cutting or tearing). There are two basic types of equipment used to grind grain that employ some or all of these methods: the hammermill and the roller mill.

In the hammermill, grain is added to a feeder that drops the grain over spinning hammers. These hammers are traveling in excess of 16,000 ft./min. When they strike the grain, it shatters, producing a variety of particle sizes. Grain particles of small enough size pass through the screen around the hammers. The primary advantage of the hammermill over the roller mill is output: more tons per hour can be ground using a hammermill. The primary disadvantage of the hammermill is that it will be difficult to produce a product of uniform particle size.

In the roller mill, grain is added to a feeder that drops grain onto the nip of rotating cylinders or rolls. Different outcomes can be achieved by varying the distance between the rolls, adjusting the speed differential between the rotating rolls, using corrugated or grooved rolls, stacking pairs of rolls, and exposing grain to steam prior to rolling. The primary advantage of the roller mill over the hammermill is product appearance. Roller mills produce a ground product of uniform particle size and can be used to crimp, crack, crumble, and flake feed. The primary disadvantage of the roller mill is output: fewer tons per hour can be ground using a roller mill than could be achieved using a hammermill.

Effect of Grain Processing on Animal Performance

Because feed cost usually represents over 50 percent of the cost of raising animals, small improvements in feed utilization will result in significant improvements in farm or ranch profitability. Grinding feed nearly always improves feed utilization. However, grinding grain to excessively small particle size has been associated with ulcers in swine (Wondra et al., 1992). The desired average grain particle size will vary with the other diet ingredients (if any), the species, the age of the animal, and the range of particle sizes that are used to calculate

Table 13–5
Suggested average particle size of complete feeds for swine and poultry

Species	Suggested Average Feed-Particle Size (microns)
Swine	700
Broilers	600
Layers	800

From Goodband, Tokach, & Nelssen, undated.

the average. Table 13–5 gives some suggested values for swine and poultry based on the information in the Kansas State Cooperative Extension publication, "The Effects of Diet Particle Size on Animal Performance" (Goodband, Tokach, & Nelssen, undated).

For ruminants, the issue for the feed mill is usually not particle size of the entire diet but particle size of the grain portion of the ration: the ration may include significant amounts of feedstuffs from nonfeed mill sources. For ruminant animals, the recommended particle size of the grain portion of the ration will depend on the amount of grain in the ration and whether this grain is fed with other ration components. The presence of grain in the manure is often taken as an indication that grain should be processed to a finer particle size. Higher-producing animals will have high dry matter intakes and rapid passage rates. For these animals, the usual philosophy regarding grain-particle size is that smaller is better. However, dusty feeds may negatively affect dry matter intake, especially if the ground grain is not fed as part of a total mixed ration.

Thermal Processing of Feedstuffs

Thermal processing methods include micronizing and popping, steam rolling and flaking, roasting, extruding and expanding, and pelleting. Grain for horses, dogs, cats, rabbits, and fish is nearly always processed using one or more types of thermal processing. Heat treatment may enhance starch digestion. In monogastrics, this is desirable because raw starches that were not digested prior to entering the large intestine may ferment there. This will result in flatulence and possibly digestive upset.

Thermal Processing Used in Pet Food Manufacture: Baking and Canning

Pet foods described as *kibbles* are baked products. A dough of the formulation is made and spread out thinly onto large sheets that are placed in an oven. After baking and cooling, the dough is cut into desired shapes and packaged.

The canning process begins with a high-moisture food formulation cooked at a high temperature. The cooked mixture is then placed in a can that is lidded and sealed. The sealed can is then sterilized in a machine called a *retort.*

Popping and Micronizing

Popping and micronizing involve rupturing the endosperm of the grain by heating the grain (using dry heat or infrared energy) until the moisture within it vaporizes. The water vapor occupies more space than the liquid, and the grain explodes or pops. When fed to livestock, popped grain shows increased starch utilization over whole grain. However, an individual kernel of popped grain occupies much more space than the unpopped kernel (consider popped corn) and this presents a problem in handling and storage. This and the fact that there are less-expensive methods of processing to get equivalent improvements in starch utilization make these processing methods rather uncommon in the feed industry.

Steam Rolling and Steam Flaking

Steam contains both heat and moisture. Steam rolling and steam flaking involve exposing grain to steam prior to passing it through the rotating cylinders of a roller mill. In steam rolling, the exposure to steam is of short duration (3 to 5 minutes), whereas in steam flaking the exposure to steam is longer (15 to 30 minutes). The effect of steam exposure is to soften the grain prior to passing it through the roller mill. When compared to a hard, dry kernel, passing a soft, moist kernel through a roller mill will produce larger particles of grain. In steam flaking, these large particles are in the form of thin flakes. Some studies have shown that processing grains such as corn, sorghum, and barley using these methods results in greater grain utilization (Theurer, Huber, & Delgado-Elorduy, 1996; Owens, Secrist, Hill, & Gill, 1997; Santos, Huber, Theurer, Nussio, Tarazon, & Santos, 1999; Theurer, Lozano, Alio, Delgado-Elorduy, Sadik, Huber, & Zinn, 1999).

Roasting

Roasting as a feed processing method is most commonly applied to whole soybeans. For ruminant animals, soybeans are roasted to change the protein from a form that degrades in the rumen to a form that is undegradable in the rumen. The recommended roaster temperature for whole soybeans is 295° F. The beans stay in the roaster for about 1 minute at this temperature and then exit the roaster, but are steeped in a box without cooling for 30 minutes. This process changes the undegradable protein content of soybeans from 26 percent of the protein prior to roasting to about 50 to 60 percent of the protein after roasting (Satter, Dhiman, & Hsu, 1994).

Extruding and Expanding

Extruders and expanders are machines that expose ground feed to moisture, high pressure, and high temperatures. In both types of machines, ground feed is conveyed through a barrel in which it is exposed to steam and the forces of shearing and pressure. Table 13–6 presents a comparison of the moisture, temperature, and pressure ranges in extruders, expanders, and pellet mills.

Of major significance in the extruding and expanding process is starch gelatinization. During starch gelatinization, the crystalline structure of the starch granules is destroyed and the starch within the mix becomes an amorphous blob. From the feed mill's point of view, this is desirable because feed containing gelatinized starch can be shaped and expanded. From the feeder's point of view, gelatinization enhances the ability of starches to absorb large quantities of water, which leads to improved digestibility in almost all cases and to improved feed conversion in many cases.

There are other feed changes that may occur during the extruding and expanding processes. Gelatinization of proteins may occur. Gelatinized proteins are denatured, which may result in improved protein digestibility, but it may also result in the destruction of some amino acids. The heat generated during extruding and expanding will also change feed proteins so they will be more resistant to microbial degradation in the rumen. Together, the heat and pressure will reduce or eliminate toxins, insects, and microbes in feed. Finally, the heat

Table 13–6
Temperature and pressure characteristics of different feed-processing equipment

Equipment	Moisture Content within Barrel (percent)	Usual Operating Temperatures (° F)	Usual Operating Pressures (psi)
Pellet mill	13–20	120–190	—
Expander	15–25	200–260	300–1,000
Extruder	20–35	230–320	400–1,000

and pressure generated during extruding and expanding will destroy some vitamins and feed additives.

As the mass of feed exits the barrel of the extruder or expander, the reduced pressure results in an expansion of the mass. This expansion creates air pockets within the pellet and a reduced-density feed product is created. This has applications in aquaculture because by changing the expansion in the extruder, it is possible to produce pellets that float or sink more slowly than would those pellets produced by a pellet mill. Trout, for example, will more readily accept a floating pellet than a sinking one.

At the exit of the extruder, the feed mass is passed through a die-and-knife assembly. Through the use of various dies, unusual product shapes can be created. This has application in semimoist and dry pet foods.

Pelleting

The pellet mill is the most common type of feed processing equipment at feed mills. The pellet mill consists of a conditioner in which the mixed feed is exposed to steam (heat and moisture) and machinery that forces the material leaving the conditioner through the holes in a pellet die. A rotating knife cuts the spaghetti-like feed emerging from the die at the prescribed pellet length.

The advantages of pellets over mash fall into two categories: feed-handling characteristics and animal-performance improvements.

Feed-handling characteristics

1. Increased bulk density. More tons of pelleted feed will fit into a given storage space than mash feed.
2. Improved flow characteristics. Pellets work well in most types of feed-delivery systems including computer feeders.
3. Reduced dustiness. Dusty feeds may be less palatable to livestock.
4. Decreased ingredient segregation during handling.

Animal-performance improvements

1. Decreased feed waste. This is probably due to the fact that pellets are more prehensible than mash.
2. Reduced selective feeding. All ingredients of the formulation should be in each pellet.
3. Destruction of insects and pathogenic organisms.
4. Thermal modification of starch and protein. Although gelatinization does not occur to the extent characteristic of extrusion, the heat of pelleting may result in improved nutrient utilization.
5. Improved feed acceptability/palatability.

The disadvantages of pellets when compared to mash are primarily of an economic nature.

1. Cost. The most significant disadvantage to pelleting feed is cost. Pellet mills are expensive and feed mills charge $10 to $15 more per ton to run mixed feed through the pellet mill.
2. Reduced feed output by the mill. Pellet mills represent a manufacturing bottleneck in feed mills. In other words, more tons of feed could be manufactured in a day if all feed were sold as mash rather than pellets.
3. Nutrient destruction. Although the thermal conditions in the conditioner are not as extreme as those in either the extruder or expander, some vitamin and additive destruction will occur. If these products must go into pellet formulation, it will sometimes be necessary to either overfortify the pelleted feed formula or else buy expensive protected forms of these products to ensure that the finished pellet contains enough to meet the targeted levels.

The alternative is to feed vitamin and additive products separately on the farm or ranch, and this represents an additional labor expense.

Pellet quality Pellet quality is also described as pellet durability or pellet integrity. Broken pellets are called *fines*. Images of good-quality pellets as well as pellets with fines are found in the file titled Images.ppt on the text's accompanying On-Line Companion. To view this file please go to http://www.agriculture.delmar.com and click on *Resources*. Click on *On-Line Resources* and select from the titles listed. Because of the additional cost, the buyer of pelleted feeds expects the product to contain very few fines and to be nearly 100 percent pelleted. However, one study suggested that the benefit of pelleted feed had less to do with pellet integrity than it did with the fact that it had been run through a pellet mill (Lanson & Smyth, 1955). Other studies suggest that diets with good pellet integrity outperform those with poor pellet integrity, which in turn outperform diets that have not been run through a pellet mill (Stark, Behnke, Hancock, & Hines, 1993).

Pellet quality is usually quantified using the pellet durability index (PDI). Measuring the PDI of a sample of pellets involves placing a sample of sieved pellets into a square box and tumbling the box for 10 minutes. The box contents are then sieved free of fines and reweighed to determine the PDI.

There are four factors that contribute to the binding force that holds a pellet together:

1. Weak hydrogen bonding
2. Covalent bonds
3. Entrapment by form-closed bonds
4. Liquid bridges

Of the four, the liquid bridges are probably the most significant forces involved in holding pellets together. For bridges to occur between particles, there must be a solvent present that can solubilize a bonding compound in the feed mix. In the conditioner, the solvent is water, and the bonding compound is starch and to a lesser extent protein. For bonding to take place, the feed particles must be forced together so that the required bridge can be formed. The pellet die is where this occurs.

Pellet quality is affected by two factors: formulation and pellet-mill operation.

The primary factor affecting pellet quality is the nature of the ingredients in the formulation. Fibrous cereal grains, minerals, and urea go through the conditioner unchanged, and their location in the finished pellet will represent weak points in the pellet structure. Fats and oils form a hydrophobic layer that interferes with liquid bridge formation. On the other hand, the inclusion of wheat middlings at 15 percent or more in the formula or the use of molasses with low inclusion rate pellet binders (3 to 11 lb. per ton) can help overcome the poor pelletability of other ingredients. Table 13–7 gives selected feedstuffs and some of their characteristics that affect pellet quality. It must be pointed out that a table such as this is subjective by nature and that the contribution to pellet quality of a particular feedstuff will vary with the pelletability of the other ingredients in the formulation.

Usually of only secondary importance in making a good quality pellet is the actual pellet mill and its manner of operation. If pellet durability problems are not due to the formulation, the conditioner is the next place to look. The mash should have enough resident time in the conditioner to allow moisture from the steam to be absorbed by feed particles. The heat of condensation from the steam should be sufficient to activate the natural adhesives found in the ingredients. In usual systems with the proper amount of steam, the moisture and heat requirements are met with a resident time of 5 to 15 seconds in the conditioner.

Table 13-7
Selected feedstuffs and several of their characteristics that affect pellet quality

Feedstuff	lb./ft^3	% Fat	% NDF	Abrasiveness	Moisture Absorbability	Pelletability
Oat hulls	11–12	1.8	78	High	Low	Very low
Alfalfa meal, dehydrated	18–22	2.4	55.4	High	High	Medium
Wheat bran	11–16	4.4	51	Low	Medium	Low
Brewers' grains	14–15	10.8	48.7	Medium	Medium	Low
Distillers' grains	18–19	10.7	46	Medium	Low	Low
Beet pulp	11–16	0.4	44.6	Medium	Medium	Low
Corn gluten feed	26–33	3.91	36.2	Low	Low	Medium
Wheat middlings	18–25	3.2	35	Low	Medium	Very high
Rice bran	20–21	15.0	33	High	Low	Low
Ground ear corn, 45 lb./bushel.	28–35	3.7	31	Very high	Medium	Very low
Oats, 38 lb./bushel, ground	25–30	5.2	29.3	Medium	High	Medium
Cottonseed meal	37–40	1.7	28.9	Low	Medium	High
Linseed meal	31–33	1.5	25	Medium	Low	High
Citrus pulp	20.5	0.6	23	Medium	Medium	Low
Hominy feed	25–28	7.3	23	Low	High	Low
Barley, heavy, ground	24–26	2.3	18.1	Medium	High	Medium
Peanut meal	40	5.5	14	Low	Low	High
Corn meal	38–40	4.3	9	Low	Medium	Medium
Corn gluten meal	32–43	2.4	8.9	Low	Medium	Medium
Soybean meal	41–42	1.6	7.79	Low	Low	Medium
Fish meal	30–40	10.7	2	Medium	Medium	Medium
Blood meal	38.5	1.7	0.92	Low	Medium	Medium
Whey, dehydrated	35–46	1.1	0	High	Medium	Low
Urea	34–42	0	0	High	Low	Very low
Minerals	Variable— usu. 45–80	0	0	High	Low	Very low

Adapted from Kniep, 1982.

Once the hot, wet mash leaves conditioner, it is passed into the die-and-roll system. Here, feed is forced through the holes of the die (a perforated disk). If properly conditioned, the feed emerges from the die in long spaghetti strands that are cut off by rotating knives to form the pellets.

Pellets leaving the die are soft, and durability will increase after they are cooled and dried. This operation involves drawing ambient air through a bed of pellets. In general, mills will make better-quality pellets in colder, drier weather than in warmer, more humid weather because of the impact these factors have on the ability to cool and dry pellets. On leaving the cooler, pellets are screened to remove fines. Some mills will also screen pellets before they are loaded into the truck for shipment.

Crumbling

Crumbling involves deliberately breaking pelleted feed, then grading the broken pellets to achieve a desired particle size. Crumbling makes it possible to produce a product smaller than would be practical with usual pellet die sizes. Viewed another way, crumbling makes it possible for the mill to produce small, particle-sized feeds while using a large pellet die, thus increasing pellet-mill output. Crumbles are used in rations for young fish, pigs, and poultry.

MEDICATED FEED

Depending on the drug category and feed type, feed mills that manufacture and sell medicated feeds are subject to federal and state laws regarding feed-mill licensing, the labeling of medicated feeds, record keeping, and routine inspection.

Medications Used in Livestock Feeds Described by Category

Category I Drugs: These are the drugs for which no withdrawal period is required at the lowest continuous feeding level for any approved animal.

Category II Drugs: These are the drugs that either require a withdrawal period at the lowest continuous feeding level in at least one animal for which the drug is approved, or are regulated on a zero-residue basis because of suspected carcinogenic properties.

The Feed Products That Contain These Drugs, Described by Type

Type A Medicated Article: Type A medicated articles are ingredients to be used in the manufacture of rations: they are not finished feeds. They are the feedstuffs that contain the most concentrated form of the drug and as such, they are often referred to as *drug premixes.* A Type A medicated article is used in mixes to make another Type A medicated article, a Type B medicated article, or a Type C medicated article (a finished feed).

Type B Medicated Article: Type B medicated articles are ingredients to be used in the manufacture of rations; they are not finished feeds. Type B medicated articles are mixtures of a Type A medicated article and at least 25 percent (by weight) other feedstuffs. The maximum permitted concentration of a drug in a Type B medicated article is 100 times the highest continuous-use level (or highest level approved) for Category II drugs and 200 times the highest continuous-use level for Category I drugs. A Type B medicated article is used in mixes to make another Type B medicated article or a Type C medicated article (a finished feed).

Type C Medicated Article: Type C medicated articles are finished feeds containing a drug in such concentration that the animal will receive the approved dose when it consumes the feed. A Type C medicated finished feed is made from another Type C medicated finished feed, or by substantially diluting a Type A or Type B medicated article. Type C medicated finished feeds may be fed as a complete feed, and in some cases, may be offered free choice or as top-dressed supplements.

Avoiding Drug Carryover

During the manufacture of medicated feeds, drug residues must be avoided to prevent drug carryover into the next batch of feed. This can be a problem for both Category I and Category II type drugs, especially if the feed mill manufactures feeds for multiple species.

One technique that is used to prevent drug carryover is planned sequencing of feeds to be manufactured. Feeds that contain the same drug are scheduled to be manufactured consecutively with those having the higher drug levels scheduled first. The next feeds to be manufactured would be those for the same species that were not to contain the drug.

A second technique used to prevent drug carryover is flushing. Grain or ground corn cobs can be moved through the system to flush out any medicated feed. This feed is then stored in a separate bin for use in the next medicated formula or for later disposal.

A third technique used to prevent drug carryover is equipment clean-out. This is the most labor-intensive and expensive of the techniques, but in some cases, may be the only option available. Feed mill customers should try to give feed mills advance notice of when they will need feed so that inexpensive techniques can be used to avoid drug carryover.

ORGANIC FEED

Congress passed the Organic Foods Production Act in 1990. This act required that the USDA develop regulations to ensure that organically labeled products meet consistent national standards. To this end, the USDA established the National Organic Program (NOP).

The regulations of NOP generally require that products labeled as organic be produced using only natural substances. NOP developed a list of approved and unapproved substances for the production of organic crops and for organic production of livestock. This list is referred to as the National List and is published at the NOP Web site (National Organic Program, 1990). In addition to the prohibited substances on the list, organic feed for livestock may not be produced using genetic engineering, ionizing radiation, or sewage sludge. Any farm or feed mill that wants to sell an agricultural product as organically produced must adhere to the NOP standards.

A feed crop may be labeled as organic if it has been grown on soil that has been free of substances prohibited on the National List for at least 3 years and was produced in a manner consistent with NOP rules.

Organically raised livestock must be fed only organic feedstuffs and must have access to the outdoors, including pasture for ruminants. A dairy farm may market its milk as organic after feeding 80 percent organic feed for 9 months, followed by 3 months of 100 percent organic feed.

FEED PACKAGING

The feed packaging cost center begins with the finished feed in the supply bins above the bagging equipment, and ends when the filled bags are stacked on pallets and conveyed to the warehouse (or vice versa).

Packaging feed in bags involves considerable investment by feed mills, either in labor or in automation equipment. For this reason, feed mills will charge considerably more for a ton of bagged feed compared to a ton of bulk feed. The packaging of a bag of feed involves weighing the feed with scales dedicated to the bagging operation, conveying the weighed amount into the bag, sealing the bags, and attaching a feed tag to the bag.

There are several types of bags currently in use in the feed industry. They vary in material and design, and are chosen based on their ability to keep feed fresh, their durability, cost, and ease of use by the mill's labor or equipment.

WAREHOUSE AND LOADOUT

Bagged feeds are stacked on pallets and moved into the warehouse. From the warehouse, the bags are loaded on trucks or rail cars for shipment.

Bulk feed that is ready for sale will be elevated to bins above the truck scales from which it can be loaded directly onto the truck. Bulk feed is seldom shipped by rail car because farms and ranches generally do not have access to a rail spur for receiving feed.

TRUCKING

The profit made by the mill from a given customer is affected by how far the operation is from the mill. However, mills generally do not give discounts to customers located closer to the mill.

Another factor that affects the mill's profit from a given customer will be the size of the load that is shipped to the customer. As an example, assume the cost per mile to operate a feed truck is $1.00. This is the cost of operating the truck, whether it is full or empty. If the truck is carrying its capacity of 24 tons and it travels 100 miles round trip, the $100.00 trucking expense is money well spent because it will result in the profits made on 24 tons of feed. If the truck has the capacity to carry 24 tons and it is carrying only the 3 tons of feed ordered by one customer whose farm is 100 miles away, the same $100.00 trucking expense will only result in the profits on 3 tons of feed. Unlike the distance factor, feed mills usually do offer big load discounts to their customers.

Effective trucking management can improve profitability not only by ensuring that trucks are full when they leave the mill, but also by carefully planning truck routes. Feed mill trucks usually have multiple bins, each with a capacity matched to the mill's minimum bulk purchase requirement. Six bins of 4-ton capacity is common. By carefully choosing which deliveries go on which trucks, the operations at the outskirts of the mill's territory can be serviced without excessive trucking expense. This is only possible, of course, as long as customers do not wait until the last minute before ordering feed.

SUMMARY

Chapter 13 describes the various types of equipment used at the feed mill. These include conveying equipment, storage equipment, processing equipment, mixing equipment, and packaging equipment. Trucking issues are also discussed. Molds and mycotoxins are discussed in the context of the feed mill's quality assurance program. Feed processing options and their effect on feeding value are discussed. Feed mills may include drugs in their finished feeds. These drugs are categorized as Category I and II, and the feeds that include them as Type A, B, and C. How the feed mill avoids drug carryover is discussed.

END-OF-CHAPTER QUESTIONS

1. What percentage of the cost of making most finished feeds does the cost of ingredients represent?
2. Name five types of feed-conveying equipment and describe an application for each.
3. Describe each of the following: commodity, quasi-commodity, and branded product. Why is it important that the ingredient purchaser from the feed mill understand the differences among these three?
4. List five quality-assurance operations that may be applied to incoming ingredients at the feed mill.
5. List three techniques that are helpful in preventing mold growth on/in stored feedstuffs.

6. Compare and contrast batch mixing systems with in-line mixing systems.
7. Compare the hammermill with the roller mill in terms of product(s) produced, initial cost, and cost of operation.
8. Give five thermal methods of processing feedstuffs and describe their primary applications.
9. To what does the term *pellet quality* refer? Discuss the factors that affect pellet quality.
10. Describe the two categories of drugs that may be included in feeds. Describe the three types of medicated feeds.
11. Give three management strategies that may be used by the feed mill to avoid drug carryover.

REFERENCES

Brown, R. W., Pier, A. C., Richard, J. L., & Krogstad, R. E. (1981). Effects of dietary aflatoxin on existing bacterial intramammary infections of dairy cows. *American Journal of Veterinary Research. 42*, 927.

Goodband, R. D, Tokach, M. D., & Nelssen, J. L. (n.d.). The effects of diet particle size on animal performance. Manhattan, KS: Cooperative Extension Service, Kansas State University.

Hawk, A. L. (2005). Mycotoxins. *Grain Elevator and Processing Society (GEAPS).* Retrieved 6/22/2005 from http://www.geaps.com/proceedings/2004/Hawk.cfm

Herrman, T. J. & Trigo-Stockli, D. (2002). Mycotoxins in feed grains and ingredients. Kansas State University, MF-2061.

Indiana state chemist office. 1996. Retrieved 12/1/2003 from: http://www.isco.purdue.edu/mycotoxins.htm

Kniep. (1982). *Pellet Mill Operator Manual.* American Feed Manufacturers' Association, Inc. Arlington, Virginia.

Lanson, R. K., & Smyth, J. R. (1955). Pellets vs. mash plus pellets vs. mash for broiler feeding. *Poultry Science. 34*, 234–235.

National Organic Program. (1990). Retrieved 10/31/2003 from: http://www.ams.usda.gov/nop/NationalList/FinalRule.html

Jones, F. T., Genter, M. B., Hagler, W. M., Hansen, J. A., Mowrey, B. A., Poore, M. H., Whitlow, L. W. (1994). Understanding and coping with effects of mycotoxins in livestock feed and forage. North Carolina State University, Department of Agricultural Research Service publication number ARS-83.

Owens, F. N., Secrist, D. S., Hill, W. J., & Gill, D. R. (1997). The effect of grain source and grain processing on performance of feedlot cattle: A review. *Journal of Animal Science. 75*, 868–879.

Santos, J. E. P., Huber, J. T., Theurer, C. B., Nussio, L. G., Tarazon, M., & Santos, F. A. P. (1999). Response of lactating dairy cows to steam-flaked sorghum, steam-flaked corn, or steam-rolled corn and protein sources of differing degradability. *Journal of Dairy Science. 82*, 728–737.

Satter, L. D., Dhiman, T. R., & Hsu, J. T. (1994, October 18–20). Use of heat processed soybeans in dairy rations. In *Proc. Cornell nutrition conference for feed manufacturers* (pp. 19–28). Rochester, NY.

Stark, C. R., Behnke, K. C., Hancock, J. D., & Hines, R. H. (1993). Pellet quality affects growth performance of nursery and finishing pigs. *Swine Day 1993 Report of Progress No. 695.* Manhattan, KS: AES, Kansas State University.

Theurer, C. B., Huber, J. T., & Delgado-Elorduy, A. (1996, October 22–24). Steam-flaking improves starch utilization and milk production parameters. In *Proc. Cornell nutrition conference for feed manufacturers* (pp. 121–130). Rochester, NY.

Theurer, C. B., Lozano, O., Alio, A., Delgado-Elorduy, A., Sadik, M., Huber, J. T., & Zinn, R. A. (1999). Steam-processed corn and sorghum grain flaked at different densities alter ruminal, small intestinal, and total tract digestibility of starch by steers. *Journal of Animal Science. 77*, 2824

Traylor, S. L., Behnke, K. C., Hancock, J. D., and Hines, R. H. (1996). Mix uniformity and pellet quality affect animal performance. In: *Proc. Feed Manufacturing Short Course* (Manhattan, KS.), June 2–13.

United States Food and Drug Administration. (2000). Industry activities staff booklet. Action levels for poisonous or deleterious substances in human food and animal feed. Retrieved 6/21/2005 from http://www.cfsan.fda.gov/~Lrd/fdaact.html.

United States Food and Drug Administration. (2001). Background paper in support of fumonisin levels in animal feed: executive summary of this scientific support document. Retrieved 6/21/2005 from http://www.cfsan.fda.gov/~dms/fumonbg4.html.

Vilgalys, R. (2002). Duke University. Quoted in radio program: Pulse of the planet. Jim Metzner Productions, Inc.

Wondra, K. J., McCoy, R. A., Hancock, J. D., Behnke, K. C., Hines, R. H., Fahrenholz, C. H., & Kennedy, G. A. (1992). Effect of diet form (pellet vs. meal) and particle size on growth performance and stomach lesions in finishing pigs. *Journal of Animal Science. 70*(supplement1), 239.

CHAPTER 14

FEEDING SWINE

A child's hand and a pig's trough must always be full.

SWISS PROVERB

The nutritional phases in the swine production cycle, illustrated in Figure 14–1, include:

Sow: lactation, breeding, gestation
Pig: suckling/creep feed, nursery, grower, finisher

A sample growth rate is plotted in Figure 14–2.

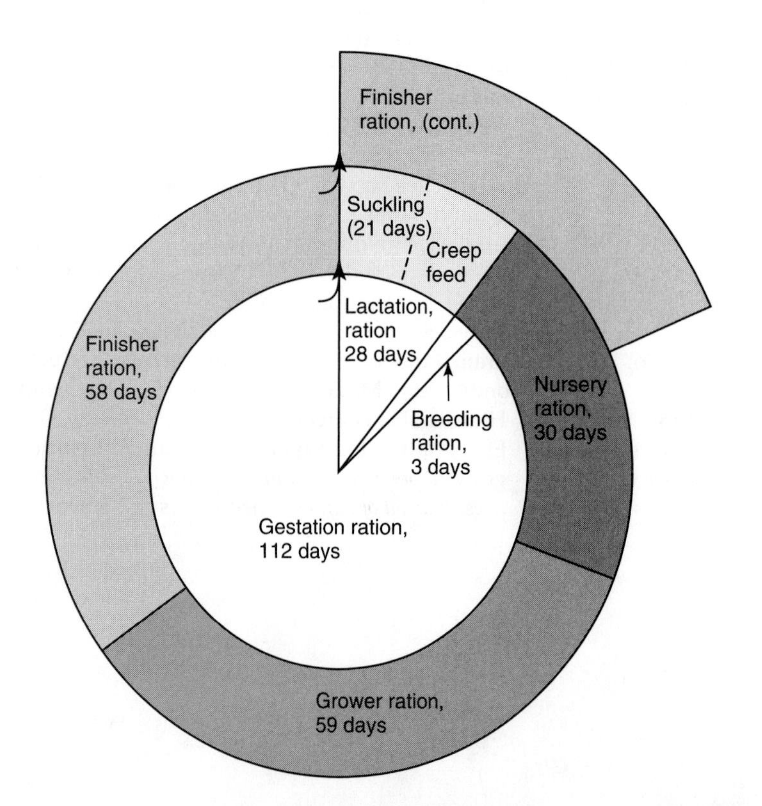

Figure 14–1
Nutritional phases in the production cycle: swine Move clockwise starting at top of innermost circle. Continue through next cycle or move to next shell after 360°.

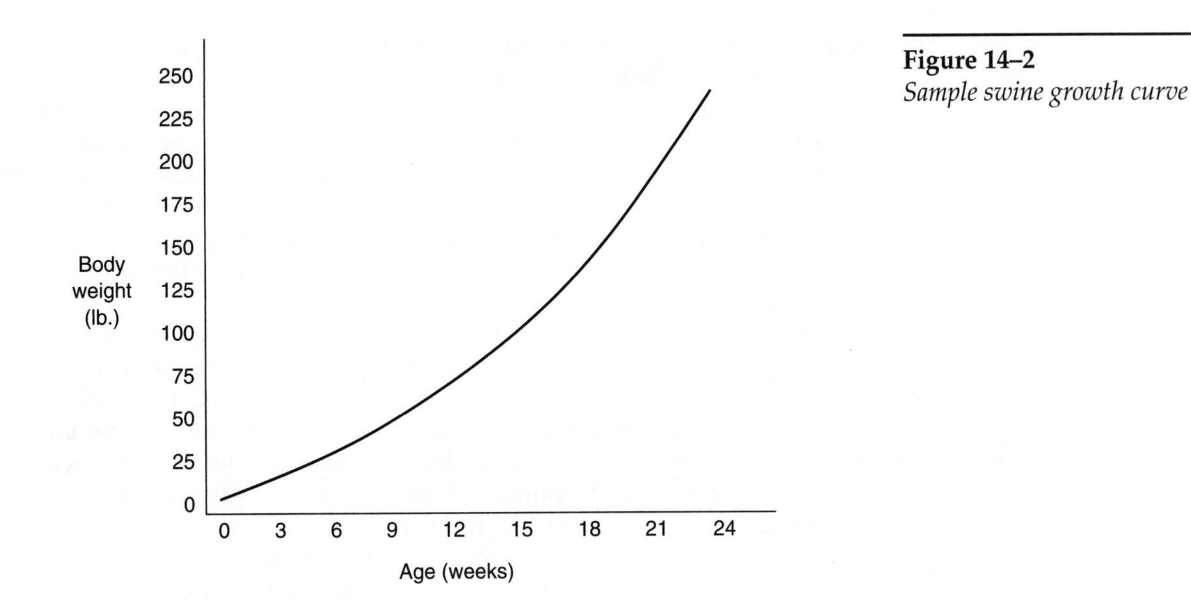

Figure 14–2
Sample swine growth curve

Table 14-1
Milk composition, as sampled basis

Constituent	Sow	Cow—Holstein	Cow—Jersey	Goat	Mare
Fat (%)	7.2	3.7	5.1	3.8	1.5
Lactose (%)	4.8	4.9	4.9	4.1	6.0
Protein (%)	6.1	3.1	3.9	3.0	2.0

Holstein and Jersey data from Ensminger, 1993; Sow data from Boland, 1966; Goat data from Haenlein & Ace, 1984; Mare data from Lewis, 1982.

BREEDING HERD

Lactating Sows

The lactating sow has the highest nutrient requirements of any type of swine. Within a few days of farrowing, the sow should be given free-choice access to a high-energy diet. The lactating sow will generally eat an amount of feed equal to 5 lb. plus 1 lb. for each pig she is nursing. Even when properly fed, the sow will probably not be able to consume enough energy to meet the needs of lactation, and will therefore lose weight. Sows are capable of producing about 20 lb. daily of milk that is the most nutrient-dense of all domesticated animals (Table 14–1).

As with all livestock, swine have the ability to store some nutrients during periods of excess intake. At times when nutrient use exceeds intake, these nutrient stores may be withdrawn to help maintain health and performance. During lactation, for example, nutrient use will often exceed intake for several nutrients. Calcium may be withdrawn from reserves in bone and made available for use in milk production (Giesemann, Lewis, Miller, & Akhter, 1998). Body fat and body protein may be mobilized to help supply the energy and amino acids used in lactation (McNamara & Pettigrew, 2002). Lactating sows deficient in these nutrients for short periods can maintain normal performance by mobilizing nutrient reserves, but a deficiency throughout lactation will result in reduced performance and health.

Livestock have little or no ability to store most other nutrients, so will require daily intakes in order to maintain health and performance. Diets inadequate in these nutrients will soon result in poor performance and symptoms of deficiency. Sows fed inadequate lysine during lactation, for example, may have a prolonged weaning-to-estrus interval (Yang, Pettigrew, Johnson, Shurson, Wheaton, White, Koketsu, Sower, & Rathmacher, 2000).

The milk of swine has considerably higher nutrient density than milk from other species (Table 14–1). However, the sow's milk does not contain an adequate amount of iron for the nursing piglet. Iron is not efficiently transferred to the milk, so manipulation of the sow's diet will not markedly affect the iron content of the milk. For this reason, and the fact that iron is not transferred to the fetal piglet in adequate amounts, the neonatal piglet should receive supplemental iron via intramuscular injection.

The structure of the sow's placenta is described as *epitheliolchorial.* This type of placenta does not allow antibody immunoglobulins to pass from the sow into the blood of the fetal piglets. The newborn piglet, therefore, has no antibody-mediated immunity to pathogens in the environment. By ingesting colostrum as soon as possible after birth, the neonatal pig acquires passive immunity from the antibody immunoglobulins in the colostrum.

The most critical nutrient for the lactating sow is energy, and these animals will generally eat to meet their requirement for energy. A sow that farrows with excess body condition will draw on the energy she has in storage to meet the energy demands of milk production, but energy from body fat is used much less efficiently than energy from the organic compounds in ingested feed. This means fat sows will have an impact on farm profitability even if they are able to perform well during the lactation.

Breeding Swine

The nutrient needs of boars, gestating sows, and sows to be bred are similar. In practice, swine farms generally use feed from the same bin for these three types of hogs.

Replacement breeding animals are usually selected from the animals raised for market within a month of marketing or at 5 to 6 months of age. Up until this time, growing animals have been fed free choice. The replacement animals are removed from the market hogs and managed separately. Their diet will be limit fed (about six lb. daily or 70 percent of full feed) to achieve breeding weight (260 to 300 lb.) at 7 to 8 months of age. Replacement gilts are usually bred after two or three estrous cycles. In practice, the diet used for replacement boars and gilts is often also the boar/gestation/breeding diet.

Most sows will come into heat within about 3 to 8 days after their pigs are weaned and may be bred at this time. However, nutritional and management factors can affect the length of the weaning-to-estrus interval. Sows that do not eat well during their lactation or are fed diets that are deficient in energy or protein may have a weaning-to-estrus interval that is longer than 8 days. Such nutritional deficiencies disrupt the normal production of reproductive hormones and cyclic function after weaning is impaired. Segregated early weaning may also result in a lengthened weaning-to-estrus interval.

Flushing is a practice whereby gilts or sows are fed an enhanced diet 2 weeks prior to and through the breeding period. The enhancement may be a greater amount of the previous diet or it may be a more nutrient-dense diet. Flushing may result in increased eggs ovulated and increased litter size. Flushing will probably be beneficial to gilts and sows that come into the breeding period in thin condition, and will probably not be beneficial to gilts and sows that are in good condition. Flushing may do more harm than good when practiced on overweight gilts and sows.

The productive life of animals in the breeding herd is dependent on management of body condition. If boars and sows are allowed to become fat, their libido will be impaired and it may be difficult to get them bred. Likewise, thin animals may not cycle properly until they receive adequate nutrition.

Gestating Swine

Nutrients ingested by the gestating sow are used for three functions: to maintain the sow, to accrete mammary tissue in late gestation, and to provide for development of the fetuses. The swine NRC software and the companion application to this text present a single requirement value for each nutrient that is the sum of what both the sow and her fetuses need.

A major challenge on many swine farms is to keep sows from getting too fat. This is because sow weight generally increases from parity to parity according to the following general pattern: gestation weight gain, 80 lb.; farrowing weight loss, 40 lb.; lactation weight loss, 25 lb. The result is a predicted weight gain of 15 lb. with each succeeding litter. The formulas in the companion application to this text predict sow weight change during gestation based on diet consumed. To manage gestation weight gain, sows should be limit fed 3 to 5 lb., depending on the energy density of the diet. An alternative to limit feeding is to use time restricted meal feeding or interval feeding. In this type of feeding management, animals receive free-choice access to feed for 2 to 6 hours only every third day.

To help prevent constipation in the lactating sow, it is advisable to increase the bulk in the diet of the late-gestation sow. Feedstuffs such as wheat bran, oat grain, dehydrated alfalfa, and beet pulp can be used in place of a portion of the corn grain in the diet. This high-bulk diet should be fed starting a few days before farrowing until a few days after farrowing.

MARKET HOGS

The Suckling Piglet

With the exception of iron, the sow's milk provides a balanced diet for the suckling piglet. As noted earlier, an iron injection should be given to the piglet shortly after birth to prevent the neonate from becoming anemic. Nutritional anemia results when an iron deficiency leads to inadequate hemoglobin synthesis. The anemic pig becomes pale and weak. The neonatal pig is not going to receive adequate iron through the diet because its only ingredient—the sow's milk—is inadequate in iron content. Supplementing the lactating sow's diet with iron does not affect the milk's iron content. The neonatal pig, therefore, must receive an intramuscular injection of an iron–carbohydrate complex such as iron dextran. A single injection of 150 mg or 2 cc of iron dextran given at 1 to 3 days of age is usually adequate to prevent anemia. Some producers will give a second iron injection at 3 weeks of age.

The structure of the sow placenta is such that the large antibodies circulating in the sow's blood are not able to cross into the blood of the developing fetuses. As a result, pigs are born with no antibody-mediated defense against environmental pathogens. The sow's first milk (colostrum) contains the antibodies from the sow's blood, and pigs receiving colostrum will be protected from the diseases to which the sow has become immune until the pig is able to make its own antibodies at about 3 to 4 weeks of age. Obviously, it is important that the neonatal pig receive colostrum as soon as possible after birth in order to receive protection. The fact that the piglet's gut is only capable of absorbing intact antibodies for a short period of time makes it even more urgent that colostrum be ingested promptly. Although it is possible that some antibody absorption takes place as late as 3 days after birth, antibody absorption is maximal at birth and declines soon thereafter.

To achieve maximum growth rate and to facilitate a successful weaning, piglets should be given access to a "creep feed," or prestarter at about 1 week of age. Creep feed is feed that is intended solely for the consumption of the suckling piglet. As such, it should be placed in a location that restricts access to only the piglets. The most important characteristic of creep feed is that it be palatable. A palatable creep feed will ensure that the piglets will consume enough solid feed to make a smooth transition to the feed available after weaning. Palatability is enhanced by the fact that creep feed is usually produced as a crumble, and its formula usually includes dried skim milk, molasses, sucrose, and/or fat. Unless the producer is weaning pigs at less than 2 weeks of age, creep feed will be the most expensive feed on the hog farm.

Traditionally, pigs have been weaned at between 3 to 4 weeks. A practice called *segregated early weaning* (SEW) involves weaning pigs much earlier. SEW takes advantage of the fact that pigs are temporarily protected from environmental pathogens after intake of sow colostrum. These protected pigs are removed from the sow barn and placed in an environment that does not contain pathogens (they are segregated). SEW pigs should never have to use feed nutrients to mount their own immune response. When compared to pigs weaned at 3 to 4 weeks, SEW pigs are healthier, and because their immune system is not stimulated by pathogens, they convert feed to pork more efficiently. The disadvantage to SEW is that it requires more intensive management. The advantages of early weaning have been summarized by Veum (1991).

The Nursery Pig

Pigs will wean themselves at 6 to 8 weeks of age. Early weaning is generally considered to be weaning at an age of less than 4 weeks. Segregated early weaning involves weaning shortly after the piglet has received its colostrum from the sow, as early as 5 days of age. Early weaning requires intensive nutritional management and it may require special equipment capable of handling liquid feeds.

The Growing Hog

In farrow-to-finish swine farms, growing pigs are usually fed three different diets: the nursery diet (weaning to about 40-lb. body weight), the grower diet (40- to about 110-lb. body weight), and the finisher diet (about 110 to 250 lb.). Feeder pig producing farms will sell pigs leaving the nursery at 40 lb. to feeder pig finishing operations. Feeder pig finishing operations will purchase 40-lb. pigs and feed them through finishing.

To achieve maximum gain, pigs in these facilities should be on full feed (they should receive feed free choice) and a primary goal of management should be to maximize feed/dry matter intake. Pigs of high-quality genotype that have high intakes of balanced rations will respond with maximum lean gains and will be ready for market at the earliest possible time.

Beginning at about 66 lb. body weight, the nutritional requirements of barrows and gilts begin to differ (NRC, 1998). Barrows will consume more feed than gilts. Also, because the carcass of gilts will have a higher lean content than the carcass of barrows, gilts have a higher requirement for dietary protein (percentage) than barrows. In practice, gilts and barrows are not fed separately, so the swine NRC software and the companion application to this text give requirement values that are prorated based on the numbers and types of animals penned together.

The Finishing Hog

Finishing hogs will be fed on both farrow-to-finish and feeder pig finishing operations. The concentration of nutrients in the ration for the finishing hogs will be lower than that for the growing hogs, but the dry matter/feed intake per hog in the finishing group will be second only to the dry matter/feed intake of the lactating sow. Sixty percent of the cost of pork production is due to feed purchases, and because of both the high intake per animal and the large number of animals, the finishing hogs will consume the largest tonnage of feed at the farrow-to-finish farm. Energy feedstuffs are the largest component of the finishing hogs' diet. Corn grain is usually the most economical energy feedstuff fed to hogs in the Americas.

Finisher hogs should be managed to achieve maximum dry matter/feed intake. If, for example, feeder space is suboptimal, some animals in the group may not consume maximally. This will result in increased variation of body weight within the group. In those pork-production systems in which an entire group of hogs is marketed at the same time, this variation in body weight may result in a financial penalty.

Carcass quality can be affected by substituting high-fiber concentrates for a portion of the corn in finishing diets. Such diets may produce a leaner carcass at the expense of feed efficiency and rate of gain. The optimal feed formulation for the finishing hogs will be determined by the added value of a leaner carcass as compared to the expenses associated with the price of high-fiber feedstuffs and longer days to market.

Finishing hogs are poor masticators of feed (Allee, 1983). For this reason, these animals show poor utilization of diets that contain unprocessed grains. Although young pigs are apparently better chewers than are finishing hogs, grinding of the cereal grain for either age improves feed efficiency.

Hogs are generally shipped to market at 250-lb. body weight. Feed conversion by swine from 40 lb. to market weight is generally 3 to 4 lb. of feed per pound of gain.

SWINE FEEDING AND NUTRITION ISSUES

Nutrient Supplementation in Swine Nutrition

Microbes in the digestive tract of all livestock use low-quality protein and nonprotein sources of nitrogen in ingested feed to build high-quality protein in microbial cells. Ruminant animals are able to digest these microbial cells because the site of microbial production (the rumen) is upstream from the primary site of absorption (the small intestine). Most microbial production in the digestive tract of swine takes place in the large intestine, downstream from the primary site of absorption. Therefore, the pig must be fed high-quality sources of protein. High-quality protein sources are those that not only contain all the amino acids found in pork, but contain these amino acids in proportions that are similar to those found in pork.

Microbes in the digestive tract also produce the B vitamins. Again, in the pig, these nutrients are synthesized downstream from the primary site of absorption so the pig does not have access to these vitamins unless it is allowed access to its manure. In addition, the usual energy and protein sources fed to swine contain low concentrations of the B vitamins. For these reasons, swine diets are usually fortified with the B vitamins.

Unlike the B vitamins, there is no mineral synthesis taking place in the digestive tract of livestock. In terms of the need for mineral supplementation of swine diets, if there is inadequate bioavailable mineral in the energy and protein sources chosen, a supplemental source of mineral will need to be added.

The Energy Requirement

In the swine NRC (1998) and the companion application to this text, the energy requirement is calculated based on information describing the pig and its environment. Animal information includes the rate of gain, which is the sum of lean gain and fat gain. Environmental information includes temperature and space per pig.

Pigs in thermal environments below their optimal temperature will use feed energy as fuel to generate the heat needed to maintain body temperature. Cold pigs will, therefore, require additional energy and their appetite will increase. However, the appetite is poorly matched to the increased need for energy due specifically to the cold stress. If the cold pig has free-choice access to feed, its appetite will result in increased metabolic energy available for all functions including fat deposition.

Crowding of growing swine has a negative impact on performance. Because performance is reduced, nutrient needs are reduced. The swine NRC and the companion application to this text show a reduced energy requirement in crowded swine. The space requirements are established as the minimum needed to avoid having a negative impact on performance. For pigs weighing less than 44 lb., the minimum space required is 4.4 ft^2 per pig. For pigs weighing 45 to 110 lb., the minimum space required is 11.4 ft^2 per pig, and for pigs over 110 lb., the minimum space required is 11.8 ft^2 per pig.

Protein in Swine Nutrition

In corn-based diets, protein is the nutrient most likely to be deficient. This is because corn is low in protein and what little protein there is in corn is of low quality. In addition, high-quality protein sources are expensive. A deficiency in protein results in reduced feed conversion, increased days to market, and a fatter carcass.

When it is said that the quality of protein in corn grain is poor, it means that the amino acid profile in corn protein matches up poorly with the animal's need for amino acids to make lean gain (Chapter 7). In selecting a protein source to mix with the corn grain, it is, therefore, important to find one where amino acid profile will make good the deficiencies of corn's amino acid profile. Soybean meal is one such protein source. Mixtures of corn meal and soybean meal have an amino acid profile that is relatively close to that required by swine to make lean gain. It is not a perfect match, however. Lysine is the amino acid that will be supporting the lowest amount of lean gain in corn–soy mixtures. Lysine is said to be the first limiting amino acid. Corn–soy diets can be made to support higher levels of lean gain by increasing the proportion of soybean meal or by adding a much smaller amount of lysine itself. Which solution is most economical will depend on the price of soybean meal and lysine.

In the swine NRC (1998) and in the companion application to this text, the lysine requirement is calculated from information describing the pig and its environment. Using the ideal protein concept (Chapter 7), swine nutritionists have established ratios that relate the requirement for each essential amino acid to the lysine requirement. Different ratios are applied to determine the essential amino acid requirements for maintenance, lean gain, and milk synthesis.

True Ileal Digestibility

Swine use amino acids to make body proteins. The primary source of these amino acids is feed protein. However, some of the amino acids in feed protein sources will not be absorbed as the feed passes through the pig's digestive tract. There are bioavailability differences among amino acids contained in the various feed protein sources. To address this issue, researchers work with ileal-cannulated swine, and based on the difference between the amino acids consumed and those that are collected at the ileum, they are able to assess amino acid bioavailability. Apparent ileal digestibility is calculated as the difference between what was consumed and what was collected at the terminal ileum. True ileal digestibility is apparent ileal digestibility corrected for the endogenous sources of amino acids that are collected at the ileum. The companion application to this text uses true ileal digestibility in predicting feedstuff amino acid value. It should be noted that the authors of the swine NRC prefer to use the term *ileal digestibilities* rather than bioavailability to describe this information because there will be some amino acids that are absorbed in a nonusable form. Table 14–2 presents three methods of expressing the amino acid content of feedstuffs: percent in feed dry matter, percent apparent ileal digestible in feed dry matter, and percent true ileal digestible in feed dry matter.

Relating Anticipated Carcass Traits to Nutrient Requirements

The two primary measurements that indicate desired qualities in a market hog are backfat thickness and loin eye area. The loin eye is the muscle in the loin region that gives rise to the pork chop product. The grade, quality, and yield of pork carcasses and wholesale cuts are related to these two indices. The desired loin eye area and fat depth at the 10th rib are carcass targets that can be used to calculate necessary carcass fat-free lean gain in the hog. The carcass fat-free lean gain can, in turn, be used to predict the amino acid levels that must be in the diet. These two indices are inputs in the companion application to this text. Most market hogs will have a loin eye area of 4 to 7 in^2 and a fat depth at the 10th rib of 0.4 to 1.6 in.

Predicting Performance

In general, performance is predicted by finding the difference between the total nutrient needed for maintenance functions and subtracting that from the total nutrient available in the diet. The difference is the amount of nutrient that can be applied to performance activity. The level of performance can be estimated by dividing the amount of nutrient that can be applied to performance by the nutrient cost for performance (Figure 14–3).

Feedstuffs Used in Swine Diets

The choice of which feedstuffs to use in formulating swine diets will depend on three factors:

1. What works to balance the ration
2. Feedstuff cost and availability
3. Feedstuff palatability

Low palatability may be an inherent characteristic of the ingredient, but palatability may also be influenced by moisture content and contaminants such as molds and toxins. Palatability may also be affected by the physical condition of the feedstuff, such as fineness of particle size or dustiness. Comments at the

Nutrient amount that can be applied to performance:
= (total nutrient available in the diet) − (total nutrient needed for maintenance functions)
Level of performance:
= (nutrient amount that can be applied to performance) / (nutrient cost for performance

Figure 14–3
Prediction of performance

Table 14–2
Methods of expressing amino acid value in feed used in swine nutrition. All values expressed as percentages.

Feedstuff	Dry Matter Basis	Apparent Ileal Digestibility	True Ileal Digestibility
Lysine			
Corn	0.29	66	78
Wheat	0.43	73	81
Barley	0.22	68	79
Oats	0.45	70	76
Soybean meal	3.36	85	90
Tryptophan			
Corn	0.07	64	84
Wheat	0.30	81	90
Barley	0.12	70	80
Oats	0.16	72	78
Soybean meal	0.72	81	90
Threonine			
Corn	0.33	69	82
Wheat	0.44	72	84
Barley	0.39	66	81
Oats	0.49	59	71
Soybean meal	2.06	78	87
Methionine			
Corn	0.19	86	90
Wheat	0.25	85	90
Barley	0.22	80	86
Oats	0.25	79	84
Soybean meal	0.74	86	91

From NRC, 1998.

feed table in the swine ration formulation application accompanying this text give suggested feedstuff limits in swine rations.

Strictly nutrition-related questions regarding the suitability of feedstuffs for swine are best answered by work with the various feedstuffs in ration development. With accurate information on animal nutrient requirements and feedstuff nutrient value, the strengths and weaknesses of feedstuffs become apparent as nutrient balance is improved with successive feedstuff selections.

In large swine operations in the United States, swine diets are generally made by fortifying a corn meal and soybean meal mix with minerals and vitamins. Swine do not require corn meal and soybean meal, however. These two feedstuffs just happen to be relatively inexpensive and available in this country; they are palatable and they complement each other well in meeting swine nutrient requirements. On small operations and in other countries, balanced swine rations are formulated using a greater variety of feedstuffs. The companion application to this text calculates a level of performance that can be expected in a typical corn–soy diet for comparison with the ration developed with selected feedstuffs.

Electrolyte Balance

The optimal electrolyte balance for swine has been suggested to be 250 mEq/kg of excess cations, calculated using the sodium and potassium cations and the chloride anion (Austic & Calvert, 1981). The 250 mEq/kg is an as fed value, which, assuming a ration of 90 percent dry matter, would be 278 mEq/kg on a dry matter basis. Sodium, potassium, and chloride are the primary dietary ions that influence acid–base status in animals. The electrolyte balance may have application in the formulation of swine diets to maintain performance when animals are heat stressed (Haydon, West, & McCarter, 1990).

Phytase

Phytase addition to swine diets has been shown to improve the utilization of dietary phosphorus (Cromwell, Stahly, Coffey, Monegue, & Randolph, 1993; Young, Leunissen, & Atkinson, 1993). This is important for two reasons. First, there may be an economic benefit: the use of phytase will enable swine to acquire the phosphorus from corn and soybean meal, and thereby eliminate the need to purchase supplemental sources of phosphorus. Second, there is an environmental benefit: the improved utilization of dietary phosphorus will reduce the excretion of phosphorus.

Feeding Antibiotics at Subtherapeutic Levels

Antibiotics are often fed at subtherapeutic levels to swine to improve feed efficiency and rate of gain. Because the use of antibiotics in livestock feed may lead to widespread microbial resistance and because such resistance may affect later therapeutic value of those antibiotics, management strategies to reduce resistance acquisition by bacteria may be warranted (Mathew, Upchurch, & Chattin, 1998). A discussion of issues related to use of antibiotics in livestock feed is found in Chapter 3.

Repartitioning Agents

The development of repartitioning agents for swine is an active area of research. Repartitioning agents redirect the use of nutrients with the result that the animal's metabolism is more focused on productive activities.

END-OF-CHAPTER QUESTIONS

1. Why are day-old piglets routinely given a shot of iron dextran?
2. Describe segregated early weaning. What are the advantages and disadvantages of SEW?
3. For what nutrient is the anticipated loin eye area used to predict the requirement?
4. What classes of swine are usually given free choice access to feed?
5. What is the name given to the solid feed that is made available to the piglet while it is still with the sow?
6. On a farrow-to-finish operation, growing/finishing pigs are usually fed at least three different rations: the nursery ration, the grower ration, and the finisher ration. Give the approximate weight ranges characteristic of pigs fed each of these rations.
7. The ideal protein concept in swine nutrition establishes the requirement for each essential amino acid based on the pig's predicted requirement for what?

8. In the United States, swine rations are usually made by fortifying two commodities with minerals and vitamins. What are these two commodities?
9. What is the difference between apparent ileal digestibility of an amino acid and true ileal digestibility of an amino acid?
10. At about what body weight do the nutritional requirements of barrows and gilts begin to differ?
11. Discuss the impact of crowding on nutrient requirements of penned swine.

REFERENCES

Allee, G. L. (1983). The effect of particle size of cereal grains on nutritional value for swine. In *First international symposium on particle size reduction in the feed industry* (pp. D1–D9). Kansas State University (edited), Department of Grain Science and Industry.

Austic, R. E., & Calvert, C. C. (1981). Nutritional interrelationships of electrolytes and amino acids. *Federation Proceedings. 40,* 63–67.

Boland, J. P. (1966). In L. K. Bustad et al. (Editors), *Swine in biomedical research.* Seattle, WA: Frayn Printing Co.

Cromwell, G. L., Stahly, T. S., Coffey, R. D., Monegue, H. J., & Randolph, J. H. (1993). Efficacy of phytase in improving the bioavailability of phosphorus in soybean meal and corn-soybean meal diets for pigs. *Journal of Animal Science. 71,* 1831–1840.

Ensminger, M. E. (1993). *Dairy cattle science* (3rd edition) (p 408). Danville, IL: Interstate Publishers, Inc.

Giesemann, M. A., Lewis, A. J., Miller, P. S., & Akhter, M. P. (1998). Effects of the reproductive cycle and age on calcium and phosphorus metabolism and bone integrity of sows. *Journal of Animal Science. 76,* 796–807.

Haenlein, G. F., & Ace, D. L. (1984). Extension goat handbook (section E-1, p.3). Washington, DC: United States Department of Agriculture.

Haydon, K. D., West, J. W., & and McCarter, M. N. (1990). Effect of dietary electrolyte balance on performance and blood parameters of growing-finishing swine fed in high ambient temperatures. *Journal of Animal Science. 68*(issue 8), 2400–2406.

Lewis, L. D. (1982). *Feeding and care of the horse.* Philadelphia: Lea and Febiger.

Mathew, A. G., Upchurch, W. G., & and Chattin, S. E. (1998). Incidence of antibiotic resistance in fecal *Escherichia coli* isolates from commercial swine farms. *Journal of Animal Science. 76,* 429–434.

McNamara, J. P., & Pettigrew, J. E. (2002). Protein and fat utilization in lactating sows: I. Effects on milk production and body composition. *Journal of Animal Science. 80,* 2442–2451.

National Research Council. (1998). *Nutrient requirements of swine* (10th revised edition). Washington, DC: National Academy Press.

Veum, T. L. (1991). In E. R. Miller et al. (Editors), *Swine nutrition.* Boston, MA: Butterworth-Heinemann.

Yang, H., Pettigrew, J. E., Johnson, L. J., Shurson, G. C., Wheaton, J. E., White, M. E., Koketsu, Y., Sower, A. F., & Rathmacher, J. A. (2000). Effects of nutrient intake during lactation on weaning to estrus interval. *Journal of Animal Science. 78,* 1001–1009.

Young, L. G., Leunissen, M., & Atkinson, J. L. (1993). Addition of microbial phytase to diets of young pigs. *Journal of Animal Science. 71,* 2147–2150.

CHAPTER 15

SWINE RATION FORMULATION

TERMINOLOGY

Workbook A workbook is the spreadsheet program file including all its worksheets.

Worksheet A worksheet is the same as a spreadsheet. There may be more than one worksheet in a workbook.

Spreadsheet A spreadsheet is the same as a worksheet.

Cell A cell is a location within a worksheet or spreadsheet.

Comment A comment is a note that appears when the mouse pointer moves over the cell. A red triangle in the upper right corner of a cell indicates that it contains a comment. Comments are added to help explain the function and operations of workbooks.

Input box An input box is a programming technique that prompts the workbook user to type information. After typing the information in the input box, the user clicks **OK** or strikes **ENTER** to enter the typed information.

Message box A message box is a programming technique that displays a message. The message box disappears after the user clicks **OK** or strikes **ENTER**.

EXCEL SETTINGS

Security: Click on the **Tools** menu, **Options** command, **Security** tab, **Macro Security** button, **Medium** setting.

Screen resolution: This application was developed for a screen resolution of 1024 × 768. If the screen resolution on your machine needs to be changed, see Microsoft Excel Help, "Change the screen resolution" for instructions.

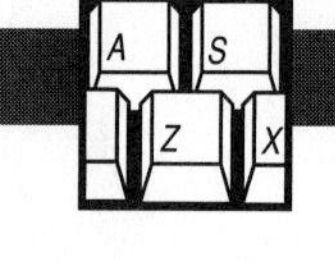

HANDS-ON EXERCISES

SWINE RATION FORMULATION

Double-click on the **SwineRation** icon. The message box in Figure 15–1 displays.

Figure 15–1

Macros may contain viruses. It is advisable to disable macros, but if the macros are legitimate, you may lose some functionality.
Disable Macros or Enable Macros or More Info

Click on **Enable macros**. The message box in Figure 15–2 displays.

Figure 15–2

Function keys F1 to F8 are set up. You may return to this location from anywhere by striking **ENTER**, then the **F1** key. Workbook by David A. Tisch. The author makes no claim for the accuracy of this application and the user is solely responsible for risk of use. You're good to go. TYPE ONLY IN THE GRAY CELLS! Note: This Workbook is made up of charts and a worksheet. The charts and worksheet are selected by clicking on the tabs at the bottom of the display. Never save the Workbook from a chart; always return to the worksheet before saving the Workbook.

Click **OK**.

Input Pig Procedure

Click the **Input Pig** button. The input box shown in Figure 15–3 displays.

Figure 15-3

1. Growing/finishing pig
2. Gestating sow
3. Lactating sow

The nutrient requirements of boars and gilts/sows to be bred are similar to that of gestating sows—these animals are usually fed the ration formulated for the gestating animals. ENTER THE APPROPRIATE NUMBER.

Click **OK** to choose the default input of #1, the Growing/finishing pig. The input box shown in Figure 15–4 displays.

Figure 15–4

Enter measured Dry Matter Intake (pounds) if available or click **OK** to accept the predicted value:

Click **OK** to accept the predicted value. The message box shown in Figure 15–5 displays.

ENTER THE APPROPRIATE CELL INPUTS.

Figure 15–5

Click **OK**.

The Cell Inputs

Table 15–1 presents the cell input values for this example. Comments associated with the cell inputs are presented in Figures 15–6 through Figure 15–10.

Table 15–1
Inputs for the swine ration formulation example

154	Current average body weight, in pounds per pig (7–230 lb.)
180	Carcass weight, Gilts (lb.)
180	Carcass weight, Barrows (lb.)
180	Carcass weight, Boars (lb.)
6	Expected 10th rib loin eye area, Gilts (in^2)
6	Expected 10th rib loin eye area, Barrows (in^2)
6	Expected 10th rib loin eye area, Boars (in^2)
0.8	Fat depth at the 10th rib, Gilts (in.)
0.8	Fat depth at the 10th rib, Barrows (in.)
0.8	Fat depth at the 10th rib, Boars (in.)
1	Number of gilts in the pen
1	Number of barrows in the pen
0	Number of boars in the pen
10	Space per pig, (ft^2) (1–160)
72	Pig's environmental temperature (° F)

Note that the finish weight is assumed to be 250 lb. for all market hogs.

Note also that the optimum temperature for the pen is calculated to be 71.1° F.

The comment behind the input, *Current average body weight, in pounds per pig (7–230 lb.)* is shown in Figure 15–6.

The value entered here is used to establish the lean gain for the pig. On a "typical" farm, pigs leave the farrowing house and enter the nursery at 3 weeks of age and 12 lb. body weight. They leave the nursery and enter the growing/finishing house at about 8 weeks of age and 45 lb. body weight. They are shipped to market at about 23 weeks of age and 250 lb. body weight.

Figure 15–6

The comment behind the three *Carcass weight* inputs is shown in Figure 15–7.

A 250-lb. live hog should yield a 180-lb. carcass. This represents a yield of about 72 percent.

Figure 15–7

The comment behind the three *Expected 10th rib loin eye area* inputs is shown in Figure 15–8.

Figure 15–8

The loin eye is an area of carcass muscle that is used to assess total carcass muscle (lean) content. Most market hogs will have a loin eye area of 4 to 7 in^2. This value and the fat depth at the 10th rib are used to predict carcass fat-free lean gain, which is used to predict the dietary amino acid levels required.

The comment behind the three *Fat depth at the 10th rib* inputs is shown in Figure 15–9.

Figure 15–9

This value, along with the loin eye area, is used to predict carcass fat-free lean gain, which is used to predict the dietary amino acid levels required. Most market hogs will have a fat depth of 0.4 to 1.6 in.

The comment behind the *Enter the space per pig, in square feet (1-160)* input is shown in Figure 15–10.

Figure 15–10

1 ft^2 = 0.093 m^2; 1 m^2 = 10.76 ft^2

Strike **ENTER** and **F1**.

Select Feeds Procedure

Click on the **Select Feeds** button. The message box shown in Figure 15–11 displays.
Click **OK**.

Figure 15–11

Nutrient content expressed on a dry matter basis. Feedstuffs are listed first by selection status, then within the same selection status by decreasing protein content, then, (within the same protein content), alphabetically.

Click **OK**.
Explore the table. Note the nutrients listed as column headings. Note also that the table ends at row 200. You select feedstuffs for use in making two different products: a blend to be mixed and sold bagged or bulk, and a ration to be fed directly to livestock.
Select the feedstuffs in Table 15–2 by placing a 1 in the column to the left of the feedstuff name. All unselected feedstuffs should have a 0 value in the column to the left of the feedstuff name. If you wish to group feedstuffs but not select them, you would place a value between 0 and 1 to the left of the feedstuff name.

Table 15–2
Feedstuffs to select for inclusion in the swine B vitamin blend

Rice bran
Biotin
Calcium pantothenate
Choline chloride, 60%
Folate (folacin, folic acid)
Niacin
Pyridoxine HCl
Riboflavin
Thiamin HCl
Vitamin B_{12} cyanocobalamin

Strike **ENTER** and **F2**.
Selected feedstuffs and their analyses are copied to several locations in the workbook.

Blend Feedstuffs Procedure

Click on the **Blend Feedstuffs** button. The message box in Figure 15–12 displays.

Figure 15–12

YOU MUST HAVE ALREADY SELECTED THE FEEDS YOU WANT TO BLEND. When your analysis is acceptable, strike **ENTER** and **F3** to name and file the blend.

Click **OK**.
In the gray area to the left of the feed name, enter the amounts to blend shown in Table 15–3.

Note that when these amounts are entered in the gray area, the blue column to the left calculates the equivalent amount in a ton of mix. Feed mills prefer to have the formula expressed on a per ton basis because the capacity of their mixer(s) is expressed in tons.

Table 15–3
Inclusion rates for feedstuffs in the swine B vitamin blend

Rice bran	150
Biotin	0.005
Calcium pantothenate	3
Choline chloride 60%	25
Folate (folacin, folic acid)	75
Niacin	1
Pyridoxine HCl	2
Riboflavin	1
Thiamin HCl	3
Vitamin B_{12} cyanocobalamin	1

Strike **ENTER** and **F3**. The input box shown in Figure 15–13 displays.

Figure 15–13

ENTER THE NAME OF THE BLEND (names may not be composed of only numbers):

Type the blend name, **Bvitpremix,** and click **OK**. The message box in Figure 15–14 displays.

Figure 15–14

The new blend has been filed at the bottom of the feed table.

Click **OK**.

View Blends Procedure

Click on the **View Blends** button to confirm that the Bvitpremix formula and analysis have been filed. The message box in Figure 15–15 displays.

Figure 15–15

Cursor right to view the blends. Cursor down for more nutrients. DO NOT TYPE IN THE BLUE AREAS.

Click **OK**. After viewing the Bvitpremix, strike **F1**.

Selecting a Blended Feed with Other Feedstuffs

Select Feeds *Procedure*

Click on the **Select Feeds** button again and select the feeds shown in Table 15–4. Unselect all other feedstuffs. Remember the Bvitpremix is located at the bottom of the feed table (row 200).

Table 15–4
Feedstuffs and blend to select for the swine ration formulation example

Feedstuffs
Soybean meal, 49%
Corn grain, ground
Dicalcium phosphate
Limestone, ground
Potassium iodide
Salt (sodium chloride)
Sodium selenite
Vitamin A supplement
Vitamin D supplement
Vitamin E supplement
Vitamin K (MSB premix)
Zinc oxide
Bvitpremix

Strike **ENTER** and **F2**.

Make Ration Procedure

Click on the **Make Ration** button. The message box shown in Figure 15–16 displays.

Figure 15–16

ENTER POUNDS TO FEED IN COLUMN B. TOGGLE BETWEEN NUTRIENT WEIGHTS AND CONCENTRATIONS USING THE **F5** KEY, RATION AND FEED CONTRIBUTIONS USING THE **F6** KEY, AND PERFORMANCE PREDICTIONS USING THE **F7** KEY. WHEN DONE, STRIKE **F1**. Cell is highlighted in red if nutrient provided is poorly matched with nutrient target. The lower limit is taken as 94 to 98 percent of target, depending on the nutrient. Except for calcium, the upper limit of acceptable mineral is taken from National Research Council, 1980, *Mineral Tolerance of Domestic Animals*: *National Academy Press.* This publication gives safe upper limits of minerals as salts of high bioavailability. The upper limit for calcium is taken as 2 percent. The upper limit of vitamin is taken from the NRC, *Vitamin Tolerance of Animals*, 1987. For other nutrients, the upper limit is based on unreasonable excess and the expense of unnecessary supplementation. See Table 15-6 for specifics.

Click **OK**.

The message box in Figure 15–17 displays.

Figure 15-17

The Goal Seek feature may be useful in finding the pounds of a specific feedstuff needed to satisfy a particular nutrient target:
1. Select the red cell highlighting the deficient nutrient
2. From the menu bar, select **Tools**, then **Goal Seek**
3. In the text box, "To Value:" enter the target to the right of the selected cell
4. Click in the text box, "By changing cell:" and then click in the gray "Pounds fed" area for the feedstuff to supply the nutrient
5. Click **OK**. You may accept the value found by clicking **OK** or reject it by clicking **Cancel**.

WARNING: Using Goal Seek to solve the unsolvable (e.g., asking it to make up an iodine shortfall with iron sulfate) may result in damage to the Workbook.
IMPORTANT: If you return to the Feedtable to remove more than one feedstuff from the selected list, you will lose your chosen amounts fed in the developing ration.

Click **OK**.

In the gray area to the right of the feedstuff name, enter the pound values shown in Table 15–5.

Table 15–5
Inclusion rates for feedstuffs and blend in the swine ration formulation example

Soybean meal, 49%	0.9
Corn grain, ground	4.6
Bvitpremix	0.006
Dicalcium phosphate	0.05
Limestone, ground	0.04
Potassium iodide	0.000002
Salt (sodium chloride)	0.012
Sodium selenite	0.000001
Vitamin A supplement	0.0055
Vitamin D supplement	0.00086
Vitamin E supplement	0.0055
Vitamin K (MSB premix)	0.00016
Zinc oxide	0.00021

The Nutrients Supplied *Display*

The application highlights ration nutrient levels, expressed as amount supplied per pig per day, that fall outside the acceptable range.

The lower limit for dry matter intake is 94 percent of the target. The lower limit for digestible energy (DE) and metabolizable energy (ME) is 94 percent of the target. The lower limit for crude protein is 95 percent of the target. The lower limit for all other ration nutrients is taken as 98 percent of the target. Table 15–6 shows the upper limits of the acceptable range for the various nutrients.

Table 15–6
Upper limits for ration nutrients used in the swine ration application

	Upper Limit	Source
DMI	6% over the predicted requirement	Author
DE	6% over the predicted requirement	Author
ME	6% over the predicted requirement	Author
n-6 fatty acids	No upper limit	—
Crude protein	20% over the predicted requirement	Author
Arginine	No upper limit	—
Histidine	No upper limit	—
Isoleucine	No upper limit	—
Leucine	No upper limit	—
Lysine	No upper limit	—
Methionine	No upper limit	—
Methionine + Cystine	No upper limit	—
Phenylalanine	No upper limit	—
Phenylalanine + Tyrosine	No upper limit	—
Threonine	No upper limit	—
Tryptophan	No upper limit	—
Valine	No upper limit	—
Calcium	2% of ration dry matter	Author

	Upper Limit	Source
Phosphorus, total	No upper limit	—
Phosphorus, available	1.5% of ration dry matter	NRC: Mineral Tolerance of Domestic Animals (1980)—swine
Sodium	3.1% of ration dry matter	NRC: Mineral Tolerance of Domestic Animals (1980)—swine based on limit for salt (sodium chloride)
Chloride	4.9% of ration dry matter	NRC: Mineral Tolerance of Domestic Animals (1980)—swine based on limit for salt (sodium chloride)
Potassium	2% of ration dry matter	NRC: Mineral Tolerance of Domestic Animals (1980)
Magnesium	0.3% of ration dry matter	NRC: Mineral Tolerance of Domestic Animals (1980)
Copper	250 mg/kg or ppm of ration dry matter	NRC: Mineral Tolerance of Domestic Animals (1980)—swine
Iodine	400 mg/kg or ppm of ration dry matter	NRC: Mineral Tolerance of Domestic Animals (1980)—swine
Iron	3,000 mg/kg or ppm of ration dry matter	NRC: Mineral Tolerance of Domestic Animals (1980)—swine
Manganese	400 mg/kg or ppm of ration dry matter	NRC: Mineral Tolerance of Domestic Animals (1980)—swine
Selenium	2 mg/kg or ppm of ration dry matter	NRC: Mineral Tolerance of Domestic Animals (1980)—swine
Zinc	1,000 mg/kg or ppm of ration dry matter	NRC: Mineral Tolerance of Domestic Animals (1980)—swine
Vitamin A	4 × requirement	NRC: Vitamin Tolerance of Animals, 1987
Vitamin D	4 × requirement	NRC: Vitamin Tolerance of Animals, 1987
Vitamin E	20 × requirement	NRC: Vitamin Tolerance of Animals, 1987
Vitamin K	1,000 × requirement	NRC: Vitamin Tolerance of Animals, 1987
Biotin	10 × requirement	NRC: Vitamin Tolerance of Animals, 1987
Choline	10 × requirement	Author
Folacin	1,000 × requirement	Author
Niacin, available	350 mg/kg body weight	NRC: Vitamin Tolerance of Animals, 1987
Pantothenic acid	100 × requirement	Author
Riboflavin	20 × requirement	Author
Thiamin B_1	1,000 × requirement	NRC: Vitamin Tolerance of Animals, 1987
Pyridoxine B_6	50 × requirement	NRC: Vitamin Tolerance of Animals, 1987
Vitamin B_{12}	300 × requirement	NRC: Vitamin Tolerance of Animals, 1987

The *Nutrients Supplied* display shows nutrient amounts in the diet and nutrient targets for the inputted pig. All nutrient levels appear to be within acceptable ranges as established by the application.

The cost of this ration using initial $/ton values is $0.52 per head per day.

1st limiting amino acid: *lysine*

The comment behind the cell containing this label is shown in Figure 15–18.

Figure 15–18

This is the amino acid that exists in the ration at a level that is farthest from the level required, or the amino acid whose requirement is most narrowly met.

Ca:P_{total} ratio: *1.04*

The comment behind the cell containing this label is shown in Figure 15–19.

Figure 15–19

When the Ca:P ratio is based on total phosphorus, the recommendation is to have 1.1 to 1.25 times as much calcium as phosphorus. Excess calcium reduces phosphorus absorption.
If the Ca:P_{avail} ratio is within acceptable limits, the Ca:P_{total} ratio can be ignored.

Ca:P_{avail} ratio: *2.37*

The comment behind the cell containing this label is shown in Figure 15–20.

Figure 15–20

When the Ca:P ratio is based on available phosphorus, the recommendation is to have 2 to 3 times as much calcium as phosphorus. Excess calcium reduces phosphorus absorption.

The Nutrient Concentration *Display (strike* `F5`*)*

The nutrients in the ration and the predicted nutrient targets here are expressed in terms of concentration. That is, the nutrients provided by the ration are divided by the amount of ration dry matter, and the nutrient targets are divided by the target amount of ration dry matter. Concentration units include percent, milligrams per kilogram or parts per million, calories per pound, and international units per pound.

Electrolyte balance: *169*

The comment behind the cell containing this label is found in Figure 15–21.

Figure 15–21

The electrolyte balance is used to assess the impact of the ration's mineral content on the body's efforts to regulate blood pH. It is calculated as mEq of excess cations: (Na + K − Cl)/kg of diet *dry matter*. The optimal electrolyte balance in the diet for pigs has been suggested to be 250, as fed (Austic and Calvert, 1981) or 278 *dry matter*. However, optimal growth has been found to occur over the range of 0 to 667 mEq/kg of *dry matter* diet. The electrolyte balance is of no value if requirements for Na, K and Cl have not been satisfied.

Strike **F5**.

The Feedstuff Contributions *Display (strike* `F6`*)*

Shown here are the nutrients contributed by each feedstuff in the ration. This display is useful in troubleshooting problems with nutrient excesses.

Strike **F6**.

The Predicted Performance *Display (strike* `F7`*)*

The *Predicted Performance* display is shown in Table 15–7.

Table 15–7
The predicted performance display for the swine ration example

GROWING/FINISHING PIGS	Corn/Soy Benchmark	Current Ration
Feed-to-gain ratio		
Gilts and boars	2.85	—
Barrows	2.90	—
Pen	2.87	2.74
Body weight gain (lb./day)		
Gilts and boars	1.89	—
Barrows	2.15	—
Pen	2.02	2.05
Protein tissue gain (lb./day)[1]		
Gilts and boars	0.92	—
Barrows	0.92	—
Pen	0.92	0.97
Fat tissue gain (lb./day)[2]		
Gilts and boars	0.87	—
Barrows	1.11	—
Pen	0.99	0.96

[1]This value is calculated from carcass inputs.
[2]This value is calculated from energy intake.

Strike **ENTER** and **F1**.

The Graphic *Display*

At the bottom of the home display are tabs. The current tab selected is the "Worksheet" tab. Other tabs are graphs based on the current ration. Note that you may have to click the leftmost navigation button at the lower-left corner to find the first tab.

DMI

Click on this tab to display a graph titled Nutrient Status: Dry Matter Intake

E&F

Click on this tab to display a graph titled Nutrient Status: Energy & Linoleic Acid

P&AA

Click on this tab to display a graph titled Nutrient Status: Crude Protein and Amino Acids

Min

Click on this tab to display a graph titled Nutrient Status: Minerals

Vit

Click on this tab to display a graph titled Nutrient Status: Vitamins

FTG

Click on this tab to display a graph titled Feed-to-Gain Ratio

DBWG

Click on this tab to display a graph titled Daily Body Weight Gain

ProtG

Click on this tab to display a graph titled Daily Protein Tissue Gain

FatG

Click on this tab to display a graph titled Daily Fat Tissue Gain
Click on the **Worksheet** tab.

Making a Medicated Complete Feed

Acquiring Drug Concentration and Dosage Information

Blend Feedstuffs Procedure

Click on the **Blend Feedstuffs** button. The same ration amounts used in the ration are shown in a blend for the selected feedstuffs. To the left of the ration amounts are feedstuff amounts expressed in pounds per ton.

We want to use arsanilic acid as an additive to aid in feed efficiency, to promote growth, and to prevent diarrhea. Click on the **Additives** button for help with including additive premixes in blends. Click on the **FDA** button to go to the FDA-approved Web site. In the search box, type arsanilic acid. This is a drug approved for use in growing swine. Click on the **Search** button. Click on **Browser View** for the first item listed. We will use the Pro-Gen Plus Feed Supplement, a Fleming Laboratories product. This product comes in three different concentrations. We will use the 50 percent arsanilic acid concentration. Note that this is a Category II (withdrawal period required), Type A medicated article (see Chapter 13). Under CFR indications, find the swine information. The approved level for growing swine is 45 to 90 g/ton. We will use an 80 g/ton level.

The concentration of this additive in this additive premix is expressed as a percentage. The dose of this additive in this additive premix is expressed in grams per ton.

Determining the Appropriate Amount of Additive Premix to Make 1 Ton of Medicated Complete Feed

Close the Web site. Look for a Type that matches our additive's concentration and dose. The match is found at Type #5. In the gray areas, enter the 50 concentration value and the 80 dose value. The calculated pounds of additive premix containing 50 percent additive to add to 1 ton of finished feed to deliver the dose of 80 g is 0.35.

Entering the Additive Premix into the Feed Table and Selecting It

Scroll up and click on the **Feed Table** button to go to the feed table to enter and select the additive premix. Enter the feed name **Pro-Gen Plus** in an empty row under FEEDNAME. Enter a cost if known. Use 20,000 for this example. Enter a DM% of **99**. Select this feedstuff along with the other components of the grain. Strike **ENTER** and **F2**.

Blending the Additive Premix into the Complete Feed

Blend Feedstuffs Procedure

Click on the **Blend Feedstuffs** button. Make sure that the amounts shown in Table 15–8 have been entered.

Table 15–8
Inclusion rates for feedstuffs and blend in the swine ration formulation example (as yet undetermined amount of drug)

Soybean meal, 49%	0.9
Corn grain, ground	4.6
Bvitpremix	0.006
Dicalcium phosphate	0.05
Limestone, ground	0.04
Potassium iodide	0.000002
Salt (sodium chloride)	0.012
Sodium selenite	0.000001
Vitamin A supplement	0.000055
Vitamin D supplement	0.0000086
Vitamin E supplement	0.00055
Vitamin K (MSB premix)	0.00016
Zinc oxide	0.00021
Pro-Gen Plus[1]	?

[1]Pro-Gen Plus Feed Supplement, a product of Fleming Laboratories.

Enter an amount of Pro-Gen Plus in the blend that converts to a value of 0.35 lb./ton or less. Recall inclusion of 0.35 lb. of this additive premix in a ton of feed will result in an arsanilic acid concentration of 80 g/ton. A value that works is 0.0009 (Table 15–9).

Table 15–9
Inclusion rates for feedstuffs, blend and drug in the G/FMedicated blend

Soybean meal, 49%	0.9
Corn grain, ground	4.6
Bvitpremix	0.006
Dicalcium phosphate	0.05
Limestone, ground	0.04
Potassium iodide	0.000002
Salt (sodium chloride)	0.012
Sodium selenite	0.000001
Vitamin A supplement	0.000055
Vitamin D supplement	0.0000086
Vitamin E supplement	0.00055
Vitamin K (MSB premix)	0.00016
Zinc oxide	0.00021
Pro-Gen Plus	0.0009

Strike **ENTER** and **F3**. Enter the name **G/FMedicated.**

View Blends Procedure

Click on the **View Blends** button and Click **OK**. Confirm that the formula and analysis of the G/FMedicated blend has been filed.

Strike **F1**.

Using the Medicated Complete Feed in the Ration

Select Feeds Procedure

Click on the **Select Feeds** button and select only the G/FMedicated feed. Unselect all other feedstuffs by entering a 0 to the left of the feedstuff name.

Strike **ENTER** and **F2**.

Make Ration Procedure

Click on the **Make Ration** button. Enter the amount to feed (Table 15–10).

Confirm that the ration is balanced as it was before the ingredients were blended. The cost of the ration is now $0.53 per head per day.

Table 15–10
Feeding rate for G/FMedicated blend in the swine ration formulation example

G/FMedicated	5.62

Strike **ENTER** and **F1**.

Print Ration or Blend Procedure

Make sure your name is entered at cell C1. Click on the **Print Ration or Blend** button.

The input box shown in Figure 15–22 is displayed.

Click **OK** to accept the default input of 1. A two-page printout will be produced by the machine's default printer.

Figure 15–22

Are you printing a swine ration evaluation or a blend formula and analysis? (**1**-RATION; **2**-BLEND):

Click on the **Print Ration or Blend** button. Type the number 2 to print a blend. The message box in Figure 15–23 displays.

Figure 15–23

Click on the green number above the blend you want to print and press **F4**. Scroll right to see additional blends.

Click **OK**.

Find the G/FMedicated blend, click on the green number above it, and strike **F4**. A one-page printout will be produced by the machine's default printer. Also print the blend named Bvitpremix.

ACTIVITIES AND WHAT-IFS

In the Forms folder on the companion CD to this text is a SwineInput.doc file that may be used to collect the necessary inputs for use of the SwineRation.xls file. This form may be printed out and used during on-farm visits to assist in ration evaluation activities.

1. Remake the ration described in Table 15–5 for the growing/finishing hog described in this chapter. Add oat grain to the selection of feedstuffs for this ration. Substitute oat grain for corn grain in the ration. What effect does this substitution have on the ration balance? What effect does this substitution have on the following performance parameters (viewed at **F7**): feed-to-gain ratio for the pen, body weight gain (lb.), protein tissue gain (lb./day), and fat tissue gain (lb./day)?
2. Remake the ration described in Table 15–5 for the growing/finishing hog described in this chapter. At *Input pig,* change carcass weights for gilts, barrows and boars from 180 lb. to 150 lb. What effect does this change have on the ration balance? What impact does this have on the following performance parameters for the pen's corn/soy benchmark (viewed at **F7**): feed-to-gain ratio, body weight gain, protein tissue gain (lb./day), and fat tissue gain (lb./day)? Explain these impacts.
3. Remake the ration described in Table 15–5 for the growing/finishing hog described in this chapter. Change the expected 10th rib loin eye area for gilts, barrows and boars from 6 to 7 in^2. What effect does this change have on the ration balance? What impact does this have on the following performance parameters for the pen's corn/soy benchmark (viewed at **F7**): feed-to-gain ratio, body-weight gain, protein tissue gain (lb./day), and fat tissue gain (lb./day)? Explain these impacts.

REFERENCES

Austic, R.E, and C.C. Calvert. 1981. Nutritional interrelationships of electrolytes and amino acids. *Federation Proceedings. 40,* 63–67.

National Research Council. 1980. *Mineral Tolerance of Domestic Animals.* Washington DC: National Academy Press.

National Research Council. 1987. *Vitamin Tolerance of Animals.* Washington DC: National Academy Press.

Pro-Gen Plus Feed Supplement, a product of Fleming Laboratories.

CHAPTER 16

FEEDING BEEF

The word "cattle" is derived from the words "chattel" and "capital."

—J. RIFKIN, 1993

The nutritional phases in the beef production cycle, illustrated in Figure 16–1, include:

Brood cow: first trimester, second trimester, third trimester, postpartum period
Market animal Route #1: forage-based program
Market animal Route #2: grain-based program

A sample growth rate is plotted in Figure 16–2.

THE BREEDING HERD

The goals for the nutrition program of the beef cow herd include:

- Maintaining productive, healthy brood cows.
- Achieving conception in cows by 80 days postcalving to maintain a 12-month calving interval.
- Producing strong, healthy, neonatal calves.

To achieve these goals, lactating cows should be fed to make adequate milk to support calf growth. If the feed resources are deficient—as during a drought—the lactating cow may not make enough milk to support desired calf growth. In this case, creep feed may be appropriate. If the brood cow's body condition needs to be adjusted, the adjustment should be made toward the end of the first and throughout the second trimester.

Profitability in the brood-cow operation is largely determined by the producer's ability to manage the forage resource to meet the animals' nutritional needs. The usual forages include pastures and native rangelands, but preserved forages such as hay and corn silage are also used.

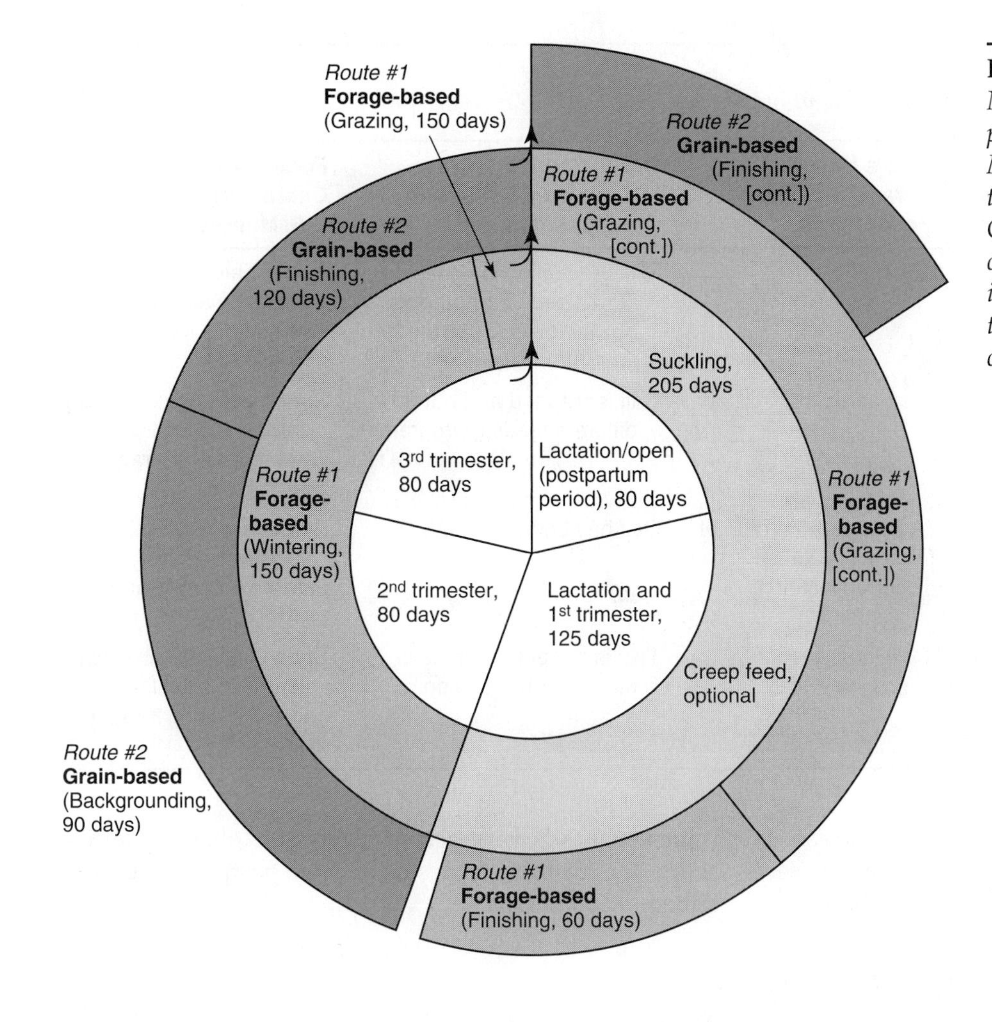

Figure 16–1
Nutritional phases in the production cycle: beef Move clockwise starting at top of innermost circle. Continue through the next cycle or move to next shell as indicated. Note that there are two possible routes after the calf is weaned.

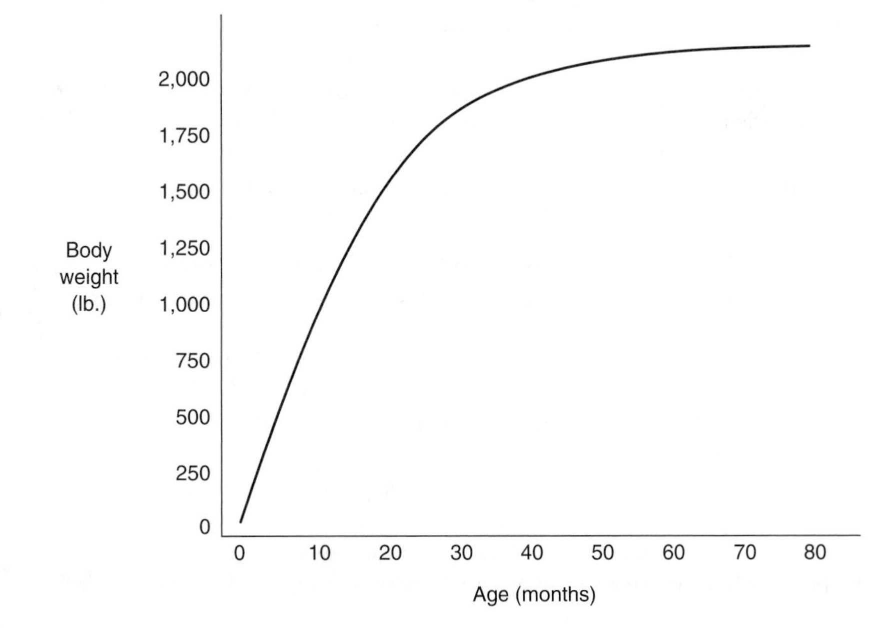

Figure 16–2
Growth curve for beef bull, intermediate frame size

Table 16-1
Brood cow phases with spring and fall calving programs

Phase/Season	Cow's Need for Forage to Support Fetal Growth	Cow's Need for Forage to Support Calf Growth through Lactation	Forage Availability Characteristic of the U.S. Midwest
1st trimester, Spring calving: early June Fall calving: early December	Minimal	Milk production is declining. Cow's need for nutrients is declining while calf's need for nutrients is increasing.	Spring calving: maximum forage availability Fall: minimal forage availability
2nd trimester, Spring calving: early September Fall calving: early March	Increasing, but still low	Calf is weaned and put on a different feeding program.	Spring calving: decreasing forage availability Fall: increasing forage availability
3rd trimester, Spring calving: early December Fall calving: early June	Maximal—90% of fetal growth occurs during last 40% of gestation.	Calf has been weaned.	Spring calving: minimal forage availability Fall: maximum forage availability
Postpartum period, Spring calving: Mid-March Fall calving: Mid-September	Cow is open (not pregnant).	This is the period of maximum milk production.	Spring calving: increasing forage availability Fall: decreasing forage availability

The brood cow's nutrient needs vary according to four phases: the first trimester, the second trimester, the third trimester, and the postpartum period. These phases are described in Table 16–1.

First Trimester

The cow nurses a calf throughout the first trimester. The first trimester begins when the nursing cow settles (becomes pregnant). It ends when the calf is weaned. Creep feeding is a management tool that impacts the nutritional programs of both the calf and the cow during the first trimester and beyond. Creep feed is nutrient-rich feed that is placed in a location so that only the calves have access to it. It is often placed near where the cows bed down. Meyers, Faulkner, Ireland, Berger, & Parrett (1999) describe many of the potential benefits of early weaning made possible through creep feeding. There are many issues that must be considered in deciding whether or not to creep feed calves:

1. A brood cow whose calf is creep fed will rebreed more quickly and require less feed to rebuild her body reserves if she lost weight during the time she was with the calf.
2. Creep fed calves will gain more weight by weaning than calves that are not creep fed.
3. The calf crop is more uniform in weight: calves of poor milkers eat more creep feed.
4. Creep feeding reduces weaning stress so weaning shrink (weight loss) is reduced.
5. Creep feed increases the cost of the brood-cow operation.
6. Heifers that are to be kept as replacements are usually not creep fed.
7. The price of calves in relation to the price of feed will determine profitability of creep feeding.
8. Creep feed is most likely to be economical when forage availability is low or of poor quality.

If creep feeding is used, it is usually started when the calf is 3 weeks old. The creep feeder should be located close to water, shade, and salt. Only 1 or 2 lb. of creep feed per calf should be placed in the feeder until the calves start to eat. Consumption will be approximately 1 lb. of creep feed per day for each 100 lb. of body weight. Feed must be kept fresh. The creep feeder should provide 4 inches of feeder length per calf (Table 12–5).

The needs of the fetal calf will not have an impact on the cow's nutrient needs during the first trimester. Toward the end of the first trimester, cow milk production decreases. As a result, calves will consume increasing amounts of forage. Cow nutrient requirements are relatively low toward the end of the first trimester. Adjustments of cow body condition are made most efficiently while the cow is still lactating toward the end of the first trimester.

Second Trimester

When the calf is weaned, the cow's nutrient requirements are at their lowest level. Although cattle replenish body-fat reserves with greater energetic efficiency while lactating, from a management perspective, the second trimester is usually the easiest time to make adjustments in the cow's body condition.

The body condition scoring system is used as a means of assessing body fat reserves and overall energy status. For beef cattle, the system uses scores from 1 to 9 as described in the following chapter, in Figure 17-9.

Cows should calve with a body-condition score between 4 and 7, and ideally, animals are fed to achieve this level of body fat by the end of the second trimester and are managed throughout the third trimester to maintain it.

Third Trimester

This is a period of increasing nutrient demand due to the developing fetus. Most of the birth weight of the calf is achieved during the third trimester of pregnancy. Failure to meet the cow's nutrient needs at this time may result in a reduced percentage calf crop: fewer of the brood cows will successfully produce a live and healthy calf. It may also result in a cow that has difficulty rebreeding.

In a spring calving program, the third trimester coincides with the winter season (Table 16-1). Cold temperatures will increase the cow's energy requirement if the environmental temperature falls below the animal's lower critical temperature (LCT). The LCT is the environmental temperature below which the animal will be unable to maintain body temperature through passive means. Below the LCT, feed energy will be repartitioned with a greater portion being oxidized to maintain body temperature and a reduction in energy available for other functions. The companion application to this text predicts the LCT given inputs regarding the animal's body size, body-condition score, hide thickness, hair coat, and energy intake. Adjustments in energy requirements are made if the inputted environmental temperature is below the animal's LCT. The third trimester ends at parturition.

The Postpartum Period

The postpartum period includes the time from parturition until the cow settles (conceives). In order for the operation to maintain a 12-month calving interval, the postpartum interval must be kept to no longer than about 80 days. This is because the gestation period is about 282 days and with a postpartum (open) period of much more than 80 days, it will take more than 365 days to complete the reproductive phases for production of a single calf.

Body-condition scoring is used to assess the amount of body fat available for mobilization in times of insufficient energy intake (Figure 17–9). Cows calving with a body condition score (BCS) of less than 4 will likely have an increased postpartum interval: the days to first estrus following calving. This delay will reduce the chances that these cows will be bred within the 80-day goal. Cows and especially heifers receiving inadequate nutrition during their pregnancy have poor reproductive performance in general (Randel, 1990). These animals show increased incidence of calving difficulty (dystocia), lowered conception rate, increased embryonic mortality, and decreased neonatal survival. The effects of inadequate nutrition during pregnancy are more severe in first-calf heifers than in older cows. At the other extreme, cows calving in fat condition (BCS over 7) are likely to have reduced conception rates.

After calving, the cow begins to lactate and will peak in milk production (11 to 27 lb./day, depending on the breed) within the postpartum period. The cow's greatest need for nutrients occurs during the postpartum period. Cows should be managed so that there are adequate forage resources available during this time. If fresh forage is not available, it will be necessary to feed preserved forages and perhaps a purchased energy supplement. Grazing cattle should receive a mineral supplement.

There are several ways to provide a supplemental source of nutrition to grazing cattle. Energy supplements may be hand- or limit fed. Hand feeding should be done during midday when cattle are not actively grazing so as not to disrupt normal grazing behavior. Also, it is important to be aware of dominance hierarchies in the cattle herd. First-calf heifers will be lower in dominance and will probably not receive any of a hand-fed supplement if they are not fed separately from the cows. Mineral supplements may be limit fed or fed free choice in the form of a loose mineral or a block. Free-choice liquid supplements can be formulated to contain energy, protein, mineral, or vitamin sources.

Some method of consumption control is necessary with free-choice supplements. A high level of salt or other substance is often added to alter palatability and limit consumption. When liquid feeds are offered through a lick tank, consumption may be controlled somewhat by adjustment of the wheel that delivers the liquid in the tank.

The postpartum period is often the time of greatest feed wastage. In a spring calving program, muddy conditions may result in feed losses of 40 percent or more of hand-fed supplements.

Beef Brood Cow Feeding and Nutrition Issues

Bloat

Bloat or *tympanites* of the rumen occurs in ruminants when the gases produced during fermentation cannot be expelled through eructation. The gases causing the rumen distension in bloat are the usual products of bacterial decomposition of carbohydrates and proteins, primarily methane and carbon dioxide with small amounts of hydrogen sulfide and others. In severe bloat, the distension of the rumen pushes the diaphragm forward, making breathing difficult. In addition, the large veins of the abdominal cavity may be compressed, interfering with general circulation. Symptoms of bloat include swelling at the left flank above the rumen, arched back with feet drawn under the abdomen, staggering gait, labored breathing, and suffocation.

There are many factors involved in the etiology of bloat. Categories of factors include those associated with the feedstuffs consumed and those associated with the animal.

The feedstuff factors that may be involved in bloat are summarized in Figure 16–3. Whereas pasture bloat is associated with the consumption of legumes

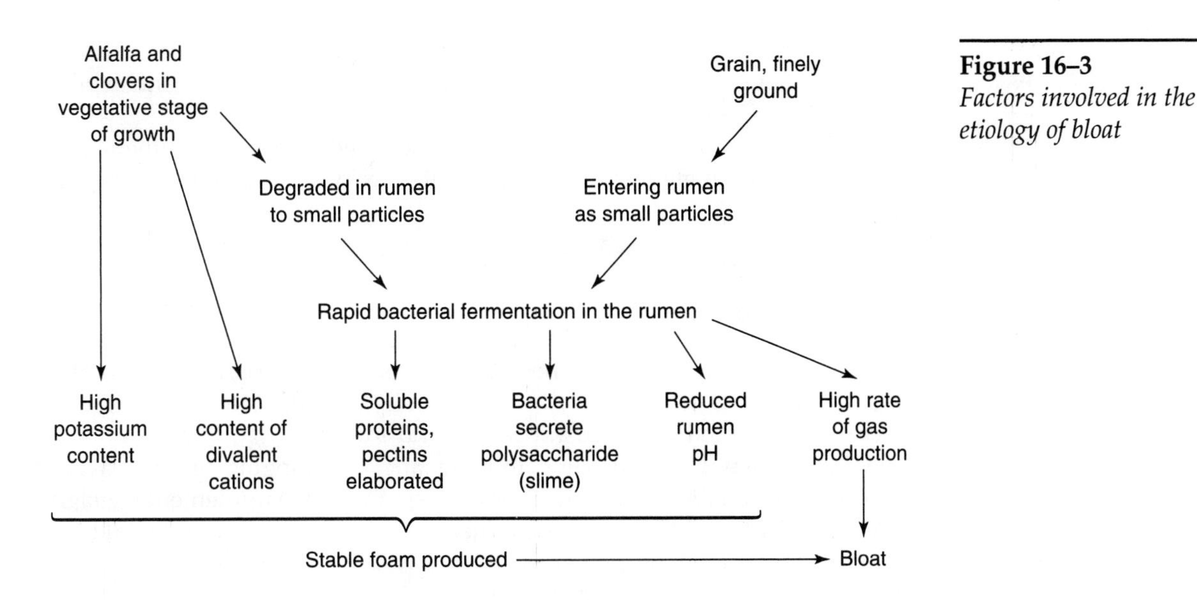

Figure 16–3
Factors involved in the etiology of bloat

such as alfalfa and clover in a vegetative stage of growth, feedlot bloat is associated with the consumption of grain that has been finely ground.

The primary factor associated with bloat in pastured cattle is the legume content of pasture. The legumes are rapidly fermented by bacteria in the rumen. This rapid fermentation results in a bacterial bloom in the rumen, leading to a high rate of gas production and polysaccharide secretion by rumen bacteria. This polysaccharide is usually described as *slime* (Blood & Radostits, 1989). In addition, the rapid fermentation of legume plants leads to the generation of a large amount of small particles in the rumen. Bacteria attach to these small particles. The foam that prevents eructation is apparently produced by some interaction between the slime and the small particles with attached bacteria. The cation content of legume has also been implicated, as positively charged ions increase the stability of foam containing negatively charged proteins.

In terms of animal factors, some animals may have a genetic predisposition to bloat. Genetic factors that influence rate of feed passage and saliva composition and/or production could influence the incidence of bloat. Animals that are bloat-susceptible should be culled from the breeding herd.

Bloat risk can be minimized in animals that are to graze pastures containing legume by waiting until the legumes have attained a more advanced stage of maturity (bloom stage or later). Also, initial turnout should be delayed until midday when pastures are dry. By managing pastures to maintain no more than 50 percent legume, bloat risk can be reduced. However, selective grazing may result in animals consuming a greater proportion of legume than is present in the pasture. It is also helpful to provide a full feed of coarse hay prior to turning animals out to a pasture containing legume. Finally, pastures containing legume should be introduced gradually, and once transitioned, animals with unrestricted access to pasture will have fewer problems with bloat than those that receive intermittent access to pasture.

Feed additives that are used in bloat prevention generally are designed to reduce the production and stability of the foam that is preventing gas release through eructation. Detergents, surfactants, and vegetable oils have been used successfully in reducing the risk of bloat. Poloxalene, marketed as Bloat Guard, is a nonionic surfactant that acts as an antifoaming agent. Poloxalene is available in meal form to be mixed with grain. It may also be fed through blocks and liquid feeds, but because consumption of such feed sources is voluntary, the effectiveness of bloat prevention is reduced. Poloxalene is also available as a drench for emergency treatment of bloat. The category of antibiotics known as ionophores has been shown to have some value in reducing the incidence of bloat (Bagley & Feazel, 1989).

Environmental factors as they affect plant growth, changing rumen microbial species, and changing rumen microbial activity may also be contributing factors to bloat.

Grass Tetany

Grass tetany may occur in cattle that are grazing on grass pastures without appropriate mineral supplements available. Grass tetany is triggered by inadequate intake of bioavailable magnesium. The disorder is described in Chapter 9 (Magnesium, Signs of deficiency). It is especially prevalent in the spring when cattle are first turned out to lush pasture, and the challenge of prevention involves the logistics of providing grazing and free-ranging animals with a source of bioavailable magnesium. For these animals, supplemental magnesium is usually provided in a loose mineral fed free choice or a mineral block. The problem can also be prevented by intravenous injection of calcium and magnesium in the gluconate form.

Nitrate Toxicity

Nitrate toxicity may occur in cattle that graze stressed pasture crops. Such stresses as cloudy weather, drought, and frost cause the plant to accumulate nitrate in the lower third of the stalk (nitrate poisoning is also called "corn stalk poisoning"). High levels of fertilizer application and herbicide treatment may also cause plants to accumulate nitrate. Suspected forage should be sent to a laboratory for analysis.

If absorbed, nitrate (NO_3^-) in the blood may cause kidney damage. In ruminant animals, however, consumption of forages containing high levels of nitrate usually results in microbial reduction of nitrate to nitrite (NO_2^-). If absorbed into the blood, nitrite converts hemoglobin to methemoglobin, which cannot transport oxygen. The symptoms of the usual nitrate poisoning in ruminants are really those of nitrite poisoning, which are those associated with asphyxiation (accelerated respiration and pulse rate), weakness, and trembling.

Crops that have accumulated nitrate may be safely fed as silage because the fermentation process chemically reduces some nitrates to gaseous nitrogen oxides. If ensiling is not an option, crops that have accumulated nitrate may usually be safely harvested for feeding by leaving a 12-inch stubble in the field. When feeding high-nitrate crops, the following precautions should be followed:

1. Feed a balanced ration (high dietary levels of readily fermentable carbohydrate increase the ruminant animal's tolerance to nitrate by increasing the efficiency of utilization of nitrogen by bacteria).
2. Feed smaller amounts of the high-nitrate feedstuff.
3. Multiple feedings of small amounts of the high-nitrate feedstuff will reduce losses from nitrate.

Water can also be an important source of nitrate. Water nitrate usually indicates surface pollution from fertilizers, manure, and/or sewage. As with forages, water suspected of nitrate contamination should be tested.

Tall Fescue Toxicosis

Beef animals that consume pasture or hay containing tall fescue that has been infected with an endophyte are susceptible to tall fescue toxicosis, which has various effects on health, production, and reproduction. Tall fescue toxicosis is discussed in Chapter 24. The causative agents involved in fescue toxicosis have been reviewed by Porter (1995).

Breeding Herd Replacements

Heifers

The heifers that are kept as herd replacements are still growing and have different nutritional needs than the cows. As with the brood cows, replacement heifers use nutrients for maintenance. In addition to maintenance, replacement heifers will use nutrients for both growth and reproduction. The diets fed to replacement heifers must supply a level of nutrition that supports all active functions. For this reason and because they will be lower on the dominance hierarchy, they should be fed and managed separately from the cows.

Replacement heifers should be managed to have their first calf by 2 years of age. The calving period for this group should be restricted to 45 days or less. To achieve these goals, the replacement heifer group must be fed to grow uniformly so that the group attains puberty in time to breed at 13 to 14½ months of age.

For heifers of typical beef breeds, including Angus, Hereford, Charolais, and Limousin, puberty is attained when animals reach 60 percent of their mature body weight. Use of ionophores in replacement heifer diets has resulted in reduced age at puberty (Bagley, 1993). Heifers of the dual-purpose or dairy breeds including Holstein, Brown Swiss, Braunvich, and Gelbvieh, should be fed to reach 55 percent of their mature body weight by breeding. With good management, these body weights correspond to 13 to 16 months of age. Because the gestation period is about 9 months (282 days), heifers bred at 13 to 16 months of age will calve at 22 to 25 months. To improve the chances for dropping a live healthy calf, bred first-calf heifers should be fed adequate nutrition during the gestation period to support the gain necessary to achieve 85 percent of mature weight by calving.

As an example, an Angus heifer's mature weight is 1,100 lb. Puberty is expected at 14 months (420 days of age) when she weighs about 60 percent of her mature weight ($0.60 \times 1{,}100 = 660$ lb.). If weaning weight at 200 days of age is 450 lb., the heifer should be fed to gain $660 - 450 = 210$ lb. for the period from weaning to breeding.

At weaning, she is 200 days old so she has $420 - 200 = 220$ days from weaning to achieve her 660-lb. breeding weight. The Angus heifer needs to gain 210 lb. in 220 days, which translates to a rate of gain of $210/220 = 0.95$ lb. per day.

In addition to the nutrient demands of pregnancy, heifers are growing animals. In our example, the first calf heifer should weigh 935 lb. (85 percent of her mature body weight) at calving. In our example, the heifer was bred at 420 days so her age at calving will be gestation length + 420 or $282 + 420 = 702$ days at calving. This translates to a gain from breeding to calving of $935 - 660 = 275$ lb. in $702 - 420 = 282$ days. The necessary rate of gain from breeding to calving is $275/282 = 0.98$ lb./head/day (Table 16–2).

Nutritional management of breeding herd replacements has been reviewed (Bagley, 1993).

Table 16–2 *Approximate relationship between age and weight of beef calves*

	Age (days)	Weight (lb.)
Weaning	200	450
Breeding	420	660
Difference weaning to breeding	220	210
Daily gain necessary—weaning to breeding	210/220 = 0.95 lb./day	
Calving	702	935
Difference breeding to calving	282	275
Daily gain necessary—breeding to calving	275/282 = 0.98 lb./day	

Bulls

Compared to growing/finishing cattle of similar body weights and rates of gain, growing bulls require 15 percent more Mcal of net energy for maintenance. Likewise, mature bulls require 15 percent more Mcal of NEm compared to cows. Other nutrient requirements are similar in comparable bulls and growing/finishing cattle. Note that the bull grows to a body weight that is 67 percent greater than the cow. Both underfeeding and overfeeding bulls will result in poor breeding performance.

GROWING/FINISHING CATTLE

The goal with weaned calves that are not intended to be kept as breeding herd replacements is to develop them into a product that may be profitably marketed. Generally, this means attaining a quality grade of Choice. Quality grade is based on the level and distribution of fat within muscle tissue, usually referred to as *marbling*. Choice quality grade is just below Prime at the top of the USDA grade standards for steers and heifers.

There are many systems for developing cattle into a product that attains a Choice carcass. Figure 16–1 illustrates two basic systems described as forage based and grain based. All systems share some features of these two basic systems. Generally, the forage-based system will require more days to achieve a quality grade of Choice than will the grain-based system. However, the grain-based system will require more purchased feed to achieve a quality grade of Choice. Generally, the heifer or steer in the forage-based system will achieve Choice at a larger body weight than the animal in the grain-based system. The decision of which system to use will be based on anticipated beef prices, availability of forage resources, and grain prices.

The finishing phase is common to many forage-based and grain-based systems. It is the final feeding phase before slaughter and it is largely responsible for the degree of marbling present in the carcass. In the finishing phase, cattle are fed a diet containing 85 to 100 percent grain, and the rate of gain is frequently greater than 3.5 lb./head/day. The companion application to this text predicts daily gain supported by both the levels of metabolizable energy and protein in the ration.

To prevent digestive problems in the finishing phase, a transition or "receiving program" is usually implemented. Although receiving programs vary, they all share the goals of acclimating cattle to bunk feeding and gradual replacement of forage with grain. If calves arrive at the finishing phase from a backgrounding or grower phase, many of the goals of the receiving program will have already been met. Some receiving diets will contain relatively high levels of antibiotics as prophylactic medication. These levels will be reduced or antibiotic feeding will be eliminated at the conclusion of the receiving period in 2 to 3 weeks. Many antibiotics have withdrawal periods and cattle must receive a diet free from these products for the duration of the withdrawal period immediately prior to slaughter. The transition from the receiving diet to the final finishing diet generally takes 2 to 3 weeks.

When cattle reach the final finishing diet, they are said to be on full feed. The grain of the finishing diet will usually be corn, milo, wheat, or barley grain. These grains may be fed dry or fermented, whole or processed. Grain is usually processed before feeding (Chapter 13). The effects of grain source and processing method on feedlot cattle performance has been reviewed (Owens, Secrist, Hill, & Gill, 1997)

The phenomenon of compensatory gain is frequently exhibited by cattle that arrive at the finishing phase from a forage-based system. The growth of such cattle has been suppressed due to poor forage quality and/or availability.

On the finishing diet, these cattle gain faster, and in some cases, more efficiently than others of the same weight.

Urea

Urea is a common component of the finishing diet. Urea is a form of nonprotein nitrogen (NPN) as opposed to preformed plant or animal protein. Urea in the rumen is first hydrolyzed by microbial urease to ammonia. This conversion is relatively rapid and requires little energy. The nitrogen in the ammonia is then incorporated into the bacteria as amino acids in cell proteins. This incorporation is relatively slow and requires considerable energy.

Urea should not be fed to a ruminant animal at a level greater than what the microbes can use. The limiting factor is the level of fermentable carbohydrate in the diet. The fermentable carbohydrate yields the energy needed by the microbes to fully utilize urea. Diets containing only low-quality forage and no grain should not contain urea because, although urea will be converted to ammonia, incorporation will not occur and the ammonia may be absorbed, leading to toxic effects. Based on usual grain levels, one rule of thumb for urea use in growing/finishing rations is to limit it to a level of urea that supplies no more than one third of the total ration nitrogen (Taylor, 1994). Given the many different types of growing/finishing diets, the application of such rules of thumb is risky. The best way to determine the upper limit for urea in a given ration is to find that level beyond which no further increase in microbial growth is supported. To this end, the companion application for ruminants predicts the bacterial growth from degradable protein sources such as urea. The companion application also calculates the percentage of ration nitrogen from nonprotein nitrogen and the equivalent crude protein from nonprotein nitrogen when using urea in a blend of feedstuffs.

The protein that the microbes make from urea is composed of amino acids. Some of these amino acids contain the mineral sulfur. Sulfur may not be present in adequate amounts in finishing diets to make efficient use of urea and may require supplementation. The usual recommendation for the ratio of nitrogen to sulfur is from 10:1 to 12:1 for ruminant diets containing urea (Qi, Lu, & Owens, 1993). The companion application to this text displays the nitrogen-to-sulfur ratio as ruminant rations are developed.

Feed Additives

A variety of additives are approved for use in finishing cattle. Feed additives are used to achieve many economically important effects including the improvement of feed efficiency, daily gain, the increase of carcass weight at Choice, and the reduction of incidence of some health problems such as coccidiosis, laminitis, acidosis, bloat, and liver abscesses. Approved additives include antibiotics, probiotics, and chemotherapeutic agents. Some products designed to improve feed utilization by cattle are administered as an implant beneath the skin rather than as a feed additive. Anabolic ear implants have been shown to increase dry matter intake (DMI) and body-weight gain, and improve feed conversion (Rumsey, Hammond, & McMurtry, 1992). Melengestrerol acetate (MGA) is a feed additive with similar effects. The primary use of these anabolic products is to increase rate of gain through protein and fat accretion. Implant use or no use is an input in the companion application to this text. When no implant is used, the application reduces predicted DMI by 6 percent. When an implant is used, DMI is not reduced and the slaughter weight at the chosen level of finish should be increased. Likewise, ionophore use or no use is an input in the companion application to this text. When an ionophore is used, the application reduces predicted DMI by 6 percent and applies a credit to the feed net energy for maintenance

(NEm) value, increasing the ration NEm concentration by 12 percent. These adjustments are made based on recommendations in the NRC (2000).

Approved feed additives are found in the Feed Additive Compendium (2004), and these as well as implants are described at the Web site titled *FDA Approved Animal Drug Products* located at http://dil.vetmed.vt.edu/NADA/default.cfm. The companion application to this text is linked to this Web site in the feedstuff blending section. A discussion of feed additives is found in Chapter 3.

Growing/Finishing Cattle Feeding and Nutrition Issues

Bloat in Feedlot Cattle

Though less common than in pastured cattle, bloat may also occur in growing/finishing cattle. Although the diets of these two types of beef animals are obviously different, the characteristics of feedstuffs associated with bloat in feedlot cattle are similar to those that cause bloat in pastured cattle (Figure 16–3). In addition, the highly fermentable carbohydrate in feedlot diets results in the elaboration of acid that may reduce rumen pH. A reduced rumen pH may result in reduced rumen motility, allowing gasses to accumulate. The incidence of bloat in feedlot cattle may be influenced by the type of grain fed, as well as the type and extent of grain processing (Cheng, McAllister, Popp, Hristov, Mir, & Shin, 1998).

Polioencephalomalacia

Polioencephalomalacia (PEM) is a disorder of sheep and cattle that appears to be caused by the ingestion of substances that create a thiamin deficiency. For ruminants, the source of thiamin is microbial activity in the rumen. Thiaminase enzymes can cause PEM by destroying thiamin in the rumen. Thiaminases are found in raw fish products, bracken fern *(Pteridum aquilinum),* field horsetail *(Equisetum arvense),* and others. PEM can also be caused if dietary factors encourage the growth of thiaminase-producing bacteria in the rumen. Such dietary factors are not well defined, but are associated with high-grain diets. Finally, PEM can be caused by ingestion of excessive sulfur-containing compounds, including excess sulfates in water. This may lead to the production of a thiamin analog, which, though it serves none of the thiamin functions, appears to replace thiamin in important metabolic reactions.

Enterotoxemia

Calves, like lambs and kids, are subject to enterotoxemia (overeating disease) if stressed or poorly transitioned from a high-forage to a high-grain diet. Enterotoxemia is discussed in Chapter 20.

Acidosis and Related Disorders

Acidosis is an ailment that frequently afflicts cattle on finishing diets. Acidosis in this text is discussed in Chapter 20. Reviews of acidosis in finishing cattle have been made (Owens, Secrist, Hill, & Gill, 1998; Schwartzkopf-Genswein, Beauchemin, Gibb, Crews, Hickman, Streeter, & McAllister, 2003).

In the companion application to this text, rumen pH is predicted from the level of physically effective neutral detergent fiber (NDF) in the ration (Figure 18–4). Beef cattle that are continually challenged with low rumen pH are more likely to have liver abscesses. Damage to the rumen epithelium caused by excessive acid production allows pathogens to leave the rumen and enter the liver. Affected animals have reduced DMI[UN1] and feed efficiency and decreased carcass yield. Control measures generally involve feeding antibiotics. A description of the causes and options for control of liver abscesses has been reviewed (Nagaraja & Chengappa, 1998).

Like liver abscesses, laminitis is associated with the acidosis complex. Factors that predispose cattle to laminitis have been reviewed (Vermunt & Greenough, 1994).

Repartitioning Agents

As with swine, the use of repartitioning agents is an active area of research in the beef industry. Such agents have been reported to improve the efficiency of production of lean beef (Moloney, Allen, Ross, Olson, & Convey, 1990).

END-OF-CHAPTER QUESTIONS

1. Discuss the pros and cons of creep feeding beef calves.
2. What mineral is associated with grass tetany?
3. What causes pasture bloat? What causes feedlot bloat?
4. Define *lower critical temperature.*
5. When can an increase in the brood cow's body condition be achieved most efficiently?
6. Explain how the nitrogen-to-sulfur ratio is applied to urea feeding. Why is the amount of urea that can be used by beef animals related to the amount of fermentable carbohydrate in the diet?
7. Explain why ruminant consumption of forages high in nitrates may result in poisoning due to nitrites.
8. What is the target weight, expressed as a percentage of mature weight, to breed Hereford and Angus replacement heifers? What percent of body weight should they reach by calving?
9. Describe the differences between a forage-based and a grain-based feeding program for beef cattle.
10. The following question pertains to conditions in the northern hemisphere. In a spring calving program, maximum forage availability corresponds with the first trimester of the brood cow. In a fall calving program, maximum forage availability corresponds with the third trimester of the brood cow. Discuss the advantages and disadvantages of each type of calving program.

REFERENCES

Bagley, C. P. (1993). Nutritional management of replacement beef heifers: a review. *Journal of Animal Science. 71,* 3155–3163.

Bagley, C. P., & Feazel, J. I. (1989). Influence of a monensin ruminal bolus on the performance and bloat prevention of grazing steers. *Nutrition Reports International. 40,* 707–716.

Blood, D. C., & Radostits, O. M. (1989). *Veterinary medicine. A textbook of the diseases of cattle, sheep, pigs, goats and horses* (7th edition). London: Baillere Tindall.

Cheng, K. J., McAllister, T. A., Popp, J. D., Hristov, A. N., Mir, Z., & Shin, H. T. (1998). A review of bloat in feedlot cattle. *Journal of Animal Science. 76,* 299–308.

Feed Additive Compendium (2004). Minnetonka, MN: Miller Publishing Co.

FDA Approved Animal Drug Products, Online Database System. (n.d.) Retrieved July 8, 2005 from: http://dil.vetmed.vt.edu/NADASecond/NADA.cfm

Moloney, A. P., Allen, P., Ross, D. B., Olson, G., & Convey, E. M. (1990). Growth, feed efficiency and carcass composition of finishing Fresian steers fed the β-adrenergic agonist L-644,969. Journal of Animal Science. *68,* 1269–1277.

Myers, S. E., Faulkner, D. B., Ireland, F. A., Berger, L. L., & Parrett, D. F. (1999). Production systems comparing early weaning to normal weaning with or without creep feeding for beef steers *Journal of Animal Science. 77,* 300–310.

Nagaraja, T. G. & Chengappa, M. M. (1998). Liver abscesses in feedlot cattle: a review. *Journal of Animal Science. 76,* 287–298.

National Research Council. (2000). *Nutrient requirements of beef cattle* (7th revised edition). Washington, DC: National Academy Press.

Owens, F. N., Secrist, D. S., Hill, W. J., & Gill, D. R. (1997). The effect of grain source and grain processing on performance of feedlot cattle: A review. *Journal of Animal Science. 75,* 868-879.

Owens, F. N., Secrist, D. S., Hill, W. J., & Gill, D. R. (1998). Acidosis in cattle: A review. *Journal of Animal Science. 76,* 275–286.

Porter, J. K. (1995). Analysis of endophyte toxins: Fescue and other grasses toxic to livestock. *Journal of Animal Science. 73,* 871–880.

Qi, K., Lu, C. D., & Owens, F. N. (1993). Sulfate supplementation of growing goats: Effects on performance, acid-base balance, and nutrient digestibilities. *Journal of Animal Science. 71,* 1579–1587.

Randel, R. D. (1990). Nutrition and postpartum rebreeding in cattle. *Journal of Animal Science. 68,* 853–862.

Rifkin, J. (1993). *Beyond Beef.* New York: The Penguin Group.

Rumsey, T. S., Hammond, A. C., & McMurtry, J. P. (1992). Response to reimplanting beef steers with estradiol benzoate and progesterone: Performance, implant absorption pattern, and thyroxine status. *Journal of Animal Science. 70,* 995–1001.

Schwartzkopf-Genswein, K. S., Beauchemin, K. A., Gibb, D. J., Crews, D. H., Jr., Hickman, D. D., Streeter, M., & McAllister, T. A. (2003). Effect of bunk management on feeding behavior, ruminal acidosis and performance of feedlot cattle: A review. *Journal of Animal Science. 81,* E149–158E.

Taylor, R. E. (1994). *Beef Production and Management Decisions* (2nd edition). New York: Macmillan Publishing Company.

Vermunt, J. J., & Greenough, P. R. (1994). Predisposing factors of laminitis in cattle. *British Veterinary Journal. 150,* 151–164.

CHAPTER 17

BEEF RATION FORMULATION

TERMINOLOGY

Workbook A workbook is the spreadsheet program file including all its worksheets.

Worksheet A worksheet is the same as a spreadsheet. There may be more than one worksheet in a workbook.

Spreadsheet A spreadsheet is the same as a worksheet.

Cell A cell is a location within a worksheet or spreadsheet.

Comment A comment is a note that appears when the mouse pointer moves over the cell. A red triangle in the upper right corner of a cell indicates that it contains a comment. Comments are added to help explain the function and operations of workbooks.

Input box An input box is a programming technique that prompts the workbook user to type information. After typing the information in the input box, the user clicks **OK** or strikes **ENTER** to enter the typed information.

Message box A message box is a programming technique that displays a message. The message box disappears after the user clicks **OK** or strikes **ENTER**.

EXCEL SETTINGS

Security: Click on the **Tools** menu, **Options** command, **Security** tab, **Macro Security** button, **Medium** setting.

Screen resolution: This application was developed for a screen resolution of 1024 × 768. If the screen resolution on your machine needs to be changed, see Microsoft Excel Help, "Change the screen resolution" for instructions.

HANDS-ON EXERCISES

BEEF RATION FORMULATION

Double click on the **BeefRation** icon. The message box in Figure 17-1 displays.

Figure 17–1

> Macros may contain viruses. It is advisable to disable macros, but if the macros are legitimate, you may lose some functionality.
> Disable Macros or Enable Macros or More Info

Click on **Enable macros**

The message box in Figure 17–2 displays.

Figure 17–2

> Function keys **F1** to **F8** are set up. You may return to this location from anywhere by striking **ENTER**, then the **F1** key. Workbook by David A. Tisch. The author makes no claim for the accuracy of this application and the user is solely responsible for risk of use. You're good to go. TYPE ONLY IN THE GRAY CELLS! Note: This Workbook is made up of charts and a worksheet. The charts and worksheet are selected by clicking on the tabs at the bottom of the display. Never save the Workbook from a chart; always return to the worksheet before saving the Workbook.

Click **OK**.

Input Beef Animal Procedure

Click the **Input Beef Animal** button. The input box shown in Figure 17-3 displays.

Figure 17–3

> 1. Growing/finishing
> 2. Brood cow, lactating
> 3. Brood cow, gestating, not lactating
> 4. Replacement heifer
>
> Note: Bulls constitute 3 to 5 percent of the cattle herd and this application does not compute bull requirements. The nutrient requirements for growing bulls may be as much as 20 percent higher than that for replacement heifers. Mature bulls are usually maintained on the cow ration.
> ENTER THE APPROPRIATE NUMBER:

Click **OK** to choose the default input of #1, the Growing/finishing beef animal. The input box shown in Figure 17–4 displays.

Enter measured dry matter intake (pounds) if available or click **OK** to accept the predicted value at CX26.

Figure 17–4

Click **OK** to accept the predicted dry matter intake (DMI). The message box shown in Figure 17–5 displays.

Enter the appropriate cell inputs in column CX and finish entering inputs by clicking on the buttons below.

Figure 17–5

Click **OK**.

The Cell Inputs

The Cell Inputs, Animal Inputs are shown in Table 17–1. Enter these values in appropriate cells on the worksheet.

Table 17–1
Animal inputs for the beef ration formulation example

1	Breeding adjustment factor for DMI
1	Body fat adjustment factor for DMI
1	Breed effect on NEm
1060	Current body weight (lb).
	N/A
5	Body condition score, 1–9 (used to compute energy needs and lower critical temperature).
	N/A
	N/A
	N/A
	N/A
	N/A
	N/A
	N/A
	N/A
	N/A
1054	Standard reference weight (lb.)
1300	The mean final weight of the group of animals under consideration.

The comment behind the input, *Breeding adjustment factor for dry matter intake,* is shown in Figure 17–6.

Breed Effect	Input Value
Holstein X holstein	1.08
Holstein X beef	1.04
Beef X beef	1.00

Figure 17–6

The comment behind the input, *Body fat adjustment factor for dry matter intake,* is shown in Figure 17–7.

Figure 17–7

Body Fat Percentage (and weight)	Input Value
21.3 (772 lb.)	1.00
23.8 (882 lb.)	0.97
26.5 (992 lb.)	0.90
29.0 (1,102 lb.)	0.82
31.5 (1,213 lb.)	0.73

The comment behind the input, *Breed effect on NEm* (net energy for maintenance) is shown in Figure 17–8.

Figure 17–8

Breed	Breed Effect on NEm Requirement
Angus	1.00
Braford	0.95
Brahman	0.90
Brangus	0.95
Braunvieh	1.20
Charolais	1.00
Chianina	1.00
Devon	1.00
Galloway	1.00
Gelbvieh	1.00
Hereford	1.00
Holstein	1.20
Jersey	1.20
Limousin	1.00
Longhorn	1.00
Maine Anjou	1.00
Nellore	0.90
Piedmontese	1.00
Pinzgauer	1.00
Polled Hereford	1.00
Red Poll	1.00
Sahiwal	0.90
Salers	1.00
Santa Gertudis	0.90
Shorthorn	1.00
Simmental	1.20
South Devon	1.00
Tarentaise	1.00

The comment behind the input, *Condition score, 1–9* (used to compute energy needs and lower critical temperature) is shown in Figure 17–9.

Condition Score

1. Extremely emaciated and listless
2. Somewhat emaciated
3. Individual ribs visible
4. Individual ribs not obvious
5. Can feel fat cover over ribs and tailhead
6. Pressure needed to feel spine
7. Some fat in brisket, feels spongy over ribs
8. Very fleshy, brisket full, cannot feel spine
9. Extremely fleshy and blocky

Figure 17–9

The comment behind the input, *Standard reference weight (lb.)* is shown in Figure 17–10.

Standard Reference Body Weight

This reference weight is the weight that corresponds to a degree of fatness for the cattle type. It is used along with the mean final weight to develop a scaling system in which the inputted animal can be compared to a medium-framed steer for calculation of nutrient requirements.

1,054 for small marbling (27.8% body fat) and replacement heifers
1,019 for slight marbling (26.8% body fat)
959 for animals finishing at trace marbling (25.2% body fat)

Figure 17–10

The comment behind the input, *The mean final weight of the group of animals under consideration* is shown in Figure 17–11.

For growing and finishing animals, this weight should reflect the weight at which the grade chosen is attained. The average final body weight for U.S. steers is 1,224 lb.
Note that use of an anabolic implant (Management Inputs) will result in increased mean final body weights at the chosen degree of fatness (see *Standard Reference Weight*).
For replacement heifers and beef cows, this value should reflect the mean weight of animals greater than 60 months of age at a condition score of 5.

Figure 17–11

Click on the **`Management Inputs`** button and enter the values in Table 17–2.

Table 17–2
Management inputs for the beef ration formulation example

yes	Anabolic implant use (yes/no)
yes	Ionophore use (yes/no)
no	Pasture use (yes/no)
	N/A
	N/A
	N/A
	N/A
	N/A
	N/A

The comment behind the cell containing the Anabolic implant use label is shown in Figure 17–12.

Figure 17–12

Anabolic implants are products implanted in the ear. Melengestrerol acetate (MGA) is a similar product delivered through feed. These products are used primarily to enhance rate of gain. They also may increase DMI.

The comment behind the Anabolic implant input cell is shown in Figure 17–13.

Figure 17–13

Unlike ionophores, anabolic agents have minimal impact on nutrient utilization. Their primary effect is to increase rate of gain, which can be accounted for by adjusting the mean final weight of the group of animals under consideration at Animal Inputs.

The comment behind the cell containing the Ionophore use label is shown in Figure 17–14.

Figure 17–14

Ionophores are products that improve the efficiency with which the animal is able to use feed energy to meet maintenance needs. Most studies suggest that there is a reduction in dry matter intake with ionophore use.

The comment behind the ionophore input cell is shown in Figure 17–15.

Figure 17–15

Ionophore Use	Adjustment Applied to Dry Matter Intake	Adjustment Applied to Feed NEm Concentration
Yes	decreased by 6%	increased by 12%
yes + implant use	no adjustment	increased by 12%
No	no adjustment	no adjustment

If ionophore is used, it is assumed that it is included at 27.5 to 33 mg/kg (ppm) diet dry matter.

Click on the **Environment Inputs** button and enter the information shown in Table 17–3.

Table 17–3
Environmental inputs for the beef ration formulation example

30	Current temperature (in ° F)
40	Average temperature for the previous month (in ° F)
Yes	Is there significant relief from the heat at night? (Heat stress is unlikely—entry should probably be "yes").
5	Wind speed (mph)
0.5	Hair coat effective depth (in.)
1	Mud adjustment factor for external insulation and DMI (see comment)
1	Hide adjustment factor for external insulation (0.8: thin; 1: average; 1.2: thick).
1	Heat stress unlikely—input should probably be "1" (see comment)

The information immediately below the Environment Cell input section is presented in Figure 17–16.

Figure 17–16

5.6 Surface area, m^2 (***not an input***)
9.0 Tissue insulation, ° C/Mcal/m^2/day (***not an input***)
8.2 External insulation, ° C/Mcal/m^2 (***not an input***)
The values above, along with energy intake, are used to calculate the animal's lower critical temperature (LCT). The LCT is the environmental temperature below which the animal's tissue insulation (hide and body fat) and external insulation (hair coat) will be unable to maintain body temperature. Below the LCT, feed energy will be repartitioned, with a greater portion being oxidized to maintain body temperature and a reduction in energy available for other functions.

The comment behind the cell containing the Mud adjustment factor for external insulation and DMI label is shown in Figure 17–17.

Figure 17–17

1 Dry & clean
2 Some mud on lower body
3 Wet & matted
4 Covered with wet snow or mud

The comment behind the cell containing the Heat Stress label is shown in Figure 17–18.

Figure 17–18

Enter **1** if no heat stress, **2** for rapid shallow panting, **3** for open-mouth panting.

Strike **ENTER** and **F1**.

Select Feeds Procedure

Click on the **Select Feeds** button. The message box in Figure 17–19 displays.

Figure 17–19

Nutrient content expressed on a dry matter basis. Feedstuffs are listed first by selection status, then, within the same selection status, by decreasing protein content, then, within the same protein content, alphabetically.

Click **OK**.

Explore the table. Note the nutrients listed as column headings. Note also that the table ends at row 200. You select feedstuffs for use in making two different products: (1) a blend to be mixed and sold bagged or bulk, and (2) the ration to be fed directly to livestock.

Select the feedstuffs in Table 17–4 by placing a 1 in the column to the left of the feedstuff name. All unselected feedstuffs should have a 0 value in the column to the left of the feedstuff name. If you wish to group feedstuffs but not select them, you would place a value between 0 and 1 to the left of the feedstuff name.

Table 17–4
Feedstuffs to select for the beef ration formulation example

Feedstuff
Barley grain, heavy
Barley silage
Copper sulfate
Limestone, ground
Potassium iodide
Salt (sodium chloride)
Vitamin A supplement
Vitamin D supplement
Vitamin E supplement
Zinc oxide

Strike **ENTER** and **F2**.

Selected feedstuffs and their analyses are copied to several locations in the workbook.

Make Ration Procedure

Click on the **Make Ration** button. The message box shown in Figure 17–20 displays.

Figure 17–20

ENTER POUNDS TO FEED IN COLUMN B. TOGGLE BETWEEN NUTRIENT WEIGHTS AND CONCENTRATIONS USING THE **F5** KEY, RATION AND FEED CONTRIBUTIONS USING THE **F6** KEY. WHEN DONE STRIKE **F1**. Cell is highlighted in red if nutrient provided is poorly matched with nutrient target. The lower limit is taken as 95 to 98 percent of target, depending on the nutrient. The upper limit of acceptable mineral is taken from NRC, 1980. This publication gives safe upper limits of minerals as salts of high bioavailability. The upper limit of vitamin is taken from the NRC, 1987. For other nutrients, the upper limit is based on unreasonable excess and the expense of unnecessary supplementation. See Table 17–6 for specifics.

Click **OK**.
The message box in Figure 17–21 displays.

The Goal Seek feature may be useful in finding the pounds of a specific feedstuff needed to reach a particular nutrient target:

1. Select the red cell highlighting the deficient nutrient
2. From the menu bar, select **Tools**, then **Goal Seek**
3. In the text box, 'To Value:' enter the target to the right of the selected cell
4. Click in the text box, "By changing cell:" and then click in the gray "Pounds fed" area for the feedstuff to supply the nutrient
5. Click **OK**. You may accept the value found by clicking **OK** or reject it by clicking **Cancel**

WARNING: Using Goal Seek to solve the unsolvable (e.g., asking it to make up an iodine shortfall with iron sulfate) may result in damage to the Workbook. IMPORTANT: If you return to the Feedtable to remove more than one feedstuff from the selected list, you will lose your chosen amounts fed in the developing ration.

Figure 17–21

Click **OK**.
In the gray area to the right of the feedstuff name, enter the pound values shown in Table 17–5.

Table 17–5
Inclusion rates for feedstuffs in the beef ration formulation example

Feedstuff	Pounds
Barley grain, heavy	18
Barley silage	27
Copper sulfate	0.00068
Limestone, ground	0.26
Potassium iodide	0.000018
Salt (sodium chloride)	0.06727
Vitamin A supplement	0.00898
Vitamin D supplement	0.00146
Vitamin E supplement	0.01917
Zinc oxide	0.00044

The Nutrients Supplied *Display*

The application highlights ration nutrient levels, expressed as amount supplied per beef animal per day, that fall outside the acceptable range.

The lower limit for dry matter intake is 95 percent of the target. The lower limit for net energy for maintenance (NEm) is 100 percent of the target. The lower limit for sodium is 97 percent of the target. The lower limit for all other ration nutrients is taken as 98 percent of the target. Table 17–6 shows the upper limits of the acceptable range for the various nutrients.

Table 17–6
Upper limits for ration nutrients used in the beef ration software

	Upper Limit	Source
DMI	5% over the predicted requirement	Author
NEm	Excess beyond maintenance is used to predict gain	
MP	Excess beyond maintenance is used to predict gain	
Ca	2% of ration dry matter	NRC: Mineral Tolerance of Domestic Animals (1980)—cattle
P	1% of ration dry matter	NRC: Mineral Tolerance of Domestic Animals (1980)—cattle
Magnesium	0.5% of ration dry matter	NRC: Mineral Tolerance of Domestic Animals (1980)—cattle
Potassium	3% of ration dry matter	NRC: Mineral Tolerance of Domestic Animals (1980)—cattle
Sulfur	0.4% of ration dry matter	NRC: Mineral Tolerance of Domestic Animals (1980)
Sodium	3.5% nonlactating, 1.6% lactating	NRC: Mineral Tolerance of Domestic Animals (1980)—cattle, based on limit for salt (sodium chloride)
Salt	9% nonlactating, 4% lactating	NRC: Mineral Tolerance of Domestic Animals (1980)—cattle
Iron	1,000 mg/kg or ppm of ration dry matter	NRC: Mineral Tolerance of Domestic Animals (1980)—cattle
Manganese	1,000 mg/kg or ppm of ration dry matter	NRC: Mineral Tolerance of Domestic Animals (1980)—cattle
Copper	100 mg/kg or ppm of ration dry matter	NRC: Mineral Tolerance of Domestic Animals (1980)—cattle
Zinc	500 mg/kg or ppm of ration dry matter	NRC: Mineral Tolerance of Domestic Animals (1980)—cattle
Iodine	50 mg/kg or ppm of ration dry matter	NRC: Mineral Tolerance of Domestic Animals (1980)—cattle
Cobalt	10 mg/kg or ppm of ration dry matter	NRC: Mineral Tolerance of Domestic Animals (1980)—cattle
Selenium	2 mg/kg or ppm of ration dry matter	NRC: Mineral Tolerance of Domestic Animals (1980)
Vitamin A	30 × the predicted requirement	NRC: Vitamin Tolerance of Animals, 1987
Vitamin D	4 × the predicted requirement	NRC: Vitamin Tolerance of Animals, 1987
Vitamin E	20 × the predicted requirement	NRC: Vitamin Tolerance of Animals, 1987

The *Nutrients Supplied* display shows nutrient amounts supplied in the diet and nutrient targets for the inputted growing/finishing beef animal. All nutrient levels appear to be within acceptable ranges as established by the application.

The cost of this ration using initial $/ton values is $1.54 per head per day.

Gain supported by dietary energy (lb.): 3.95

The comment behind the cell containing this label is shown in Figure 17–22.

Figure 17–22

If the energy in the diet is in excess of maintenance needs, gain can be supported. This gain is on a per-day basis.
Note that the urea cost is a maintenance energy cost and that it has been added to the maintenance energy requirement. Because maintenance is the top priority, any urea cost will reduce the energy available for gain and other functions.

Gain supported by metabolizable protein (lb.): 4.87

The comment behind the cell containing this label is shown in Figure 17–23.

If the metabolizable protein exceeds maintenance needs, gain can be supported. Metabolizable protein includes bypass protein (RUP) and microbial protein. This gain is on a per-day basis.

Figure 17–23

The comment behind the cell with this calculation is shown in Figure 17–24.

Ration protein and energy content should support similar levels of performance, otherwise nutrients are wasted. This waste may be acceptable if it does not adversely affect profitability, the environment, or animal health.

Figure 17–24

Ca:P: *1.83*

The comment behind the cell containing this label shown in Figure 17–25.

Calcium-to-phosphorus ratio. Ratios between 1:1 and 7:1 result in similar performance provided that the phosphorus requirement is met. Source: Beef NRC, 2000.

Figure 17–25

Percent forage: *32*
N:S: *10.3*

The comment behind the cell containing this label is shown in Figure 17–26.

The dietary nitrogen-to-sulfur ratio should be between 10:1 and 12:1 for efficient utilization of nonprotein nitrogen (urea). (Bouchard & Conrad, 1973; Qi et al., 1993). This is because the rumen microbes need a source of sulfur with nitrogen to synthesize the sulfur containing amino acids methionine and cysteine. Nitrogen was calculated from crude protein using 16 percent as the average nitrogen content of crude protein. This ratio may be ignored if urea or a similar source of NPN is not being fed.

Figure 17–26

UREA cost: *0.000*

The comment behind this cell is shown in Figure 17–27.

The urea cost is the cost in Mcal of metabolizable energy to dispose of the nitrogen in the ration's unused protein. This amount of energy becomes part of the maintenance requirement at the expense of other functions.
The value here has been included in the calculation of the maintenance energy requirement.

Figure 17–27

Predicted rumen pH: *6.12*

The comment behind this cell is shown in Figure 17–28.

Figure 17–28

DMI becomes variable at a ruminal pH of less than 6.2.
At a ruminal pH of less than 6.0, the ability of bacteria to derive energy from forages declines.
For finishing beef animals, a ruminal pH as low as 5.8 and possibly even lower may be acceptable for a short period of time.

Cost/lb. gain—energy:	*0.39*
Cost/lb. gain—protein:	*0.32*
Predicted cost/lb. gain:	*0.39*

The comment behind the Predicted cost/lb. gain label is shown in Figure 17–29.

Figure 17–29

The predicted cost per unit gain is calculated as the ration cost divided by the pounds gain supported by dietary energy or metabolizable protein—whichever supports the lower gain.

The Nutrient Concentration *Display (strike* `F5`*)*

The nutrients in the ration and the predicted nutrient targets are expressed here in terms of concentration. That is, the nutrients provided by the ration are divided by the amount of ration dry matter and the nutrient targets are divided by the target amount of ration dry matter. Concentration units include percent, milligrams per kilogram or parts per million, calories per pound, and international units per pound.

Lower critical temperature (LCT) (° F): *14.2*

The comment behind the cell containing this label is shown in Figure 17–30.

Figure 17–30

The LCT is the environmental temperature below which the animal's tissue insulation (hide and body fat) and external insulation (hair coat) will be unable to maintain body temperature. Below the LCT, feed energy will be repartitioned, with a greater portion being oxidized to maintain body temperature and a reduction in energy available for other functions.
Calculated using formulas from the dairy NRC (2001) as well as the beef NRC (2000).

	Pounds	**Percent**
RDP	2.320	73.7

The comment behind the cell containing this label is shown in Figure 17–31.

Figure 17–31

Rumen degradable protein. Also called *degradable intake protein* (DIP). RDP is a subset of crude protein.

The comment behind the cell containing the percent RDP value is shown in Figure 17–32.

The percentage shown here is the percentage of RDP in crude protein.

Figure 17–32

	Pounds	Percent
RUP	0.830	26.3

The comment behind the cell containing this label is shown in Figure 17–33.

Rumen undegradable protein. Also called *undegradable intake protein* (UIP) and *bypass protein*. RUP is a subset of crude protein.

Figure 17–33

The comment behind the cell containing the percent RUP value is shown in Figure 17–34.

The percentage shown here is the percentage of RUP in crude protein.

Figure 17–34

	Pounds	Percent
MP-bacteria	1.262	61.5

The comment behind the cell containing this label is shown in Figure 17–35.

Metabolizable protein from rumen bacteria. The rumen bacteria grow on RDP and carbohydrate. This represents the metabolizable portion of bacterial protein. The formula used in this calculation is based on that published in the dairy NRC, 2001. Economical beef rations will usually supply more metabolizable protein through bacterial growth than from RUP.

Figure 17–35

The comment behind the percent MP-bacteria value is shown in Figure 17–36.

MP-bacteria percentage—percentage of total metabolizable protein.

Figure 17–36

	Pounds	Percent
MP-RUP	0.664	32.4

The comment behind the cell containing this label is shown in Figure 17–37.

Metabolizable Protein from RUP (rumen undegradable protein). RUP is also called bypass protein and UIP (undegradable intake protein). This value represents the metabolizable portion of RUP. Economical beef rations will usually supply more metabolizable protein through bacterial growth than from RUP.

Figure 17–37

The comment behind the cell containing the percent MP-RUP value is shown in Figure 17–38.

Figure 17–38

MP-RUP percentage—percentage of total metabolizable protein.

	Pounds	**Percent**
MP-endogenous	0.126	6.1

The comment behind the cell containing this label is shown in Figure 17–39.

Figure 17–39

MP from endogenous sources. MP-endogenous sources include metabolizable protein coming from sloughed epithelial cells in the digestive and respiratory systems as well as enzyme secretions into the abomasum. Calculated from formulas in the dairy NRC, 2001.

The comment behind the cell containing the percent MP-endogenous value is shown in Figure 17–40.

Figure 17–40

MP-endogenous percentage—percentage of total metabolizable protein.

	Pounds	**Percent**
MP total	2.052	7.7

The comment behind the cell containing this label is shown in Figure 17–41.

Figure 17–41

Metabolizable protein total. Maximizing MP total is an important goal in feeding ruminants and requires an understanding of the impact of different feeding management strategies on the supplies of metabolizable protein from bacteria and metabolizable protein from RUP.

The comment behind the cell containing the percent MP total value is shown in Figure 17–42.

Figure 17–42

Total metabolizable protein percentage—percentage of DMI.

	Pounds	**Percent**
CP total	3.150	11.8

The comment behind the cell containing this label is shown in Figure 17–43.

Figure 17–43

Crude protein is measured as 6.25 × (the feed nitrogen content). Because other feed components besides protein contain nitrogen, it is described as *crude* protein.

The comment behind the cell containing the percent CP total value is shown in Figure 17–44.

Total crude protein percentage—percentage of DMI.

Figure 17–44

Strike **F5**.

The Feedstuffs Contributions *Display (strike* F6*)*

Shown here are the nutrients contributed by each feedstuff in the ration. This display is useful in troubleshooting problems with nutrient excesses.

Strike **F1**.

The Graphic *Display*

At the bottom of the home display are tabs. The current tab selected is the Worksheet tab. Other tabs are graphs based on the current ration. Note that you may have to click the leftmost navigation button at the lower-left corner to find the first tab.

DMI

Click on this tab to display a graph titled Nutrient Status: Dry Matter Intake

NE

Click on this tab to display a graph titled Nutrient Status: Partitioning of Net Energy

MP

Click on this tab to display a graph titled Nutrient Status: Partitioning of Metabolizable Protein

RUP

Click on this tab to display a graph titled Source and Supply of Metabolizable Protein

Gain

Click on this tab to display a graph titled Energy and Protein Support of Gain (Growing/Finishing Animals and Replacement Heifers)

Macro

Click on this tab to display a graph titled Nutrient Status: Macrominerals

Micro

Click on this tab to display a graph titled Nutrient Status: Microminerals

Vit

Click on this tab to display a graph titled Nutrient Status: Vitamins A, D & E

Click on the **Worksheet** tab.

Blend Feedstuffs Procedure

Click on the **Blend Feedstuffs** button. The message box in Figure 17–45 displays.

Figure 17–45

YOU MUST HAVE ALREADY SELECTED THE FEEDS YOU WANT TO BLEND. When your analysis is acceptable, strike **ENTER** and **F3** to name and file the blend.

Click **OK**.

In the gray area to the left of the feedstuff name, the amounts entered in the ration are entered as the amounts to blend. This farm wants the feed mill to blend all ingredients except the barley grain and the barley silage. Place a 0 in the gray area to the left of these feedstuffs (Table 17–7). Note that when these amounts are entered in the gray area, the blue column to the left calculates the equivalent amount in a ton of mix. Feed mills prefer to have the formula expressed on a per-ton basis because the capacity of their mixer(s) is expressed in tons.

Table 17–7
Inclusion rates for feedstuffs in the beef G/FMin/Vit blend

Barley grain, heavy	0
Barley silage	0
Copper sulfate	0.00068
Limestone, ground	0.26
Potassium iodide	0.000018
Salt (sodium chloride)	0.06727
Vitamin A supplement	0.00898
Vitamin D supplement	0.00146
Vitamin E supplement	0.01917
Zinc oxide	0.00044

Strike **ENTER** and **F3**. The input box in Figure 17–46 displays.

Figure 17–46

ENTER THE NAME OF THE BLEND (names may not be composed of only numbers):

Type the blend name, **G/FMin/Vit,** and click **OK**. The message box shown in Figure 17–47 displays.

Figure 17–47

The new blend has been filed at the bottom of the feed table.

Click **OK**.

View Blends Procedure

Click on the **View Blends** button. The message box in Figure 17-48 displays.

Figure 17–48

Cursor right to view the blends. Cursor down for more nutrients. DO NOT TYPE IN THE BLUE AREAS.

Click **OK**. Confirm that the G/FMin/Vit formula and analysis have been filed.
Strike **F1**.

Using the Blended Feed in the Balanced Ration

Select Feeds *Procedure*

Click on the **Select Feeds** button and select the feeds shown in Table 17–8. Unselect all other feedstuffs. Remember that the G/FMin/Vit blend is located at the bottom of the feed table at row 200.

Barley grain, heavy
Barley silage
G/FMin/Vit

Table 17–8
Feedstuffs and blend to select for the beef ration formulation example

Strike **ENTER** and **F2**.

Make Ration *Procedure*

Click on the **Make Ration** button. Enter the ration shown in Table 17–9.

Barley grain, heavy	18
Barley silage	27
G/FMin/Vit	0.36

Table 17–9
Feeding rates for feedstuffs and blend in the beef ration formulation example

The amount of barley grain and barley silage to feed has already been established. The amount of the G/FMin/Vit to feed is the total amount of its component ingredients in the balanced ration. That value is:

$$0.00068 + 0.26 + 0.000018 + 0.06727 + 0.00898 + 0.00146 + 0.01917 + 0.00044 = 0.36 \text{ lb.}$$

This value is recorded at View Blends under Formula, as entered. The ration is balanced as it was when the components of G/FMin/Vit were fed unmixed.

Strike **ENTER** and **F1**.

Print Ration or Blend Procedure

Make sure your name is entered at cell C1. Click on the **Print Ration or Blend** button. The input box in Figure 17–49 displays.

Are you printing a beef animal ration evaluation or a blend formula and analysis? (**1**-RATION, **2**-BLEND):

Figure 17–49

Click **OK** to accept the default input of 1. A two-page printout will be produced by the machine's default printer.
Click on the **Print Ration or Blend** button. Type the number 2 to print a blend. The message box in Figure 17–50 displays.

Figure 17–50

Click on the green number above the blend you want to print and press **F4**. Scroll right to see additional blends.

Click **OK**.

Find the G/FMin/Vit blend, click on the green number above it, and strike **F4**. A one-page printout will be produced by the machine's default printer.

ACTIVITIES AND WHAT-IFS

In the Forms folder on the companion CD to this text is a BeefInput.doc file that may be used to collect the necessary inputs for use of the BeefRation.xls file. This form may be printed out and used during on-farm visits to assist in ration-evaluation activities.

1. Remake the ration described in Table 17–5 for the growing/finishing beef animal described in this chapter. What is the gain supported by the energy of this ration? Change the ionophore use to **no** (Input Beef Animal/Management Inputs). What is the gain supported by the energy content of this ration now? Explain why the change occurred.
2. Remake the ration described in Table 17–5 for the growing/finishing beef animal described in this chapter. Record the following values in this ration:

   Gain supported by dietary energy and the cost per unit gain—energy.
   Gain supported by dietary metabolizable protein and the cost per unit gain—protein.
   The actual gain realized will be the lower of the two gain-supported values.

   What is the predicted cost per unit gain for this ration? Change implant use to **no** (Input Beef Animal/Management Inputs). Change the final body weight from 1,300 to 1,200 (Animal Inputs) to reflect the consequence of no implant use. Reduce the amount of silage fed to meet the new dry matter intake target. Now reevaluate the cost for gain. Which produces the cheaper gains: animals with anabolic implants or animals without anabolic implants?
3. Remake the ration described in Table 17–5 for the growing/finishing beef animal described in this chapter. Record the gain supported by dietary energy and the cost per unit gain—energy. Record the gain supported by dietary metabolizable protein and the cost per unit gain—protein. The actual gain realized will be the lower of the two gain-supported values. What is the cost per unit gain for this ration? Add bakery waste to the selection of feedstuffs for this ration. Use the bakery waste in place of a portion of the barley grain, heavy. What are the amounts of barley grain and bakery waste that result in a balanced ration that produces the least expensive cost per unit of gain?

REFERENCES

Bouchard, R., Conrad, H. R. (1973). Sulfur requirement of lactating dairy cows. I. Sulfur balance and dietary supplementation. *Journal of Dairy Science. 56*, 1276–1282.

National Research Council. (1980). *Mineral Tolerance of Domestic Animals*. Washington DC: National Academy Press.

National Research Council. (1987). *Vitamin Tolerance of Animals*. Washington DC: National Academy Press.

National Research Council. (2000). *Nutrient Requirements of Beef Cattle* 7th revised edition. Washington DC: National Academy Press.

National Research Council. (2001). *Nutrient Requirements of Dairy Cattle*, 7th revised edition. Washington DC: National Academy Press.

Qi, K., Lu, C. D., & Owens, F. N. (1993). Sulfate supplementation of growing goats: effects on performance, acid-base balance, and nutrient digestibilities. *Journal of Animal Science. 71*, 1579–1587.

CHAPTER 18

FEEDING DAIRY

. . . all the really good ideas I'd ever had came to me while I was milking a cow.

G. WOOD

The nutritional phases in the dairy production cycle, illustrated in Figure 18–1, include:

Cows: Fresh, mid-lactation, late lactation, early-mid dry, close-up dry
Calves and heifers: Liquid feeding/calf starter, calf starter, prebred heifer, postbred heifer

A sample growth rate is plotted in Figure 18–2.

COWS

Fresh Cows

Fresh cows are cows that have recently transitioned from pregnancy through the process of parturition and on to lactation. The health and productivity of the fresh cow is dependent on the success of this transition. The period of time from the last 3 weeks of the dry period to the first 3 weeks of lactation is described as the *transition* or *periparturient* period. Transition cows are cows that are changing from the hormonal and metabolic status characteristic of the non-lactating (dry) cow to the hormonal and metabolic status of the lactating cow. There are many problems that may occur if the transition proceeds poorly. These include:

Ketosis
Fatty liver syndrome
Displaced abomasum
Retained placenta
Mastitis
Metritis
Milk fever
Acidosis

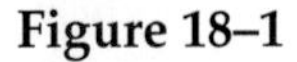

Figure 18–1
Nutritional phases in the production cycle: dairy. Move clockwise starting at top of innermost circle. Continue through the next cycle or move to next shell after 360°.

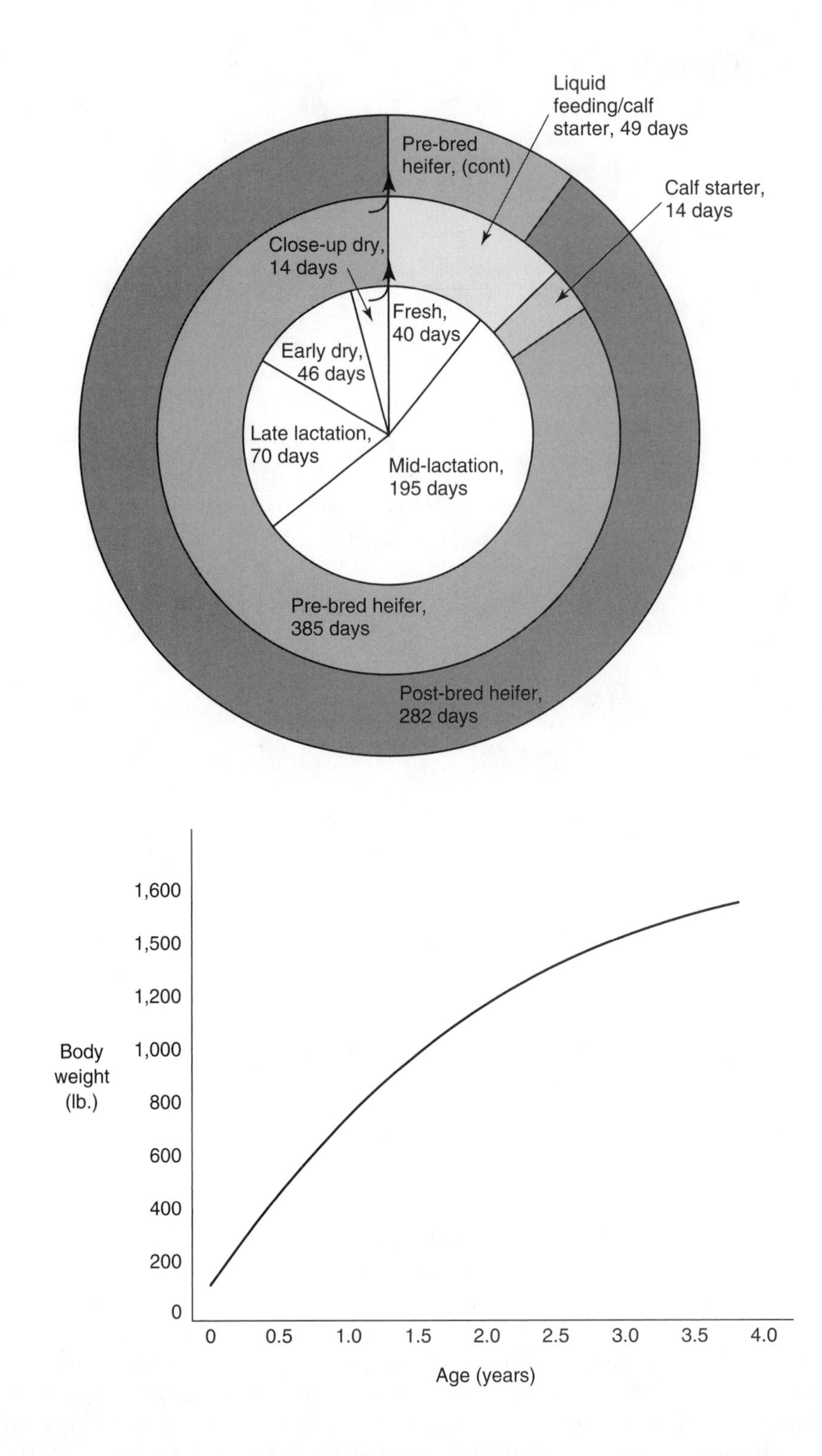

Figure 18–2
Sample dairy bovine growth curve (large breed)

The primary nutritional consideration during the transition period is meeting the cow's energy requirement. This is because the cow's need for energy is increasing at a time when her appetite is declining (Figure 18–3). Normally, cow dry matter intake (DMI) falls by approximately 30 percent during the transition period (Hayirli, Grummer, Nordheim, & Crump, 2002). With the exception of milk fever and acidosis, the energy deficit of the transition cow is a contributing factor to all of the problems that may be experienced by the transition cow.

Recall from the discussion of additives in Chapter 3 that ionophores are products that increase the efficiency with which ruminants are able to use feed energy to meet their maintenance energy needs. These products may, therefore,

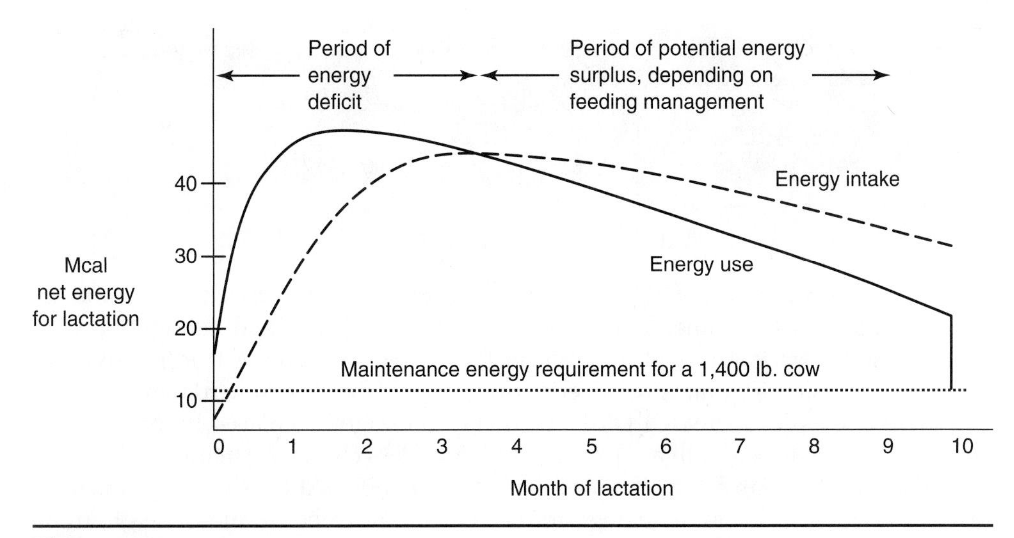

Figure 18–3
Energy status during lactation assuming a 10-month lactation period

be useful in transition cow diets to help minimize the energy deficit and alleviate all the problems associated with it. However, most studies have shown that use of ionophores results in reduced dairy matter intake, and this would deepen the energy deficit. One ionophore has recently been approved for use in dry and lactating dairy cows in the United States. The choice of whether to use an ionophore in transition cows and lactating cows will depend on whether the improvement in energy efficiency more than offsets the reduction in DMI. Transition cow problems are discussed later in this chapter under "Dairy Feeding and Nutrition Issues."

Dairy cattle, like beef cattle, are usually managed to maintain a 12-month calving interval. To do this, cows must become pregnant (settle) within about 80 days of calving. This is because the gestation length is about 282 days. If cows are open much more than 80 days, it will take more than 365 days to complete the reproductive cycle. The period of 80 days postcalving corresponds to the period of time when the cow is peaking in milk production and continues to run an energy deficit. Following calving, the cow's ovaries remain inactive (ovulation does not occur) until energy balance progresses beyond its most negative value (the nadir) and is returning toward balance. Specifically, the cow's first postpartum ovulation occurs 1 to 2 weeks after the negative energy balance nadir is reached (Canfield & Butler, 1991).

The most important factor determining the productivity and profitability of fresh cows is DMI. Feeding and management decisions that lead to increased DMI by cows in early lactation will always result in increased profits. DMI is limited by the resident time of feed in the digestive tract. DMI, then, is potentially increased as the rate of feed removal from the rumen increases. Feed material is removed from the rumen by one of four means:

1. Feed may be colonized and fermented by rumen microbes, which incorporate feed nutrients into microbial cells. These cells are regularly washed out of the rumen.
2. Microbial activity may process a portion of the feed into gases, which are eructated.
3. Microbial activity may turn feed material into valuable compounds that are absorbed through the rumen wall.
4. Finally, feed material may be reduced in size to small and dense particles that pass out of the reticulum and rumen into the omasum.

Figure 18–4
Predicting rumen pH using peNDF

Rumen pH = 5.425 + (0.04229 × peNDF), where peNDF is expressed as a percentage of ration dry matter.
Example: Given peNDF in ration dry matter of 20%.
$5.425 + (0.04229 \times 20) = 5.425 + 0.846$
$= 6.27$, predicted rumen pH

Diets composed of high levels of coarse forage are processed slowly and accumulate in the rumen. At some point, the gut is full and eating stops. As coarse forage is replaced by grain in these diets, the rate of feed removal increases because grain is removed more rapidly from the rumen than is forage. This allows for increased DMI. However, as forage is replaced by grain, a decreased amount of saliva is produced because less chewing is required. Saliva is the primary source of buffer for the rumen microbes. At the same time, increasing amounts of acid are being produced as the microbes ferment the grain. As forage continues to be replaced by grain, the rumen pH will begin to fall. In the CNCPS v.5 (Fox, Tylutki, Tedeschi, Van Amburgh, Chase, Pell, & Overton., 2004) and in the companion application to this text, the rumen pH is predicted from the amount of physically effective neutral detergent fiber (peNDF) in the ration (Figure 18–4).

The rumen bacteria that are most sensitive to decreases in pH are those that work on the forages. When these microbes begin to suffer, DMI will decline because forages are not being removed from the rumen as rapidly. Further decreases in rumen pH may result in systemic acidosis, which is a threat to the well-being of the cow. Acidosis is discussed in Chapter 20.

Mid-Lactation and Late Lactation

Lactating cows should be fed a ration that matches their production potential. In general, a nutrient deficiency during lactation will result in reduced milk production rather than a reduced level of that nutrient in the milk produced. Both deficiencies and excesses will have a negative impact on farm profitability. Meeting the nutritional requirements for lactating cows as they progress through their lactation is achieved by establishing groups on the farm, developing rations for those groups, and moving animals into the appropriate group at the proper time.

A simplified daily time budget for lactating dairy cattle (Grant, 2003) would include 9 to 14 meals eaten during 3 to 5 hours per day. This budget would also include 10 to 12 hours lying down and resting, 7 to 10 hours ruminating, 30 minutes drinking, and 2 to 3 hours in the milking parlor. It has been suggested that for each hour additional rest beyond 7 hours and up to 14 hours, cows may be expected to produce 2 lb. additional milk daily (Albright, 2001). Studying the time budgets of lactating dairy cows shows how animal management can trump animal nutrition when it comes to animal performance.

Energetic efficiency would be greatest if cow body condition did not cycle; that is, if the requirement for energy was always matched by the intake of feed energy. However, during early lactation, the high-producing cow will be in an energy-deficit situation. To maintain high levels of milk production, the cow will need to draw on stored reserves of body fat as a secondary source of energy. During lactation, cow body condition will change as she uses these energy reserves and loses weight during the first 70 days or so, then replenishes these reserves and gains weight as milk production declines. As with beef cattle, a body condition scoring system is used with dairy cattle to assess body fat reserves and overall energy status. The dairy system uses scores of 1 to 5 as described in the following chapter in Table 19-2.

Ideally, feeding management is such that weight loss during lactation corresponds to no more than one body condition score (BCS). Given that the target body condition score for the fresh cow is 3.5 to 4.0, most cows should not fall below a BCS of 2.5 during their lactation. Late lactation is the best time to make feeding adjustments to increase or decrease cow body condition in preparation for the next lactation.

Early-Mid Dry and Close-Up Dry

Metabolic and nutritional disorders such as ketosis, milk fever, retained placenta, and displaced abomasum (DA) all have their root cause in the nutritional management during the months leading up to calving (the dry cow period). In the dairy industry, there is currently a great deal of interest in the possibility of reducing the dry period from the traditional 60-day duration and even eliminating it altogether (Annen, Collier, McGuire, & Vicini, 2004). Although more research is needed to consider the biological processes that occur within the mammary gland during the dry period (Bachman & Schairer, 2003), there is a paucity of research on what effect a shortened dry period would have on the incidence of metabolic disorders (Grummer & Rastani, 2004). It is likely that with proper management, a shortened dry period will not necessarily be associated with increased incidence of metabolic disorders.

A successful dry cow feeding program consists of the following.

1. Providing proper nutrition for the development of the fetus. About two thirds of the weight of the newborn calf is gained during the last 2 months of gestation. The fetus of a calf weighing 90 lb. at birth would, therefore, increase in weight about a pound a day during a 60-day dry period. A shortage of nutrients at this time will result in the birth of weak or dead calves.
2. Preparing the digestive system for the upcoming lactation. The stress of feed change can be minimized by preparing the cow during the last portion of the dry period. This involves the introduction of grain to the diet to develop the proper microbial population. During the 2 weeks prior to freshening, dry cows and heifers should be fed increasing amounts of grain up to a maximum of 0.5 percent of the body weight (NRC, 1989). It is inadvisable to use the lactation ration as part of the dry cow diet, even for only the last few weeks of the dry period. This is primarily due to the fact that the mineral and vitamin needs of lactating cows are dramatically different from those of dry cows. In addition, lactation rations often contain buffers to combat the acids generated from the high-grain diets they are fed. Dry cows will not have the high-grain diet and so will not have high acid production. As a result, dry cows fed buffers may have undesirably high rumen pH values. Some minerals, particularly magnesium, are absorbed with decreasing efficiency as rumen pH increases. Magnesium deficiency has been implicated in some postcalving disorders.
3. Retooling the cow's metabolism for the challenges of lactation. Many metabolic pathways must be established and redirected to support lactation. Included are pathways that involve the generation of glucose needed for milk lactose synthesis, the mobilization of body fat stores for milk fat synthesis, and the management of body mineral stores to ensure that the cow's health is maintained, even as large amounts of mineral leave the body in milk.
4. Maintaining the proper body condition. In most situations, the focus of dry-cow feeding is not so much rebuilding body stores as it is avoiding overconditioning. Management of the cow's body weight is an important goal because cows that are dried off fat are more likely to have difficulties at the end of their dry period when they calve. Ideally, cows are dried off at the same BCS as that at which they will calve, and there is no weight change during

the dry period. The recommended BCS for cows at calving is 3.5–4.0 (Table 19–2). To achieve the 3.5–4.0 BCS in the cow at dry-off time means careful management of her feeding program, and has implications on grouping strategies used at the farm.

CALVES AND HEIFERS

The objective of the calf-raising program is to keep the calf alive, healthy, and growing at a rate at which she will be large enough to breed by 14 months of age.

Liquid Feeding/Calf Starter

The structure of the cow's placenta is described as epitheliolchorial. This type of placenta does not allow antibody immunoglobulins to pass from the cow into the blood of the fetal calf. These antibody immunoglobulins are acquired by the neonatal calf through colostrum. Colostrum, or first milk, is secreted by the cow the first few days following calving. As is the case with other mammalian livestock, colostrum should be consumed by the neonatal calf as soon as possible. If calves are left with the dam to suckle, many will not receive adequate colostrum, so colostrum consumption should be a closely managed activity. An esophageal feeder may be used to get colostrum into calves if they will not drink by themselves. It is possible to store excess high-quality colostrum from older cows by freezing. Before use, frozen colostrum should be thawed in warm water. The importance of colostrum to the neonatal calf cannot be overestimated. Colostrum is discussed in Chapter 20.

The recommendations for colostrum consumption for all ruminants are that (1) neonates should receive an amount of colostrum that is equivalent to 10 percent of the birth weight or 1.5 ounces per pound of body weight within 12 hours of birth, and that (2) half this amount should be ingested within 2 hours of birth, the sooner the better. For an 80-lb. calf, this amounts to about 1 gallon (4 quarts) of high-quality colostrum within the first 12 hours of birth and at least 2 quarts of colostrum within 2 hours of birth. Colostrum should be fed for the first 3 days of life at a rate of 4 quarts per day.

When the calf is 4 and 5 days of age, milk or milk replacer should be fed at a rate of 1 quart twice daily. On day 6 through weaning, calves should be fed 5 percent of body weight twice daily for a maximum of 10 percent per day (milk and milk replacer weigh about 8 pounds per gallon). All milk-feeding equipment should be kept clean and sanitized.

Milk replacers are intended to replace whole milk as calf feed. The reason for feeding milk replacer is an economic one: the cost of the milk replacer is less than the value of whole milk.

Feed mills have available several types of milk replacers. The most expensive contain protein from skim milk, casein, and whey. The least expensive generally contain untreated soybean protein. There are also milk replacers available that are soybean based but have been treated to improve the feeding value for calves. Unlike milk replacers made from milk protein sources, those made from untreated soybean protein contain enzyme inhibitors, and for some calves, allergens. The enzyme inhibitors will reduce protein digestion and retard growth in calves. Another concern with soybean-based milk replacers is the amino-acid balance. Compared to milk protein, soybean protein is somewhat deficient in methionine and, therefore, soybean-based milk replacers are often fortified with this amino acid. Antibiotics are also sometimes added to milk replacers to aid in growth promotion and prevention of bacterial scours.

When the calf is 4 days old, a calf starter should be offered fresh daily along with clean, fresh water. A high-quality calf starter should be highly palatable and nutritionally balanced. Feedstuffs often included in calf starters include cracked or flaked corn, oats and oat products, beet pulp, molasses, soybean meal, and milk products. Although the functional rumen is renowned for its ability to ferment forage, it is the volatile fatty acids from the digestion of the grain in calf starters that stimulate early rumen development.

The following guidelines will help determine when to wean the calf.

1. The calf should be at least 4 weeks old.
2. The calf should be eating 1.5 to 2 lb. daily of calf starter.
3. The above rate of calf starter consumption should have occurred for 3 consecutive days.
4. The size and health of the calf should be considered.

On most farms, calves will be weaned at 4 to 8 weeks of age.

Calf Starter

By 8 weeks of age or about 2 weeks after weaning, calves should be eating 6 to 8 lb. of calf starter. The calf now has a functional rumen and is capable of rumination. At this time, high-quality forages should be introduced to the calf's diet. Calf starters generally contain an ionophore. Ionophores help control the protozoan disease coccidiosis. Use of an ionophore in dairy calves and heifers also consistently improves feed efficiency (Chapter 3).

Prebred Heifer and Postbred Heifer

The objective of a heifer program is to produce a well-grown heifer ready to breed early, calve early, and begin to contribute profits at an early age.

Within 2 weeks following weaning, calves should be consuming 6 to 8 lb. of starter grain. At this time, herd replacement calves should be changed to a heifer feeding program that will include forages. The heifer feeding program should include a ration balanced for all nutrients to support proper heifer growth and development.

Top-quality hay is always desirable for young heifers. If poor hay must be utilized in a dairy operation, it is best fed to dry cows or to heifers more than 1 year old. In addition to hay, hay crop silage, corn or sorghum silage, and pasture can all be useful in building balanced heifer rations. It is always important to know the nutrient content of the forages, whether of good or poor quality, so that proper supplementation can be applied to the ration(s) of which they are a part.

To the extent possible, heifers should be grouped according to nutritional need. Also, the grouping strategy used should minimize the effect of dominance hierarchies on animal access to feed. In growing heifers, both nutritional need and position in the dominance hierarchy correspond well to animal size. In practice, then, animals are grouped according to body size.

Two weeks prior to calving, the heifer should be treated and fed differently, and a separate prefresh group is desirable. At this time, the feeding strategy is designed not only to meet nutritional requirements, but also to prepare the microbes of the rumen for the high-grain diet of the lactating group(s). During the 2 weeks prior to freshening, animals should be fed increasing amounts of grain up to a maximum of 0.5 percent of the body weight (NRC, 1989). Unless the animal has an additional need for nutrients, grain beyond this level provides no additional benefit and may result in undesirable fat deposition.

The period during the 3 weeks prior to freshening to about 3 weeks after freshening is called the *transition period*. More is said about this critical period in the section on "Fresh Cows" and in the section on "Dairy Feeding and Nutrition Issues."

Lactating heifers have different nutrient requirements than mature cows because they are growing in addition to producing milk. Also, heifers are generally of lower rank in the herd and are less aggressive. Ideally, lactating heifers are fed separately from the cows.

DAIRY FEEDING AND NUTRITION ISSUES

Transition Cow Problems

Energy and ketosis (or acetonemia) and fatty liver syndrome

Most animals evolved in an energy-scarce environment. The milk made by mammals is, therefore, rich in energy to help ensure newborn survival. Cows acquire the energy to make this high-energy product from two sources: the energy contained in ingested nutrients from the diet and stored body fat. The energetic efficiency is higher when milk is produced using feed energy rather than the energy in stored fat, and this is one reason that high-producing cows should be managed to maximize DMI. Another reason is that when cows depend excessively on body fat to meet their energy requirement, they become ketotic.

Body fat is a good source of energy for the transition cow, but it is not a versatile source. Body fat is effectively used to make milk fat, but it is poorly used to meet the cow's other energy needs. This is important because although a significant portion of the energy cost of making milk comes from fat production (NRC, 2001), there are other energy costs involved in milk production, and these other energy costs are met with glucose, not mobilized body fat.

Where does the cow get the glucose it needs to meet the portion of its energy requirement that cannot be met by fat? The carbohydrate in feed material is a source of glucose for nonruminants. For these animals, carbohydrate is digested to monosaccharides, including glucose, which are absorbed into the blood. However, carbohydrate in the ruminant goes into the rumen, where it is used as an energy source by the microbes. Glucose of dietary origin typically makes little net contribution to the glucose supply of ruminants (Reynolds, Harman, & Cecava, 1994) because little carbohydrate passes out of the rumen.

Most of the glucose needed by the ruminant animal must be manufactured by the animal during gluconeogenesis [*gluco* (glucose), *neo* (new), *genesis* (creation)]. Gluconeogenesis is carried out primarily by the liver. A representation of the cow's sources and uses for glucose is found in Figure 18–5. Because glucose is used to make lactose (milk sugar) and because lactose content in milk is relatively constant (NRC, 2001), a deficiency of lactose will result in a reduction in milk production. Glucose is made from substances called *glucogenic precursors*. Glucogenic precursors include:

- Fermentation products (mostly propionate).
- Feed amino acids.
- Muscle amino acids.

Note that the first two sources of glucose come from the digestive tract. However, during the transition period, the cow's appetite is usually depressed, so transition cows are usually glucose deficient to some degree. A glucose deficiency is essentially the same as an energy deficiency. The energy deficiency of transition cows is probably a contributing factor to all of the transition cow problems except milk fever and acidosis.

Cows in an energy deficit are mobilizing body fat to help meet the body's energy needs. As evidence of this fact, samples of blood from transition cows in energy deficit show high levels of circulating nonesterified fatty acids (NEFA). In order to use body fat as an energy source, lipolysis must occur during which

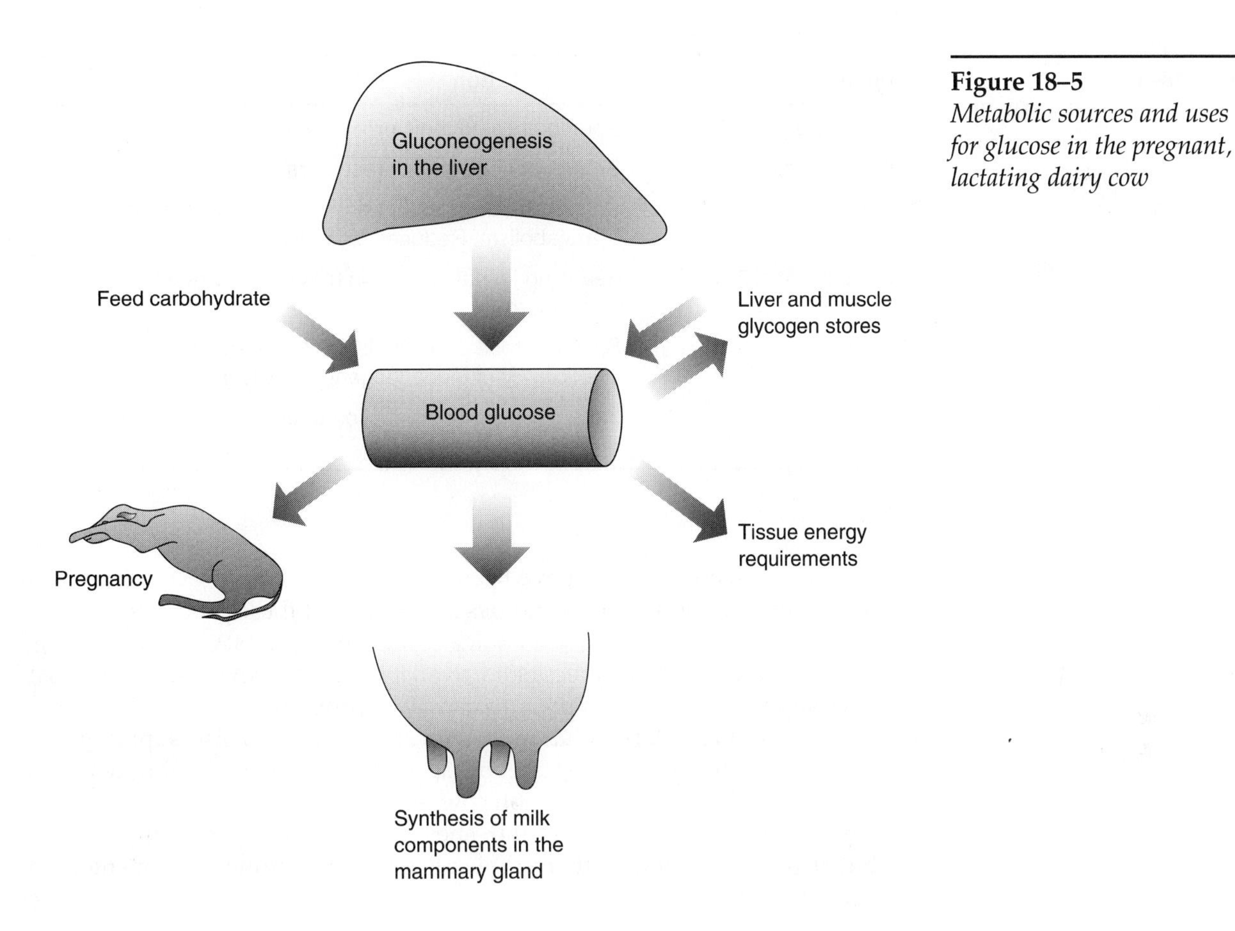

Figure 18–5
Metabolic sources and uses for glucose in the pregnant, lactating dairy cow

the ester bonds of triacylglycerols in fat tissue are broken. This releases glycerol and NEFA into the blood in the proportion of 1:3. Most of the fat's energy resides in the NEFA. NEFA may be used directly by the mammary gland to make milk fat.

The liver takes up NEFA in proportion to its level in the blood. This is a factor in fatty liver syndrome because imported NEFA are reesterified back into triglycerides in the liver. The triglycerides remain in the liver until they can be oxidized in liver cell mitochondria or repackaged for export. Because the export process is slow relative to the import of NEFA, triglyceride may accumulate in the liver, leading to fatty liver syndrome. Symptoms of fatty liver syndrome are due to impaired liver function. Because of the wide variety of activities carried out by the liver, symptoms are nonspecific and include general weakness, poor health, reduced production, and depression.

Oxidation of liver triglyceride involves degrading triglyceride back to NEFA and sending the NEFA to the tricarboxylic acid (TCA) cycle in cell mitochondria. However, in order for NEFA to be completely oxidized in the TCA cycle, glucogenic precursors (most amino acids, propionate, and glycerol from lipolysis) must be available.

If glucogenic precursors are in short supply, as will be the case in an animal with a poor appetite, the NEFA will be incompletely oxidized. The incomplete oxidation of NEFA by the liver produces ketones (also called *ketone bodies*), energy-containing compounds that may be used by a few tissues of the body. Most tissues cannot use ketones, however, and their presence in the blood further depresses appetite. Because ketones are removed from the body via the lungs, they may be detected on the breath of the ketotic animal as the sweetish smell of acetone. Because ketones are also excreted via the kidneys, this same sweetish smell may be detected in urine. Other symptoms of ketosis include dullness, depression, incoordination, rapid weight loss, and a drop in milk production.

Table 18–1
Products and modes of action for additives that have been used to prevent ketosis in dairy cattle

Product	Mode of Action
Propylene glycol	Source of glucogenic precursors
Calcium propionate	Source of glucogenic precursors
Niacin	Specific mode of action is unknown—involved in energy metabolism. Reduces blood NEFA.
Conjugated linoleic acid	Reducing the fat content in milk, thereby reducing the need for energy
Choline	Specific mode of action is unknown—possibly improves efficiency of gluconeogenesis in liver
Monensin	Increases rumen production of propionate, a glucogenic precursor.

Numerous feed products have been developed to help prevent ketosis in transition cows. Table 18–1 lists and describes some of these products.

During the transition period in dairy cows, feeders must carefully manage the supply of glucogenic precursors available to the cow. The most cost-effective supply of glucogenic precursors will be propionate from carbohydrate (both NFC and NDF) fermentation in the rumen. Proteins also supply glucogenic amino acids. Maintaining a good appetite, therefore, should always be the first strategy to prevent transition cow problems.

The energy deficit that occurs at freshening is responsible not only for ketosis, but also for most of the other problems that occur during the transition period in dairy cows.

Energy and Displaced Abomasum

The displaced abomasum (DA) is strongly and consistently associated with energy nutrition. Although the nature of the link is not well understood, minimizing the energy deficit for the transition cow is the best strategy to prevent DA.

Energy and Mastitis, Metritis, and Retained Placenta

A significant component of dietary energy and protein is used to replenish the supply of the cow's white blood cells, specifically, the neutrophils. An 1,800-lb. cow has approximately 1.4×10^{11} neutrophils in its blood (Kehrli, 2001). In the cow, half this number must be replaced every 6 hours and this replacement requires a significant amount of energy. Transition cows in energy deficit, therefore, experience immunosuppression. Given the fact that the energy demand of lactation is largely responsible for the energy deficit, it is not surprising that mastectomized cows showed a much shorter period of immunosuppression following calving than did intact lactating cows (Kimura, Goff, Kehrli, Harp, & Nonnecke, 1997). Immunosuppression predisposes cows to infections such as mastitis and metritis, and it may contribute to retained placentas (Goff & Horst, 1997) due to diminished attack by neutrophils on placental tissues following parturition.

Excess Protein

Dietary protein that is degraded in the rumen becomes ammonia. This ammonia is utilized by the rumen microbes as a source of nitrogen with which to build microbial cell proteins. If a ruminant animal is fed more degradable protein than can be incorporated into cell protein, the ammonia will accumulate and may be absorbed into the blood through the rumen wall. This absorbed ammonia is converted to urea by the liver.

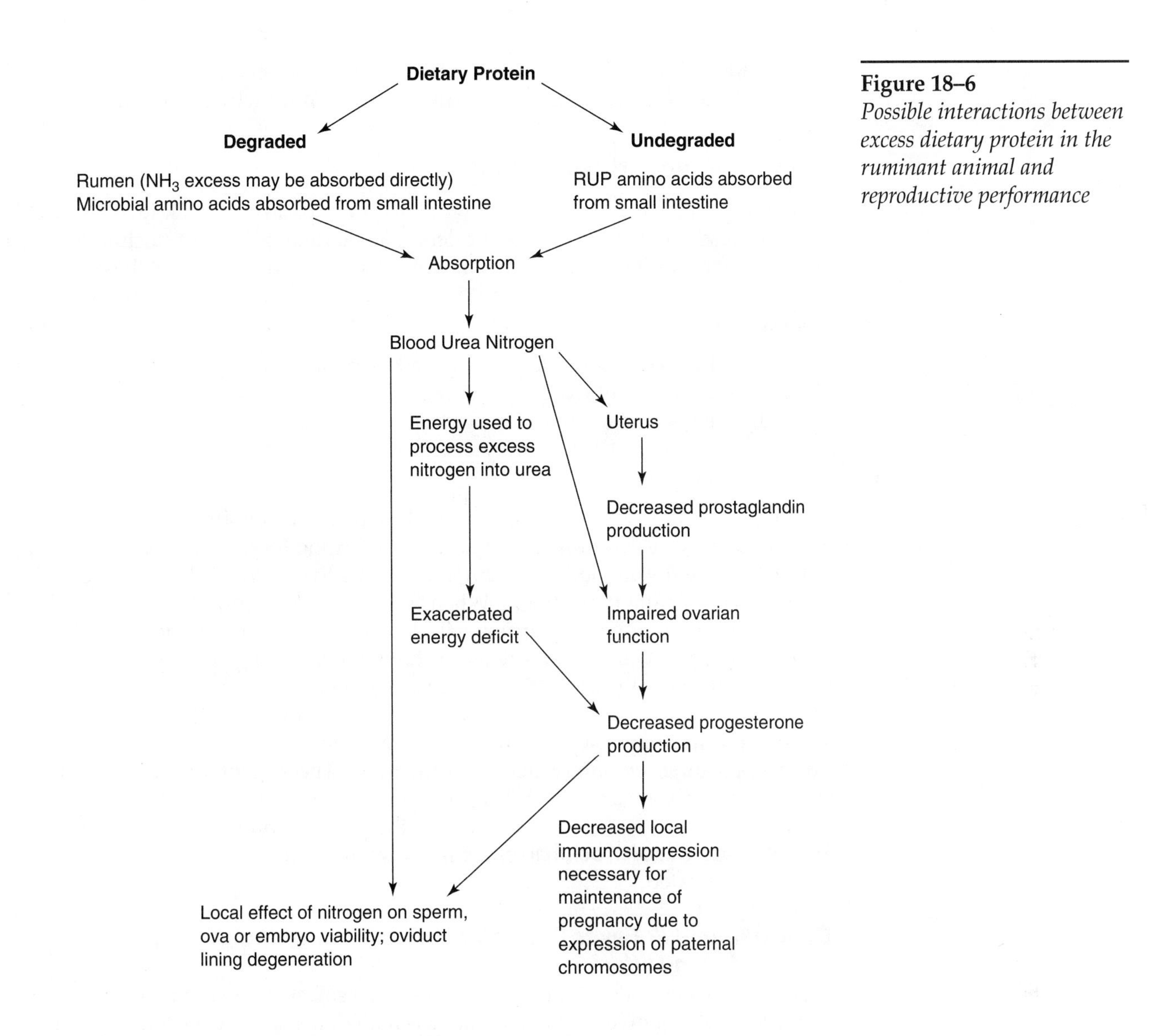

Figure 18–6
Possible interactions between excess dietary protein in the ruminant animal and reproductive performance

Dietary protein that passes through the rumen undegraded arrives at the small intestine. If it is digestible, most of the component amino acids will be absorbed (Chapter 7). If the animal cannot use the absorbed amino acids to make body proteins, either because the level of protein in the diet is excessive or because the quality of the protein is poor, the nitrogen is removed from the excess amino acids and it, too, is processed into urea.

Much of the blood urea is filtered out by the kidneys and excreted. A small amount of the urea in the blood is recycled in saliva. Some will exit the body in milk. Excess dietary protein may be assessed by examining blood and milk urea nitrogen levels (BUN and MUN). Protein excesses are indicated by BUN or MUN levels above 19 mg/dl. (Butler, 1998).

Excess dietary protein has been associated with reproductive problems (McCormick, French, Brown, Cuomo, Chapa, Fernandez, Beatty, & Blouini, 1999; Butler, 1998; Elrod & Butler, 1993; Canfield, Sniffen, & Butler, 1990; Bruckental, Dori, Kaim, Lehrer, & Folman, 1989; Jordan, Chapman, Holtan, & Swanson, 1983; Jordan & Swanson, 1979). The relationship between blood urea levels and reproductive performance is complex. Figure 18–6 summarizes relevant theories.

The manufacture of urea from excess dietary protein requires energy. The companion application to this text calculates the metabolizable energy that will be needed to process any excess protein into urea. This amount of energy is referred to as the *urea cost*. Because this is an essential metabolic process, the urea cost is added to the animal's maintenance energy requirement, leaving less energy available to support growth, productive and reproductive functions.

Milk Fever or Parturient Paresis

Lactating animals may be afflicted with parturient paresis or milk fever. Milk fever is caused by a reduced blood calcium level. Reduced blood calcium also leads to poor muscle tone, and this may predispose cattle to DA, retained placenta, and mastitis (Goff, 2000). The drop in blood calcium is caused by a combination of three factors.

1. The drain on blood calcium imposed by milk production
2. The inability of the digestive system (because of reduced appetite) to deliver enough calcium to replace that which is used for lactation
3. The inability of the parathyroid gland to direct mobilization of stored calcium reserves

Blood calcium is normally maintained within a narrow range, even when dietary intake of calcium is inadequate to match the loss of calcium in milk. This is because the endocrine system is very sensitive to small changes in concentration of calcium in the blood. When serum calcium is low, the parathyroid gland secretes parathyroid hormone (PTH), which facilitates removal of calcium from bone. When serum calcium is high, the thyroid gland secretes calcitonin, which facilitates deposition of calcium in bone.

Milk fever occurs when blood calcium falls below the level necessary to support normal physiology. It usually occurs within 3 days of parturition. Cows stop eating and are unsteady on their feet. The eyes of affected animals are dull and the pupils are dilated. Fever is not a symptom of milk fever; in fact, body temperature may be subnormal. As the problem progresses, the animal will fall down and will be unable to rise. Cows may become comatose and will die if left untreated.

Controlling Milk Fever through Manipulation of the Dietary Cation–Anion Difference

According to Stewart's Theory (see Chapter 9), a solution must remain neutral. If an excess of negatively charged anions is added, the solution becomes acidic rather than negatively charged: the anions replace the OH^- in the solution. A reduction in OH^- concentration means an excess of H^+ or acid. If an excess amount of anions are fed and absorbed into the blood, the blood is at risk of becoming acidic. The body reacts by excreting acid, primarily through action of the kidneys. In addition, the body mobilizes stored cations to counteract the anions. The primary cation that the body has in storage is calcium in the bone. Calcium is mobilized from the bone through the action of the parathyroid gland's PTH. Another mineral that is important in this discussion is magnesium. Magnesium is essential for proper function of the PTH.

For cows that are unable to maintain blood calcium level through the transition period, a manipulation of the dietary cation–anion difference (DCAD) has been an effective prophylactic, provided the cow absorbs adequate magnesium. By feeding more anions (Table 18–2) and thereby reducing the cation–anion difference for the 2 weeks prior to freshening, the cow is put into a similar situation as she is in at calving. She is forced to mobilize the cation she has in storage, which is calcium. This means that the parathyroid gland is forced to secrete PTH prior to freshening. It is important that these cows be put on lactation diets with normal cation–anion difference levels immediately following freshening to prevent the withdrawal of excessive levels of calcium from bone.

In a sense, by reducing the DCAD of the diet, an idle parathyroid gland is toned up and readied for active duty. Dry-cow feeding programs that predis-

Table 18–2
Important cations and anions in animal nutrition

Cations	Anions
Sodium	Chloride
Potassium	Sulfur
Calcium	Phosphorus
Magnesium	

pose transition cows to a "lazy" parathyroid gland are those with excess cations. The two cations that appear to be most important are calcium and potassium. The intake of calcium and potassium by dry cows should be restricted, depending on the incidence of transition cow problems. Overall intake of calcium should be limited to 50 to100 g per cow per day and potassium content of forages for dry cows should be less than 2 percent (Chase, 2001). High levels of potassium in dry-cow diets may also interfere with magnesium absorption. Buffers should not be fed to dry cows because buffers in the dry cow's rumen may create conditions that reduce the efficiency with which magnesium is absorbed.

Acidosis

Because fresh cows are likely to be on high-grain diets, acidosis is a potential problem during this period, although it could be a problem throughout the lactation. To avoid acidosis problems in dairy cows, the rumen pH should be maintained above 6.28. Acidosis is discussed in Chapter 20.

Milk-Fat Depression

Diets that result in a fall of rumen pH below 6.0 are associated with milk-fat depression in dairy cows (NRC, 2001). Milk-fat depression is a reduction in the normal fat component of milk. This may directly affect the market value of the milk, but it is also an indication that the rumen ecology is unsettled. The fat in the milk and tissues of ruminant animals is highly saturated, regardless of the type of fat in the diet. This is because—in the absence of oxygen—rumen microbes use available unsaturated fats as oxidizing agents. As unsaturated fatty acids become reduced or "biohydrogenated," they become saturated fatty acids. These saturated fatty acids are passed on to the small intestine where absorption takes place. In an acid rumen, the process of biohydrogenation is apparently altered, resulting in an increased microbial production of trans fatty acids (Chapter 8). These trans fatty acids made in the rumen are incorporated into milk fat, but apparently with reduced efficiency, resulting in milk-fat depression. Correction of milk-fat depression involves maintaining a rumen pH above 6.28 by feeding adequate forage. Dietary buffers are also helpful.

Prussic Acid Poisoning

Under certain conditions, some grasses, particularly forage sorghums, may accumulate hydrocyanic acid (prussic acid) in their tissues. Prussic acid poisoning is discussed in Chapter 20.

Tall Fescue Toxicosis

Dairy animals that consume pasture or hay containing tall fescue that has been infected with an endophyte are susceptible to tall fescue toxicosis, which has various effects on health, production, and reproduction. Tall fescue toxicosis is discussed in Chapter 24.

END-OF-CHAPTER QUESTIONS

1. What is a transition cow?
2. What does a high blood level of nonesterified fatty acids in transition cows indicate?
3. What class of compounds is produced as a result of the incomplete oxidation of NEFA by the liver? What should be the first management strategy to prevent ketosis?
4. Ketosis, fatty liver, DA, retained placenta, mastitis, and metritis are all problems that share the same root cause. What is it?
5. What ration component will do the most to stimulate rumen development in the young calf?
6. How should the BCS of cows at the end of their lactation compare to the BCS of cows when they calve?
7. Explain how feeding anionic salts can prevent milk fever.
8. What is the fate of unused protein in dairy cow diets? Give three negative consequences of feeding excess protein to dairy cows.
9. What are two effects of ionophore when fed to dairy animals?
10. What ration constituent is used in the formula that predicts rumen pH?

REFERENCES

Albright, J. L. (2001). Personal communication to R. J. Grant.

Annen, E. L., Collier, R. J., McGuire, M. A., & Vicini, J. L. (2004). Effects of dry period length on milk yield and mammary epithelial cells. *Journal of Dairy Science. 87,* E66–E76.

Bachman, K. C., & Schairer, M. L. (2003). Invited review: Bovine studies in optional lengths of dry periods. *Journal of Dairy Science. 86,* 3027–3037.

Bruckental, I, Dori, D., Kaim, M., Lehrer, H., & Folman, Y. (1989). Effects of source and level of protein on milk yield and reproductive performance of high-producing primiparous and multiparous dairy cows. *Animal Production. 48,* 319–329.

Butler, W. R. (1998.) Review: Effect of protein nutrition on ovarian and uterine physiology in dairy cattle. *Journal of Dairy Science. 81,* 2533–2539.

Canfield, R. W., Sniffen, C. J., & Butler, W. R. (1990). Effects of excess degradable protein on postpartum reproduction and energy balance in dairy cattle. *Journal of Dairy Science. 73,* 2342–2349.

Canfield, R. W., & Butler, W. R. (1991). Energy balance, first ovulation and the effects of naloxone on LH secretion in early postpartum dairy cows. *Journal of Animal Science. 69,* 740–746.

Chase, L. E. (2001, November 16–17). DCAD and mineral updates for the transition cow. In: *Proceedings transition cow conference.* Canadaigua, NY.

Elrod, C. C., & Butler, W. R. (1993). Reduction of fertility and alteration of uterine pH in heifers fed excess ruminally degradable protein. *Journal of Animal Science. 71,* 694–701.

Fox, D. G., Tylutki, T. P., Tedeschi, L. O., Van Amburgh, M. E., Chase, L. E., Pell, A. N., & Overton, T. R. (2004). Cornell net carbohydrate and protein system (ver. 5). Ithaca, NY: Cornell University.

Goff, J. P., & Horst, R. L. (1997). Physiological changes at parturition and their relationship to metabolic disorders. *Journal of Dairy Science. 80,* 1260–1268.

Goff, J. P. (2000). Factors to concentrate on to Prevent Periparturient disease in the dairy cow. In *The dairy professional program at Cornell, transition cows.* Ithaca, NY: Cornell University.

Grant, R. J. (2003, October 21–23). Taking advantage of dairy cow behavior: Cost of ignoring time budgets. In *Proc. Cornell nutrition conference.* Syracuse, NY.

Grummer, R. R., & Rastani, R. R. (2004). Why reevaluate dry period length? *Journal of Dairy Science. 88,* E77–E85.

Hayirli, A., Grummer, R. R., Nordheim, E. V., & Crump, P. M. (2002). Animal and dietary factors affecting feed intake during the prefresh transition period in Holsteins. *Journal of Dairy Science. 85,* 3430–3443.

Jordan, E. R. & Swanson, L. V. (1979). Effect of crude protein on reproductive efficiency, serum total protein, and albumin in the high producing dairy cow. *Journal of Dairy Science. 62,* 58–63.

Jordan, E.R., Chapman, T.E., Holtan, D.W., & Swanson, L. V. (1983). Relationship of dietary crude protein to composition of uterine secretions and blood in high-producing dairy cows. *Journal of Dairy Science. 66,* 1854–1862.

Kehrli, M. E., Jr. (2001, November 16–17). Pfizer global research & development, veterinary medicine pharmaceutical discovery, immunological dysfunction in periparturient cows—What role does it play in postpartum infectious diseases? In *Proc. transition cow conference.* Canadaigua, NY.

Kimura, K. J., Goff, J. P., Kehrli, M. E., Jr., Harp, J. A., & Nonnecke, B. J. (1997). Effect of mastectomy on phenotype and function of leukocytes in periparturient dairy cows. *Journal of Dairy Science. 80,* 264.

McCormick, M. E., French, D. D., Brown, T. F., Cuomo, G. J., Chapa, A. M., Fernandez, J. M., Beatty, J. F., & Blouini, D. C. (1999). Crude protein and rumen undegradable protein effects on reproduction and lactation performance of Holstein cows. *Journal of Dairy Science. 82,* 2697–2708.

National Research Council. (1989). *Nutrient requirements of dairy cattle* (6th revised edition). Washington, DC: National Academy Press.

National Research Council. (2001). *Nutrient requirements of dairy cattle* (7th revised edition). Washington, DC: National Academy Press.

Reynolds, C. K., Harmon, D. L., & Cecava, M. J. (1994). Absorption and delivery of nutrients for milk protein synthesis by portal-drained viscera. *Journal of Dairy Science. 77,* 2787–2808.

Wood, G. (1936). Aritist's Odyssey. *Literary Digest. 18.* April.

CHAPTER 19

Dairy Ration Formulation

TERMINOLOGY

Workbook A workbook is the spreadsheet program file including all its worksheets.

Worksheet A worksheet is the same as a spreadsheet. There may be more than one worksheet in a workbook.

Spreadsheet A spreadsheet is the same as a worksheet.

Cell A cell is a location within a worksheet or spreadsheet.

Comment A comment is a note that appears when the mouse pointer moves over the cell. A red triangle in the upper right corner of a cell indicates that it contains a comment. Comments are added to help explain the function and operations of workbooks.

Input box An input box is a programming technique that prompts the workbook user to type information. After typing the information in the input box, the user clicks **OK** or strikes **ENTER** to enter the typed information.

Message box A message box is a programming technique that displays a message. The message box disappears after the user clicks **OK** or strikes **ENTER**.

EXCEL SETTINGS

Security: Click on the **Tools** menu, **Options** command, **Security** tab, **Macro Security** button, **Medium** setting.

Screen resolution: This application was developed for a screen resolution of 1024 × 768. If the screen resolution on your machine needs to be changed, see Microsoft Excel Help, "Change the screen resolution" for instructions.

HANDS-ON EXERCISES

DAIRY RATION FORMULATION

Double click on the **DairyRation** icon. The message box in Figure 19–1 displays.

Figure 19–1

> Macros may contain viruses. It is advisable to disable macros, but if the macros are legitimate, you may lose some functionality.
> Disable Macros or Enable Macros or More Info

Click on **Enable macros**. The message box in Figure 19–2 displays.

Figure 19–2

> Function keys F1 to F10 are set up. You may return to this location from anywhere by striking **ENTER**, then the **F1** key. Workbook by David A. Tisch. The author makes no claim for the accuracy of this application and the user is solely responsible for risk of use. You're good to go. TYPE ONLY IN THE GRAY CELLS! Note: This workbook is made up of charts and a worksheet. The charts and worksheet are selected by clicking on the tabs at the bottom of the display. Never save the workbook from a chart; always return to the worksheet before saving the Workbook.

Click **OK**.

Input Dairy Animal Procedure

Click on the **Input Dairy Animal** button. The input box shown in Figure 19–3 displays.

Figure 19–3

> 1. Lactating
> 2. Dry cow
> 3. Replacement heifer.
>
> ENTER THE APPROPRIATE NUMBER:

Click **OK** to choose the default input of #1, the Lactating dairy animal. The input box shown in Figure 19–4 displays.

Figure 19–4

> Enter measured DMI (pounds) if available or click **OK** to accept the predicted value at CX26.

Click **OK** to accept the predicted dry matter intake. The message box shown in Figure 19–5 displays.

Enter the appropriate Cell Inputs in column CX; then click the **ENVIRONMENT INPUTS** button to continue.

Figure 19–5

Click **OK**.

The Cell Inputs

Enter the cell inputs in Table 19–1. Comments associated with the cell inputs follow.

Animal Inputs

Table 19–1
Animal inputs for the dairy ration formulation example

Value	Input
65	Age (months). If a lactating animal was chosen, this entry must be >21 months.
1400	Body weight (lb.)
no	Ionophore inclusion (yes/no)
70	Days pregnant. For dry cows, this value should be 220–282. For heifers <55% mature weight, this value should be 0
3	Condition Score (1–5)
150	Days in milk (DIM). DIM should be greater than the number of days pregnant by the number of days past calving that you wait to rebreed.
24	Age at first calving (months)
12	Calving interval (months)
4	Breed (1—Ayrshire, 2—Brown Swiss, 3—Guernsey, 4—Holstein, 5—Jersey, 6—Milking Shorthorn)
3.5	Milk fat (%)
3.3	Milk protein (% true protein)
4.8	Milk lactose (%)
12	Milk price, dollars per cwt (100 lb.)
24,000	Rolling herd average (used with lactation # to create a lactation curve for the herd)
3	Lactation # (used with rolling herd average to create a lactation curve for the herd)

The comment behind the cell containing the Condition Score label is shown in Table 19–2.

Table 19–2
Body condition score descriptions for dairy cattle

BCS	1	2	3	4	5
Tailhead	Deep cavity	Shallow cavity	No cavity	Folds of fat	Buried in fat
Pelvic bones	Sharp and easily felt	Prominent and easily felt	Smooth but felt with pressure	Rounded but felt with firm pressure	Cannot be felt
Processes of lumbar vertebrae	Distinct, giving a sawtooth appearance	Ends rounded, upper surfaces felt with slight pressure	Fat covering upper surfaces but still felt with pressure	Cannot be felt	Rounded and covered with fat
Area between processes of lumbar vertebrae	Deep depression	Depression still obvious	Slight depression	Nearly flat across back	Rounded

The comment behind the cell containing the Breed label is shown in Figure 19–6.

Figure 19–6

Breed effects on nutrient requirements are minimal. The breed selected will determine mature weight and calf birth weight, which will affect growth requirements.

The information below the input section is shown in Figure 19–7. Note associated comments behind these cells.

Figure 19–7

84 Cow Milk Potential (lb.) based on where the cow is on the herd's lactation curve. NOT AN INPUT.
0.0 Target ADG (lb.) to meet growth targets. NOT AN INPUT.

Click on the button at the bottom of the *Animal Inputs* display to go to the *Environment Inputs* section and enter the values shown in Table 19–3.

Table 19–3
Environmental inputs for the dairy ration formulation example

60	Current temperature (° F)
60	Temperature average for the previous month (° F)
1	Wind speed (mph)
no	Grazing (yes/no)
	N/A
	N/A
	N/A
0.25	Hair depth, inches (in.)
2	Coat condition (1—clean/dry, 2—some shud on legs, 3—shud on lower body, 4—covered with shud)
1	Heat stress unlikely—input should probably be 1 (see comment)
Yes	Is there significant relief from the heat at night? (Heat stress unlikely—input should probably be *yes.*)

The comment behind the cell containing the Heat Stress label is shown in Figure 19–8.

Figure 19–8

> Enter **1** if no heat stress, **2** for rapid/shallow panting, **3** for open-mouth panting.

Strike **ENTER** and **F1** to return to the home display.

Select Feeds Procedure

Click on the **Select Feeds** button. The note in Figure 19–9 displays.

Figure 19–9

> Nutrient content expressed on a dry matter basis. Feedstuffs are listed first by selection status, then, within the same selection status, by decreasing protein content, then, within the same protein content, alphabetically. Feedstuff DE and NDFdig are calculated values—strike **F10** to calculate these values after entering new analysis values.

Click **OK**.

Explore the table. Note the nutrients listed as column headings. Note also that the table ends at row 200. You select feedstuffs for use in making two different products: (1) a blend to be mixed and sold bagged or bulk, and (2) the ration to be fed directly to livestock.

Select the feedstuffs in Table 19–4 by placing a 1 in the column to the left of the feedstuff name. All unselected feedstuffs should have a 0 value in the column to the left of the feedstuff name. If you wish to group feedstuffs but not select them, you would place a value between 0 and 1 to the left of the feedstuff name.

Table 19–4 *Feedstuffs to select for the dairy ration formulation example*

Feedstuff
Blood meal, flash dried
Soybean meal, 49%
Soybeans, roasted
Mixed grass/legume silage, mid maturity
Mixed grass/legume hay, mid maturity
Corn grain, ground
Corn silage, normal
Calcium carbonate
Cobalt carbonate
Copper sulfate
Dicalcium phosphate
EDDI (Ethylenediamine dihydroiodide)
Salt (sodium chloride)
Sodium selenite
Vitamin A supplement
Vitamin D supplement
Vitamin E supplement
Zinc oxide

Strike **ENTER** and **F2**.

Selected feedstuffs and their analyses are copied to several locations in the workbook.

Make Ration Procedure

Click on the **Make Ration** button. The message box shown in Figure 19–10 displays.

Figure 19–10

ENTER **POUNDS TO FEED** IN COLUMN B. TOGGLE BETWEEN NUTRIENT WEIGHTS AND CONCENTRATIONS USING THE **F5** KEY, RATION AND FEED CONTRIBUTIONS USING THE **F6** KEY, AMINO ACID FLOWS USING THE **F7** KEY, AND ENERGY CALCULATIONS USING THE F9 KEY. WHEN DONE STRIKE **F1**. Cell is highlighted in red if nutrient provided is poorly matched with nutrient target. The lower limit is taken as 94 to 98 percent of target, depending on the nutrient. The upper limit of acceptable mineral is taken from National Research Council. 1980. *Mineral Tolerance of Domestic Animals.* National Academy Press. This publication gives safe upper limits of minerals as salts of high bioavailability, which corresponds to the values calculated here. Vitamin limits are taken from the NRC, 1987. For other nutrients, the upper limit is based on unreasonable excess and the expense of unnecessary supplementation. See Table 19–6 for specifics.

Click **OK**.

The message box in Figure 19–11 displays.

Figure 19–11

The Goal Seek feature may be useful in finding the pounds of a specific feedstuff needed to reach a particular nutrient target:

1. Select the red cell highlighting the deficient nutrient
2. From the menu bar, select **Tools**, then **Goal Seek**
3. In the text box, "To Value:" enter the target to the right of the selected cell
4. Click in the text box, "By changing cell:" and then click in the gray "Pounds fed" area for the feedstuff to supply the nutrient
5. Click **OK**. You may accept the value found by clicking **OK** or reject it by clicking **Cancel**.

WARNING: Using Goal Seek to solve the unsolvable (e.g., asking it to make up an iodine shortfall with iron sulfate) may result in damage to the Workbook.

IMPORTANT: If you return to the Feedtable to remove more than 1 one feedstuff from the selected list, you will lose your chosen amounts fed in the developing ration.

Click **OK**.

In the gray area to the right of the feedstuff name, enter the pound values shown in Table 19–5.

Table 19–5
Inclusion rates for feedstuffs in the dairy ration formulation example

Feedstuff	Pounds
Blood meal, flash dried	0.2
Soybean meal, 49%	5.5
Soybeans, roasted	1.5
Mixed grass/legume silage, mid maturity	38
Mixed grass/legume hay, mid maturity	3
Corn grain, ground	16
Corn silage, normal	38
Calcium carbonate	0.25
Cobalt carbonate	0.000015
Copper sulfate	0.00062
Dicalcium phosphate	0.03681
EDDI (Ethylenediamine dihydroiodide)	0.0000314
Salt (sodium chloride)	0.28
Sodium selenite	0.0000237
Vitamin A supplement	0.02392
Vitamin D supplement	0.00848
Vitamin E supplement	0.02238
Zinc oxide	0.002

The Nutrients Supplied *Display*

The application highlights ration nutrient levels, expressed as amount supplied per dairy animal per day, that fall outside the acceptable range.

The lower limit for DMI is 94 percent of the target. With the exception of energy and protein, the lower limit for all ration nutrients is taken as 98 percent of the target. Energy and protein levels are used to predict performance so no limits are set. Table 19–6 shows the upper limits of the acceptable range for the various inputs.

Table 19–6
Upper limits for ration nutrients used in the dairy ration application.

	Upper Limit	Source
DMI	6% over predicted requirement	Author
Energy	No upper limit. Excess beyond maintenance is used to predict performance.	
Metabolizable protein	No upper limit. Excess beyond maintenance is used to predict performance.	
Calcium—bioavailable	2% of diet dry matter	NRC: Mineral Tolerance of Domestic Animals (1980)—cattle
Phosphorus—bioavailable	1%	NRC: Mineral Tolerance of Domestic Animals (1980)—cattle
Magnesium—bioavailable	0.5%	NRC: Mineral Tolerance of Domestic Animals (1980)—cattle
Potassium—bioavailable	3%	NRC: Mineral Tolerance of Domestic Animals (1980)—cattle
Sulfur—bioavailable	0.4%	NRC: Mineral Tolerance of Domestic Animals (1980)
Sodium—bioavailable	1.6% lactating, 3.5% nonlactating	NRC: Mineral Tolerance of Domestic Animals (1980)—cattle, based on limit for salt (sodium chloride)
Chloride—bioavailable	2.5 lactating, 5.5% nonlactating	NRC: Mineral Tolerance of Domestic Animals (1980)—cattle, based on limit for salt (sodium chloride)
Iron—bioavailable	1000 mg/kg or ppm	NRC: Mineral Tolerance of Domestic Animals (1980)—cattle
Manganese—bioavailable	1000 mg/kg or ppm	NRC: Mineral Tolerance of Domestic Animals (1980)—cattle
Copper—bioavailable	100 mg/kg or ppm	NRC: Mineral Tolerance of Domestic Animals (1980)—cattle
Zinc—bioavailable	500 mg/kg or ppm	NRC: Mineral Tolerance of Domestic Animals (1980)—cattle
Iodine—bioavailable	50 mg/kg or ppm	NRC: Mineral Tolerance of Domestic Animals (1980)—cattle
Cobalt—bioavailable	10 mg/kg or ppm	NRC: Mineral Tolerance of Domestic Animals (1980)—cattle
Selenium—bioavailable	2 mg/kg or ppm	NRC: Mineral Tolerance of Domestic Animals (1980)
Vitamin A	30 times the requirement	NRC: Vitamin Tolerance of Animals (1987)
Vitamin D	4 times the requirement	NRC: Vitamin Tolerance of Animals (1987)
Vitamin E	20 times the requirement	NRC: Vitamin Tolerance of Animals (1987)

The *Nutrients Supplied* display shows nutrient amounts supplied in the diet and nutrient targets for the inputted lactating dairy animal. All nutrient levels appear to be within acceptable ranges as established by the application.

The cost of this ration using initial dollar-per-ton values is $3.49 per head per day.

Milk supported by energy (lb.): *87*

The comment behind the cell containing the Milk Supported by Energy label is shown in Figure 19–12.

Figure 19–12

Note that the urea cost is a maintenance energy cost and that it has been added to the maintenance energy required.

The comment behind the cell showing the predicted pounds milk supported by energy is shown in Figure 19–13.

> This is the milk production (in pounds) that is supported by the level of energy in the diet.

Figure 19–13

Milk supported by protein (lb.): *82*

The comment behind the cell containing the Milk Supported by Protein label is shown in Figure 19–14.

> Ration protein and energy content should support similar levels of performance—otherwise nutrients are wasted. This waste may be acceptable if it does not adversely affect profitability, the environment, or animal health.
> In early lactation, it may be desirable to formulate the ration so that dietary protein will support slightly more milk production than dietary energy, because the animal can use energy reserves in body fat in addition to feed energy to support milk production. In later lactation, the animal should receive a ration that allows her to restore her energy reserves in body fat.

Figure 19–14

The comment behind the cell showing the predicted pounds milk supported by protein is shown in Figure 19–15.

> This is the milk production (in pounds) that is supported by the level of protein in the diet. The level of milk supported by the protein content should be within 15 percent of that supported by the energy content.

Figure 19–15

Milk supported by methionine: *119*

The comment behind the cell showing the predicted pounds milk supported by the first limiting amino acid is shown in Figure 19–16.

> This is the milk production (in pounds) that is supported by the level of the first limiting amino acid in the diet (see notes at **F7**).

Figure 19–16

Max milk supported: *82*

The comment behind the cell containing this label is shown in Figure 19–17.

> If the max milk supported is significantly greater than the cow milk potential, cows may not be able to realize the milk production supported by this ration. Excess nutrients will be excreted.

Figure 19–17

Cow milk potential: *84*

The comment behind the cell containing this label is shown in Figure 19–18.

Figure 19–18

This is the predicted amount of milk that this cow is capable of making based on the inputs.

IOF (Income Over Feed) cost using max milk: *$6.40*

The comment behind the cell containing this label is shown in Figure 19–19.

Figure 19–19

IOF (Iincome Over Feed) cost is calculated here as follows:

[(inputted milk price/100) × max milk supported] − ration cost

Note that the IOF cost using max milk increases with increasing max milk supported. However, if the max milk supported is greater than the cow milk potential, the excess nutrients will go into storage or be excreted.

IOF (Income Over Feed) cost using cow milk potential: *$6.56*

The comment behind the cell containing this label is shown in Figure 19–20.

Figure 19–20

IOF cost is calculated here as follows:

(inputted milk price/100 × cow milk potential) − ration cost

Because cow milk potential is a fixed value based on animal and environment inputs, IOF cost using cow milk potential decreases with increasing ration cost.

	Pounds	**Percent**
RDP	6.13	64.02

The comment behind the cell containing this label is shown in Figure 19–21.

Figure 19–21

RDP is rumen degradable protein. It is the portion of feed protein that rumen microbes can use for food. In most diets, RDP should be greater than RUP because RDP is cheaper and usually the microbes grown on RDP deliver a better-quality protein to the mammary gland.

The comment behind the cell containing the percent RDP value is shown in Figure 19–22.

Figure 19–22

Percentage of crude protein.

	Pounds	Percent
RUP	3.45	35.98

The comment behind the cell containing this label is shown in Figure 19–23.

RUP is rumen undegradable protein. This is the portion of feed protein that bypasses the rumen microbes and goes directly to the small intestine. If digestible and metabolizable, it then is delivered to the tissues and mammary gland.

Figure 19–23

The comment behind the cell containing the percent RUP value is shown in Figure 19–24

Percentage of crude protein.

Figure 19–24

	Pounds	Percent
MP-bacteria	2.87	48.15

The comment behind the cell containing this label is shown in Figure 19–25.

MP-Bacteria is the metabolizable protein from the bacterial cells grown in the rumen on the feed's RDP (rumen degradable protein). 45 to 80 percent of the nitrogen reaching the small intestine is of microbial origin.

Figure 19–25

The comment behind the cell containing the percent MP-bacteria value is shown in Figure 19–26.

MP-bacteria percentage: percentage of total metabolizable protein.

Figure 19–26

	Pounds	Percent
MP-RUP	2.83	47.59

The comment behind the cell containing this label is shown in Figure 19–27.

MP-RUP is the portion of the feed protein that is not degraded in the rumen (rumen undegradable protein), yet is digestible and metabolizable. It is also known as *bypass protein.*
Though most nutritionists recommend that most of the total MP come from MP-bacteria, higher producers will generally have to have a greater proportion of their total MP coming from RUP.

Figure 19–27

The comment behind the cell containing the percent MP-RUP value is shown in Figure 19–28.

Figure 19–28

MP-RUP percentage: percentage of total metabolizable protein. In most rations, this should not be as great as the MP produced in the rumen by bacteria.

	Pounds	Percent
MP-Endogenous	0.25	4.27

The comment behind the cell containing this label is shown in Figure 19–29.

Figure 19–29

MP-endogenous includes metabolizable protein coming from sloughed epithelial cells in the digestive and respiratory systems as well as enzyme secretions into the abomasum.

The comment behind the cell containing the percent MP-endogenous value is shown in Figure 19–30.

Figure 19–30

MP-endogenous percentage: percentage of total metabolizable protein.

	Pounds	Percent
MP total	5.95	11.06

The comment behind the cell containing this label is shown in Figure 19–31.

Figure 19–31

MP is Metabolizable Protein, and the total level provided is the sum of MP from bacteria, MP from RUP and MP from endogenous sources. Maximizing MP total is an important goal in feeding ruminants and requires an understanding of the impact of different feeding management strategies on the supply of MP-Bacteria and MP-RUP.

The comment behind the cell containing the percent MP total value is shown in Figure 19–32.

Figure 19–32

MP total percentage: percentage of DMI.

	Pounds	Percent
CP total	9.58	17.79

The comment behind the cell containing this label is shown in Figure 19–33.

Figure 19–33

Crude protein is measured as 6.25 × (the feed nitrogen content). Because other feed components besides protein contain nitrogen, it is described as *crude* protein.

The comment behind the cell containing the percent CP total value is shown in Figure 19–34.

CP total percentage: percentage of DMI.

Figure 19–34

	Mcal
UREA cost	0.00

The comment behind the cell containing this label is shown in Figure 19–35.

The urea cost is the cost in Mcal of ME to dispose of the nitrogen in the ration's unused protein. This amount of energy becomes part of the maintenance requirement at the expense of other functions. Approximately 0.52 Mcal of ME is used to make 1 lb. of milk containing 3.5 percent fat. Therefore, if energy is limiting production, each multiple of 0.52 ME used in protein excretion results in a pound of lost milk. The value here has been included in the calculation of the maintenance energy (ME) requirement.

Figure 19–35

% Forage: 60.7%

The comment behind the cell containing this label is shown in Figure 19–36.

This is the percentage of forage in the feed dry matter. It is one of the oldest ways of assessing effective fiber in the diet. The usual recommendation is that the ration should contain more than 40 percent forage.

Figure 19–36

Predicted rumen pH: 6.46

The comment behind the cell containing this label is shown in Figure 19–37.

Dry matter intake (DMI) begins to fluctuate at a ruminal pH below 6.2. At a ruminal pH below 6.0, the ability of bacteria to derive energy from forages declines.
Note that although it is advisable to feed buffer to help manage ruminal pH on high-grain diets, buffer is not considered here in the ruminal pH prediction.

Figure 19–37

The Nutrient Concentration *Display (strike* `F5`*)*

The nutrients in the ration and the predicted nutrient targets are expressed in terms of concentration. That is, the nutrients provided by the ration are divided by the amount of ration dry matter and the nutrient targets are divided by the target amount of ration dry matter. Concentration units include percentage, milligrams per kilogram or parts per million, calories per pound, and international units per pound.

DCAD: 137

The comment behind the cell containing this label is shown in Figure 19–38.

Figure 19–38

Dietary Cation–Anion Difference (DCAD). DCAD is used to assess the impact of the ration's mineral content on the body's efforts to regulate blood pH. It can also be manipulated to prevent milk fever. Feeding anionic salts to dry cows for the 2 weeks prior to freshening reduces the DCAD, forcing cows to mobilize calcium. This prepares cows for early lactation in which they also must mobilize calcium. Because anionic salts are generally unpalatable, there is a risk that feed intake will be negatively impacted and, therefore, manipulation of DCAD should probably only be considered where milk fever is a significant problem. The equation used here is:

$$(mEq\ (Ca+Mg+Na+K) - mEq\ (Cl+S+P))/kg\ DMI$$

(Goff, J. 2000.)

To prevent milk fever, a goal would be to achieve a −40 to +50 DCAD. The Goff reference is a personal communication reported in D. G. Fox, T. P. Tylutki, M. E. Van Amburgh, L. E. Chase, A. N. Pell, T. R. Overton, L. O. Tedeschi, C. N. Rasmussen, & V. M. Durbal. 2000. The Net Carbohydrate and Protein System for Evaluating Herd Nutrition and Nutrient Excretion, CNCPS version 4.0. Animal Science Department Mimeo 213. Cornell University.

Ca:P(abs) *1.25*

The comment behind the cell containing this label is shown in Figure 19–39.

Figure 19–39

Calcium-to-phosphorus ratio. This ratio is of little significance if requirements for phosphorus and calcium are met. No differences in milk yield, persistency of milk production, milk composition, or reproductive performance were found in cows during early lactation fed diets with calcium to phosphorus ratios ranging from 1:1 to 8:1.

Dairy NRC, 2001.

N:S(abs) *14.2*

The comment behind the cell containing this label is shown in Figure 19–40.

Figure 19–40

The dietary nitrogen-to-sulfur ratio should be between 10:1 and 12:1 for efficient utilization of nonprotein nitrogen (urea). (Bouchard, R. and H. R. Conrad. 1973. *J. Dairy Sci. 56,* 1276–1282; Qi et al., 1993. *J. Anim. Sci.*). This is because the rumen microbes need a source of sulfur with nitrogen to synthesize the sulfur containing amino acids. This recommendation is based on total sulfur, not bioavailable sulfur. However, since most sulfur sources are 100 percent bioavailable, little error is incurred when using bioavailable sulfur. Nitrogen was calculated from crude protein using 16 percent as the average nitrogen content of protein. This ratio may be ignored if urea is not being fed.

LCT (° F) *−39.46*

The comment behind the cell containing this label is shown in Figure 19–41.

LCT (lower critical temperature) is calculated from body insulation and heat production. Since heat production is dependent upon the energy content of the ration, the LCT will change with the energy content of the ration. Animals in environments below the LCT are cold stressed and will use feed energy to maintain their body temperature, leaving fewer nutrients available for other functions.

Figure 19–41

Daily weight change, lb. (cows) *0.44*

The comment behind the cell containing this label is shown in Figure 19–42.

This calculation is based on how the ration's energy status is affecting daily body weight change. If there is excess energy, the animal will be replenishing reserves and gaining weight; if there is an energy shortage, the animal will be mobilizing body fat and losing weight.
For dry cows, this value should be within 20 percent of the target average daily gain (ADG).
For close-up dry cows, which will be on the ration for only a short time, it may be acceptable to have a high predicted rate of gain because the goal is to prepare the rumen for the high-grain ration of a lactating group.

Figure 19–42

Days to gain 1 CS *471*

The comment behind the cell containing this label is shown in Figure 19–43.

Based on whether dietary energy is deficient or in excess of requirement, the animal may be gaining or losing body weight. The value shown is the number of days to lose or gain enough body-fat reserves to change a BCS (5 point scale). To calculate what one BCS equates to in terms of pounds, multiply the days to gain/lose 1 BCS by the daily weight change.
A cow's energy deficit should not be so deep that she is predicted to lose a BCS in less than 100 days.
For close-up dry cows, see the comment under Daily weight change.

Figure 19–43

Strike **F5**.

The Feedstuffs Contributions *Display (strike* F6*)*

Shown here are the nutrients contributed by each feedstuff in the ration. This display is useful in troubleshooting problems with nutrient excesses.

Strike **F6**.

The Amino Acid Flows and Production Potentials Display *(strike* `F7`*)*

Shown here are the essential amino acids, their predicted flows to the small intestine, the percent of MP this represents, and the pounds of milk this could support. These calculations do not consider how shifting metabolic pathways affect amino acid supply and demand.

Strike **`F1`**.

The Graphic *Display*

At the bottom of the home display are tabs. The current tab selected is the Worksheet tab. Other tabs are graphs based on the current ration. Note that you may have to click the leftmost navigation button at the lower-left corner to find the first tab.

DMI

Click on this tab to display a graph titled Nutrient Status: Dry Matter Intake

Energy

Click on this tab to display a graph titled Nutrient Status: Energy

MP

Click on this tab to display a graph titled Sources of Metabolizable Protein (MP)

Macro

Click on this tab to display a graph titled Nutrient Status: Macrominerals

Micro

Click on this tab to display a graph titled Nutrient Status: Microminerals

Vitamins

Click on this tab to display a graph titled Nutrient Status: Vitamins A, D & E

Milk

Click on this tab to display a graph titled Nutrient Support for Milk Production

Click on the **`Worksheet`** tab.

Blend Feedstuffs Procedure

Click on the **`Blend Feedstuffs`** button. The note in Figure 19–44 displays.

Figure 19–44

YOU MUST HAVE ALREADY SELECTED THE FEEDSTUFFS YOU WANT TO BLEND. When your analysis is acceptable, strike **ENTER** and **F3** to name and file the blend. NOTE: The formulas in this Workbook use measured protein digestion rates, %/hr (Kd), mineral bioavailability and fat, CP, RUP, NFC, & NDF digestibilities for individual feedstuffs. When a blend is made and used in a ration, the blend is "decomposed" and the characteristics of the blend's component feedstuffs are used. For this reason, DO NOT use one blend as a component of another blend.

Click **OK**.

We will plan on having one mineral/vitamin blend.

To the left of the feedstuff name, enter an amount of 0 for all ingredients except the mineral and vitamin sources as shown in Table 19–7.

Table 19–7
Inclusion rates for feedstuffs in the dairy MinVit blend

Feedstuff	Amount
Blood meal, flash dried	0
Soybean meal, 49%	0
Soybeans, roasted	0
Mixed grass/legume silage mid maturity	0
Mixed grass/legume hay mid maturity	0
Corn grain, ground	0
Corn silage, normal	0
Calcium carbonate	0.25
Cobalt carbonate	0.000015
Copper sulfate	0.00062
Dicalcium phosphate	0.03681
EDDI (Ethylenediamine dihydroiodide)	0.0000314
Salt (sodium chloride)	0.28
Sodium selenite	0.0000237
Vitamin A supplement	0.02392
Vitamin D supplement	0.00848
Vitamin E supplement	0.02238
Zinc oxide	0.002

Notice that as the amounts entered change in the gray area, a calculation is made of the appropriate pounds per ton in the blue area. Because feed mill mixers have capacities rated in tons, it is necessary that formulas to be mixed be expressed in pounds per ton.

Strike **ENTER** and **F3**. The input box in Figure 19–45 displays.

Figure 19–45

The name of this blend is BLEND1. This name is accessed in the application to assign nutrient value; therefore, the name must not be altered.

Click **OK**. The message box in Figure 19–46 displays.

Figure 19–46

The new blend has been filed at the bottom of the feed table. When used in a ration, the blend will be "decomposed" and the digestion characteristics of its component feedstuffs will be used in calculations.

Click **OK**.

View Blends Procedure

Click on the **View Blends** button to confirm that the BLEND1 formula and analysis have been filed. The note in Figure 19–47 displays.

Figure 19–47

Cursor down for more nutrients. DO NOT TYPE IN THE BLUE AREAS.

Click **OK**.
After viewing BLEND1, strike **F1** to return to the home display.

Using the Blended Feed in the Balanced Ration

Select Feeds *Procedure*

Click on the **Select Feeds** button and select the feeds shown in Table 19–8. Unselect all other feedstuffs. Remember that the BLEND1 blend is located at the bottom of the feed table (row 200).

Table 19–8
Feedstuffs and blend to select for the dairy ration formulation example

Blood meal, flash dried
Soybean meal, 49%
Soybeans, roasted
Mixed grass/legume silage, mid maturity
Mixed grass/legume hay, mid maturity
Corn grain, ground
Corn silage normal
BLEND1

Strike **ENTER** and **F2**.

Make Ration *Procedure*

Click on the **Make Ration** button. Enter the ration shown in Table 19–9.

Table 19–9
Feeding rates for feedstuffs and blend in the dairy ration formulation example

Blood meal, flash dried	0.2
Soybean meal, 49%	5.5
Soybeans, roasted	1.5
Mixed grass/legume silage, mid maturity	38
Mixed grass/legume hay, mid maturity	3
Corn grain, ground	16
Corn silage, normal	38
BLEND1	0.62

The amount of blood meal, soybean meal 49%, soybeans roasted, mixed grass/legume silage mid maturity, mixed grass/legume hay mid-maturity, corn grain ground, and corn silage have already been established. The amount of the BLEND1 to feed is the total amount of its component ingredients in the balanced ration. That value is:

0.25 + 0.000015 + 0.00062 + 0.03681 + 0.0000314 + 0.28 + 0.0000237 + 0.02392 + 0.00848 + 0.02238 + 0.002 = 0.62 lb.

This value is recorded at View Blends under Formula, as entered. The ration is balanced as it was when the components of BLEND1 were fed unmixed.

Strike **ENTER** and **F1**.

Print Ration or Blend Procedure

Make sure your name is entered at cell C1.
Click on the **Print Ration or Blend** button. The input box in Figure 19–48 displays.

Figure 19–48

> Are you printing a dairy ration evaluation or a blend formula and analysis? (1-RATION, 2 BLEND)

Click **OK** to accept the default input of 1. A two-page ration printout will be produced by the machine's default printer.
Click on the **Print Ration or Blend** button. This time enter a value of 2 to print a blend. Click **OK**. The message box shown in Figure 19–49 displays.

Figure 19–49

> Click on the green number above the blend you want to print and press **F4**.

Click **OK**.
Find the blend named BLEND1. Click on the green number above it and strike **F4**. A one-page printout of the blend formula and analysis will be produced by the machine's default printer.

ACTIVITIES AND WHAT-IFS

In the Forms folder on the companion CD to this text is a DairyInput.doc file that may be used to collect the necessary inputs for use of the Dairy Ration.xls file. This form may be printed out and used during on-farm visits to assist in ration evaluation activities.

1. Remake the ration described in Table 19–5 for the lactating cow described in this chapter. Blend all feedstuffs. If there is not extensive sorting of the feed fed to the lactating cows, this blend represents the nutrient value of the refusals. Input a dry cow at 250 days pregnant. From the feed table, select the blend that represents the refusals. Meet the dry cow's DMI with the refusals blend. How many days are needed to gain a condition score for this dry cow? What problems might you expect due to this situation? What is the urea cost in this ration? Explain why this value indicates an excess of dietary protein.
2. Remake the ration described in Table 19–9 for the lactating cow described in this chapter. The answers to the following questions are found by striking **F9**. Given that metabolizable energy is used with similar efficiencies for the functions of maintenance and milk production, feed energy content is expressed as net energy for lactation (NEl) whether discussing maintenance or lactation energy needs. How many megacalories of NEl are required to meet the maintenance needs of this cow? How many pounds of feed dry matter are needed to deliver this number of megacalories? How many additional pounds of feed dry matter is this animal eating beyond that needed to supply maintenance energy? How many megacalories of NEl will these additional pounds provide? Given the megacalories of NEl in one pound of milk as shown at F9, how many pounds of milk is the energy beyond maintenance in this ration capable of supporting? Based on the animal and environmental inputs, what is this cow's milk production potential? What is the difference between the pounds of milk that the energy content of the ration can support and the cow's milk production potential? How many megacalories of NEl does this difference represent? Given the megacalories NEl per pound reserve change shown at F9, how many pounds of reserves will the cow replenish with the extra megacalories of NEl?
3. Remake the ration described in Table 19–9 for the lactating cow described in this chapter. What amount of milk is supported by the energy and protein levels of the ration? Strike **F1** and click on **Select Feeds**. What is the digestible energy, Mcal/lb. of the corn silage? Change the NDF content for the corn silage from 45 to 35 percent. Strike **ENTER** and **F10** to recalculate digestible energy and digestible NDF. What is the new digestible energy, Mcal/lb. for corn silage? Strike **F2** to select this new corn silage. What are the new levels of milk supported by the ration? Back at the feed table, return the corn silage NDF to 45 percent. Strike **ENTER** and **F10**. In the mixed grass/legume silage, record the DE, Mcal/lb. and the digestibility of the NDF (NDFdig) lb./lb. Change the lignin percentage in the mixed grass/legume silage from 5.9 to 4 percent. Strike **F10**. What is the new NDFdig, lb./lb.? What is the new DE, Mcal/lb.? Strike **F2** to select the new hay crop silage. What is the effect of this changed analysis on pounds of milk supported by the ration? Explain how changes in forage NDF and lignin affect digestible energy content and NDF digestibility.

REFERENCES

Bouchard, R. & Conrad, H. R. (1973). Sulfur requirement of lactating dairy cows. I. Sulfur balance and dietary supplementation. *Journal of Dairy* Science. *56,* 1276–1282.

Fox, D. G., Tylutki. T. P., Van Amburgh, M. E., Chase, L. E., Pell, A. N., Overton, T. R., Tedeschi, L. O., Rasmussen, C. N., & Durbal, V. M. (2000). The Net Carbohydrate and Protein System for Evaluating Herd Nutrition and Nutrient Excretion, CNCPS version 4.0. Animal Science Department Mimeo 213. Cornell University.

National Research Council. 1980. *Mineral Tolerance of Domestic Animals.* Washington DC: National Academy Press.

National Research Council. 1987. *Vitamin Tolerance of Animals.* Washington DC: National Academy Press.

Qi, K., Lu, C. D., & Owens, F. N. (1993). Sulfate supplementation of growing goats: effects on performance, acid-base balance, and nutrient digestibilities. *Journal of Animal Science. 71,* 1579–1587.

CHAPTER 20

FEEDING SHEEP

The mountain sheep are sweeter,
But the valley sheep are fatter;
We therefore deem it meter
To carry off the latter.

THOMAS LOVE PEACOCK, *1785–1866*

The nutritional phases in the sheep production cycle, illustrated in Figure 20–1, include:

Ewe: Lactation, flushing/breeding, early-mid gestation, late gestation
Lamb: Liquid feeding/creep feed, growing/finishing

A sample growth rate is plotted in Figure 20–2.

THE EWE AND EWE LAMB

Lactation

Ewes achieve their peak milk production within 2 to 4 weeks after lambing, and 70 percent of their total milk production will be made in the first 8 weeks. After 8 weeks, lactation will have minimal impact on nutrient requirements of the ewe. Ewes nursing twins will make 20 to 50 percent more milk than ewes nursing singles. Information about when the milk is produced and how much milk is produced should be applied to ewe nutritional management. To the extent possible, ewes in similar production status should be grouped together so that they can be fed diets targeted to meet their nutritional needs. The primary nutrient requirement that will vary among lactating ewes will be the need for energy. Ewes will lose body weight if the energy demand for milk production is greater than energy intake. Ewes should be managed to lose no more than one body condition score (BCS) during the first month and should gain this back during the second month of lactation (Snowder & Glimp, 1991).

Lambs can be weaned as early as 4 weeks if they are eating creep feed at a rate of 0.5 lb. daily. It is more common to wean lambs at 6 to 10 weeks. Absent the suckling lamb(s), the mammary glands cease to produce milk and the nutrient requirements of the ewe are reduced. If ewes have not yet regained the

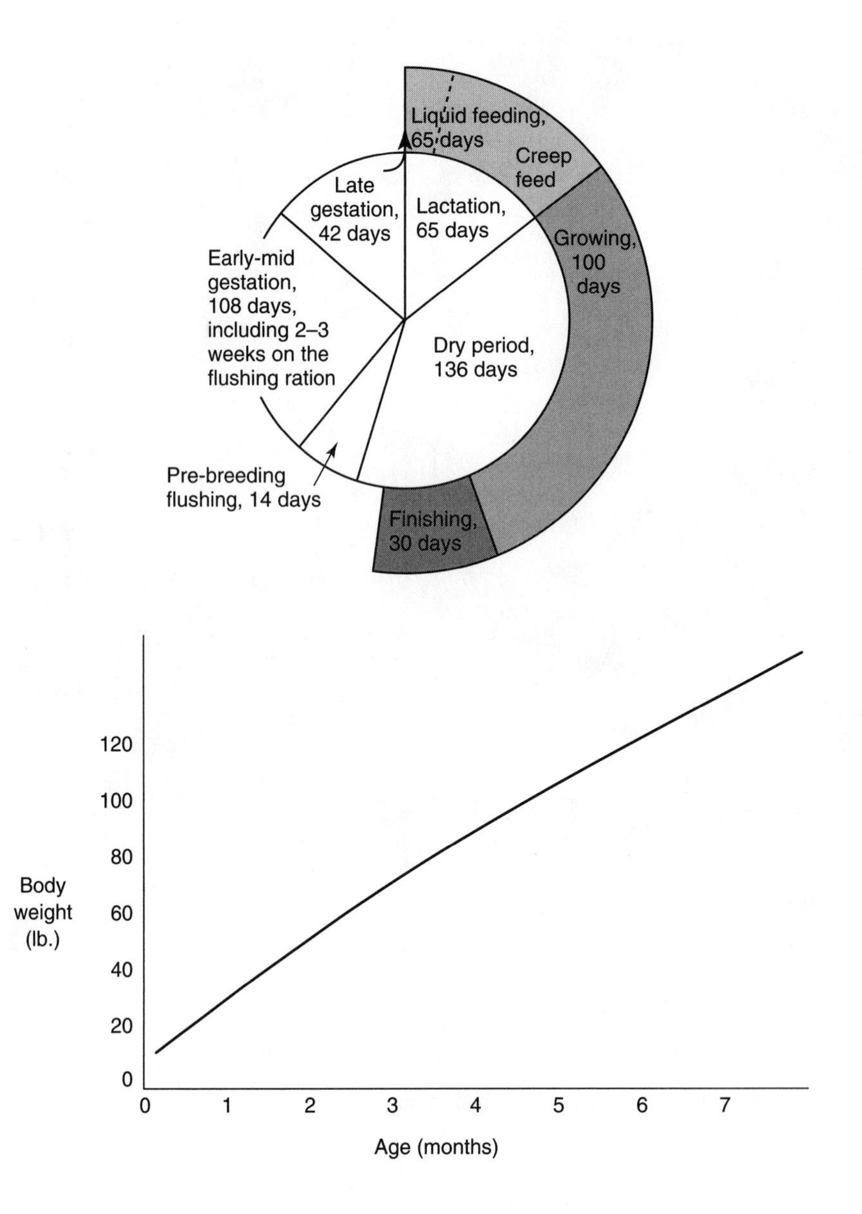

Figure 20–1
Nutritional Phases in the Production Cycle—Sheep. Move clockwise starting at top of innermost circle. Continue through the next cycle or move to next shell after 360°.

Figure 20–2
Sample growth curve for sheep

weight lost during the late gestation and lactation phases, they should be fed a diet to support gain after weaning.

Ewe Lambs

Ewe lambs are young ewes that are be bred to lamb at 12 to 14 months of age. These animals are selected from the market sheep at 3 to 4 months of age (80 to 90 lb.) and are usually managed as a separate flock. Success with ewe lambs requires a high level of nutritional management. Their diet should support growth such that they reach 65 percent of their mature body weight by breeding time at 7 to 9 months, but they should not be allowed to get fat. Lactation is a critical period for ewe lambs because these animals are still growing and will not have the stored reserves of nutrients that mature ewes have. Ewe lambs are generally not bred when sheep are managed under range conditions because of the difficulty in managing nutrient intake on the range.

Flushing/Breeding

Flushing is the practice of increasing the nutrient intake (primarily energy but also protein) around the time of breeding to improve ovulation rate and, hence,

Condition 1 (Emaciated)
Spinous processes are sharp and prominent. Loin eye muscle is shallow with no fat cover. Transverse processes are sharp; one can pass fingers under ends. It is possible to feel between each process.
Condition 2 (Thin)
Spinous processes are sharp and prominent. Loin eye muscle has little fat cover but is full. Transverse processes are smooth and slightly rounded. It is possible to pass fingers under the ends of the transverse processes with a little pressure.
Condition 3 (Average)
Spinous processes are smooth and rounded and one can feel individual processes only with pressure. Transverse processes are smooth and well covered, and firm pressure is needed to feel over the ends. Loin eye muscle is full with some fat cover.
Condition 4 (Fat)
Spinous processes can be detected only with pressure as a hard line. Transverse processes cannot be felt. Loin eye muscle is full with a thick fat cover.
Condition 5 (Obese)
Spinous processes cannot be detected. There is a depression between fat where spine would normally be felt. Transverse processes cannot be detected. Loin eye muscle is very full with a very thick fat cover.

Figure 20–3
Sheep body condition scoring (Thompson & Meyer)

the lambing rate. Lambing rate would be 100 percent if each ewe in the flock gives birth to a single lamb; it would be 200 percent if each ewe has twins. Flushing has been shown to result in a 10 to 20 percent improvement in lambing rate (Cole & Cupps, 1997) in ewes of moderate body condition (Gunn, 1983). Where it is effective, the increased nutrient intake from flushing has an impact on the developing follicles of the ovary.

Flushing is accomplished by moving ewes onto a high-quality pasture or by feeding grain, usually at a rate of 0.25 to 0.5 lb. The period of flushing usually starts 2 weeks before breeding and may continue for several weeks into the breeding season. The reason for continuing flushing into the breeding season is to decrease embryonic death loss. Flushing is effective in improving reproductive performance in ewes that are to be bred either early or late in the breeding season when ovulation rates are less than optimal. Because flushing is of short duration, it will usually not have a significant impact on the ewe's body condition.

Body Condition Scoring

Body condition scoring in sheep is generally based on a 5-point scale similar to the dairy system (Figure 20–3). The 9-point system used in the beef industry is also sometimes used for sheep (Figure 17–9). Moderate body condition corresponds to a body condition score (BCS) within the range of 2.0 to 3.5 on a 5-point scale. Outside this range of body condition, flushing appears to have less of an impact on ovulation rate than does the body condition.

Early/Mid-Gestation

The gestation length of a ewe is approximately 5 months (146 to 150 days). The embryo becomes attached to the uterine wall at about 45 days postbreeding, the development of the placenta is completed by 90 days, and up until about 100 days into the pregnancy, the embryo increases only slightly in size. Severe nutrient deficiencies can affect embryonic development, but nutrients during this time will be used primarily for maintenance functions, fleece production, and, if the pregnant animal is a ewe lamb, growth. Due to the relatively low nutrient demands of early/mid-gestation, this is a period in which ewes can be fed to achieve the targeted BCS at lambing. As with dairy cows, the target score at parturition is 3.5.

Late Gestation

Two thirds of the fetal growth occurs during the last third of gestation, and the nutrition of the ewe during this period will have a significant impact on the survivability of her lamb(s). Lambs born to ewes that were fed poorly during this period will have reduced energy reserves as evidenced by low birth weights, and in the case of multiple births, dissimilar birth weights. Such ewes will also give birth to lambs with a reduced number of functional wool follicles. This may have an impact on lamb survivability, particularly if the lamb is born in a cold environment. The ewe that is fed poorly during the last part of gestation will have a decreased milk production potential because this is the time when mammogenesis occurs. Even if the ewe is well fed once lactation begins, if mammogenesis has not progressed properly, there will be less milk for the lamb(s).

The nutrient requirements for the last 6 weeks of gestation will be considerably greater for the ewe carrying twins than for the ewe carrying a single fetus. If ewes do not receive adequate energy from the feed they are eating during this period, they will mobilize body fat to meet their energy needs. An excessive reliance on body fat to meet energy needs results in a problem called *pregnancy toxemia* in sheep and goats. As body fat is mobilized, ketone bodies accumulate in the blood. Pregnancy toxemia is also called *ketosis.* A ewe with ketosis lags behind the flock, shows labored breathing, staggers, and finally becomes paralyzed. Ketosis is described in Chapter 18. For sheep, goats, and dairy cattle, ketosis is the result of an energy deficit. In cows, ketosis is more likely to occur during early lactation, but in ewes and does, ketosis is more likely to occur during late gestation, especially if the ewe or doe is carrying twins or triplets.

Accelerated Lambing

In accelerated lambing programs, animals are managed to produce more than one lambing per year. In these programs, it is possible to have three lamb crops per ewe in 2 years or five lamb crops per ewe in 3 years. The decision to use an accelerated lambing program will have a significant impact on the choice of sheep breed, marketing options, and labor inputs, as well as on nutritional management. Ewes must be fed to return to breeding weight (BCS of 3.5) sooner than if they were to lamb only once per year. Early weaning of lambs may be useful in accelerated lambing programs.

THE LAMB

Colostrum

The structure of the ewe's placenta is described as epitheliolchorial. This type of placenta does not allow antibody immunoglobulins to pass from the ewe into the blood of the fetal lamb(s). The ewe's first milk or colostrum is rich in these antibody immunoglobulins and the newborn ruminant receives passive immunity by intestinal absorption of the antibodies present in colostrum. Antibodies are made of protein and protein is normally digested in the intestine into its component amino acids. Fortunately, this does not happen in the newborn ruminant. If it did, the antibodies would be destroyed. For a short time after birth, the neonatal ruminant is able to absorb the antibodies in colostrum intact. However, this ability begins to decline within minutes of birth. Colostrum is also a laxative and contains higher levels of some nutrients than does regular ewe milk. To ensure prompt and adequate colostrum intake, neonatal ruminants are often bottle-fed colostrum.

The recommendations for colostrum consumption for all ruminants are (1) neonates should receive an amount of colostrum that is equivalent to 10 percent of the birth weight or 1.5 ounces per pound of body weight within 12 hours of birth; and (2) half this amount should be ingested within 2 hours of birth, the sooner the better. A 10-lb. lamb should, therefore, receive about 1 lb. of colostrum. One pound of colostrum occupies a volume of approximately 0.5 quarts or 15 ounces.

If the ewe is unable to provide her lamb with colostrum, colostrum may be provided from another ewe or from frozen ewe or cow colostrum, warmed to body temperature and bottle-fed.

Creep Feed

Creep feed is feed that is placed in a location that makes it accessible to the lambs but inaccessible to the ewes. The location should be warm, well lit, and not far from where the ewes rest. Creep feed for lambs may contain antibiotics to improve gain and feed efficiency, and also to provide some protection against low-level infection. The issues that must be considered in deciding whether or not to creep feed lambs are the same as those for beef animals (Chapter 16). In addition, creep feeding lambs will reduce the stress on ewes with twins and triplets; this may be especially important with accelerated lambing programs.

Early Weaning

At birth, the lamb's rumen is nonfunctional. Until the rumen becomes functional, needed nutrients must be supplied through milk or milk replacer. Early weaning necessitates the use of milk replacer.

Traditionally, lambs stay with the ewe for about 6 to 10 weeks, though weaning can be successful as early as 4 weeks if lambs are eating creep feed at a rate of 0.5 lb. daily. Early weaning involves removing lambs from ewes shortly after they have received colostrum and feeding them milk replacer until weaning. Early weaning involves more labor and feed expense than leaving lambs with ewes, but it offers the following advantages:

- Increased weight gains
- Lambs are ready for market at a younger age
- Pasture-sparing effect—more ewes can be supported on a given pasture
- Less stress on ewes, which may help them to return to breeding condition more quickly (as would be necessary in accelerated lambing management)
- Fewer problems with parasitism and predators

When lambs are becoming accustomed to milk replacer, it should be fed warm. However, it is recommended that eventually milk replacer be self-fed from multiple nipple pails at a cold temperature. Cold milk replacer is consumed in smaller but more frequent meals than warm milk replacer, and is thereby utilized more efficiently.

Lambs can be weaned from milk replacer as small as 25 lb. and as young as 4 weeks of age. To be ready for weaning at this age, lambs should be well accustomed to creep feed, eating about 0.5 lb. daily. For this to happen, the creep feed should be made available to lambs at 1 week of age. Although week-old lambs will not eat significant amounts of creep feed, the small amounts ingested are important to establish a functioning rumen at an early age.

Growing and Finishing

Weaned lambs may be sent directly to a feedlot for finishing, or they may be placed on pasture or crop aftermath as part of a growing phase prior to finishing.

To be ready for most markets, lambs should have attained a degree of finish corresponding to a backfat depth of 0.1–0.2 in. Under some systems of management, lambs may reach this goal while still with the ewe, but most lambs in the United States will need to be placed on a finishing diet to reach this target. Larger-frame sheep will attain this level of finish at higher body weights than smaller frame sheep. Continuing to feed lambs that have attained 0.2 in. of backfat results in reduced profitability because these lambs are converting increasing amounts of feed energy into fat and this conversion is relatively inefficient.

At the feedlot, lambs should be given immediate access to free-choice water. The finishing diet will be a high-concentrate, energy-dense ration. Such rations may contain as much as 80 percent grain. Care should be taken to ensure that the lambs arriving at the feedlot are transitioned properly to the finishing diet. All changes of diet must be gradual to allow for proper adaptation of the ruminal microbes and the lamb. A gradual transition to the finishing diet will help prevent problems such as acidosis, enterotoxemia, and diarrhea. In warm weather, dry matter intakes (DMI) may improve if lambs are sheared soon after arrival at the feedlot.

Feeder space for self-fed lambs should be 3 to 4 in. per lamb; feeder space for limit fed lambs should be 9 to 12 in. per lamb (Table 12–5). Limit fed lambs should be fed at least twice daily.

It is important to know the growth potential of the lambs. Feeding a diet that would result in maximal gains in lambs with rapid growth potential to lambs with only moderate growth potential will increase fat deposition and reduce profitability. Ram lambs have higher feed intakes and higher nutrient requirements than do ewe lambs, and should be fed accordingly for maximum profitability.

SHEEP FEEDING AND NUTRITION ISSUES

Nutrition and Wool Production

Energy is the primary nutrient impacting wool production. However, energy-deficient sheep will be growing fewer high-protein microbes in the rumen. It may be this decreased ruminal protein production that most directly affects wool production. Except in the case of severe nutritional deficiencies, the impact of nutrition on wool production is quantitative; that is, mild deficiencies will result in reduced wool yield but will have no impact on wool quality.

Nutritional deficiencies will have different impacts on wool production depending on when they occur. If the gestating ewe is fed a deficient diet, her fetus will not fully develop all its potential wool follicles. If the neonatal lamb is fed a deficient diet, not all of its developed wool follicles will be capable of producing a wool fiber. This damage to the wool follicle will be permanent. If a mature sheep is fed a deficient diet, it may result in a temporary disruption of the normal wool follicle growth cycle.

Blocks for Pastured/Range Sheep

Under pasture and especially range conditions, there is limited opportunity to alter the nutrient intake of sheep. However, under some conditions, blocks that contain limiting nutrients can be provided to these animals. The nutrients most likely to be limiting for pasture- and range-fed sheep are salt, energy, crude protein, phosphorus, and vitamin A.

If energy is limiting, supplemental crude protein should be true protein from a source such as soybean meal rather nonprotein nitrogen (NPN) such as urea. This is because NPN can only be utilized if ruminal microbes have access to adequate energy in the form of fermentable carbohydrates. Salt supplements may be provided in block or in loose (granular) form. Loose salt is usually preferable because sheep may bite the blocks and damage teeth.

By changing block location, these feed supplements can be used to help manage pasture utilization.

Copper

The copper level of the total diet for sheep should be less than 25 mg/kg or ppm (National Research Council, 1980). This is because sheep accumulate copper in the liver more readily than do most animals. Eventually the copper destroys liver cells and is released into the blood, causing symptoms of anorexia, excessive thirst, and depression. Death usually follows within a few days.

The sheep's sensitivity to copper makes many products designed for species other than sheep unsuitable for sheep. Because molybdenum forms an insoluble complex with copper and thereby reduces its availability (Figure 9–1), it is possible to limit copper absorption through molybdenum supplementation. However, the Food and Drug Administration does not recognize molybdenum as a feed additive so the use of molybdenum to prevent copper toxicity in sheep will be a veterinary rather than a nutritional application. It should be emphasized that sheep do have a copper requirement and excess molybdenum may result in copper deficiency.

The Need for Vitamin D

Sheep on pasture normally will not require supplemental vitamin D because unless they have heavy fleeces, sheep are able to manufacture vitamin D from sterol compounds in their skin tissue when exposed to sunlight.

Bloat

Bloat is a concern with pastured ruminants, particularly with pastures heavy in legumes. In bloat, the gases of fermentation are trapped and cannot be eructated. Unless the resulting pressure is relieved, the animal will die. Feeding hay before or with pasture can help prevent bloat. The subject of bloat is discussed further in Chapter 16.

Processing Grains

The processing of grains, either into a mash/meal form or into a pellet, adds expense to the ration. The expense is usually warranted if it results in increased nutrient availability over feeding the unprocessed grains. With mature sheep, this is usually not the case. Mature sheep with good teeth chew their grain thoroughly so that there is usually no benefit from feed processing. For lambs, however, this may not be the case. There may be enough improvement in feed intake and nutrient availability to make grain processing a good investment when feeding lambs. In addition, grinding grain may be necessary if excessive sorting (preferential consumption) of feed ingredients becomes a problem.

Urinary Calculi

Urinary calculi are mineral crystals that form in the urinary tract from excess dietary minerals, primarily cations. The crystals grow and can block passageways in the urinary tract. Since the source of the calculi is excess dietary mineral, a logical prevention program would be to limit excesses of minerals, which also makes nutritional, economic, and environmental sense. Reducing the dietary cation–anion difference (DCAD) through the addition of anionic salts such as ammonium chloride or ammonium sulfate is also helpful in reducing the incidence of urinary calculi. For this reason, it is common in the feed industry to formulate sheep grain that includes ammonium sulfate or ammonium chloride at a level of 0.5 percent. Animals fed excess anions excrete acids in the urine and an acidified urine discourages the formation of urinary calculi. The companion application to this text calculates the DCAD in sheep rations.

Feeder Space

When providing supplemental feed to ewes, pelleted feed or whole grains should be spread out in such a way that all animals can feed at feeding time. This is important because ewes are generally limit fed and all will be eating at the same time. When using troughs or communal feeders, it is important that adequate feeder space be provided to ewes. Table 12–5 gives feeder space recommendations for livestock.

Prussic Acid Poisoning

Under certain conditions, some grasses—particularly forage sorghums—may accumulate hydrocyanic acid in their tissues. These conditions include stresses such as frost, severe drought, or a period of heavy trampling or physical damage. The breakdown of cell structure in these stressed plants results in a mixing of plant enzymes with cyanogenic glycosides in the plant, producing hydrocyanic acid. Hydrocyanic acid is a source of cyanide and is also known as prussic acid. Consumption of such crops by sheep or any grazing animal results in a disorder called *prussic acid poisoning.* Symptoms of prussic acid poisoning include convulsions, frothing at the mouth, breathing difficulty, and sudden death. The ensiling process removes the hydrocyanic acid and so these grasses can be safely fed as silage.

Nitrate Poisoning

Nitrate poisoning is described in Chapter 16. Sheep are especially prone to consuming nitrate-containing compounds such as commercial fertilizers. The pioneer cattlemen are said to have used this characteristic of sheep to drive sheep-raisers and their flocks from an area of free range in the public domain of the western United States. "Salting the range" refers to the practice of spreading saltpeter (potassium nitrate) in places where sheep would eat it and succumb to nitrate poisoning.

Poisonous Plants

Sheep will generally avoid plants that contain potential toxins. However, during times of feed shortage, toxic plants may be ingested. Tables 3–3, 3–4, and

3–5 list plant species that are toxic to sheep, grouped according to the nature of the toxicity.

Ruminal Acidosis

In ruminant animals, the term *acidosis* is generally applied to the condition in which the rumen contents drop in pH to below 6.2. There are two forms of ruminal acidosis: acute and subacute. The difference between acute and subacute ruminal acidosis relates to the rate and quantity of acid production and how long the acid condition persists in the rumen. For purposes of discussion here, subacute ruminal acidosis will be considered to exist when ruminal pH lies within the range of 5.7 to 6.2, and acute ruminal acidosis exists when ruminal pH falls below 5.7.

In acute ruminal acidosis, the ruminal pH drops rapidly to below 5.7. Acute ruminal acidosis may occur in feedlots when the transition to a concentrate-rich diet is poorly managed. Acute ruminal acidosis may result in death of the animal. Animals that survive acute ruminal acidosis may have sustained permanent damage to the rumen wall. These animals become chronic poor-performers due to an impaired ability to properly absorb nutrients.

Subacute ruminal acidosis is a more widespread problem than is the acute form. The ruminal pH characteristic of the subacute form of ruminal acidosis is 5.7 to 6.2. Subacute ruminal acidosis may occur in feedlots but it is also common in dairy animals fed for high rates of production.

The acid in ruminal acidosis originates during the ruminal fermentation of readily fermentable carbohydrates, primarily grain. The rate of fermentation of the carbohydrate in cereal grains varies, so some grains are more likely to be associated with acidosis than others. Wheat grain is the most rapidly fermented, followed by barley grain and then corn grain. Some processing methods such as steam flaking will increase the rate of fermentation of a cereal grain. The primary acid associated with grain fermentation is propionic acid. Acids are also produced from fermentation of carbohydrate in the forage component of the diet, but more slowly than from the grain portion. The primary acid associated with forage fermentation is acetic acid. Ruminal fermentation activities also produce lesser amounts of butyric and other organic acids.

The microbes that ferment the major carbohydrate in grains are the amylolytic bacteria, and those that ferment the major carbohydrate in forages are the cellulolytic bacteria. The primary product of the amylolytic bacteria is propionic acid. The primary product of the cellulolytic bacteria is acetic acid.

The cellulolytic bacteria are more sensitive to a reduction in ruminal pH than are the amylolytic bacteria. As a result, during acidosis, the proportion of acetate falls as the production of propionate rises. Figure 20–4 displays these relationships. Acidosis, therefore, leads to a decreased acetate-to-propionate molar ratio in the rumen. Because acetate is used to build milk fat, one theory as to the cause of milk-fat depression in ruminants suggests that acidosis results in an acetate deficiency. The more widely accepted theory faults the increased production of trans fatty acids in the acidic rumen. Milk-fat depression is discussed in Chapter 18.

Acetic and propionic acids produced by the ruminal bacteria are absorbed through the rumen wall into the portal blood that carries absorbed material to the liver. These acids are an important source of energy for the ruminant. The butyric acid made in the rumen is utilized as an energy source by the epithelial cells lining the rumen. Butyric acid that is not utilized by the rumen's epithelial cells is converted to ketones before absorption.

If too much grain is fed without appropriate forage or if there is an insufficient adaptation period when transitioning to a high-grain diet, the bacterial populations and the fermentation processes will be altered (Figure 20–5).

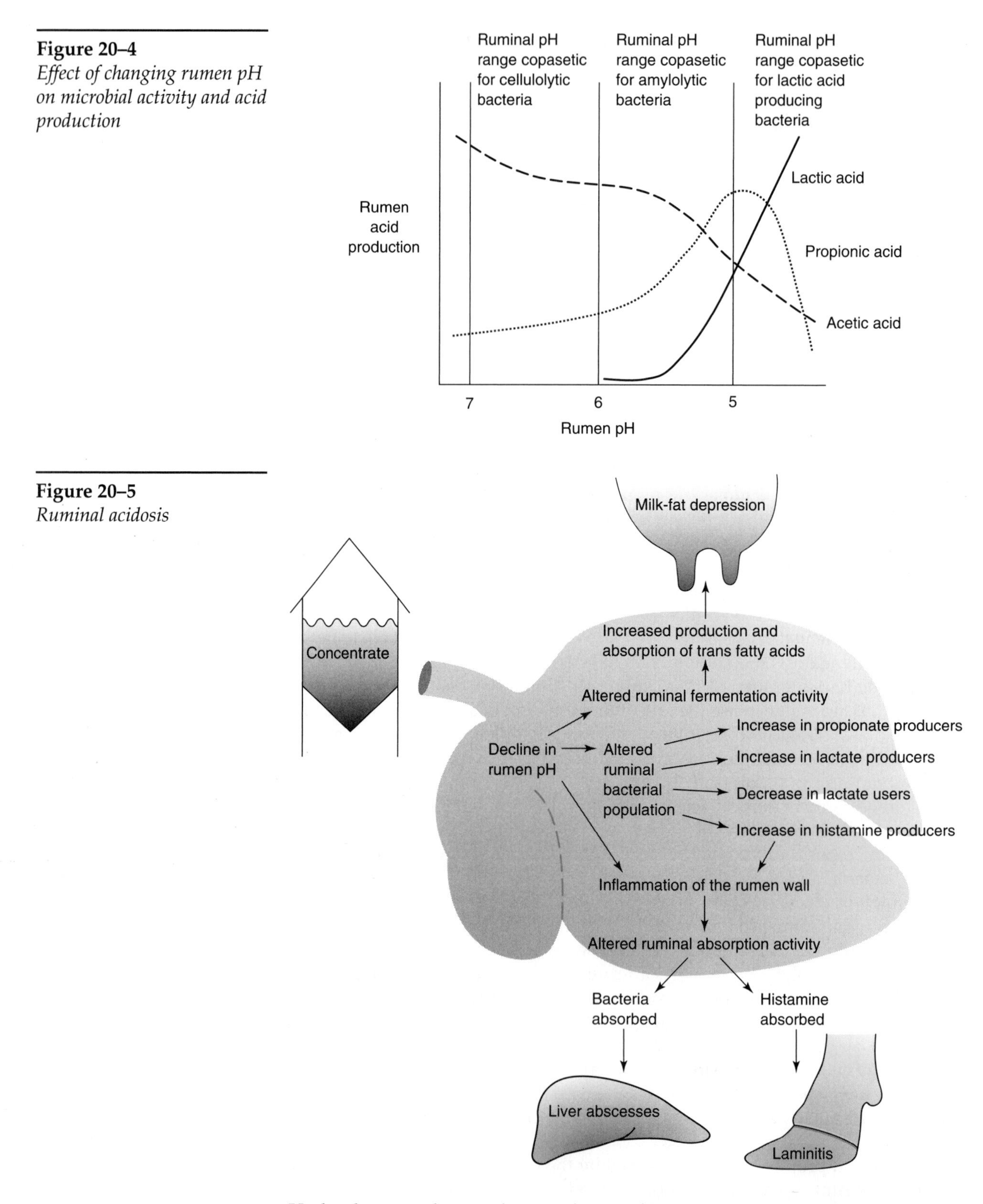

Figure 20–4
Effect of changing rumen pH on microbial activity and acid production

Figure 20–5
Ruminal acidosis

Under these conditions, the population of lactic acid–producing bacteria in the rumen increases and there is a concomitant reduction in the use of lactic acid by other bacteria in the rumen. Lactic acid is a stronger acid than the acids discussed thus far: the pKa value is 3.85 for lactic acid compared to 4.76 for acetic acid, 4.87 for propionic acid, and 4.85 for butyric acid. These pKa values mean that in terms of its ability to reduce ruminal pH, lactic acid is about 10 times as strong as the volatile fatty acids normally produced in the rumen.

Animals with a ruminal pH of between 6.2 and 5.7 show variable DMI. In the rumen itself, there is reduced microbial growth and fiber fermentation. At a ruminal pH of less than 6.0, the ability of bacteria to derive energy from forages declines (Schwartzkopf-Genswein, Beauchemin, Gibb, Crews, Hickman, Streeter, & McAllister, 2003). DMI may be dramatically reduced if ruminal pH drops to less than 5.7. In the companion application to this text, ruminal pH is predicted from the level of physically effective NDF in the ruminant ration (Figure 18–4).

A reduction of ruminal pH leads to several conditions that contribute to the systemic problems associated with acidosis.

1. There is a change in microbial activity, leading to increased production of trans fatty acids.
2. There is a change in the population of microorganisms inhabiting the rumen, increasing the number of those that decarboxylate the amino acid histidine to create histamine.
3. Ruminal histamine concentration increases.
4. The rumen wall becomes inflamed, and histamine—along with bacteria—passes into the portal blood (Garner, Hay, Guard, & Russell, 2003)

The increased production and absorption of trans fatty acids leads to milk fat depression (Chapter 18).

The bacteria that are absorbed into the portal blood of an animal with ruminal acidosis cause liver abscesses. Liver abscesses result in significant economic losses from liver condemnation, reduced animal performance, and reduced carcass yield (Nagaraja & Chengappa, 1998).

An increase in the level of histamine in the blood causes changes in the pattern of blood circulation, especially in the hoof. Histamine is a powerful inflammatory agent. As such, it increases the permeability of blood vessels, allowing fluids to escape from the vessels and seep out into extracellular space. This seepage results in inflammation and blood stagnation. As a result of this stagnation, oxygen is not delivered in adequate amounts to the differentiating epithelial cells that are destined to become hoof wall. This cell damage in the hoof is evidenced as laminitis.

Ruminal acidosis combined with excess sulfur intake may lead to a disorder called *polioencephalomalacia,* which is discussed in Chapter 9.

Ruminal acidosis has been reviewed (Schwartzkopf-Genswein, Beauchemin, Gibb, Crews, Hickman, Streeter, & McAllister, 2003; Owens, Secrist, Hill, & Gill, 1998).

Enterotoxemia

Enterotoxemia, or overeating disease, is a common problem in finishing lambs, and also occurs in cattle, goats, horses, and rabbits. It occurs in stressed animals and/or animals whose diet has been changed without a proper transition. Enterotoxemia occurs when bacteria from the genera *Clostridium* and/or *Escherichia* proliferate in the intestine, producing toxins that are absorbed. Prevention involves ensuring that lambs have 7 to 10 days to make the transition from a high-forage diet to one containing large amounts of concentrate. Protection through vaccination is also possible, and in flocks where enterotoxemia is a constant threat, vaccinations should be routine practice. Symptoms include depression, grinding of the teeth, and death.

Grass Tetany

Grass tetany may occur in sheep turned out to lush pastures without being provided with a mineral supplement containing adequate magnesium. Grass tetany is described in Chapters 9 and 16.

White Muscle Disease

White muscle disease or stiff lamb disease is caused by a deficiency of selenium and/or vitamin E. Affected lambs are usually 3 to 8 weeks of age. The primary symptom is stiffness in the legs and death. A necropsy reveals a paleness in the heart and skeletal muscle. Prevention involves ensuring that rations meet requirements for selenium and vitamin E. These nutrients may also be provided through injection.

Milk Fever or Parturient Paresis

The lactating ewe may be afflicted with parturient paresis or milk fever. Milk fever is described in Chapter 18.

Tall Fescue Toxicosis

Sheep that consume pasture or hay containing tall fescue that has been infected with an endophyte are susceptible to tall fescue toxicosis, which has various effects on health, production, and reproduction. Tall fescue toxicosis is discussed in Chapter 24.

Polioencephalomalacia

Polioencephalomalacia (PEM) is a disorder occurring most commonly in sheep and cattle that is caused by a thiamin deficiency. It may be caused by ingestion of feedstuffs containing thiaminases, the creation of a rumen environment encouraging the growth of thiaminase-producing bacteria, or excess sulfur intake. Symptoms of PEM include lethargy, blindness, gastrointestinal stasis, muscle tremors, convulsions, and death. PEM is discussed in Chapter 16.

END-OF-CHAPTER QUESTIONS

1. In the sheep industry, when is the threat of ketosis the greatest? What is a synonym of ketosis in the sheep industry?
2. Explain why mineral mixes formulated for other livestock are unsuitable for sheep.
3. Why is it usually unnecessary to feed supplemental vitamin D to pastured sheep?
4. Adjustment of the DCAD in sheep diets is helpful in preventing what disorder?
5. Explain how the fact that amylolytic bacteria are less sensitive to an acid environment than cellulolytic bacteria contributes to the development of acidosis.
6. List three problems caused by acidosis. Explain how each problem listed is caused by acidosis.
7. Give two methods to prevent enterotoxemia.
8. What causes stiff lamb disease?
9. Describe accelerated lambing.
10. Define flushing/feeding management. Give two purported benefits to using flushing rations.

REFERENCES

Cole, H. H., & Cupps, P. T. (1997). *Reproduction in domestic animals* (3rd edition). New York: Academic Press, Inc.

Garner, M. R., Hay, A. G., Guard, C. L., & Russell, J. B. (2003). Bovine laminitis and *Allisonella histaminiformans.* In *Proceedings Cornell nutrition conference.* East Syracuse, NY.

Gunn, R. G. (1983). The influence of nutrition on the reproductive performance of ewes. In *Sheep production.* W. Haresign, England: Butterworth.

Kaufmann, W., Hagemeister, H., & Dirksen, G. (1980). Adaptation to changes in dietary composition, level and frequency of feeding. In: *Digestive Physiology and Metabolism in Ruminants,* editors Y. Ruckenbusch and P. Thivend. Westport, CT: AVI Publishing Co.

Nagaraja, T. G., & Chengappa, M. M. (1998). Liver abscesses in feedlot cattle: A review. *Journal of Animal Science. 76,* 287–298.

National Research Council. (1980). *Mineral tolerance of domestic animals.* Washington, DC: National Academy Press.

Owens, F. N., Secrist, D. S., Hill, W. J., & Gill, D. R. (1998). Acidosis in cattle: A review. *Journal of Animal Science. 76,* 275–286.

Schwartzkopf-Genswein, K. S., Beauchemin, K. A., Gibb, D. J., Crews, D. H., Jr., Hickman, D. D., Streeter, M., & McAllister, T. A. (2003). Effect of bunk management on feeding behavior, ruminal acidosis and performance of feedlot cattle: A review. *Journal of Animal Science. 81,* E149–E158.

Snowder, G. D., & Glimp, H. A. (1991). Influence of breed, number of suckling lambs, and stage of lactation on ewe milk production and lamb growth under range conditions. *Journal of Animal Science. 69,* 923–930.

Thompson, J. M. & Meyer, H. Undated. Body Condition Scoring of Sheep. Retrieved October, 2004 from: http://oregonstate.edu/dept/animal-sciences/bcs.htm

CHAPTER 21

SHEEP RATION FORMULATION

TERMINOLOGY

Workbook A workbook is the spreadsheet program file including all its worksheets.

Worksheet A worksheet is the same as a spreadsheet. There may be more than one worksheet in a workbook.

Spreadsheet A spreadsheet is the same as a worksheet.

Cell A cell is a location within a worksheet or spreadsheet.

Comment A comment is a note that appears when the mouse pointer moves over the cell. A red triangle in the upper right corner of a cell indicates that it contains a comment. Comments are added to help explain the function and operations of workbooks.

Input box An input box is a programming technique that prompts the workbook user to type information. After typing the information in the input box, the user clicks **OK** or strikes **ENTER** to enter the typed information.

Message box A message box is a programming technique that displays a message. The message box disappears after the user clicks **OK** or strikes **ENTER**.

EXCEL SETTINGS

Security: Click on the **Tools** menu, **Options** command, **Security** tab, **Macro Security** button, **Medium** setting.

Screen resolution: This application was developed for a screen resolution of 1024 × 768. If the screen resolution on your machine needs to be changed, see Microsoft Excel Help, "Change the screen resolution" for instructions.

HANDS-ON EXERCISES

SHEEP RATION FORMULATION

Double click on the **SheepRation** icon. The message box in Figure 21–1 displays.

Figure 21–1

Macros may contain viruses. It is advisable to disable macros, but if the macros are legitimate, you may lose some functionality.
Disable Macros or Enable Macros or More Info

Click on **Enable macros**.

The message box in Figure 21–2 displays.

Figure 21–2

Function keys F1 to F8 are set up. You may return to this location from anywhere by striking **ENTER**, then the **F1** key. Workbook by David A. Tisch. The author makes no claim for the accuracy of this application and the user is solely responsible for risk of use. You're good to go. TYPE ONLY IN THE GRAY CELLS! Note: This Workbook is made up of charts and a worksheet. The charts and worksheet are selected by clicking on the tabs at the bottom of the display. Never save the Workbook from a chart; always return to the worksheet before saving the Workbook.

Click **OK**.

Input Sheep Procedure

Click on the **Input Sheep** button. The input box in Figure 21–3 displays.

Figure 21–3

1. Ewes
2. Ewe lambs (pregnant or lactating)
3. Replacement ewe lambs
4. Replacement ram lambs
5. Lambs finishing (4 to 7 months old)
6. Early weaned lambs moderate growth potential
7. Early weaned lambs rapid growth potential

Enter the appropriate number:

Click **OK** to select the default input of 1, Ewes. The input box shown in Figure 21–4 displays.

Figure 21–4

1. Maintenance
2. Flushing—2 weeks prebreeding and first 3 weeks of breeding
3. Nonlactating—first 15 weeks gestation
4. Last 4 weeks gestation (130 to 150 percent lambing rate expected) or last 4 to 6 weeks lactation suckling singles
5. Last 4 weeks gestation (180 to 225 percent lambing rate expected)
6. First 6 to 8 weeks lactation suckling singles or last 4 to 6 weeks lactation suckling twins
7. First 6 to 8 weeks lactation suckling twins

Enter the appropriate number:

Enter the input of 5, Last 4 weeks gestation (180–225 percent lambing rate expected) and click **OK**.
The input box in Figure 21–5 displays.

Figure 21–5

Enter measured DMI (pounds) if available or click **OK** to accept the predicted value at CX26.

Click **OK** to accept the predicted DMI. The message box in Figure 21–6 displays.

Figure 21–6

Enter the appropriate cell inputs in column CX.

Click **OK**.

The Cell Inputs

Enter the Cell Inputs shown in Table 21–1.

Table 21–1
Inputs for the sheep ration formulation example

60	Current temperature (° F)
yes	Is there significant relief from the heat at night? (Heat stress unlikely—input should probably be *yes*)
no	Is animal shorn[2]? (yes/no). This input affects the DMI adjustment for cold stress.
170	Ewe's current body weight (110–200 lb.)
	N/A
	N/A
	N/A
	N/A
	N/A
	N/A

The comment behind the "Current temperature, degrees F" input cell reads as shown in Figure 21–7.

Figure 21–7

The temperature impact on DMI. is adjusted based on beef data.

Strike **ENTER** and **F1**.

Select Feeds Procedure

Click on the **Select Feeds** button. The message in Figure 21–8 displays.

Nutrient content expressed on a dry matter basis. Feedstuffs are listed first by selection status, then, within the same selection status, by decreasing protein content, then, within the same protein content, alphabetically.

Figure 21–8

Click **OK**.

Explore the table. Note the nutrients listed as column headings. Note also that the table ends at row 200. You select feedstuffs for use in making two different products: a blend to be mixed and sold bagged or bulk, and a ration to be fed directly to livestock.

Select the feedstuffs in Table 21–2 by placing a 1 in the column to the left of the feedstuff name. All unselected feedstuffs should have a 0 value in the column to the left of the feedstuff name. If you wish to group feedstuffs but not select them, you would place a value between 0 and 1 to the left of the feedstuff name.

Table 21–2
Feedstuffs to select for the sheep ration formulation example

Feedstuff
Soybean meal, 44%
Brome hay, mid bloom
Corn grain, ground
Molasses, cane
Calcium sulfate
EDDI (Ethylenediamine dihydroiodide)
Limestone, ground
Salt (sodium chloride)
Sodium selenite
Vitamin A supplement
Vitamin D supplement
Vitamin E supplement
Zinc oxide

Strike **ENTER** and **F2**.

Selected feedstuffs and their analyses are copied to several locations in the workbook.

Make Ration Procedure

Click on the **Make Ration** button. The message box in Figure 21–9 displays.

Figure 21–9

ENTER **POUNDS TO FEED** IN COLUMN B. TOGGLE BETWEEN NUTRIENT WEIGHTS AND CONCENTRATIONS USING THE **F5** KEY, RATION AND FEED CONTRIBUTIONS USING THE **F6** KEY. WHEN DONE STRIKE **F1**. Cell is highlighted in red if nutrient provided is poorly matched with nutrient target. The lower limit is taken as 95 to 98 percent of target, depending on the nutrient. The upper limit of acceptable mineral is taken from National Research Council. 1980. *Mineral Tolerance of Domestic Animals.* National Academy Press. This publication gives safe upper limits of minerals as salts of high bioavailability. The upper limit for vitamins is taken from the NRC, 1987. For other nutrients, the upper limit is based on unreasonable excess and the expense of unnecessary supplementation. See Table 21–4 for specifics.

Click **OK**. The message box in Figure 21–10 displays.

Figure 21–10

The Goal Seek feature may be useful in finding the pounds of a specific feedstuff needed to reach a particular nutrient target:

1. Select the red cell highlighting the deficient nutrient
2. From the menu bar, select **Tools**, then **Goal Seek**
3. In the text box, "To Value:" enter the target to the right of the selected cell
4. Click in the text box, "By changing cell:" and then click in the gray "Pounds fed" area for the feedstuff to supply the nutrient
5. Click **OK**. You may accept the value found by clicking **OK** or reject it by clicking **Cancel**.

WARNING: Using Goal Seek to solve the unsolvable (e.g., asking it to make up an iodine shortfall with iron sulfate) may result in damage to the Workbook.

IMPORTANT: If you return to the Feedtable to remove more than 1 one feedstuff from the selected list, you will lose your chosen amounts fed in the developing ration.

Click **OK**.

In the gray area to the right of the feedstuff name, enter the pound values shown in Table 21–3.

Table 21–3
Inclusion rates for feedstuffs in the sheep ration formulation example

Feedstuff	Pounds
Soybean meal, 44%	0.4
Brome hay, mid bloom	2.2
Corn grain, ground	2
Molasses, cane	0.24
Calcium sulfate	0.0086
EDDI (Ethylenediamine dihydroiodide)	0.000000084
Limestone, ground	0.0273
Salt (sodium chloride)	0.00883
Sodium selenite	0.00000089
Vitamin A supplement	0.00214
Vitamin D supplement	0.00019
Vitamin E supplement	0.00131
Zinc oxide	0.000004

The Nutrients Supplied *Display*

The application highlights ration nutrient levels, expressed as amount supplied per sheep per day, that fall outside the acceptable range.

The lower limit for DMI is 95 percent of the target. The limit for all other ration nutrients is taken as 98 percent of the target. Table 21–4 shows the upper limits of the acceptable range for the various nutrients.

Table 21–4
Upper limits for ration nutrients used in the sheep ration application

	Upper Limit	Source
Dry matter intake (DMI)	5% over the requirement	Author
Total digestible nutrients (TDN)	25% over the requirement	Author
Digestible energy (DE)	25% over the requirement	Author
Metabolizable energy (ME)	25% over the requirement	Author
Net energy for maintenance (NEm)	No upper limit	—
Crude protein (CP)	30% over the requirement	Author
Calcium (Ca)	2% of ration dry matter	NRC: Mineral Tolerance of Domestic Animals, 1980—sheep
Phosphorus (P)	0.6% of ration dry matter	NRC: Mineral Tolerance of Domestic Animals, 1980—sheep
Magnesium (Mg)	0.5% of ration dry matter	NRC: Mineral Tolerance of Domestic Animals, 1980—sheep
Potassium (K)	3% of ration dry matter	NRC: Mineral Tolerance of Domestic Animals, 1980—sheep
Sulfur (S)	0.4% of ration dry matter	NRC: Mineral Tolerance of Domestic Animals, 1980
Sodium (Na)	10 × the predicted requirement	Author
Salt	9% of ration dry matter	NRC: Mineral Tolerance of Domestic Animals, 1980—sheep
Iron (Fe)	500 mg/kg or ppm of ration dry matter	NRC: Mineral Tolerance of Domestic Animals, 1980—sheep
Manganese (Mn)	1000 mg/kg or ppm of ration dry matter	NRC: Mineral Tolerance of Domestic Animals, 1980—sheep
Copper (Cu)	25 mg/kg or ppm of ration dry matter	NRC: Mineral Tolerance of Domestic Animals, 1980—sheep
Zinc (Zn)	300 mg/kg or ppm of ration dry matter	NRC: Mineral Tolerance of Domestic Animals, 1980—sheep
Iodine (I)	50 mg/kg or ppm of ration dry matter	NRC: Mineral Tolerance of Domestic Animals, 1980—sheep
Cobalt (Co)	10 mg/kg or ppm of ration dry matter	NRC: Mineral Tolerance of Domestic Animals, 1980—sheep
Selenium (Se)	2 mg/kg or ppm of ration dry matter	NRC: Mineral Tolerance of Domestic Animals, 1980
Vitamin A	30 × the predicted requirement	NRC: Vitamin Tolerance of Animals, 1987
Vitamin D	4 × the predicted requirement	NRC: Vitamin Tolerance of Animals, 1987
Vitamin E	20 × the predicted requirement	NRC: Vitamin Tolerance of Animals, 1987

The *Nutrients Supplied* display shows the ration nutrients and the nutrient targets expressed in terms of pounds, grams, milligrams, megacalories, and international units. All nutrient levels appear to be within acceptable ranges as established by the application.

The cost of this ration using initial dollars-per-ton values is $0.28 per head per day.

Ca:P: *1.5*

The comment behind the cell containing this label is shown in Figure 21–11.

Figure 21–11

Calcium-to-phosphorus ratio. With adequate levels of calcium and phosphorus, ruminants can tolerate a calcium-to-phosphorus ratio as wide as 7:1. Sheep NRC, 1985.

NDF:NFC: *0.7*

The comment behind the cell containing this label is shown in Figure 21–12.

Figure 21–12

This is a ratio of carbohydrates in the diet. It is a ratio of fiber content to sugar and starch content. Though there are no specific recommendations, rumen health may be compromised if the ratio falls too far below 1 and productivity compromised if the ratio is too far above 1.

N:S: *10.2*

The comment behind the cell containing this label is shown in Figure 21–13.

Figure 21–13

The dietary nitrogen-to-sulfur ratio should be between 10:1 and 12:1 for efficient utilization of nonprotein nitrogen. (Bouchard & Conrad, 1973. Qi et al., 1993). This is because the rumen microbes need a source of sulfur with nitrogen to synthesize the sulfur-containing amino acids methionine and cysteine. Nitrogen was calculated from crude protein using 16 percent as the average nitrogen content of protein. This ratio may be ignored if urea is not being fed.

DCAD: *169*

The comment behind the cell containing this label is shown in Figure 21–14.

Figure 21–14

Dietary Cation–Anion Difference (DCAD) is used to assess the impact of the ration's mineral content on the body's efforts to regulate blood pH through urinary excretion. It is expressed as mEq/kg and calculated as follows:

$$mEq\ (Na + K) - mEq\ (Cl + S).$$

In sheep nutrition, the feeding of ammonium sulfate (0.5 to 1.0 percent), ammonium chloride and other compounds have been suggested as means of reducing the incidence of urinary calculi. These have the effect of acidifying the urine. This effect is measured generally as the DCAD.

Predicted rumen pH: *6.46*

The comment behind the cell containing this label is shown in Figure 21–15.

Figure 21–15

DMI becomes variable at a ruminal pH of less than 6.2. At a ruminal pH of less than 6.0, the ability of bacteria to derive energy from forages declines.
Note that although it is advisable to feed buffer to help manage ruminal pH on high grain diets, buffer is not considered in the ruminal pH prediction.
For feedlot animals, a ruminal pH as low as 5.8 may be acceptable for a short period of time.

Nutrient Concentration *Display (strike* `F5`*)*

The nutrients in the ration and the predicted nutrient targets are expressed in terms of concentration. That is, the nutrients provided by the ration are divided by the amount of ration dry matter and the nutrient targets are divided by the target amount of ration dry matter. Concentration units include percent, milligrams per kilogram or parts per million, calories per pound, and international units per pound.

	Pounds	Percent
RDP	0.45	70.31

The comment behind the cell containing this label is shown in Figure 21–16.

Figure 21–16

Rumen degradable protein (RDP). This is the portion of crude protein that is degraded and used as a food source by the rumen microbes. It is also called *degradable intake protein* (DIP). The optimal mix of degradable and undegradable protein has not been determined in sheep, though the RDP portion should never be less than 50 percent of the crude protein.

The comment behind the cell containing the percent RDP value is shown in Figure 21–17.

Figure 21–17

RDP% is the percentage of crude protein.

	Pounds	Percent
RUP	0.19	29.69

The comment behind the cell containing this label is shown in Figure 21–18.

Figure 21–18

Rumen undegradable protein (RUP). This is the portion of crude protein that is not degraded by the rumen microbes. It is also called *undegradable intake protein* (UIP) and *bypass protein*. The optimal mix of degradable and undegradable protein has not been determined in sheep, though the RUP portion should never be more than 50 percent of the crude protein.

The comment behind the cell containing the percent RUP value is shown in Figure 21–19.

Figure 21–19

RUP% is percentage of crude protein.

	Pounds	Percent
MP-bacteria	0.383	69.0

The comment behind the cell containing this label is shown in Figure 21–20.

Figure 21–20

MP-bacteria is predicted using formulas in the dairy NRC, 2001 as follows:

$$Microbial\ crude\ protein = 0.13 \times discounted\ TDN.$$

However, if RDP intake is less than 1.18 times this TDN-predicted value, then microbial crude protein is predicted as 0.85 × RDP.

$$MPbacteria = Microbial\ crude\ protein \times 0.64.$$

Economical sheep rations will usually supply more metabolizable protein through bacterial growth than from RUP.

The comment behind the cell containing the percent MP-bacteria value is shown in Figure 21–21.

Figure 21–21

MP-bacteria% is percentage of metabolizable protein.

	Pounds	Percent
MP-RUP	0.152	27.4

The comment behind the cell containing this label is shown in Figure 21–22.

Figure 21–22

MP-RUP calculated using a formula from the dairy NRC, 1989 as follows:

$$MP - RUP = RUP \times 0.80.$$

Economical sheep rations will usually supply more metabolizable protein through bacterial growth than from RUP.

The comment behind the cell containing the percent MP-RUP value is shown in Figure 21–23.

Figure 21–23

MP-RUP% is percentage of metabolizable protein.

	Pounds	Percent
MP-endogenous	0.020	3.6

The comment behind the cell containing this label is shown in Figure 21–24.

MP-endogenous includes metabolizable protein coming from sloughed epithelial cells in the digestive and respiratory systems as well as enzyme secretions into the abomasum. Calculated from formulas in the dairy NRC, 2001.

Figure 21–24

The comment behind the cell containing the percent MP-endogenous value is shown in Figure 21–25.

MP-endogenous% is percentage of metabolizable protein.

Figure 21–25

	Pounds	**Percent**
MP total	0.555	13.0

The comment behind the cell containing this label is shown in Figure 21–26.

Metabolizable protein is true protein absorbed at the intestine, supplied by microbial protein, RUP, and endogenous sources.

Figure 21–26

The comment behind the cell containing the percent MP total value is shown in Figure 21–27.

MP total% is percentage of ration dry matter.

Figure 21–27

	Pounds	**Percent**
CP total	0.64	14.95

The comment behind the cell containing this label is shown in Figure 21–28.

Crude protein is measured as 6.25 × (the feed nitrogen content). Because other feed components besides protein contain nitrogen, it is described as *crude* protein.

Figure 21–28

The comment behind the cell containing the percent CP total value is shown in Figure 21–29.

CP total% is percentage of ration dry matter.

Figure 21–29

Strike **F5**.

The Feedstuffs Contributions *Display (strike* `F6`*)*

Shown here are the nutrients contributed by each feedstuff in the ration. This display is useful in troubleshooting problems with nutrient excesses.

Strike **F1**.

The Graphic *Display*

At the bottom of the home display are tabs. The current tab selected is the Worksheet tab. Other tabs are graphs based on the current ration. Note that you may have to click the leftmost navigation button at the lower-left corner to find the first tab.

DMI

Click on this tab to display a graph titled Nutrient Status: Dry Matter Intake

Energ&Prot

Click on this tab to display a graph titled Nutrient Status: Energy and Protein

Macro

Click on this tab to display a graph titled Nutrient Status: Macrominerals

Micro

Click on this tab to display a graph titled Nutrient Status: Microminerals

Vitamins

Click on this tab to display a graph titled Nutrient Status: Vitamins A, D, & E

Click on the **Worksheet** tab.

Blend Feedstuffs Procedure

Click on the **Blend Feedstuffs** button. The message box in Figure 21–30 displays.

Figure 21–30

> YOU MUST HAVE ALREADY SELECTED THE FEEDS YOU WANT TO BLEND. When your analysis is acceptable, strike **ENTER** and **F3** to name and file the blend.

Click **OK**.

Enter a value of 0 in the gray area to the left of the feedstuff name for the one feedstuff that the mill would not handle: the hay (Table 21–5). Note that in the blue column to the left, the values are converted to a pounds-per-ton basis.

Table 21–5
Inclusion rates for feedstuffs in the EweGrain blend

Soybean meal, 44%	0.4
Brome hay, mid bloom	0
Corn grain, ground	2
Molasses, cane	0.24
Calcium sulfate	0.0086
EDDI (Ethylenediamine dihydroiodide)	0.000000084
Limestone, ground	0.0273
Salt (sodium chloride)	0.00883
Sodium selenite	0.00000089
Vitamin A supplement	0.00214
Vitamin D supplement	0.00019
Vitamin E supplement	0.00131
Zinc oxide	0.000004

Strike **ENTER** and **F3**. The input box in Figure 21–31 displays.

ENTER THE NAME OF THE BLEND (names may not be composed of only numbers):

Figure 21–31

Name the blend EweGrain and click **OK**. The message box in Figure 21–32 displays.

The new blend has been filed at the bottom of the feed table.

Figure 21–32

Click **OK**.

View Blends Procedure

Click on the **View Blends** button. The message box in Figure 21–33 displays.

Cursor right to view the blends. Cursor down for more nutrients. DO NOT TYPE IN THE BLUE AREAS.

Figure 21–33

Click **OK**. Confirm that the EweGrain formula and analysis have been filed.
Strike **F1**.

Using the Blended Feed in the Balanced Ration

Select Feeds *Procedure*

Click on the **Select Feeds** button. Unselect all feedstuffs that were in the EweGrain. Select the EweGrain at the bottom of the feed table (row 200) and the Brome hay mid (Table 21–6).

Table 21–6
Feedstuff and blend to select for the sheep ration formulation example

Brome hay, mid bloom
EweGrain

Strike **ENTER** and **F2**.

Make Ration *Procedure*

Click on the **Make Ration** button. Enter the ration shown in Table 21–7.

Table 21–7
Feeding rates for feedstuff and blend in the sheep ration formulation example

EweGrain	2.69
Brome hay, mid bloom	2.2

The amount of brome hay has already been established. The amount of EweGrain to feed is the total amount of its component ingredients in the balanced ration. That value is:

$$0.4 + 2 + 0.24 + 0.0086 + 0.000000084 + 0.0273 + 0.00883 + 0.00000089 + 0.00214 + 0.00019 + 0.00131 + 0.000004 = 2.69$$

This value is recorded at View Blends under Formula, as entered. The ration is balanced as it was when the components of EweGrain were fed unmixed.

Strike **ENTER** and **F1**.

Print Ration or Blend Procedure

Make sure your name is entered at cell C1. Click on the **Print Ration or Blend** button. The input box in Figure 21–34 displays.

Figure 21–34

Are you printing a sheep ration evaluation or a blend formula and analysis? (**1**—RATION, **2**—BLEND):

Click **OK** to accept the default input of 1. A two-page printout will be produced by the machine's default printer.
Click on the **Print Ration or Blend** button. Type **2** to print a blend. The message box in Figure 21–35 displays.

Figure 21–35

Click on the green number above the blend you want to print and press **F4**. Scroll right to see additional blends.

Click **OK**.
Find the EweGrain blend, click on the green number above it, and strike **F4**. A one-page printout will be produced by the machine's default printer.

ACTIVITIES AND WHAT-IFS

In the Forms folder on the companion CD to this text is a SheepInput.doc file that may be used to collect the necessary inputs for use of the SheepRation.xls file. This form may be printed out and used during on-farm visits to assist in ration evaluation activities.

1. Remake the ration described in Table 21–3 for the ewe described in this chapter. Strike **F5**. How many pounds of metabolizable protein are provided by this ration? Of this total, how many pounds are predicted to come from the growth of bacteria (MP-bacteria)? How many pounds are predicted to come from rumen-undegradable feed sources (MP-RUP)? How many pounds are predicted to come from endogenous sources (MP-Endogenous)? Add urea and roasted soybeans to the selection of feedstuffs for this ration. Make the ration and strike **F5**. Add 0.005 lb. of urea and evaluate the effect on the three sources of MP. Remove the urea and add 0.005 lb. of roasted soybeans to the ration. Evaluate the effect on the three sources of MP. Which feedstuff has more of an effect on MP-bacteria? Which feedstuff has more of an effect on MP-RUP? Explain these results. Why is it that neither feedstuff has an impact on the level of MP-Endogenous?
2. Remake the ration described in Table 21–3 for the ewe described in this chapter. Input a ewe, first 6 to 8 weeks lactation suckling twins. What problem is evident with feeding the lactating ewe the same ration as is fed the gestating ewe? Fix the DMI shortage by increasing the amount of brome hay, mid-bloom. What nutrients are still deficient? Describe in general how the ration for the lactating ewe should differ from that of the gestating ewe.
3. Remake the ration described in Table 21–3 for the ewe described in this chapter. What is the current ration cost? Add wheat middlings to the selection of feedstuffs for this ration. Replace as much of the corn meal and soybean meal (44 percent) with wheat middlings as possible while still keeping the ration balanced. What is the price of your new formula?

REFERENCES

Bouchard, R., Conrad, H. R. (1973). Sulfur requirement of lactating dairy cows. I. Sulfur balance and dietary supplementation. *Journal of Dairy Science. 56,* 1276–1282.

National Research Council. (1980). *Mineral Tolerance of Domestic Animals.* Washington DC: National Academy Press.

National Research Council. (1985). *Nutrient Requirements of Sheep* (6th revised edition). Washington DC: National Academy Press.

National Research Council. (1987). *Vitamin Tolerance of Animals.* Washington DC: National Academy Press.

Qi, K., Lu, C. D., & Owens, F. N. (1993). Sulfate supplemtation of growing goats: effects on performance, acid-base balance, and nutrient digestibilities. *Journal of Animal Science. 71,* 1579–1587.

CHAPTER 22
FEEDING GOATS

Beware of a servant become master, or a goat become gardener.

ESTONIAN PROVERB

The nutritional phases in the goat production cycle, illustrated in Figure 22–1, include:

Doe: lactation, flushing, lactation and gestation, dry and gestation
Growing kid Route #1: liquid feeding/creep feed, growing short duration
Growing kid Route #2: liquid feeding/creep feed, growing long duration

A sample growth rate is plotted in Figure 22–2.

THE DOE

Lactation

At kidding, the doe experiences a drain of nutrients from her body in milk, especially if the doe is a dairy breed such as Saanen, Alpine, Toggenburg, LaMancha, or Nubian. The doe's appetite cannot immediately make the adjustment necessary to replace the nutrients, so in early lactation she will draw on stored reserves of nutrients. As she does so, she will lose body weight. Doe body condition should be managed so that she has adequate but not excessive fat reserves at kidding. This corresponds to a body condition score (BCS) of 3.5. The guidelines in Figure 20–3 may also be applied to goats.

Dairy goats may be milked for 305 days or more. Meat goats and Angora goats will be dried off at the weaning of their kid(s). Depending on the breed, they may then enter an anestrus period that lasts until the start of the next breeding season.

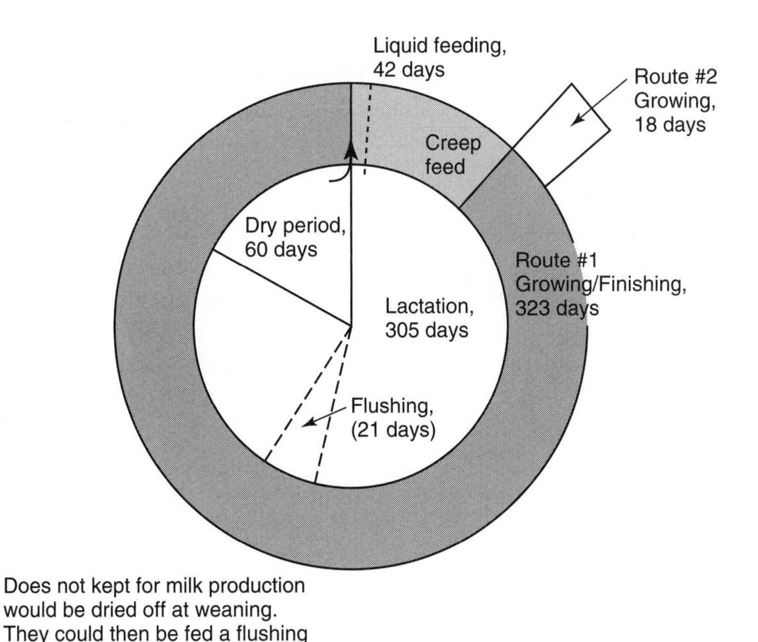

Figure 22–1
Nutritional Phases in the Production Cycle—Goats. Move clockwise starting at top of innermost circle. Continue through the next cycle or move to next shell after 360°. Note that there are two possible routes after the kid is weaned.

Body weight (lb.)
140
120
100
80
60
40
20
0.0
0 2 4 6 8 10 12
Age (months)

Figure 22–2
Sample growth curve for meat goats.

Anestrous

Strictly seasonal breeds of goats do not cycle out of season and the period of noncycling is termed the *anestrous* period. The breeding does are fed a maintenance diet during their anestrous period unless they are lactating. Many goat breeds are at least partially seasonal breeders, cycling reliably during periods of decreasing daylight (late August through January in the northern hemisphere). Dairy goats will be lactating through their anestrous period. The does in a meat goat herd that were dried off at weaning should be fed a diet supporting only maintenance or one adjusted to account for less than ideal body condition until the start of the next breeding season.

Flushing/Breeding

Goat breeds that evolved in the temperate regions of the world will be seasonal breeders, cycling most reliably during periods of decreasing day length.

Seasonal breeders will have an anestrus period during late spring. Goat breeds that evolved in tropical regions will cycle year round. To optimize market options for weaned kids sold for meat, it is best to include some of the tropical goat trait of nonseasonality in the genetics of the meat goat herd.

Beginning 2 to 3 weeks prior to breeding and continuing for a week or more into the breeding season, does are often flushed. This means that they are fed an enhanced diet containing increased nutrient levels. Flushing thin does has been shown to improve breeding performance as measured by the ovulation rate and conception rate. It also may decrease embryonic death loss. Flushing does that are in good condition may have no benefit and flushing fat does will be detrimental. Flushing is discussed in greater detail in Chapter 20.

Gestation

The gestation period for goats is 146 to 150 days. Two-thirds of the birth weight of the kid(s) is acquired during the last third of pregnancy. Dairy goats given a 60-day dry period will be lactating through the first 90 days of their gestation. Some concentrate feeding during the first half of the dry period may be necessary to achieve the targeted 3.5 BCS and during the second half of the dry period to prevent pregnancy toxemia. Gestating meat goats may also require grain supplementation during the last month of gestation to prevent pregnancy toxemia, especially if multiple births are expected.

Pregnancy toxemia is the result of a drop in the doe's blood glucose toward the end of the gestation as a result of the rapidly increasing energy demand of the fetus(es). Although it is normal for the doe to mobilize some body fat, excessive reliance on body fat to meet energy needs leads to the accumulation of ketone bodies in the blood. Pregnancy disease is also called *ketosis.* In sheep and goats, ketosis is most likely to occur during late gestation whereas in dairy cattle, ketosis is most likely to occur during early lactation. Ketosis is discussed in Chapter 18.

THE KID

Liquid Feeding/Creep Feed

Most of the world's goats are raised on pasture and range. Under these conditions, the only nutrients available to goat kids will come from colostrum, milk, and forage with no supplementation.

The structure of the doe's placenta is described as epitheliolchorial. This type of placenta does not allow antibody immunoglobulins to pass from the doe into the blood of the fetal kid(s). The neonatal kid acquires these antibody immunoglobulins through the doe's first milk or colostrum. The importance of colostrum to the neonatal kid cannot be overestimated. (Colostrum is discussed in Chapter 20.) The recommendations for colostrum consumption for all ruminants are (1) that neonates should receive an amount of colostrum equivalent to 10 percent of the birth weight, or 1.5 ounces per pound of body weight, within 12 hours of birth; and (2) that half this amount should be ingested within 2 hours of birth, the sooner the better. An 8-lb. kid should, therefore, receive about 0.8 lb. (about 12 ounces) of colostrum.

After the second day, kids may be fed goat milk, cow milk, or milk replacer at a daily rate of up to 3 pints, split into two feedings. Limiting the liquid feed available to kids to 3 pints maximum encourages them to consume a creep or starter feed. This level of feeding should be continued until weaning. A gradual transition from goat milk to cow milk or milk replacer is possible after the kid is a few days old.

Small amounts of a starter feed may be made available after the first week. This will stimulate rumen development and ease the transition at weaning. Some high-quality forage may also be given to kids prior to weaning.

Growing

Weaning of kids can be successful as early as 4 weeks of age. Success depends on how well the kid is eating solid feed before weaning. A more typical weaning age is 2 to 3 months. Newly weaned kids should be provided with the highest-quality forage available and 0.3 lb. of high-quality starter feed formulated for calves. Weaned kids to be sold for meat will be placed on either a short-duration grain-based feeding program or long-duration forage-based feeding program (Figure 22–1).

GOAT FEEDING AND NUTRITION ISSUES

State of Research into Goat Nutrient Requirements

Goats produce milk, meat, and fiber (mohair and cashmere). Because research on goat nutrition has not been as well funded as research for other ruminant species, much of the goat nutrient requirement data published in the NRC's *Nutrient Requirements of Goats: Angora, Dairy, and Meat Goats in Temperate and Tropical Countries* (1981) is based on sheep and cattle research.

In the goat industry, it is common to share grain mixtures among groups and between species. In other words, grains developed for lactating dairy cows are often used for lactating dairy goats; yearling goats are often fed the same grain used for the lactating does; the buck usually receives the same grain as the does; and calf grain is often used for kids. Nevertheless, goats do have unique nutritional requirements, and producers should be aware of how the nutrient levels being fed match up to the predictions of what the goat requires in order to efficiently troubleshoot future problems. Although the symptoms of nutrient deficiencies are often similar to those of infectious diseases, the cures are obviously different.

Feed Passage Rate

Compared to cattle, small ruminants such as goats and sheep pass feed material through their digestive systems relatively quickly. At a given particle size, the same feedstuff would be digested more completely when fed to cattle than it would when fed to goats or sheep because there is less time of exposure to digestive activity. A more rapid passage rate also means that relative to body size, the small ruminant will be capable of a higher level of dry matter intake (DMI) than would cattle because a higher passage rate means that the reticulorumen will empty out faster, making room for additional feed intake.

Selective Feeding Behavior

Goats are unique in their preferences when grazing; they will consume feedstuffs that would be avoided by grazing sheep and cattle. For this reason, goats complement these species in mixed grazing systems. This selective behavior also enables the goat grazing on pastures with a high level of plant diversity to consume a diet that has a higher nutrient concentration than the average nutrient concentration in the available vegetation. Goats grazing on land with little plant diversity, however, will consume a diet of similar nutrient content to that of sheep and cattle.

The ability of goats to select particular plants and plant parts when grazing is applied by goats to grain mixtures, usually with undesirable results. Selective consumption or sorting may lead to nutrient deficiencies if goats leave behind unpalatable components of a balanced ration. To prevent the goat's ingested ration from being significantly different in nutrient content from the fed ration, it may be necessary to pellet the nonforage portion.

Goats are sometimes used to control browse plants or brush in abandoned fields. Goats are particularly effective at this because they are agile animals, and they have prehensile lips that enable them to selectively consume the more nutritious parts of woody plants, including the leaves, buds, twigs, and shoots. A flattened underline of trees and shrubs is often evident in areas grazed by goats. It is referred to as the *browse line* and marks the accessible height of the bipedal goat. Goats used to control brush should have free-choice access to a mineral mix containing both macro- and microminerals, as well as water.

Although the nutritional value of browse should be considered in the feeding management of goats (Huston, 1978), goats do not require browse. In fact, the plants that make up browse contain an uncertain level of available nutrients. Browse crude protein may be high in acid detergent fiber insoluble protein (Chapter 4) and, therefore, unavailable to the goat (Ramirez, Loyo, Mora, Sanchez, & Chaire, 1991). Additionally, if the browse plants are sparse, range goats will have an increased energy requirement due to the increased muscular activity involved in browsing. On the other hand, the goat's ability to more effectively utilize browse makes it possible for this species to acquire its needed nutrients with less time and effort than would be the case with sheep (Kronberg & Malechek, 1997).

Although goats are the livestock of choice on arid and semiarid rangeland where browse is the primary feedstuff available, productivity will usually be improved by providing supplemental sources of energy and protein. Grains fed to goats should be processed to improve utilization. Processing options for grain are given in Chapter 13.

Fiber Production by Goats

Goats produce three types of fiber. Meat goats produce a coarse guard hair and a variable amount of fine undercoat hair called cashmere. Angora goats produce mohair. Both mohair and cashmere have market value.

The use of metabolizable energy for mohair production by Angora goats is assumed to be 33 percent efficient (NRC, 1981). This value is used in predicting the mohair produced by goats in the companion application to this text. The only nutritional requirement besides energy that is influenced by mohair production according to the NRC publication is protein. Although the goat NRC (1981) and the companion application to this text give no additional requirements for dietary sulfur due to mohair production, the sulfur-containing amino acids are significant components of mohair, and adequate sulfur intake is critical in Angora goats.

Activity Level and Nutrient Requirements

The goat NRC (1981) and the companion application to this text use four activity descriptions as inputs in the determination of nutrient requirements for goats. These are as follows.

1. Minimal activity (includes stable feeding conditions).
2. Low activity (intensive management and tropical range). Metabolizable energy requirements for goats described as low activity are 25 percent greater than the requirements for goats described as maintenance. The requirements for nonenergy nutrients are increased only slightly over maintenance.
3. Medium activity (semiarid rangeland and slightly hilly pastures). Metabolizable energy requirements for goats described as medium activity are 50 percent greater than the requirements for goats described as maintenance.

The requirements for nonenergy nutrients are increased only slightly over low-activity levels.

4. High activity (arid rangeland, sparse vegetation, and mountainous pastures). Metabolizable energy requirements for goats described as high activity are 75 percent greater than the requirements for goats described as maintenance. The requirements for nonenergy nutrients are increased only slightly over medium activity.

Feedstuffs Affecting Milk Flavor

Some crops, such as turnips, when fed to lactating does shortly before milking, can impart off flavors to the milk. If such crops are fed immediately after milking, the risk that they will affect milk flavor at the next milking is reduced.

Urinary Calculi

Goats are prone to urinary calculi. The companion application to this text calculates the dietary cation–anion difference (DCAD) in formulated rations. Keeping this value low through the addition of ammonium chloride or other anionic salts is helpful in preventing the development of urinary calculi. This disorder is discussed in Chapter 20.

Poisonous Plants

A review of plants that are poisonous to goats has been done (Ace & Hutchison, 1984). A listing of some poisonous plant species, grouped according to the nature of the toxicity, is presented in Tables 3–3, 3–4, and 3–5.

Bloat

Bloat occurs in ruminants when the gases produced during fermentation are unable to be expelled through eructation. Bloat is a concern with pastured goats, particularly with pastures rich in legume. Making hay available free choice to goats on pasture can help prevent bloat. The subject of bloat is discussed further in Chapter 16.

Acidosis

The diet for growing/finishing goats and lactating does may contain as much as 70 percent grain. It is important that ruminants moving from a high-forage diet to a high-grain diet receive an appropriate transition to allow an opportunity for the rumen microbial population to make the adjustment. A poorly managed transition may lead to a reduction in rumen pH, which in turn, may result in rumen acidosis. In the companion application to this text, rumen pH is predicted from the level of physically effective NDF in the ration (Figure 18–4). Acidosis is discussed in Chapter 20.

Prussic Acid Poisoning

Under certain conditions, some grasses, particularly forage sorghums, may accumulate hydrocyanic acid (prussic acid) in their tissues. Prussic acid poisoning is discussed in Chapter 20.

Nitrate Poisoning

Goats are susceptible to nitrate poisoning. Nitrate poisoning is caused by consumption of plants that have accumulated nitrate in their tissues. It may also be caused by ingestion of poor-quality water. Symptoms include poor appetite, general unthriftiness, and abortion. Nitrate poisoning is discussed in Chapter 16.

Parturient Paresis

High-producing dairy goats may be afflicted with parturient paresis or milk fever. Symptoms include incoordination, recumbency, and unconsciousness. Milk fever occurs as a result of a drop in blood calcium during lactation, and is described in Chapter 18.

Tall Fescue Toxicosis

Goats that consume pasture or hay containing tall fescue that has been infected with an endophyte are susceptible to tall fescue toxicosis, which has various effects on health, production, and reproduction. Tall fescue toxicosis is discussed in Chapter 24.

Enterotoxemia

Goats, like sheep and cattle, are subject to enterotoxemia (overeating disease) if stressed or poorly transitioned from a high-forage to a high-grain diet. Enterotoxemia is discussed in Chapter 20.

White Muscle Disease

White muscle disease is caused by a deficiency of selenium and/or vitamin E. Symptoms are stiffness, general difficulty in movement, and death. White muscle disease is discussed in Chapter 20.

END-OF-CHAPTER QUESTIONS

1. What is browse? What role can it play in goat nutrition?
2. Compared to other ruminant livestock, goats are better able to selectively consume the most nutritious parts of plants. Why is this the case?
3. Describe mixed grazing. What role might goats play in a mixed grazing system?
4. Describe the causes of the following problems as they relate to feeding goats: bloat, acidosis, parturient paresis, and nitrate poisoning.
5. Name a feedstuff that might cause an off-flavor milk to be produced if fed to a lactating dairy goat immediately prior to milking.
6. Give an example of a feeding environment where associated goat activity would be characterized as minimal. Give an example of a low-activity feeding environment. Give an example of a medium-activity feeding environment. Give an example of a high-activity feeding environment. What nutrient requirement adjustments are made for low, medium and high activity as compared to the minimal-activity feeding environment?
7. Does carrying multiple fetuses in late gestation that are fed only low-quality forage would be at risk for what metabolic disorder?
8. Give three functions of colostrum. Which of these functions is lost as colostrum consumption by the neonatal kid is delayed? Give recommendations for colostrum consumption by the neonatal kid.
9. How does the rate of passage of feed ingested by the goat compare with the rate of passage of feed ingested by the bovine? What impact would you expect this difference to have on the DMI, expressed as a percentage of body size? What impact would you expect this to have on the completeness with which the digestive tract is able to extract feed nutrients?
10. Discuss the status of research in goat nutrition relative to that in other ruminant livestock. Of what consequence is this to those involved in goat production?

REFERENCES

Ace, D. L., & Hutchison, L. J. (1984). Poisonous plants. Goat extension handbook, fact sheet C-4, Pennsylvania State University, University Park, Pennsylvania.

Huston, J. E. (1978). Forage utilization and nutrient requirements of the goat. *Journal of Dairy Science 61*, 988

Kronberg, S. L., & J. C. Malechek. (1997). Relationships between nutrition and foraging behavior of free-ranging sheep and goats. *Journal of Animal Science 75*, 1756–1763.

National Research Council. (1981). *Nutrient requirements of goats: Angora, dairy and meat goats in temperate and tropical countries.* Washington, DC: National Academy Press.

Ramirez, R. G., Loyo, A., Mora, R., Sanchez, E. M., & Chaire, A. (1991). Forage intake and nutrition of range goats in a shrubland in northeastern Mexico. *Journal of Animal Science 69*, 879–885.

CHAPTER 23

GOAT RATION FORMULATION

TERMINOLOGY

Workbook A workbook is the spreadsheet program file including all its worksheets.

Worksheet A worksheet is the same as a spreadsheet. There may be more than one worksheet in a workbook.

Spreadsheet A spreadsheet is the same as a worksheet.

Cell A cell is a location within a worksheet or spreadsheet.

Comment A comment is a note that appears when the mouse pointer moves over the cell. A red triangle in the upper right corner of a cell indicates that it contains a comment. Comments are added to help explain the function and operations of workbooks.

Input box An input box is a programming technique that prompts the workbook user to type information. After typing the information in the input box, the user clicks **OK** or strikes **ENTER** to enter the typed information.

Message box A message box is a programming technique that displays a message. The message box disappears after the user clicks **OK** or strikes **ENTER**.

EXCEL SETTINGS

Security: Click on the **Tools** menu, **Options** command, **Security** tab, **Macro Security** button, **Medium** setting.

Screen resolution: This application was developed for a screen resolution of 1024 × 768. If the screen resolution on your machine needs to be changed, see Microsoft Excel Help, "Change the screen resolution" for instructions.

HANDS-ON EXERCISES

GOAT RATION FORMULATION

Double click on the **GoatRation** icon. The message box in Figure 23–1 displays.

Figure 23–1

Macros may contain viruses. It is advisable to disable macros, but if the macros are legitimate, you may lose some functionality.
Disable Macros or Enable Macros or More Info

Click on **Enable macros**.

The message box in Figure 23–2 displays.

Figure 23–2

Function keys F1 to F8 are set up. You may return to this location from anywhere by striking **ENTER**, then the **F1** key. Workbook by David A. Tisch. The author makes no claim for the accuracy of this application and the user is solely responsible for risk of use. You're good to go. TYPE ONLY IN THE GRAY CELLS! Note: This Workbook is made up of charts and a worksheet. The charts and worksheet are selected by clicking on the tabs at the bottom of the display. Never save the Workbook from a chart; always return to the worksheet before saving the Workbook.

Click **OK**.

Input Goat Procedure

Click on the **Input Goat** button. The input box in Figure 23–3 is shown.

Figure 23–3

1. Goats at maintenance
2. Gestating does
3. Growing kids
4. Lactating does
5. Goats producing only mohair.

(Bucks usually receive the same kind of feedstuffs as the doe, with the amount dependent upon body condition.) Enter the appropriate number:

Click **OK** to choose the default input of #3, the Growing Kid. The input box shown in Figure 23–4 displays.

Figure 23–4

1. Minimal activity—includes stable feeding conditions
2. Low activity—intensive management and tropical range
3. Moderate activity—semiarid rangeland and slightly hilly pastures
4. High activity—arid rangeland, sparse vegetation and mountainous pastures

Enter the appropriate number:

Click **OK** to accept the default value of 1. The input box in Figure 23–5 displays. Click **OK** to accept the predicted dry matter intake (DMI).

Figure 23–5

Enter measured DMI (pounds) if available or click **OK** to accept the predicted value at CX26.

Click **OK**. The message box in Figure 23–6 displays.

Figure 23–6

Enter the appropriate Cell Inputs in column CX.

Click **OK**.

The Cell Inputs

Enter the cell inputs shown in Table 23–1.

Table 23–1
Inputs for the goat ration formulation example

70	Current temperature (° F)
yes	Is there significant relief from the heat at night[1]?
30	Goat's current body weight (22–200 lb.)
	N/A
0.22	Anticipated weight gain, lb./day (0–0.55)
	N/A
	N/A
	N/A

Strike **`ENTER`** and **`F1`**.

Select Feeds Procedure

Click on the **`Select Feeds`** button. The message in Figure 23–7 displays.

Figure 23–7

Nutrient content expressed on a dry matter basis. Feedstuffs are listed first by selection status, then, within the same selection status, by decreasing protein content, then, within the same protein content, alphabetically.

Click **OK**.

Explore the table. Note the nutrients listed as column headings. Note also that the table ends at row 200. You select feedstuffs for use in making two different products: a blend to be mixed and sold bagged or bulk, and a ration to be fed directly to livestock.

Select the feedstuffs in Table 23–2 by placing a 1 in the column to the left of the feedstuff name. All unselected feedstuffs should have a 0 value in the column to the left of the feedstuff name. If you wish to group feedstuffs but not select them, you would place a value between 0 and 1 to the left of the feedstuff name.

Table 23–2
Feedstuffs to select for the goat ration formulation example

Feedstuff
Brewers dried grains
Corn grain, ground
Timothy hay, full bloom
Molasses, cane
Limestone, ground
Salt (sodium chloride)
Sodium selenite
Vitamin A supplement
Vitamin D supplement
Vitamin E supplement

Strike **ENTER** and **F2**.

Make Ration Procedure

Click on the **Make Ration** button. The message box in Figure 23–8 displays.

Figure 23–8

> ENTER **POUNDS TO FEED** IN COLUMN B. TOGGLE BETWEEN NUTRIENT WEIGHTS AND CONCENTRATIONS USING THE **F5** KEY, RATION AND FEED CONTRIBUTIONS USING THE **F6** KEY. WHEN DONE STRIKE **F1**. A cell is highlighted in red if nutrient provided is poorly matched with the nutrient target. The lower limit is taken as 95 to 98 percent of target, depending on the nutrient. Except for copper, the upper limit of acceptable mineral is taken from National Research Council. (1980). *Mineral Tolerance of Domestic Animals.* National Academy Press. Maximum tolerable copper is taken as 1.6 × the value given for sheep. The upper limit for vitamins is taken from the NRC, 1987. For other nutrients, the upper limit is based on unreasonable excess and the expense of unnecessary supplementation. See Table 23–4 for specifics.

Click **OK**. The message box in Figure 23–9 displays.

Figure 23–9

The Goal Seek feature may be useful in finding the pounds of a specific feedstuff needed to reach a particular nutrient target:
1. Select the red cell highlighting the deficient nutrient
2. From the menu bar, select **Tools**, then **Goal Seek**
3. In the text box, "To Value:" enter the target to the right of the selected cell
4. Click in the text box, "By changing cell:" and then click in the gray "Pounds fed" area for the feedstuff to supply the nutrient
5. Click **OK**. You may accept the value found by clicking **OK** or reject it by clicking **Cancel**.

WARNING: Using Goal Seek to solve the unsolvable (e.g., asking it to make up an iodine shortfall with iron sulfate) may result in damage to the Workbook.
IMPORTANT: If you return to the Feedtable to remove more than one feedstuff from the selected list, you will lose your chosen amounts fed in the developing ration.

Click **OK**.
In the gray area to the right of the feedstuff name, enter the pound values shown in Table 23–3.

Table 23–3
Inclusion rates for the feedstuffs in the goat ration formulation example

Brewers dried grains	0.2
Corn grain, ground	0.4
Timothy hay, full bloom	1
Molasses, cane	0.1
Limestone, ground	0.00121
Salt (sodium chloride)	0.00742
Sodium selenite	0.000000049
Vitamin A supplement	0.00018
Vitamin D supplement	0.000048
Vitamin E supplement	0.00032

The Nutrients Supplied *Display*

The application highlights ration nutrient levels, expressed as amount supplied per goat per day, that fall outside the acceptable range.

The lower limit for dry matter intake is 95 percent of the target. The lower limit for all other ration nutrients is taken as 98 percent of the target. Table 23–4 shows the upper limits of the acceptable range for the various nutrients.

Table 23–4
Upper limits for ration nutrients used in the goat ration application

	Upper Limit	Source
Dry matter intake (DMI)	5% over the requirement	Author
Total digestible nutrients (TDN)	No upper limit	—
Digestible energy (DE)	No upper limit	—
Metabolizable energy (ME)	No upper limit	—
Net energy	An energy excess beyond that needed for maintenance, activity, and pregnancy is used to predict performance	—
Crude protein (CP)	A protein excess beyond that needed for maintenance, activity, and pregnancy is used to predict performance	—
Calcium (Ca)	2% of ration dry matter	NRC: Mineral Tolerance of Domestic Animals, 1980—sheep
Phosphorus (P)	0.6% of ration dry matter	NRC: Mineral Tolerance of Domestic Animals, 1980—sheep
Magnesium (Mg)	0.5% of ration dry matter	NRC: Mineral Tolerance of Domestic Animals, 1980—sheep
Potassium (K)	3% of ration dry matter	NRC: Mineral Tolerance of Domestic Animals, 1980—sheep
Sulfur (S)	0.4% of ration dry matter	NRC: Mineral Tolerance of Domestic Animals, 1980
Sodium (Na)	3.5% or ration dry matter	NRC: Mineral Tolerance of Domestic Animals, 1980—sheep, based on limit for salt (sodium chloride)
Salt	9% of ration dry matter	NRC: Mineral Tolerance of Domestic Animals, 1980—sheep
Iron (Fe)	500 mg/kg or ppm of ration dry matter	NRC: Mineral Tolerance of Domestic Animals, 1980—sheep
Manganese (Mn)	1,000 mg/kg or ppm of ration dry matter	NRC: Mineral Tolerance of Domestic Animals, 1980—sheep
Copper (Cu)	40 mg/kg or ppm of ration dry matter	Author
Zinc (Zn)	300 mg/kg or ppm of ration dry matter	NRC: Mineral Tolerance of Domestic Animals, 1980—sheep
Iodine (I)	50 mg/kg or ppm of ration dry matter	NRC: Mineral Tolerance of Domestic Animals, 1980—sheep
Cobalt (Co)	10 mg/kg or ppm of ration dry matter	NRC: Mineral Tolerance of Domestic Animals, 1980—sheep
Selenium (Se)	2 mg/kg or ppm of ration dry matter	NRC: Mineral Tolerance of Domestic Animals, 1980
Vitamin A	30 × the predicted requirement	NRC: Vitamin Tolerance of Animals, 1987
Vitamin D	4 × the predicted requirement	NRC: Vitamin Tolerance of Animals, 1987
Vitamin E	20 × the predicted requirement	NRC: Vitamin Tolerance of Animals, 1987

The *Nutrients Supplied* display shows nutrient amounts supplied in the diet and nutrient targets for the inputted goat. All nutrient levels appear to be within acceptable ranges as established by the application.

The cost of this ration using initial $/ton values is $0.09 per head per day.

Energy allowable gain (lb./head/day): *0.19*

The comment behind the cell containing the value for Energy Allowable Gain (lb./head/day) is shown in Figure 23–10.

Figure 23–10

If this cell is highlighted in yellow, it is because the energy content of this ration supports a rate of gain that is more than 30 percent in excess of the target growth rate as indicated in the Input section.

Protein allowable gain (lb./head/day): *0.74*

The comment behind the cell containing the value for Protein Allowable Gain (lb./head/day) is shown in Figure 23–11.

Figure 23–11

Calculated from crude protein.

Predicted cost per lb. gain: *$0.47*

The comment behind the cell containing this label is shown in Figure 23-12.

Figure 23–12

Calculated as the ration cost divided by the pounds gain supported by dietary net energy or crude protein—whichever supports the lower gain.

Ca:P: *1.4*

The comment behind the cell containing this label is shown in Figure 23–13.

Figure 23–13

Calcium-to-phosphorus ratio. Generally, the ratio for goats should be maintained above 1.2. High levels of phosphorus have been implicated in urolithiasis (urinary calculi). (Goat NRC, 1981.)

NDF:NFC: *1.3*

The comment behind the cell containing this label is shown in Figure 23–14.

Figure 23–14

This is a ratio of carbohydrates in the diet. It is essentially a ratio of fiber content to sugar and starch content. Though there are no specific recommendations, rumen health may be compromised if the ratio falls too far below 1 and productivity compromised if the ratio is too far above 1.

N:S: *9.6*

The comment behind the cell containing this label is shown in Figure 23–15.

Figure 23–15

The dietary nitrogen-to-sulfur ratio should be between 10:1 and 12:1 for efficient utilization of nonprotein nitrogen (urea). (Bouchard & Conrad, 1973; Qi et al., 1993). This is because the rumen microbes need a source of sulfur with nitrogen to synthesize the sulfur-containing amino acids methionine and cysteine. Nitrogen was calculated from crude protein using 16 percent as the average nitrogen content of protein. This ratio may be ignored if urea is not being fed.

DCAD: *144*

The comment behind the cell containing this label is shown in Figure 23–16.

Figure 23–16

Dietary Cation–Anion Difference (DCAD) is used to assess the impact of the ration's mineral content on the body's efforts to regulate blood pH. It is expressed as mEq/kg and calculated as follows:

$$mEq\,(Na + K) - mEq\,(Cl + S).$$

In goat nutrition, the feeding of ammonium sulfate (0.5 to 1.0 percent), ammonium chloride, and other compounds have been suggested as means of reducing the incidence of urinary calculi. These have the effect of acidifying the urine. This effect is measured generally as the DCAD.

Predicted rumen pH: *6.46*

The comment behind the cell containing this label is shown in Figure 23–17.

Figure 23–17

DMI becomes variable at a ruminal pH of less than 6.2. At a ruminal pH of less than 6.0, the ability of bacteria to derive energy from forages declines.
Note that although it is advisable to feed buffer to help manage ruminal pH on high-grain diets, buffer is not considered in the ruminal pH prediction. For feedlot animals, a ruminal pH as low as 5.8 may be acceptable for a short period of time.

The Nutrient Concentration *Display (strike* `F5`*)*

The nutrients in the ration and the predicted nutrient targets are expressed in terms of concentration. That is, the nutrients provided by the ration are divided by the amount of ration dry matter and the nutrient targets are divided by the target amount of ration dry matter. Concentration units include percent, milligrams per kilogram or parts per million, calories per pound, and international units per pound.

	Pounds	Percent
RDP	0.090	56.25

The comment behind the cell containing this label is shown in Figure 23–18.

Figure 23–18

> Rumen degradable protein (RDP). This is the portion of crude protein that is degraded and used as a food source by the rumen microbes. It is also called *degradable intake protein* (DIP). The optimal mix of degradable and undegradable protein has not been determined in goats, though the RDP portion should never be less than 50 percent of the crude protein.

The comment behind the cell containing the percent RDP value is shown in Figure 23–19.

Figure 23–19

> RDP% is percentage of crude protein.

	Pounds	**Percent**
RUP	0.070	43.75

The comment behind the cell containing this label is shown in Figure 23–20.

Figure 23–20

> Rumen undegradable protein (RUP). This is the portion of crude protein that is not degraded by the rumen microbes. It is also called *undegradable intake protein* (UIP) and *bypass protein*. The optimal mix of degradable and undegradable protein has not been determined in goats, though the RUP portion should never be more than 50 percent of the crude protein.

The comment behind the cell containing the percent RUP value is shown in Figure 23–21.

Figure 23–21

> RUP% is percentage of crude protein.

	Pounds	**Percent**
MP-bacteria	0.077	54.79

The comment behind the cell containing this label is shown in Figure 23–22.

Figure 23–22

> MP-bacteria is predicted using formulas in the dairy NRC, 2001 as follows:
>
> *Microbial crude protein* = *0.13* × *discounted TDN.*
>
> However, if RDP intake is less than 1.18 times this TDN-predicted value, then microbial crude protein is predicted as 0.85 × RDP.
>
> *MPbacteria* = *Microbial crude protein* × *0.64.*
>
> Economical goat rations will usually supply more metabolizable protein through bacterial growth than from RUP.

The comment behind the cell containing the percent MP-bacteria value is shown in Figure 23–23.

Figure 23–23

MP-bacteria% is percentage of metabolizable protein.

	Pounds	**Percent**
MP-RUP	0.056	40.11

The comment behind the cell containing this label is shown in Figure 23–24.

Figure 23–24

MP-RUP calculated using a formula from the dairy NRC, 1989 as follows:

$$MP\text{-}RUP = RUP \times 0.80.$$

Economical goat rations will usually supply more metabolizable protein through bacterial growth than from RUP.

The comment behind the cell containing the percent MP-RUP value is shown in Figure 23–25.

Figure 23–25

MP-RUP% is percent age of metabolizable protein.

	Pounds	**Percent**
MP—endogenous	0.007	5.10

The comment behind the cell containing this label is shown in Figure 23–26.

Figure 23–26

MP-endogenous includes metabolizable protein coming from sloughed epithelial cells in the digestive and respiratory systems as well as enzyme secretions into the abomasum. Calculated from formulas in the dairy NRC, 2001.

The comment behind the cell containing the percent MP-endogenous value is shown in Figure 23–27.

Figure 23–27

MP-endogenous% is percentage of metabolizable protein.

	Pounds	**Percent**
MP total	0.140	9.25

The comment behind the cell containing this label is shown in Figure 23–28.

Figure 23–28

Metabolizable protein is true protein absorbed at the intestine, supplied by microbial protein, RUP, and endogenous sources.

The comment behind the cell containing the percent MP total value shown in Figure 23–29.

Figure 23–29

MP total% is percentage of ration dry matter.

	Pounds	Percent
CP total	0.160	10.60

The comment behind the cell containing this label is shown in Figure 23–30.

Figure 23–30

Crude protein is measured as 6.25 × (the feed nitrogen content). Because other feed components besides protein contain nitrogen, it is described as *crude* protein.

The comment behind the cell containing the percent CP total value is shown in Figure 23–31.

Figure 23–31

CP total% is percentage of ration dry matter.

Strike **`F5`**.

The Feedstuffs Contributions *Display (strike* `F6`*)*

Shown here are the nutrients contributed by each feedstuff in the ration. This display is useful in troubleshooting problems with nutrient excesses.

Strike **`ENTER`** and **`F1`**.

The Graphic *Display*

At the bottom of the home display are tabs. The current tab selected is the Worksheet tab. Other tabs are graphs based on the current ration. Note that you may have to click the leftmost navigation button at the lower-left corner to find the first tab.

DMI

Click on this tab to display a graph titled Nutrient Status: Dry Matter Intake

Energ&Prot

Click on this tab to display a graph titled Nutrient Status: Energy and Protein

Macro

Click on this tab to display a graph titled Nutrient Status: Macrominerals

Micro

Click on this tab to display a graph titled Nutrient Status: Microminerals

Vitamins

Click on this tab to display a graph titled Nutrient Status: Vitamins

Gain

Click on this tab to display a graph titled Energy and Protein Allowable Gain

Milk

Click on this tab to display a graph titled Energy and Protein Allowable Milk

Click on the **Worksheet** tab.

Making a Medicated Complete Feed

Acquiring Drug Concentration and Dosage Information

***Blend Feedstuffs* Procedure**

Click on the **Blend Feedstuffs** button. The message box in Figure 23–32 displays.

Figure 23–32

YOU MUST HAVE ALREADY SELECTED THE FEEDS YOU WANT TO BLEND. When your analysis is acceptable, strike **ENTER** and **F3** to name and file the blend.

Click **OK**. We want to use monensin to control coccidiosis in our goat herd. Click on the **Additives** button for help with including additive premixes in blends. Click on the **FDA** button to go to the FDA-approved Web site. In the search box, type **monensin**. This is a drug approved for use in confined goats, not lactating. Monensin is used in preventing coccidiosis. Click on the **Search** button. Find the product, Rumensin® and click on **Browser** to view that product. We will use a product that contains 45 g monensin sodium/lb. Note that this is a Category I (no withdrawal period required), Type A medicated article (see Chapter 13). Under CFR indications, find the goat information. The amount to use is 20 g/ton.

The concentration of this additive in this additive premix is expressed in grams per pound.

The dose of this additive in this additive premix is expressed in grams per ton.

Determining the Appropriate Amount of Additive Premix to Make 1 Ton of Medicated Complete Feed

Close the Web site. Look for a Type that matches our additive's concentration and dose. The match is found at Type 6. In the gray area beside the cell containing the label, "Enter g/lb. additive in additive premix," enter **45**. In the gray area beside the cell containing the label, "Enter animal dose in g/ton feed," enter **20**. The result is shown across from the label, "Pounds of additive premix containing additive to add to 1 ton of blend to deliver dose." The result is 0.44.

Entering the Additive Premix into the Feed Table and Selecting It

Scroll up and click on the **Feed Table** button to go to the feed table to enter and select the additive premix. Enter the feed name **Rumensin** in an empty row under FEEDNAME. Enter a cost if known. Use 20,000 for this example. Enter a DM percent of 99. Select this feedstuff along with the other components of the grain. Strike **ENTER** and **F2**.

Blending the Additive Premix into the Complete Feed

Blend Feeds Procedure

Click on the **Blend Feeds** button and confirm that the amounts of feedstuffs are those in Table 23–5. Note that the amount of hay will be 0 in this blend.

Table 23–5
Inclusion rates for feedstuffs in the goat blend (as yet undetermined amount of drug)

Brewers dried grains	0.2
Corn grain, ground	0.4
Timothy hay, full bloom	0
Molasses, cane	0.1
Limestone, ground	0.00121
Salt (sodium chloride)	0.00742
Sodium selenite	0.000000049
Vitamin A supplement	0.00018
Vitamin D supplement	0.000048
Vitamin E supplement	0.00032
Rumensin[1]	?

[1]Rumensin®, a product of Elanco Animal Health.

Enter an amount in the blend that translates to a value of 0.44 lb./ton or less. Recall that inclusion of 0.44 lb. of this additive premix in a ton of feed will result in a monensin concentration of 20 g/ton. A value that works is 0.00015 (Table 23-6).

Table 23–6
Inclusion rates for feedstuffs in the medicated goat blend

Brewers dried grains	0.2
Corn grain, ground	0.4
Timothy hay, full bloom	0
Molasses, cane	0.1
Limestone, ground	0.00121
Salt (sodium chloride)	0.00742
Sodium selenite	0.000000049
Vitamin A supplement	0.00018
Vitamin D supplement	0.000048
Vitamin E supplement	0.00032
Rumensin	0.00015

Strike **ENTER** and **F3**. The message box in Figure 23–33 displays.

Figure 23–33

ENTER THE NAME OF THE BLEND (names may not be composed of only numbers):

Enter the name **Goatmedicated**. Click **OK**. The message box in Figure 23–34 displays.

> The new blend has been filed at the bottom of the feed table.

Figure 23–34

Click **OK**.

***View Blends* Procedure**

Click on the **View Blends** button to verify that the formula and analysis of the Goatmedicated blend has been filed. The message box in Figure 23–35 displays.

> Cursor right to view the blends. Cursor down for more nutrients. DO NOT TYPE IN THE BLUE AREAS.

Figure 23–35

Click **OK**.
Strike **F1**.

Using the Medicated Complete Feed in the Ration

***Select Feeds* Procedure**

Click on the **Select Feeds** button and select the feeds shown in Table 23–7. Unselect all others. Remember the Goatmedicated blend is located at the bottom of the feed table (row 200).

Goatmedicated
Timothy hay, full bloom

Table 23–7
Feedstuff and blend to select for the goat ration formulation example

Strike **ENTER** and **F2**.

***Make Ration* Procedure**

Click on the **Make Ration** button. Enter the ration shown in Table 23–8.

Goatmedicated	0.71
Timothy hay, full bloom	1

Table 23–8
Feeding rates for the feedstuff and blend in the goat ration formulation example

The ration is balanced as it was when the components of the grain were fed unmixed. With the monensin product, the ration now costs $0.10 per head per day.

Strike **ENTER** and **F1**.

Print Ration or Blend Procedure

Make sure your name is entered at cell C1. Click on the **Print Ration or Blend** button. The input box in Figure 23–36 displays.

> Are you printing a goat ration evaluation or a blend formula and analysis? (**1**—RATION, **2**—BLEND):

Figure 23–36

Click **OK** to accept the default input of 1. A two-page printout will be produced by the machine's default printer.
Click on the **Print Ration or Blend** button. Type the number 2 to print a blend. The message box in Figure 23–37 displays.

Figure 23–37

Click on the green number above the blend you want to print and press **F4**. Scroll right to see additional blends.

Click **OK**.
Find the Goatmedicated blend, click on the green number above it, and strike **F4**. A one-page printout will be produced by the machine's default printer.

ACTIVITIES AND WHAT-IFS

In the Forms folder on the companion CD to this text is a GoatInput.doc file that may be used to collect the necessary inputs for use of the GoatRation .xls file. This form may be printed out and used during on-farm visits to assist in ration evaluation activities.

1. Remake the ration described in Table 23–3 for the goat described in this chapter. Add ammonium sulfate to the selection of feedstuffs for this ration. Add enough to reduce the DCAD to 50. What percentage of the ration does this represent? For what reason is ammonium sulfate added to goat rations at this level?
2. Remake the ration described in Table 23–7 for the goat described in this chapter. Add alfalfa hay, early vegetative to the list of selected feedstuffs for this ration. Replace the timothy hay, full bloom with an equivalent amount of dry matter from alfalfa hay, early vegetative. What happens to predicted rumen pH? Rumen pH is predicted by the following formula:

$$5.435 + 0.04229 \times \text{effective NDF}$$

   Explain why a substitution of timothy hay, full bloom, for alfalfa hay, early vegetative, caused a predicted reduction in rumen pH.
3. Remake the ration described in Table 23–7 for the goat described in this chapter. Input a growing kid on slightly hilly pastures (moderate activity). The ration for the growing kid at minimal activity is not a balanced ration for the growing kid on slightly hilly pasture. Meet the DMI with hay and note the ration improvements. Return to 1 lb. timothy hay and meet the DMI with the Goatmedicated blend. Which does a better job of improving the ration balance for this moderate-activity kid: adding more hay or adding more grain? Explain why this is the case.

REFERENCES

Bouchard, R., Conrad, H. R. (1973). Sulfur requirement of lactating dairy cows. I. Sulfur balance and dietary supplementation. *Journal of Dairy Science. 56,* 1276–1282.

National Research Council. (1980). *Mineral Tolerance of Domestic Animals.* Washington DC: National Academy Press.

National Research Council. (1981). *Nutrient Requirements of Goats: Angora, Dairy, and Meat Goats in Temperate and Tropical Countries.* Washington DC: National Academy Press.

National Research Council. (1987). *Vitamin Tolerance of Animals.* Washington DC: National Academy Press.

National Research Council. (1989). *Nutrient Requirements of Dairy Cattle* (6th revised edition). Washington DC: National Academy Press.

National Research Council. (2001). *Nutrient Requirements of Dairy Cattle* (7th revised edition). Washington DC: National Academy Press.

Qi, K., Lu, C. D., & Owens, F. N. (1993). Sulfate supplemtation of growing goats: effects on performance, acid-base balance, and nutrient digestibilities. *Journal of Animal Science. 71,* 1579–1587.

Rumensin®, a product of Elanco Animal Health, A Division of Eli Lilly & Co.

CHAPTER 24

FEEDING HORSES

A horse is dangerous at both ends and uncomfortable in the middle.

IAN FLEMING, 1966

The nutritional phases in the horse production cycle, illustrated in Figure 24–1, include:

Brood mare: early lactation, late lactation, maintenance that includes anestrous and the first 8 months of gestation, the 9th month of gestation, the 10th month of gestation, and the 11th month of gestation
Growing horse: suckling, weaning, yearling, long yearling, 2-year-old

A sample growth rate is plotted in Figure 24–2.

THE BREEDING HERD

The Brood Mare

Lactation

Parturition is an extremely stressful event for mammals. As she recovers from the activities of foaling, the mare's appetite is usually depressed. At the same time, her mammary glands begin producing nutrient-rich colostrum. A primary goal after parturition will be to manage the mare so that she regains a healthy appetite. Mares are often fed a warm mash made of wheat bran and water to stimulate appetite at this time. Wheat bran has the added benefit of being a mild laxative. Generally mares should be gradually introduced to the lactating diet over a period of 7 to 10 days.

Aside from the short-duration muscular effort of the working horse, lactation requires the greatest amount of energy in the horse production cycle. In addition, the manufacture and secretion of milk increases the mare's requirement for other nutrients.

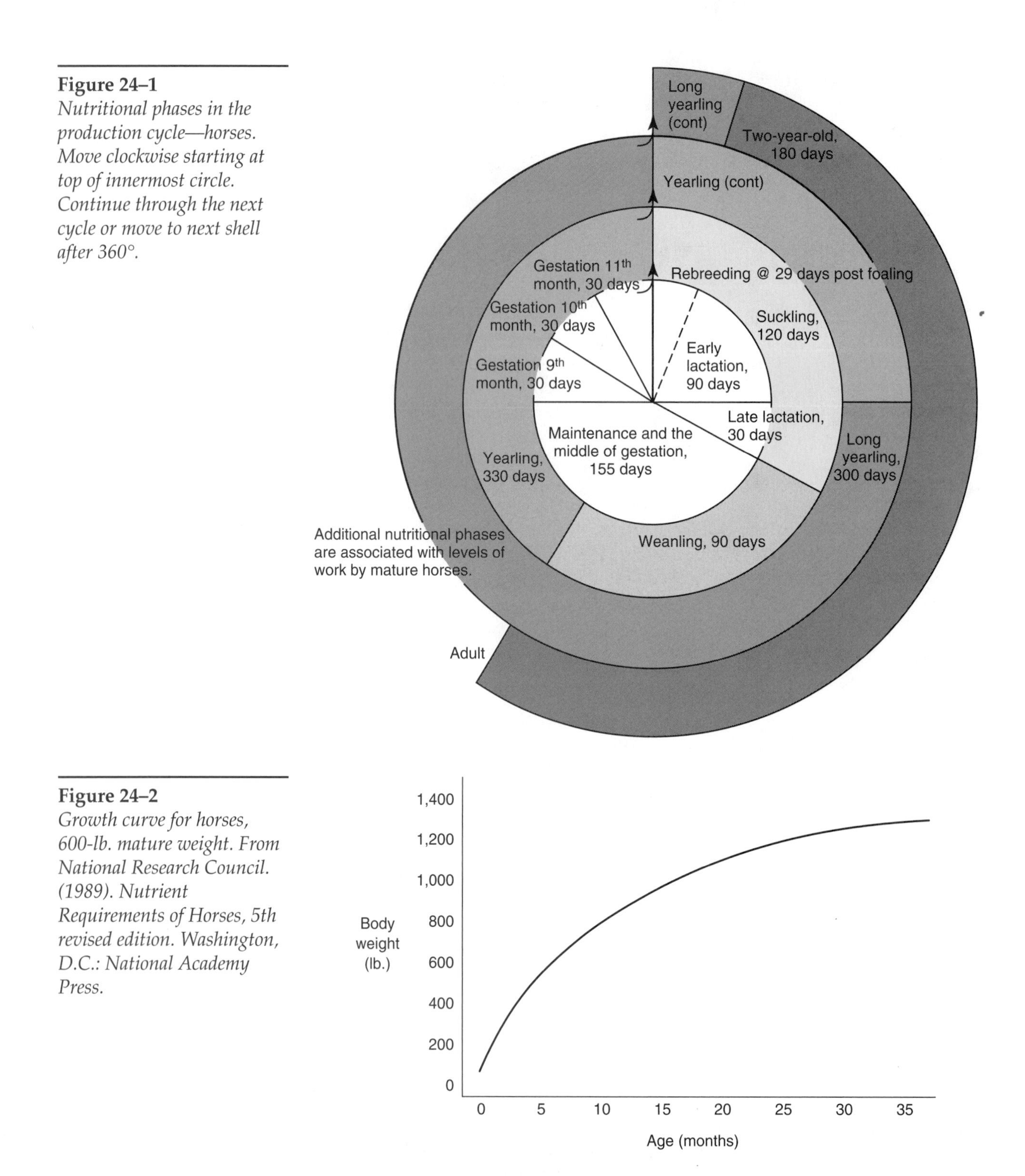

Figure 24–1
Nutritional phases in the production cycle—horses. Move clockwise starting at top of innermost circle. Continue through the next cycle or move to next shell after 360°.

Figure 24–2
Growth curve for horses, 600-lb. mature weight. From National Research Council. (1989). Nutrient Requirements of Horses, 5th revised edition. Washington, D.C.: National Academy Press.

Maintenance/Early Gestation

An idle, mature horse and a mare in the first three quarters of gestation are animals that use feed nutrients only for maintenance functions. Once these animals achieve the desired body condition score (BCS) of 5 or 6, they should receive diets that support only maintenance. On maintenance diets, they will neither gain nor lose weight; they will be consuming just enough nutrients to maintain the status quo. Balancing rations for idle horses can be as challenging as balancing rations for horses with higher nutrient requirements. A diet based on early-cut grass hay or alfalfa hay cut at an early bloom stage of maturity will often contain energy, protein, and calcium nutrients in excess of these animals' requirements.

Environments that increase the loss of body heat will result in an increased maintenance energy requirement. In areas with extreme winters, it may be desirable to feed mares in mid-gestation excess energy to increase tissue insulation (BCS of 6 or higher). The companion application to this text considers five determinants of body heat loss: environmental temperature, night cooling, wind speed, external insulation (hair coat), and tissue insulation (body fat). Based on inputted values, adjustments are made on predicted dry matter intake (DMI) and the energy requirement.

A well-managed pasture, water source, and appropriate mineral supplement available free choice should be able to support the nutritional needs of one idle horse per acre. On pastures that are not managed, and during times of drought, it may take four or more acres to support the nutritional needs of one idle horse.

Horses described as "easy keepers" appear to have reduced maintenance energy requirements. Easy keepers appear to maintain or gain body weight when fed the diets that cause hard keepers to lose weight.

Gestation, Months 9 to 11

Nutrient requirements increase rapidly during the last 3 months of gestation because 60 percent of the neonatal foal's body weight is gained during this period. These requirements must be met without excesses to ensure that the foal's birth weight is acceptable and that the mare's body condition is maintained.

In late gestation, the appetite of mares will usually be somewhat depressed. This means that it may not be possible to meet her increased nutrient requirements with increased pounds of the maintenance diet. Ration nutrient density will have to be increased. Alfalfa hay cut at an early bloom stage will likely be useful in balancing rations for mares in late gestation, as will grains and other concentrated sources of nutrients.

The Stallion

The stallion should be fed a diet that is matched to his activity. Idle stallions in good body condition should be fed a diet that supports maintenance only. During the breeding season, the stallion's daily requirements for nutrients are 25 to 50 percent higher than the maintenance requirements, according to the horse NRC (1989).

THE GROWING HORSE

Suckling

The structure of the mare's placenta is described as epitheliolchorial. This type of placenta does not allow antibody immunoglobulins to pass from the mare into the blood of the fetal foal. These antibody immunoglobulins are acquired by the neonatal foal through colostrum. Colostrum or first milk, is secreted by the mare the first few days following foaling. It is different from ordinary milk in that it is a more concentrated source of nutrients and has a laxative effect, which aids in the passing of the meconium (the fecal material that accumulates in the large intestine during prenatal growth). Most important, however, colostrum contains antibodies. Colostrum should be consumed by the neonatal foal as soon as possible. Colostrum is discussed in Chapter 20. Foals too weak to nurse should receive colostrum by stomach tube. The quantity of colostrum needed will range from 1 to 6 pints (1 pint equals approximately 1 pound), depending on the quality of the colostrum and how many hours have passed since foaling.

If the mare is well fed, her milk should provide all the nutrients needed by the foal for the first 3 months. Many owners will make the weanling grain available to the suckling foal beginning at about 1 to 2 months of age. Doing so will ease the transition at weaning. The foal is usually weaned at 4 to 6 months, and at this age it will be eating 5 to 8 lb. of the weanling grain if it has been available. In addition to grain, the foal should be offered hay free choice.

Weanling

At 2 months of age, the foal is usually interested in supplementing milk with other sources of nutrients. By 6 months of age, milk will play a minor role in supplying needed nutrients if the foal has been offered grain and hay. Foals are generally weaned at 4 to 6 months, and the stress of weaning can be minimized if the foal has been introduced to grain starting at 1 to 2 months of age.

Weanling foals should be fed balanced diets that supply enough nutrients to support a reasonable rate of gain. Generally, this means that grain is fed at no more than 10 lb. and hay is fed to achieve satiety. Weanlings sometimes tend to sort through a grain mix, consuming only the more palatable feedstuffs. Sorting can be prevented by using feed in pelleted form.

Yearling, Long Yearling, and 2-Year-Old

The proper development of the skeletal system is perhaps more critical in horses than in other species because horses are used for work and athletic activity. Calcium and phosphorus are the two primary minerals that make up the skeleton and their levels in the diet must be carefully managed. Magnesium is also an important constituent of bone.

The bioavailability of calcium in feed for horses ranges from 50 to 70 percent, whereas the bioavailability of phosphorous is about 30 to 55 percent. Bioavailability is a function of mineral source, and also declines with the age of the horse. The calcium requirement published by the horse NRC (1989) assumes that the feedstuffs supplying calcium are 50 percent bioavailable. The phosphorus requirement published by the NRC assumes that the feedstuffs supplying phosphorus are 45 percent bioavailable for lactating mares and growing horses, and 35 percent for all other classes of horses. The higher value for the lactating mares and growing horses is used because it is assumed that these animals will be receiving a significant amount of their phosphorus in the more bioavailable, inorganic form. The only other mineral for which the NRC uses predicted bioavailability in determining requirements is magnesium. It is assumed that magnesium is 40 to 60 percent bioavailable. The NRC uses the 40 percent value in determining requirements. These assumed bioavailabilities are applied by the NRC as safety factors in predicted requirements; the companion application to this text uses these same values.

Energy requirements for growing horses are given in the NRC nutrient requirement tables based on whether the expected growth rate is moderate or rapid. The energy requirement for growing horses is calculated more precisely in the text of the NRC publication. Rather than use qualitative inputs of growth potential, the text presents a formula that contains a variable of average daily gain; this formula is used in the companion application to this text.

As the horse grows, dry matter/feed intake increases and the required nutrient density in the diet declines. Rations need to be reformulated periodically during growth because excess nutrient levels may result in a rate of growth that negatively impacts soundness and longevity. This issue is discussed later in this chapter under the topic developmental orthopedic disease.

HORSE FEEDING AND NUTRITION ISSUES

The Working Horse

Adult horses must be fed a diet that will support maintenance. Nutrient levels beyond maintenance will be necessary if the horse is used for work. In the horse NRC (1989) and in the companion application to this text, horse work levels are described as light, moderate, or intense. Light work examples include pleasure riding, bridle path hack, and equitation. Moderate work examples include ranch work, roping, cutting, barrel racing, and jumping. Intense work examples include race training and polo. Draft horses doing heavy work should be described as light if work is less than 3 hours, moderate if 3 to 6 hours, and intense if greater than 6 hours daily. The NRC formulas and those in the companion application to this text increase the requirement for digestible energy above maintenance by 25 percent, 50 percent, and 100 percent for light, moderate, and intense work, respectively.

Because of the high need for energy, horses at intense work may not be able to maintain body condition on a 50:50 forage-to-concentrate diet. Under such conditions, horses may benefit from added fat in the diet (Taylor, Ferrante, Kronfeld, & Meacham, 1995). It will also be best to offer some hay before the grain and to feed at least three times daily.

Working horses may lose significant amounts of electrolytes in sweat. The most important electrolytes contain sodium, chloride, and potassium. These mineral ions are not stored by the body and must be replenished daily to prevent problems with muscle function and fluid balance. The best way to replenish them is in small, frequent feedings. Endurance horses competing in long distance races should be given small amounts of a supplement containing electrolytes throughout the competition.

Sweat also contains water but in order to minimize the risk of water founder, hot, tired horses should not be given free access to cold water. These animals should get all the water they desire, but it should be made available in small quantities, offered at 10-minute intervals.

Fermentation in the Large Intestine

The large intestine provides a favorable environment for the bacterial fermentation of material that was not digested and absorbed in the stomach and small intestine. Available carbohydrates including fiber are broken down by the bacteria inhabiting the large intestine into volatile fatty acids (VFA), primarily acetic acid (acetate) and propionic acid (propionate), which may then be absorbed through the wall of the large intestine and utilized by the various tissues as sources of energy (Bergman, 1990; Hintz, 1983). For horses at maintenance, the VFA from fiber fermentation, together with the small amount of nonfiber carbohydrate and other organic compounds in forage, may satisfy the entire energy requirement, so these animals may not require grain.

Amino acids are made by the microbes inhabiting the large intestine of the horse. These amino acids become the protein in microbial cells. Whereas in ruminant animals the digestive processes of the true stomach (the abomasum in ruminants) and small intestine are downstream from the site of microbial protein production, the digestive processes are upstream in the horse. The microbes growing in the horse's large intestine are never exposed to the digestive processes. Neither do they pass through the small intestine, the primary site of amino acid absorption. It is, therefore, unclear how much of the microbial protein that is produced in the horse's large intestine becomes available to the horse.

The B vitamins and vitamin K are also formed during the microbial activity in the large intestine of the horse. As with microbial protein, a large portion of these synthesized nutrients end up in the manure because the large intestine does not absorb vitamins efficiently (Hintz, 1983). Many of the B vitamins function in energy metabolism, and it is reasonable to assume that horses with high energy requirements would have an increased need for B vitamins. Whether or not usual diets for heavily worked horses would be improved by supplementation of these nutrients is unknown.

Given the uncertainty regarding availability of the nutrients synthesized in the horse's large intestine, it would be prudent to feed a high-quality source of protein as the primary source of protein to growing horses. Regarding the B vitamins and vitamin K, feed the highest quality forage on hand to the growing horses. Avoid feeding diets to any horse that would be totally devoid of these nutrients, such as those consisting exclusively of overly matured, bleached hay.

Because the capacity of the horse's digestive tract is relatively small, a large meal has a relatively rapid rate of passage. A rapid passage rate may result in incomplete digestion in the stomach and small intestine. Incompletely digested feed material arriving at the large intestine has the potential to cause problems in the horse. Colic may result (colic is discussed later in this section). It is theorized that a disruption of the microbial populations may occur when large amounts of starch arrive at the large intestine, and that this may lead to the release of endotoxins. These endotoxins may play a part in the etiology of founder or laminitis (founder is also discussed later in this chapter).

Group Feeding

Ideally, pastured horses that must receive supplemental grain to balance the ration are brought into separate stalls at feeding time. If this is not practical, horses should be grouped according to nutrient need, and the grain can be offered to the group in such a way as to minimize fighting. Table 12–5 gives recommendations on feeder spacing when feeding grain to horses on pasture. If pasture is not providing adequate forage, it may be necessary to feed hay to the group. Because of increased waste, it will be necessary to feed about 20 percent more hay than the animals are expected to eat when hay is fed to pastured horses.

General Feeding Guidelines

All feed-related health problems in horses, including colic, could be prevented by providing a constant and consistent flow through the digestive tract of wholesome feed and water in an amount that satisfies the horse's appetite and thirst, and results in the absorption of nutrients at the level required by the tissues. In striving to achieve this ideal nutritional management, the following practical feeding guidelines should be followed:

1. Keep feed boxes clean.
2. Make feed changes, especially carbohydrate changes, gradually. An overload of carbohydrate has often been implicated in the etiology of digestive disturbances in the horse such as colic and laminitis. There may not be a single carbohydrate component that is responsible for the overload in all cases. The hydrolyzable carbohydrate in "sweet feed" may be the agent involved in carbohydrate overload, but the nonhydrolyzable, rapidly fermentable carbohydrate of spring pasture may also cause carbohydrate overload in horses (Hoffman,Wilson, Kronfeld, Cooper, Lawrence, Sklan, & Harris, 2001). Avoiding carbohydrate overload is accomplished by adapting the horse's digestive system to changing feed carbohydrates and feeding frequently.

3. Feed frequently. Horses on pasture will spend at least half the day eating and meals will be fairly evenly spaced throughout the day. Stabled horses should be fed at least twice a day, preferably three times. Frequent feedings will help prevent colic and reduce the chances of the horse developing stable vices such as wood chewing and cribbing.
4. Always feed at the same time of day. Regularity in feeding programs helps avoid feeding problems.
5. Feed by weight, not volume. Oat grain is notoriously variable in density. A 1-quart container filled with oats of high-hull content may weigh 0.7 lb., whereas a quart container filled with oats of low-hull content may weigh more than a pound. A 1-quart container filled with corn grain weighs 1.8 lb.
6. For those horses that tend to bolt their feed, make an effort to slow down their rate of consumption. Strategies include:
   a. placing baseball-sized stones or chunks of block salt in the feeder
   b. using a large box and spreading the feed
   c. mixing chopped hay with grain in the feeder
7. Know the role of bulk in the diet. Bulk is necessary for effective performance of the digestive tract. However, it may be necessary to minimize bulk in the meal taken before heavy work.
8. When necessary, bring body condition down gradually.
9. Have fresh, clean water available at all times. Control water intake of a "hot" horse. Encourage horses to drink water before feeding.
10. Check teeth regularly to ensure that the horse can properly chew its feed.
11. Feed rations that contain appropriate levels of all required nutrients.

Colic

Generally, the term *colic* refers to the symptoms that are due to adverse conditions within the digestive tract. Such conditions accompany almost every systemic disorder of the horse. Colic, therefore, is not a symptom of a specific disease.

Horses with colic show general signs of anxiety, and may bite or kick at their abdomen, groan, or roll. Colic may be mild and the problem may resolve itself, or the problem causing colic may be life-threatening.

Listed below are problems that cause a horse to show symptoms of colic that involve feeding management.

1. Feeding unwholesome feedstuffs and feeds can cause a horse to show symptoms of colic.
2. The tendency of some horses to bolt their feed can lead to colic if such behavior is not properly managed.
3. Overfeeding is a common cause of colic.
4. Improper feeding management can contribute to constipation that will lead to colic. Providing the horse with a balanced ration and water will usually prevent constipation. A wheat bran mash is often fed to horses to prevent constipation. It should be recognized that the effectiveness of this feedstuff in preventing constipation has nothing to do with its nutritional content, but rather with its ability to soften the feces (a laxative effect).
5. An irregular feeding schedule can cause colic in some horses. Confined animals look forward to feeding time with great anticipation. Physical changes take place within the digestive tract as feeding time approaches. It is a fundamental principle of animal husbandry that animals be able to rely on the predictability of the feeding activity.
6. The sequence of component feeding will affect the completeness of digestion, and incomplete digestion may be a contributing factor to colic. Ideally, hay

and grain are fed and consumed together. If this is not possible, hay should be fed first. Hay is more effective than grain in stimulating the flow of saliva and gastric juices, and this should set the stage for a more complete digestion of subsequent feedstuffs. In addition, hay passes through the digestive tract relatively slowly. The passage of grain will be delayed by the presence of hay in the digestive tract, resulting in more complete digestion. Grain that is incompletely digested in the stomach and small intestine may disrupt microbial populations when it arrives at the large intestine, resulting in carbohydrate overload and causing colic.

7. Horses that are fed only one large meal per day will be susceptible to colic. The digestive anatomy of the horse is designed for more or less constant processing of small quantities of feed. Figure 2–4d indicates that the stomach, the primary feed storage organ, represents only 9 percent of the volume of the digestive tract. When compared with the pig's 26 percent, the dog's 63 percent, and the ruminant's 67 percent, it is apparent that the horse is not designed to consume large meals that can be stored for later processing. Add this to the fact that the bile made by the horse's liver is not stored in a gall bladder but rather is constantly secreted directly into the intestine, and it becomes apparent that horses are not designed to process large, infrequent meals. A horse fed a single meal daily will eat rapidly. Ingested feed will pass rapidly through the digestive tract, leading to incomplete digestion.
8. Diet changes that are made without an appropriate transition can result in colic. As the availability and/or economic characteristics of feed resources change, the horse's diet should change. Also, as the horse's nutrient requirements change, the nutritional content of the diet should change accordingly. Ideally, these diet adjustments involve a reformulation so that the appetite of the animal is always satisfied when it receives the required amounts of nutrients. Drastic feed changes without a proper transition will often result in colic. When minor changes in the diet are called for, the transition to the new diet should be made over a period of several days. Major changes in the diet should be made gradually over a period of 2 to 3 weeks. When feeding a thin but otherwise healthy animal, take into consideration the fact that changing from a ration that is deficient in nutrients to a ration that meets nutrient requirements is still a ration change. Any ration change needs to be made gradually to minimize the risk of indigestion or colic.
9. Unlike ruminants, horses cannot regurgitate swallowed feed and they cannot eructate gases produced during fermentation. Because they cannot regurgitate, horses must chew their feed thoroughly before swallowing. Because horses cannot eructate, gases formed during fermentation can only be passed through the rectum. If excess gases are produced in the digestive tract, distension of the gut can interfere with blood circulation and respiration. The first eight causes of colic discussed in this section, along with #12 which follows, may result in excess gas formation.
10. The availability of water relative to the availability of feed can sometimes be a contributing factor to colic. Most horses drink rapidly, and if this flow of water washes large amounts of incompletely digested feed material into the intestine, colic may result. This problem is usually prevented by making water available to the horse during meals.
11. The horse's work schedule should be developed around feeding times to avoid colic. The distribution of blood in the body changes depending on what tissues in the body have the greatest need for delivery of oxygen and nutrients and the removal of wastes. The muscles of an exhausted horse will continue to hold priority status for the blood supply for a period of time after vigorous exercise, and the digestive tract will not be capable of processing feed until after a period of rest. Small amounts of water may be

given at frequent intervals, but feed should be withheld for about 2 hours after vigorous exercise to avoid colic.

12. Conversely to number 11, if an animal has just been fed a large meal, it should be given time to digest it before being exercised. When feed is being digested, the animal's digestive muscles and glands are active. This activity is supported by an increased blood supply to the digestive tract. Hard work will divert blood to the skeletal muscles at the expense of the digestive organs. Incomplete digestion will result. Incompletely digested feed arriving at the intestine may result in colic. Another reason to avoid working a horse immediately after a meal has to do with its affect on the respiratory system. Immediately after a meal, the digestive tract is distended. The abdominal cavity occupies more space at the expense of the thoracic cavity, and the lungs are prevented from expanding to their fullest. Such a horse may experience difficult breathing during heavy exercise.

Laminitis and Founder

Laminitis is defined as an inflammation of the laminae of the feet. Laminitis has several causes. *Founder* is the term describing laminitis caused by a disturbance originating in the digestive tract.

The inflammation that characterizes founder is caused, in part, by the effects of histamine on the circulation of blood and lymph in the feet. Histamine is useful in the body when a wound or injury requires additional blood factors for repair activity. This is because histamine increases localized blood flow, which is evident as a swelling in the affected area. But the effects of histamine in the foot tissues cause pain and potentially permanent damage.

What conditions lead to an increase in the level of histamine in the foot tissues? Recall from Chapter 20 that in ruminant animals, an acid condition in the rumen leads to the establishment of a population of bacteria that convert the amino acid histidine to histamine, which is absorbed, contributing to laminitis in these species. The source of the histamine that contributes to laminitis in the horse may be the horse's own tissues, acting in response to the presence of endotoxin, as during enterotoxemia.

Enterotoxemia may be caused by starch overload. The resulting disruption in normal microbial populations of the digestive tract may cause the proliferation of undesirable bacteria of the genera *Clostridium* and *Escherichia.* If endotoxins produced by these bacteria are absorbed, enterotoxemia results. Enterotoxemia may also be caused by indigestion or by drinking cold water while the horse is still hot from exercise.

Another form of founder, called *grass founder,* occurs when horses that have spent the winter months working and consuming a high-grain diet are suddenly turned out on lush spring grass. As is likely to happen with any sudden dietary change, the horse experiences a dramatic gastrointestinal upset. In the case of grass founder, this leads to enterotoxemia, histamine release, and laminitis.

Investigations are being conducted to examine the possibility that the histamine that leads to laminitis in horses originates in the digestive tract, produced by bacteria and absorbed into the blood (Garner, Flint, & Russell, 2002). This etiology is more in line with what has been found by ruminant researchers.

Azoturia

Azoturia (also known as *paralytic myoglobinuria* and *equine exertional rhabdomyolysis*) occurs most often in the physically fit horse whose ration is balanced to satisfy the nutrients needed during intense work. The problem begins when such a horse is given a rest period of one to several days, but is still fed as if it was at work. When the horse commences physical exertion, the muscles cramp and the animal is unable to move.

During the course of azoturia, there is extensive damage to muscle tissue. In addition, there may be kidney damage and paralysis.

Azoturia can be prevented by maintaining a balance between diet and exercise. Once a horse has begun a rigid training schedule, that schedule and its corresponding daily ration must be maintained and carefully managed to ensure that changes are made gradually.

Heaves

Heaves, or alveolar pulmonary emphysema, is a respiratory condition in horses. As with ordinary emphysema, the lung alveoli are damaged and the animal has difficulty breathing. Heaves appears to have several causes. Heaves may be a complication from a severe or neglected case of infectious respiratory disease. Heaves may also be the result of a reaction to inhaled particulate matter.

The feed is one of many potential sources of airborne particulate matter (dust) in a horse's environment. The most potent source of dust in most horse diets is the hay. The source of the dust may be fine particles of hay itself, or it may be mold that is growing on the hay. If the source of dust is mold, the hay should not be fed. When feeding hay, there are several actions that can be taken to reduce its dustiness. It can be shaken outside the barn. The dustiness of hay may be minimized by moistening it or pouring diluted molasses over it. As always, the feeder should be cleaned between feedings.

If hay is a problem for the horse with heaves, there are several hay modifications and alternatives that may be used. Hay baled at a higher moisture content than the usual 12 to 15 percent will be less dusty and perhaps more desirable for the horse with heaves. A mold inhibitor such as propionic acid must be used to prevent mold growth and spontaneous combustion of hay baled at greater than 15 percent moisture. Propionic acid treatment of hay does not affect the palatability of the hay. Cubed, pelleted, or fermented forage (silage) will be less dusty than long hay. Forage alternatives such as beet pulp may be fed. During the grazing season, hay in the diet may be replaced with pasture.

Feeding management can be altered to help minimize symptoms of the horse with heaves. The chances that irritants will enter the nasal cavity will be reduced by allowing the horse to feed from the floor rather than from a hay rack. Again, it is critical that the feeding area be cleaned between feedings.

Body Condition Scoring

Body condition scoring of horses is based on a 9-point scale as shown in Table 24–1.

Prussic Acid Poisoning

Under certain conditions, some grasses, particularly forage sorghums, may accumulate hydrocyanic acid (prussic acid) in their tissues. Prussic acid poisoning is discussed in Chapter 20.

Endophyte-Infected Grasses

An endophyte is a fungus that lives inside a grass plant. The two organisms are involved in a symbiotic relationship in which the fungus receives a place to live in return for providing the plant with increased resistance to environmental stresses such as attack by insects.

Table 24–1
Body condition score description in horses

	Anatomy			
Score	**Spinous Processes**	**Ribs**	**Tailhead**	**Withers**
1	Little fat covering spinous processes	Prominent	Prominent and individual vertebrae evident	Bone structure is evident
2	Some fat covering spinous processes	Prominent	Prominent and individual vertebrae faintly discernible	Bone structure faintly discernible
3	Fat buildup about halfway on spinous processes	Slight fat over ribs	Prominent but individual vertebrae not evident	Withers lean but bone structure not apparent
4	Slight convex ridge along back from ends of spinous processes	Faint outline of ribs discernible	Fat can be felt around tailhead	Some fat apparent along withers
5	Flat over back—spinous processes buried	Ribs not visible but can be felt	Fat around tailhead feels spongy	Withers rounded over spinous processes
6	Slight concave ridge along back	Spongy fat felt over ribs	Fat around tailhead feels soft	Some fat deposited along side of withers
7	Crease along back	Area between ribs filled with fat	Fat around tailhead feels soft	Fat deposited along side of withers
8	Crease along back	Difficult to feel ribs	Fat around tailhead very soft	Area along withers filled with fat
9	Obvious crease along back	Patchy fat over ribs	Fat around tailhead is bulging	Bulging fat along withers

Consumption of endophyte-infected tall fescue, whether in pasture or hay, may result in "tall fescue toxicosis." Tall fescue toxicosis impacts health, production, and reproduction in livestock (Waller & Fribourg, n.d.), especially horses. The effects on reproduction include:

- Lower conception rates
- Abortions
- Prolonged pregnancy
- Dystocia
- Retained placenta
- Higher rates of neonatal foal death
- Agalactia or poor milk production

The nonreproduction effects of tall fescue toxicosis on livestock include:

- Increased risk of laminitis
- Loose feces
- Reduced growth rate
- Reduced dry matter intake

Because most of the problems associated with consumption of endophyte-infected tall fescue are related to reproduction, one recommended measure is to remove the source of endophyte-infected tall fescue from the brood mare diet 60 to 90 days prior to foaling (Coleman, Henning, Lawrence, & Lacefield, n.d.). Endophyte-free tall fescue seed is available for those considering pasture renovation.

Plants other than tall fescue can become infected with endophytes. Ryegrass may be infected with an endophyte, and livestock consuming infected ryegrass may show muscle system symptoms such as incoordination and tetany. This disorder is called *ryegrass staggers.*

Feedstuffs Affecting Urinary Tract Health

Cystitis is inflammation of the bladder. It is characterized by painful, frequent urination and sometimes by bloody urine. Consumption of forage sorghum, Sudan grass, and sorghum-Sudan grass hybrids have been associated with cystitis or inflammation of the bladder in horses.

Because excess dietary protein leads to increased urea filtration by the kidneys, it used to be thought that high-protein feedstuffs such as early-cut alfalfa could cause kidney damage. It is now known that healthy kidneys are not damaged by the consumption of high-protein feedstuffs.

Developmental Orthopedic Disease

Developmental orthopedic disease (DOD) is a general term for several problems associated with improper development and maturation of the horse's skeleton. Nutrition, genetics, and endocrine function all may play a role in DOD.

The increasing incidence of developmental orthopedic diseases in horses has been attributed to feeding young horses for maximal growth (Ropp, Raub, & Minton, 2003). Mineral imbalances including copper deficiency may also cause DOD.

A nutrition–hormone interaction due to infrequent feeding may result in DOD (Hintz, 1995). Infrequent feeding results in insulin spikes that may cause a reduction in thyroxine (a hormone from the thyroid gland) production. Proper thyroxine levels are necessary to ensure normal maturity of the skeleton.

Nutritional Needs of the Geriatric Horse

Many feed manufacturers market feed formulations specifically designed for the geriatric horse. Some of the characteristics of these feed products include:

- Greater attention to grain processing, assuming some deterioration in the teeth of older horses
- Reduced fiber and increased nutrient density, assuming that older horses have reduced digestive efficiency

The horse NRC (1989) does not give specific nutrient recommendations for the geriatric horse.

Feedstuffs Used in Horse Diets

The choice of which feedstuffs to use in formulating horse diets will depend on four factors.

1. What works to balance the ration
2. Feedstuff cost and availability
3. Feedstuff palatability
4. Feedstuff acceptability

Low palatability may be an inherent characteristic of the feedstuff, but palatability may also be influenced by moisture content and contaminants such as molds and toxins. Palatability may also be affected by the physical condition of the feedstuff such as fineness of particle size. The acceptability of feedstuffs may vary, depending on the condition of the individual horse. For example, rations for horses afflicted with the respiratory ailment heaves should not be made with dusty feedstuffs unless efforts have been made to contain the dust.

END-OF-CHAPTER QUESTIONS

1. Give three techniques that may be used to slow feed consumption in horses that tend to bolt their feed.
2. Give three strategies that can be used to ameliorate heaves caused by components of the diet.
3. Discuss the microbial activity that takes place in the horse's large intestine as it impacts the need to include dietary sources of B vitamins.
4. What happens to the horse afflicted with azoturia? How can it be prevented?
5. At what month of gestation do the nutrient demands of pregnancy begin to materially affect the mare's nutrient intake requirements?
6. What impact does a full digestive tract have on the ability of the horse to rapidly and fully expand the thoracic cavity as would be necessary during work or exercise?
7. Describe a horse with a BCS of 2. Describe a horse with a BCS of 5. Describe a horse with a BCS of 8.
8. What is an endophyte and what is its relevance to feeding horses?
9. Describe how to manage mineral nutrition to prevent developmental orthopedic disease.
10. Give five feeding management practices that may result in colic in horses. Give one horse behavior that may result in colic. Give one feedstuff that, when consumed by a horse, may cause colic.

REFERENCES

Bergman, E. N. (1990). Energy contributions of volatile fatty acids from the gastrointestinal tract in various species. *Physiological Reviews. 70,* 567–590.

Coleman, R. J., Henning, J. C., Lawrence, L. M., & Lacefield, G. D. (n.d.). Understanding endophyte-infected tall fescue and its effect on broodmares. Retrieved 10/30/2003 from: http://www.ca.uky.edu/agc/pubs/id/id144/id144.htm

Garner, M. R., Flint, J. F., & Russell, J. B. (2002). *Allisonella histaminiformans* gen. nov., sp. nov. A novel bacterium that produces histamine, utilizes histidine as its sole energy source, and could play a role in bovine and equine laminitis. *Systematic and Applied Microbiology 25,* 498–506.

Hintz, H. F. (1983). *Horse nutrition, A practical guide.* New York: Prentice-Hall Press, Inc.

Hintz, H. F. (1995). Horses. In *Basic Animal Nutrition and Feeding,* 4th edition. Pond, W.G., Church, D. C., Pond, K. R. New York: John Wiley & Sons.

Hoffman, R. M., Wilson, J. A., Kronfeld, D. S., Cooper, W. L., Lawrence, L. A., Sklan, D., & Harris, P. A. (2001). Hydrolyzable carbohydrates in pasture, hay, and horse feeds: Direct assay and seasonal variation. *Journal of Animal Science 79,* 500–506.

National Research Council. (1989). *Nutrient requirements of the horse* (5th rev. ed.). Washington, DC: National Academy Press.

Ropp, J. K., Raub, R. H., & Minton, J. E. (2003). The effect of dietary energy source on serum concentration of insulin-like growth factor-I, growth hormone, insulin, glucose, and fat metabolites in weanling horses. *Journal of Animal Science. 81,* 1581–1589.

Taylor, L. E., Ferrante, P. L., Kronfeld, D. S., & Meacham, T. N. (1995). Acid-base variables during incremental exercise in sprint-trained horses fed a high-fat diet. *Journal of Animal Science 73,* 2009–2018.

Waller, J. C., & Fribourg, H. A. (n.d.). Performance of steers grazing pastures of endophyte-infected and endophyte-free tall fescue with and without clover. Retrieved 10/30/2003 from: http://www.agriculture.utk.edu/ansci/pdf/Reports/performance.pdf

CHAPTER 25

HORSE RATION FORMULATION

TERMINOLOGY

Workbook A workbook is the spreadsheet program file including all its worksheets.

Worksheet A worksheet is the same as a spreadsheet. There may be more than one worksheet in a workbook.

Spreadsheet A spreadsheet is the same as a worksheet.

Cell A cell is a location within a worksheet or spreadsheet.

Comment A comment is a note that appears when the mouse pointer moves over the cell. A red triangle in the upper right corner of a cell indicates that it contains a comment. Comments are added to help explain the function and operations of workbooks.

Input box An input box is a programming technique that prompts the workbook user to type information. After typing the information in the input box, the user clicks **OK** or strikes **ENTER** to enter the typed information.

Message box A message box is a programming technique that displays a message. The message box disappears after the user clicks **OK** or strikes **ENTER**.

EXCEL SETTINGS

Security: Click on the **Tools** menu, **Options** command, **Security** tab, **Macro Security** button, **Medium** setting.

Screen resolution: This application was developed for a screen resolution of 1024 × 768. If the screen resolution on your machine needs to be changed, see Microsoft Excel Help, "Change the screen resolution" for instructions.

HANDS-ON EXERCISES

HORSE RATION FORMULATION

Double click on the **HorseRation** icon. The message box in Figure 25–1 displays.

Figure 25–1

> Macros may contain viruses. It is advisable to disable macros, but if the macros are legitimate, you may lose some functionality.
> Disable Macros or Enable Macros or More Info

Click on **Enable macros**.

The message box in Figure 25–2 displays.

Figure 25–2

> Function keys F1 to F8 are set up. You may return to this location from anywhere by striking **ENTER**, then the **F1** key. Workbook by David A. Tisch. The author makes no claim for the accuracy of this applicaiton and the user is solely responsible for risk of use. You're good to go. TYPE ONLY IN THE GRAY CELLS! Note: This Workbook is made up of charts and a worksheet. The charts and worksheet are selected by clicking on the tabs at the bottom of the display. Never save the Workbook from a chart; always return to the worksheet before saving the Workbook.

Click **OK**.

Input Horse Procedure

Click the **Input Horse** button. The input box shown in Figure 25–3 displays.

Figure 25–3

> 1. Idle horse (maintenance activity)
> 2. Stallion (breeding season)
> 3. Pregnant mare
> 4. Lactating mare
> 5. Working horse
> 6. Growing horse (age 4 to 24 months)
>
> Enter the appropriate number:

Click **OK** to choose the default input of #1, the Idle Horse. The input box shown in Figure 25–4 displays.

Figure 25–4

> Enter measured DMI (pounds) if available or click **OK** to accept the predicted value at CX22.

Click **OK** to accept the predicted dry matter intake (DMI). The message box shown in Figure 25–5 displays.

Figure 25–5

Enter the appropriate Cell inputs in column CX.

Click **OK**.

The Cell Inputs Table 25–1 presents the cell inputs for this example. Comments associated with the cell inputs are presented in Tables 25–2 through 25–5. The information used to assign body condition score (BCS) for horses is found in the comment associated with this input as well as in Chapter 23.

Table 25–1
Inputs for the horse ration formulation example

58	Temperature (° F)
Yes	Is there relief from heat at night? (heat stress unlikely—input should be "yes.")
1.03	Wind speed and hair coat (external insulation) adjustment factor
0	Animal surface mud, rain, and snow input value (used to adjust tissue insulation and DMI)
3.9	Activity (expressed as acres walked per day)
1,100	Body weight (lb.) (100–2,200)
6	BCS (1–9)
	N/A
	N/A
	N/A
	N/A
	N/A
	N/A

Table 25–2
Temperature adjustment factors. Note: These factors are based on empirical observations rather than scientific data.

Effective Temperature	Adjustment Factor Applied to DE Requirement
>95° F, no night cooling	0.65
>95° F, with night cooling	0.90
77–95° F	0.95
57–76° F	1.00
41–58° F	1.06
24–40° F	1.16
<24° F	1.20

Table 25–3
Wind speed and hair coat (external insulation) adjustment factor. Adjustment factor applied to DE requirement. Note: These factors are based on empirical observations rather than scientific data.

	Hair Coat Depth (in.)			
Wind Speed (mph)	0.2	0.4	0.6	1.2
1.0	1.05	1.03	1.02	1.00
4.0	1.10	1.08	1.03	1.00
8.0	1.15	1.10	1.05	1.00
≥ 16.0	1.20	1.15	1.10	1.00

	Input Value	Adjustment Factor Applied to DMI	Adjustment Factor Applied to Digestible Energy Requirement
No mud	0	1	1
Mildly muddy	1	0.85	1.05
Muddy	2	0.77	1.25
Extremely muddy	3	0.70	1.50

Table 25–4
Animal surface mud, rain, and snow input value (used to adjust tissue insulation and DMI). Note: These factors are based on empirical observations rather than scientific data.

Acres Walked per Day	Adjustment Factor Applied to DE Requirement
0.0–0.7	1.0
0.8–3.9	1.2
4.0–7.9	1.3
>7.9	1.5

Table 25–5
Activity, expressed as acres walked daily Note: These factors are based on empirical observations rather than scientific data. This is the best place to adjust for the increased energy requirements of "hard" keepers.

Strike **ENTER** and **F1**.

Select Feeds Procedure

Click on the **Select Feeds** button. The message box shown in Figure 25–6 displays.

Nutrient content expressed on a dry matter basis. Feedstuffs are listed, first by selection status, then, within the same selection status, by decreasing protein content, then, within the same protein content, alphabetically.

Figure 25–6

Click **OK**.

Explore the table. Note the nutrients listed as column headings. Note also that the table ends at row 200. You select feedstuffs for use in making two different products: (1) a blend to be mixed and sold bagged or bulk, and (2) a daily ration to be fed directly to livestock.

Select the feedstuffs in Table 25–6 by placing a 1 in the column to the left of the feedstuff name. All unselected feedstuffs should have a 0 value in the column to the left of the feedstuff name. If you wish to group feedstuffs but not select them, you would place a value between 0 and 1 to the left of the feedstuff name.

Feedstuff
Oats, 38 lb./bu.
Grass hay, mature, >60% NDF
Salt (sodium chloride)
Sodium selenite
Vitamin A supplement
Vitamin D supplement
Vitamin E supplement
Zinc oxide

Table 25–6
Feedstuffs to select for the horse ration formulation example

Strike **ENTER** and **F2**.

Selected feedstuffs and their analyses values are copied to several locations in the Workbook.

Make Ration Procedure

Click on the **Make Ration** button. The message box in Figure 25–7 displays.

Figure 25–7

ENTER **POUNDS TO FEED** IN COLUMN B. TOGGLE BETWEEN NUTRIENT WEIGHTS AND CONCENTRATIONS USING THE **F5** KEY, RATION AND FEED CONTRIBUTIONS USING THE **F6** KEY. WHEN DONE STRIKE **F1**. Cell is highlighted in red if nutrient provided is poorly matched with nutrient target. The lower limit is taken as 95 to 98 percent of target, depending on the nutrient. With the exception of sulfur, the upper limit of acceptable mineral is taken from National Research Council. (1980). *Mineral Tolerance of Domestic Animals.* National Academy Press. This publication gives safe upper limits of minerals as salts of high bioavailability. The upper limit for sulfur is taken as 0.4 percent of diet dry matter. The upper limit for vitamins is taken from the NRC, 1987. For other nutrients, the upper limit is based on unreasonable excess and the expense of unnecessary supplementation. See Table 25–8 for specifics.

Click **OK**. The message box in Figure 25–8 displays.

Figure 25–8

The Goal Seek feature may be useful in finding the pounds of a specific feedstuff needed to reach a particular nutrient target:

1. Select the red cell highlighting the deficient nutrient
2. From the menu bar, select **Tools**, then **Goal Seek**
3. In the text box, "To Value:" enter the target to the right of the selected cell
4. Click in the text box, "By changing cell:" and then click in the gray "Pounds fed" area for the feedstuff to supply the nutrient
5. Click **OK**. You may accept the value found by clicking **OK** or reject it by clicking **Cancel**.

WARNING: Using Goal Seek to solve the unsolvable (e.g., asking it to make up an iodine shortfall with iron sulfate) may result in damage to the Workbook.

IMPORTANT: If you return to the Feedtable to remove more than one feedstuff from the selected list, you will lose your chosen amounts fed in the developing ration.

Click **OK**.

In the gray area to the right of the feedstuff name, enter the pound values shown in Table 25–7.

Table 25–7
Inclusion rates for feedstuffs in the horse ration formulation example

Oats, 38 lb./bu.	5
Grass hay, mature, >60% NDF	20
Salt (sodium chloride)	0.126
Sodium selenite	0.0000013
Vitamin A supplement	0.0078
Vitamin D supplement	0.0015
Vitamin E supplement	0.0255
Zinc oxide	0.0005

The Nutrients Supplied *Display*

The application highlights ration nutrient levels, expressed as amount supplied per horse per day, that fall outside the acceptable range.

The lower limit for most ration nutrients is taken as 98 percent of the target. The single exception is DMI, for which the lower limit is taken as 95 percent of the target. Table 25–8 shows the upper limits of the acceptable range.

Table 25–8
Upper limits for ration nutrients used in the horse ration application

	Upper Limit	Source
DMI	5% more than predicted requirement	Author
Digestible energy	1.4 × predicted requirement	Author
Lysine	No upper limit	—
Calcium	2% of ration dry matter	NRC: Mineral Tolerance of Domestic Animals (1980)—horse
Phosphorus	1% of ration dry matter	NRC: Mineral Tolerance of Domestic Animals (1980)—horse
Magnesium	0.3% of ration dry matter	NRC: Mineral Tolerance of Domestic Animals (1980)
Potassium	3% of ration dry matter	NRC: Mineral Tolerance of Domestic Animals (1980)
Sulfur	0.4% of ration dry matter	NRC: Mineral Tolerance of Domestic Animals (1980)
Sodium	1.18% of ration dry matter	NRC: Mineral Tolerance of Domestic Animals (1980) based on limit for salt (sodium chloride)
Chloride	1.82% of ration dry matter	NRC: Mineral Tolerance of Domestic Animals (1980) based on limit for salt (sodium chloride)
Salt (sodium chloride)	3% of ration dry matter	NRC: Mineral Tolerance of Domestic Animals (1980)
Iron	500 mg/kg or ppm of ration dry matter	NRC: Mineral Tolerance of Domestic Animals (1980)
Manganese	400 mg/kg or ppm of ration dry matter	NRC: Mineral Tolerance of Domestic Animals (1980)
Copper	800 mg/kg or ppm of ration dry matter	NRC: Mineral Tolerance of Domestic Animals (1980)—horse
Zinc	500 mg/kg or ppm of ration dry matter	NRC: Mineral Tolerance of Domestic Animals (1980)
Iodine	5 mg/kg or ppm of ration dry matter	NRC: Mineral Tolerance of Domestic Animals (1980)—horse
Cobalt	10 mg/kg or ppm of ration dry matter	NRC: Mineral Tolerance of Domestic Animals (1980)
Selenium	2 mg/kg or ppm of ration dry matter	NRC: Mineral Tolerance of Domestic Animals (1980)
Vitamin A	4 × predicted requirement	NRC: Mineral Tolerance of Domestic Animals (1980)
Vitamin D	4 × predicted requirement	NRC: Mineral Tolerance of Domestic Animals (1980)
Vitamin E	20 × predicted requirement	NRC: Mineral Tolerance of Domestic Animals (1980)

The *Nutrients Supplied* display shows nutrient amounts supplied in the diet and nutrient targets for the inputted horse. All nutrient levels appear to be within acceptable ranges as established by the application.

The cost of this ration using initial $/ton values is $1.13 per horse per day.

DCAD provided: 275

The comment associated with the cell containing this label is shown in Figure 25–9.

Figure 25–9

Dietary Cation–Anion Difference (DCAD):

$$mEq\ (Na + K) - mEq\ (Cl + S)$$

DCAD is used to assess the impact of the ration's mineral content on the body's efforts to regulate blood pH. This is a useful value for some species of livestock, but its significance in horse nutrition is unknown.

Electrolyte balance: 366

The comment associated with the cell containing this value is shown in Figure 25–10.

Figure 25–10

The electrolyte balance is calculated as mEq of excess cations: (Na + K – Cl)/kg of diet dry matter. Like DCAD, it is used to assess the impact of the ration's mineral content on the body's efforts to regulate blood pH. Also like DCAD, there is little information as to the optimal electrolyte balance for horses. The optimal electrolyte balance in the diet for pigs has been suggested to be 250, as fed (Austic and Calvert, 1981) or 278 dry matter. However, optimal growth has been found to occur over the range of 0 to 667 mEq/kg of dry matter diet. The electrolyte balance is of no value if requirements for Na, K and Cl have not been satisfied.

Electrolyte balance from requirements: 50

The comment associated with the cell containing this value is shown in Figure 25–11

Figure 25–11

This is the calculated electrolyte balance based on required levels of minerals. Since the significance of the electrolyte balance in horse nutrition is unknown, this value should not be viewed as a target for ration balancing.

Ca:P: 1.20

The comment associated with the cell containing this label is shown in Figure 25–12.

Figure 25–12

Calcium-to-phosphorus ratio. The Ca:P ratio should be greater than 1:1. Ratios less than 1:1 indicate that the diet contains more phosphorus than calcium. This situation may be detrimental to calcium absorption. Ratios as high as 6:1 in diets for growing horses may not be detrimental if phosphorus intake is adequate. horse NRC, 1989.

The Nutrient Concentration *Display (strike* `F5`*)*

The nutrients in the ration and the predicted nutrient targets are expressed in terms of concentration. That is, the nutrients provided by the ration are divided by the amount of ration dry matter and the nutrient targets are divided by the target amount of ration dry matter. Concentration units include percent, milligrams per kilogram or parts per million, calories per pound, and international units per pound.

Strike **F5**.

The Feedstuff Contributions *Display (strike* `F6`*)*

Shown here are the nutrients contributed by each feedstuff in the ration. This display is useful in troubleshooting problems with nutrient excesses.

Strike **F6**.
Strike **ENTER** and **F1**.

The Graphic *Display*

At the bottom of the home display are tabs. The current tab selected is the Worksheet tab. Other tabs are graphs based on the current ration. Note that you may have to click the leftmost navigation button at the lower-left to find the first tab.

DMI

Click on this tab to display a graph titled Nutrient Status: Dry Matter Intake

Energ&Prot

Click on this tab to display a graph titled Nutrient Status: Energy & Protein

Macro

Click on this tab to display a graph titled Nutrient Status: Macrominerals

Micro

Click on this tab to display a graph titled Nutrient Status: Microminerals

Vitamins

Click on this tab to display a graph titled Nutrient Status: Vitamins A, D, & E

Click on the **Worksheet** tab.

Blend Feedstuffs Procedure

This horse farm wants the feed mill to blend all the horse ration feedstuffs except the hay and oat grain. The blended ingredients will be purchased from the feed mill in bagged form.

Click on the **Blend Feedstuffs** button. The message box in Figure 25–13 displays.

Figure 25–13

YOU MUST HAVE ALREADY SELECTED THE FEEDSTUFFS YOU WANT TO BLEND. When your analysis is acceptable, strike **ENTER** and **F3** to name and file the blend.

Click **OK**.

In the gray area to the left of the feed name, enter the amounts to blend shown in Table 25–9. Note that these are the same amounts that were used in the balanced ration, except that we have entered 0 for oats and hay. Note also that when these amounts are entered in the gray area, the blue column to the left calculates the equivalent amount in a ton of mix. Feed mills prefer to have the formula expressed on a per-ton basis because the capacity of their mixer(s) is expressed in tons.

Table 25–9
Inclusion rates for feedstuffs in the IdleMin/Vit blend

Oats, 38 lb./bu.	0
Grass hay, mature, >60% NDF	0
Salt (sodium chloride)	0.126
Sodium selenite	0.0000013
Vitamin A supplement	0.0017
Vitamin D supplement	0.0003
Vitamin E supplement	0.0286
Zinc oxide	0.0004

Strike **ENTER** and **F3**. The input box shown in Figure 25–14 displays.

Figure 25–14

ENTER THE NAME OF THE BLEND (names may not be composed of only numbers):

Type the blend name, **IdleMin/Vit**, and click **OK**. The message box shown in Figure 25–15 displays.

Figure 25–15

The new blend has been filed at the bottom of the feed table.

Click **OK**.

View Blends Procedure

Click on the View Blends button. The message box in Figure 25-16 displays.

Figure 25–16

Cursor right to view the blends. Cursor down for more nutrients. DO NOT TYPE IN THE BLUE AREAS.

Click **OK**. Confirm that the IdleMin/Vit formula and analysis have been filed.
Strike `F1`.

Using the Blended Feed in the Balanced Ration

Select Feeds *Procedure*

Click on the **Select Feeds** button and select the feedstuffs shown in Tale 25–10. Unselect all other feedstuffs. Remember that the IdleMin/Vit blend is located at the bottom of the feed table at row 200.

Oats, 38 lb./bu.
Grass hay, mature, >60% NDF
IdleMin/Vit

Table 25–10
Feedstuffs and blend to select for the horse ration formulation example

Strike `ENTER` and `F2`.

Make Ration *Procedure*

Click on the **Make Ration** button. Enter the ration shown in Table 25-11.

Oats, 38 lb./bu.	5
Grass hay, mature, >60% NDF	20
IdleMin/Vit	0.16

Table 25–11
Feeding rates for feedstuffs and blend in the horse ration formulation example

The amount of oats and hay to feed has already been established. The amount of the IdleMin/Vit to feed is the total amount of its component ingredients in the balanced ration. That value is:

$$0.126 + 0.0000013 + 0.0017 + 0.0003 + 0.0286 + 0.0004 = 0.157$$

This value is recorded at View Blends under Formula, as entered. The ration is balanced as it was when the components of IdleMin/Vit were fed unmixed.

The Feeding Directions *Display (strike* `F7`*)*

Here, weighed feed amounts are converted to volume amounts for simplicity of feeding. In each case, two realistic conversion values are used. The results are shown in Table 25–12. The differences shown here emphasize the need to recalibrate volume feeding equipment to actual feed weight with each new load of feed.

Table 25–12
Feeding directions display for the horse ration example

	Flakes to Feed		Cans to Feed		Tablespoons to Feed	
	If Hay is 2 lb./Flake	If Hay is 4 lb./Flake	If Grain is 0.8 lb./Can	If Grain is 1.4 lb./Can	If Supplement is 10 g/tbsp	If Supplement is 20 g/tbsp
Oats, 38 lb./bu.			6.3	3.6		
Grass hay, mature, >60% NDF	10	5				
IdleMin/Vit					7.25760	3.62880

Strike **ENTER** and **F1**.

Print Ration or Blend Procedure

Make sure your name is entered at cell C1. Click on the **Print Ration or Blend** button.

The input box shown in Figure 25–17 displays.

Figure 25–17

Are you printing a horse ration evaluation or a blend formula and analysis? (**1**-RATION, **2**-BLEND):

Click **OK** to accept the default input of 1. A two-page printout will be produced by the machine's default printer.
Click on the **Print Ration or Blend** button. Type the number 2 to print a blend. The message box in Figure 25–18 displays.

Figure 25–18

Click on the green number above the blend you want to print and press **F4**. Scroll right to see additional blends.

Click **OK**.
Find the IdleMin/Vit blend, click on the green number above it, and strike **F4**. A one-page printout will be produced by the machine's default printer.

ACTIVITIES AND WHAT-IFS

In the Forms folder on the companion CD to this text is a HorseInput.doc file that may be used to collect the necessary inputs for use of the HorseRation.xls file. This form may be printed out and used during on-farm visits to assist in ration evaluation activities.

1. Remake the ration described in Table 25–12 for the horse described in this chapter. Strike **F7**. If the grass hay in this balanced ration weighs 2 lb. per flake, how many flakes should be fed? How many flakes should be fed if the hay weighs 4 lb. per flake? If the grain in this balanced ration is to be fed by the coffee can and a full can weighs 0.8 lb., how many coffee cans should be fed? How many cans should be fed if a full can weighs 1.4 lb.? If the mineral supplement in this ration is to be fed by the tablespoon and a full tablespoon weighs 10 g, how many tablespoons should be fed? How many tablespoons should be fed if a full tablespoon weighs 20 g? Based on your responses, what recommendations would you make regarding feeding horses by the flake, can and tablespoon?
2. Remake the ration described in Table 25–7 for the horse described in this chapter. What is the feed cost per horse per day on this ration? Add corn grain, ground to the selection of feedstuffs for this ration. Replace the oats grain with corn grain and use Goal Seek to rebalance the ration. What is the new feed cost per horse per day on this ration? With your results in mind, discuss why oats are such a popular horse feed.
3. Remake the ration described in Table 25–12 for the horse described in this chapter. Input a growing horse, 800-lb. body weight. If fed the ration for the idle horse, for what nutrients would this growing horse be deficient? Balance the two nutrients that are not minerals or vitamins. As you do so, make sure your ration is one that the growing horse can eat: as you add something, remove something else. What are the new amounts of oats and hay? Can you think of a better approach to balance these two nutrients in the ration for the growing horse? Try balancing the mineral and vitamin requirements with the IdleMin/Vit blend. What, if any, suggestions would you make regarding the use of the IdleMin/Vit blend for the growing horse?

REFERENCES

Austic, R. E., & Calvert, C. C. (1981). Nutritional interrelationships of electrolytes and amino acids. *Federation Proceedings. 40*, 63–67.

National Research Council. (1980). *Mineral Tolerance of Domestic Animals.* Washington DC: National Academy Press.

National Research Council. (1987). *Vitamin Tolerance of Animals.* Washington DC: National Academy Press.

National Research Council. (1989). *Nutrient Requirements of Horses,* 5th revised edition. Washington DC: National Academy Press.

CHAPTER 26

FEEDING CHICKENS

. . . a hen is only an egg's way of making another egg.

SAMUEL BUTLER, 1878

THE LAYER

Layers are chickens that produce eggs intended for use as food. The dominant layer is the Leghorn-type chicken. Leghorn-type chickens are relatively small in body size but are prolific layers.

The nutritional phases in the layer production cycle, illustrated in Figure 26–1, include:

Layer: starter, developer, grower, layer phase 1, layer phase 2, layer phase 3.

A sample growth rate is plotted in Figure 26–2.

Starter

At hatching, Leghorn-type pullets weigh about 40 g or 0.09 lb. Nutritional deficiencies and imbalances during the starter period may impair growth rate and future egg laying performance. The starter feed is generally fed for the first 6 weeks, at which time the birds are switched to a grower diet. Starter diets generally contain antibiotics to promote rapid, efficient growth and reduce mortality. Coccidiostats are also included to combat the protozoan parasites causing coccidiosis.

Grower

Birds are fed the grower diet from 6 weeks to sexual maturity. By the end of the grower phase—at approximately 21 weeks—the Leghorn-type pullet has reached 1,475 g or 3.25 lb. In some cases, birds may be placed on a developer diet after about 6 weeks on the grower diet.

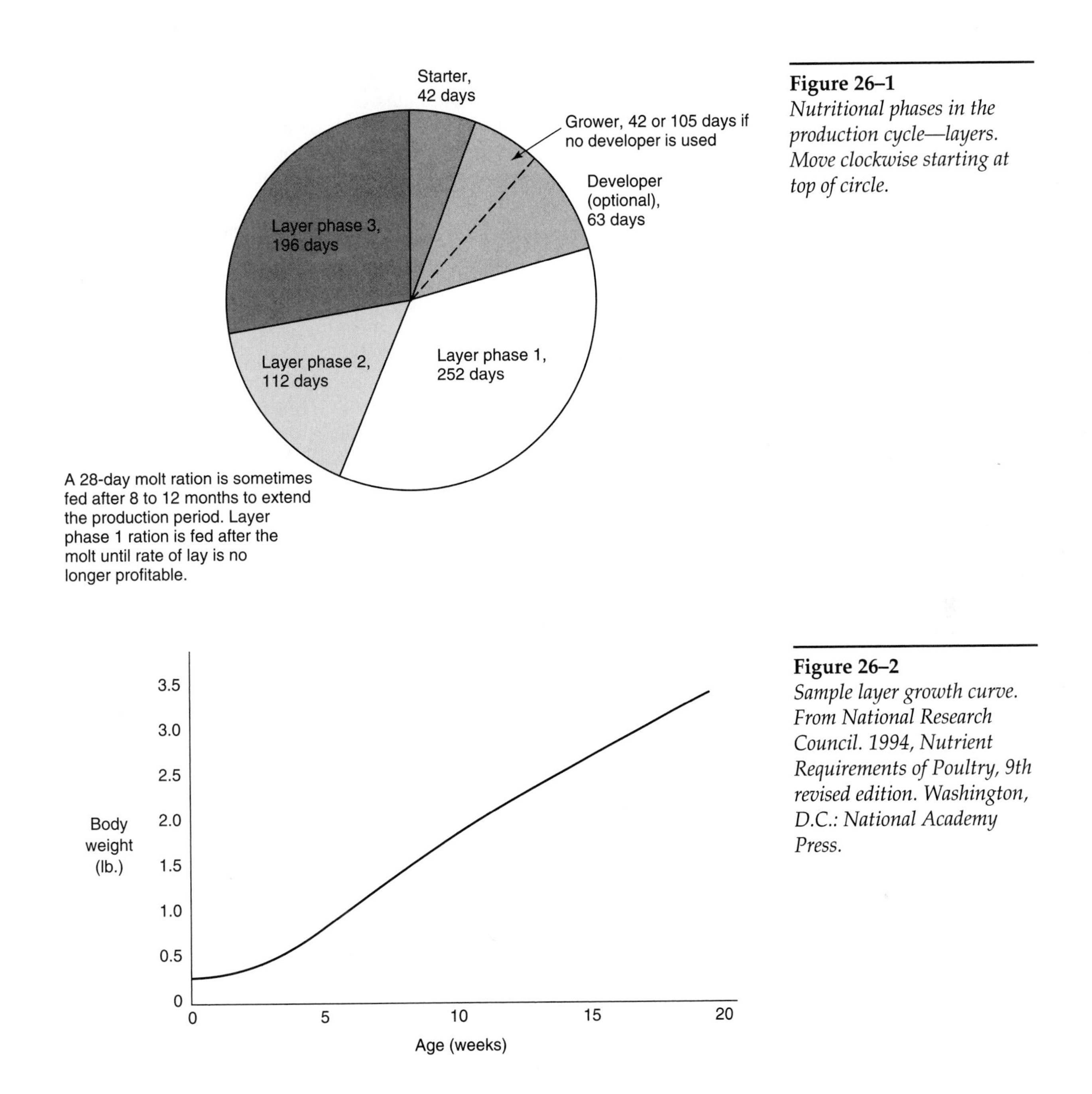

Figure 26–1
Nutritional phases in the production cycle—layers. Move clockwise starting at top of circle.

Figure 26–2
Sample layer growth curve. From National Research Council. 1994, Nutrient Requirements of Poultry, 9th revised edition. Washington, D.C.: National Academy Press.

Developer

A developer diet may be inserted between the grower and the layer diet in order to better address the changing nutritional needs of growing pullets. Although the feed for birds fed a developer diet will be less expensive than it would if they stayed on a grower diet, there will be an additional expense associated with managing these birds as a separate group. Developers are fed from 12 weeks of age until birds are up to 5 percent egg production (five eggs per 100 birds) at which time, birds are placed on the layer feed. This occurs at about the 21st week.

Egg Production

Feed for layers is provided free choice. Large amounts of calcium are required for egg-shell production by the laying hen. The calcium content of the egg shell determines its strength, and commercial eggs must be strong enough to withstand handling associated with automated collection, processing, and packaging. In addition, the calcium requirement may be affected by environmental temperature, the rate of lay, egg size, and the age of the bird. Diets for high-producing hens may contain ratios of calcium to available phosphorus as high as 12:1 (National Research Council, 1994).

Because methionine will be the first limiting amino acid in layer diets that are based on corn and soybean, methionine analog is often included in the layer ration formulation.

Phase Feeding

In a phase-feeding nutritional program for laying hens, nutrient requirements are adjusted according to expected requirements for maintenance and egg production. The adjustments amount to reductions in nutrient requirements as the laying period progresses. During the first phase, birds are growing and increasing in production, and at this time the feed formulation is at maximum nutrient density. During subsequent phases, the rate of lay declines as does nutrient density.

Layer phase 1: Phase 1 is the time from the onset of egg production until past the time of maximum egg mass output.
Layer phase 2: Phase 2 is a period following phase 1 of high (but declining) egg production and increasing egg weight. Egg production during phase 2 declines to about 65 percent of maximum.
Layer phase 3: Phase 3 is a period following phase 2 during which egg production continues to decline below 65 percent of maximum while egg weight increases only slightly.

The validity of the phase feeding system for laying hens has not been established (NRC, 1994). There is no evidence that the nutrient requirements of layers change during the period of lay. The companion application to this text does not use phase feeding to predict the nutrient requirements of laying hens. Rather, the companion application assumes that the amount of nutrient needed each day remains the same throughout a hen's production period.

Molting

Layers are sometimes molted to extend the production period. During a molt, feed and light are restricted. A molt can also be induced by feeding a diet containing a nutrient deficiency or excess. The molt may last 3 to 6 weeks. Following the molt, layers returned to a balanced diet will resume egg production, usually at a higher rate than that preceding the molt.

Growing Flock Replacements

Feed represents approximately 60 percent of the cost of raising replacement pullets in both the layer and broiler industries. Replacement pullets are chicks that are grown to produce viable eggs to hatch into chicks that will be sold as broilers or kept to produce hens that lay eggs for food. The nutrient requirement for birds producing hatching eggs is increased over that for hens producing eggs for human consumption for iodine, manganese, and zinc, and for the vitamins E, K, B_{12}, folacin, pantothenic acid, pyridoxine, and riboflavin. From about the eighth week until replacement pullets are placed on the layer feed, feed is usually restricted to a level that supports less than maximal gain to delay the onset of sexual maturity. This is necessary to optimize production of viable eggs in the adult.

THE BROILER

Originally, the broiler was a by-product of the egg industry: male pullets were grown and sold for their meat. Today, broilers are chickens that have been bred specifically for rapid and efficient gains, and a large body size. Although they may be marketed at a variety of ages and weights, broilers are usually marketed at 5 to 7 weeks of age when they weigh about 4.5 to 5.5 lb.

Broiler breeders are birds that produce eggs expected to hatch into the pullets that are grown to become meat birds. The production cycle of the meat bird—the broiler—is illustrated in Figure 26–3. It includes the phases of egg incubation, starter, grower, and finisher. Feed conversion in broiler production is about 2 lb. of feed for each pound of bird. A sample growth rate is plotted in Figure 26–4.

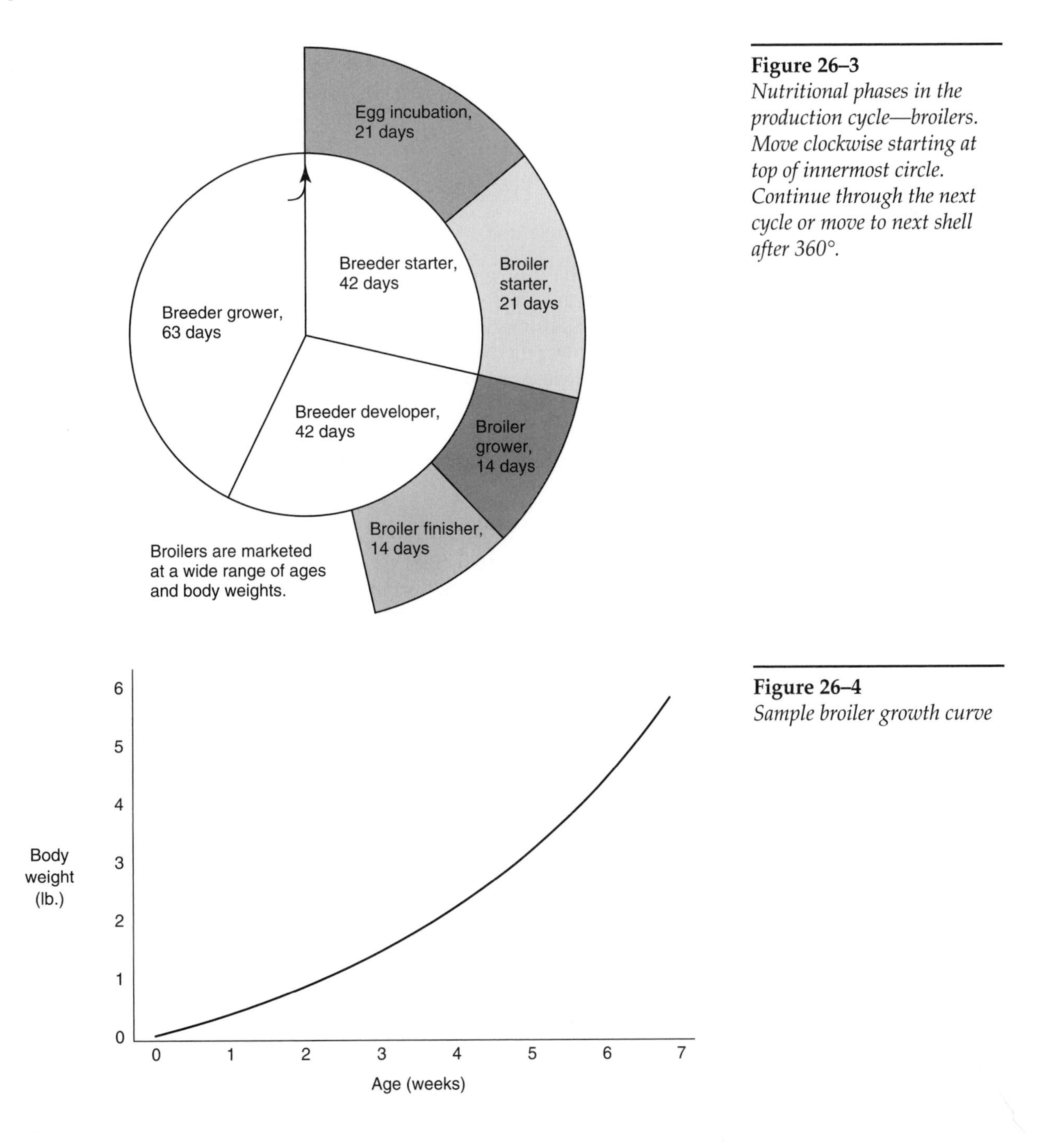

Figure 26–3
Nutritional phases in the production cycle—broilers. Move clockwise starting at top of innermost circle. Continue through the next cycle or move to next shell after 360°.

Figure 26–4
Sample broiler growth curve

Egg Incubation

Eggs hatch in 21 days.

Starter

The starter diet for broiler chicks corresponds to the starter diet for the Leghorn-type pullets. However, broiler pullets are capable of more rapid growth rates than Leghorn-type pullets, so a nutritionally balanced ration for broiler pullets will be of higher nutrient density than one for layer pullets. Antibiotics and coccidiostats are usually included in the broiler starter diet. The starter diet is usually fed for 2 to 3 weeks.

Grower

At about 3 weeks of age, broilers are switched to diets of lower nutrient density. The grower phase is inserted between the starter phase and the finisher phase to take advantage of the reduced nutrient densities required by these older birds compared to the starters. Grower diets are generally fed from 3 to 6 weeks of age.

Finisher

By the end of the grower phase—at 7 weeks—the male broiler has reached 2,590 g or 5.7 lb. and the female has reached 2,134 g or 4.7 lb. The finisher diet is fed until birds reach market weight. The finisher diet is lower in nutrient content than the grower diet and is therefore less expensive. Finisher rations do not contain antibiotic and are sometimes referred to as *withdrawal rations.*

LAYER AND BROILER FEEDING AND NUTRITION ISSUES

Dry Matter/Feed Intake

Knowledge of the dry matter/feed intake is essential in order to be able to fortify the diet appropriately. The formulas that have been developed to predict dry matter/feed intake by poultry consider the variables of environmental temperature, energy concentration in the diet, and rate of egg production. Undoubtedly, there are other factors that affect dry matter/feed intake. In the poultry industry, flow meters are sometimes used to measure actual feed consumption, replacing predictive equations.

The Energy Requirement

Through the cloaca, urine and feces are excreted together. Add to this the fact that chickens produce negligible amounts of gaseous products during digestion, and it becomes apparent why nutritionists have chosen to express chicken energy requirements and feedstuff energy values in terms of metabolizable energy: metabolizable energy excludes the energy content of feces, urine, and gas. In poultry nutrition, the metabolizable energy is expressed as "nitrogen-corrected" or MEn. The nitrogen correction involves accounting for feed protein that is retained in the body rather than metabolized for energy.

Because appetite is driven by the need for energy, the energy concentration of the diet has an impact on dry matter/feed intake (Table 26-1). Rations of high energy density should be fortified with high concentrations of amino acids, minerals, and vitamins because the intake of such diets will be relatively low.

Table 26–1
Effect of ration energy concentration on DMI in broilers

Diet Metabolizable Energy (kcal/lb., as fed basis)	Dry Matter Intake (lb.)[1]
1,452	36.6
1,315	39.2
1,179	41.1
1,043	46.4
862	52.0
726	60.9

[1]per day average over day 35(49 for 100 birds
Adapted from Leeson et al., 1996.

Conversely, rations of lower energy density should be fortified with lower concentrations of amino acids, minerals, and vitamins because the intake of such diets will be relatively high. The optimal energy density will generally depend on which formulation results in the lowest feed cost per unit of weight gain or egg production. However, the relationship between dietary energy concentration and feed intake is not precise, and for this reason, caution must be used in feeding untested or unusual formulations. Further discussion on the relationship between ration energy density and dry matter/feed intake is found in Chapter 5.

The poultry NRC (1994) gives energy concentrations based on what is typical in the poultry industry. The companion application to this text uses these typical energy concentrations as targets in ration formulation.

Male broiler chickens grow more efficiently than do females. This is because females tend to deposit more fat in the carcass and fat production requires more energy than does muscle production.

Amino Acids and Nitrogen Requirements

The amino acids that are most often first limiting in corn–soy based chicken diets include methionine, lysine, arginine, and tryptophan. Because the amino acids cysteine and tyrosine are made from the essential amino acids methionine and phenylalanine, respectively, they may become limiting with marginal levels of their precursors. Glycine and serine are amino acids that have been reported as essential for growing chickens. The two are interconvertible, so the requirement may be met in one of three ways:

1. Enough glycine to meet the need for glycine and support synthesis of serine
2. Enough serine to meet the need for serine and support synthesis of glycine
3. Adequate amounts of both glycine and serine (Akrabawi & Kratzer, 1968).

In addition to the need for essential amino acids, the chicken's diet must contain enough protein (nonspecific nitrogen sources) to allow metabolic synthesis of the nonessential amino acids.

The amino acid tryptophan can and will be converted into the B vitamin niacin when intake of niacin is inadequate. Likewise, the amino acid methionine will be converted into the B vitamin choline when intake of choline is inadequate. These conversions may be applied to diet formulation given a consideration of the cost of supplements.

While the mammalian dietary requirement for arginine is due to inadequate synthetic activity to meet metabolic needs, the poultry requirement for arginine is due to a complete lack of synthesis.

Requirements for individual amino acids, just as with all nutrients, should be viewed as targets. Missing the target through dietary excess may be just as bad as missing the target through dietary deficiency. An excess of lysine interferes with utilization of arginine and may result in a depressed appetite. A general excess of amino acid intake may create a deficiency of the first limiting amino acid.

Enzymes

Phytase addition to poultry diets has been shown to improve the utilization of dietary phosphorus (Simons, Versteegh, Jongbloed, Kemme, Slump, Bos, Wolters, Beudeker, & Verschoor, 1990). This is important for two reasons. First, there is an economic benefit: the use of phytase will enable poultry to acquire the phosphorus from corn meal and soybean meal, thereby removing the need to purchase supplemental sources of phosphorus. Second, there is an environmental benefit: the improved utilization of dietary phosphorus will reduce the excretion of phosphorus.

The addition of enzymes that act on carbohydrates has also had positive effects in poultry diets. Such enzymes have been shown to improve weight gain, feed intake, and feed efficiency in broilers (Mathlouthi, Lalles, Lepercq, Juste, & Larbier, 2002). The use of these enzymes in commercial diets is limited due to their high cost.

Xanthophylls and Carotenoids

Feeds that are rich in xanthophylls and carotenoids will produce a deep yellow color in the beak, feet, skin, shanks, fat tissue, and egg yolks of chickens. For some consumers, this coloration is desirable, and these substances are included in diet formulations in the form of additives and natural feedstuffs. Chapter 3 presents additional information on xanthophylls and carotenoids.

Electrolyte Balance

The optimal electrolyte balance for poultry has been suggested to be 250 mEq/kg of excess cations, calculated using the sodium and potassium cations and the chloride anion (Mongin, 1981). The 250 mEq/kg is an as fed value, which, assuming a ration of 90 percent dry matter, would be 278 mEq/kg on a dry matter basis. Sodium, potassium, and chloride are the primary dietary ions that influence acid–base status in animals. Dietary electrolyte balance will influence growth, bone development, egg-shell quality, and amino acid utilization in chickens.

Feeding Antibiotics at Subtherapeutic Levels

Antibiotics are often fed at subtherapeutic levels to chickens to improve feed efficiency, rate of gain, and egg production. A discussion of issues related to use of antibiotics in livestock feed is found in Chapter 3.

Grit

Poultry are sometimes fed grit to improve the grinding efficiency in the gizzard. Grit is usually composed of the ground shells of oyster, clam, or coquina, or may be made from limestone or other types of stone. Feeding grit is usually not necessary at commercial chicken farms because nutrients are efficiently released from highly processed feeds without extensive grinding in the gizzard.

END-OF-CHAPTER QUESTIONS

1. What is the relationship between feed energy density and the level of feed intake?
2. Give four economically important factors that are influenced by the electrolyte balance in chicken diets. What has been suggested as the optimum electrolyte balance for poultry, expressed as mEq/kg, as fed basis?
3. Describe a phase-feeding program for layers.
4. How does the required nutrient density for a broiler change with increasing age? Give the names that describe the rations fed during the three phases of broiler production.
5. What does the "n" in the energy value used in chicken nutrition, MEn, represent? Define MEn.
6. List two metabolic routes for meeting the chicken's requirement for each of the following: niacin, serine, and choline.
7. How are chickens different from most mammals in arginine metabolism?
8. What is grit? Describe its importance in commercial broiler and layer production.
9. What role is played by xanthophylls and carotenoids in feeding poultry?
10. Characterize the availability to chickens of phosphorus in corn meal and soybean meal. In this regard, what role can be played by phytase? Give two potential benefits of using phytase in chicken diets.

REFERENCES

Akrabawi, S. S., & Kratzer, F. H. (1968). *Journal of Nutrition. 95,* 41–48.

Butler, Samuel 1878. *Life and Habit.*

Leeson, S., Caston, L., & Summers, J. D. (1996). Broiler response to energy or energy and protein dilution in the finisher diet. *Poultry Science. 75,* 522–529.

Mathlouthi, N., Lalles, J. P., Lepercq, P., Juste, C., & Larbier, M. (2002). Xylanase and β-glucanase supplementation improve conjugated bile acid fraction in intestinal contents and increase villus size of small intestine wall in broiler chickens fed a rye-based diet. *Journal of Animal Science. 80,* 2773–2779.

Mongin, P. (1981). Recent advances in dietary cation-anion balance: Applications in poultry. *Proceedings of the Nutrition Society. 40,* 285–294.

National Research Council. (1994). *Nutrient requirements of poultry* (9th revised edition). Washington, DC: National Academy Press.

Simons, P. C. M., Versteegh, H. A. J., Jongbloed, A. W., Kemme, P. A., Slump, P., Bos, K. D., Wolters, M. G. E., Beudeker, R. F., & Verschoor, G. J. (1990). Improvement of phosphorus availability by microbial phytase in broilers and pigs. *British Journal of Nutrition. 64,* 525–540.

CHAPTER 27

CHICKEN RATION FORMULATION

TERMINOLOGY

Workbook A workbook is the spreadsheet program file including all its worksheets.

Worksheet A worksheet is the same as a spreadsheet. There may be more than one worksheet in a workbook.

Spreadsheet A spreadsheet is the same as a worksheet.

Cell A cell is a location within a worksheet or spreadsheet.

Comment A comment is a note that appears when the mouse pointer moves over the cell. A red triangle in the upper right corner of a cell indicates that it contains a comment. Comments are added to help explain the function and operations of workbooks.

Input box An input box is a programming technique that prompts the workbook user to type information. After typing the information in the input box, the user clicks **OK** or strikes **ENTER** to enter the typed information.

Message box A message box is a programming technique that displays a message. The message box disappears after the user clicks **OK** or strikes **ENTER**.

EXCEL SETTINGS

Security: Click on the **Tools** menu, **Options** command, **Security** tab, **Macro Security** button, **Medium** setting.

Screen resolution: This application was developed for a screen resolution of 1024 × 768. If the screen resolution on your machine needs to be changed, see Microsoft Excel Help, "Change the screen resolution" for instructions.

HANDS-ON EXERCISES

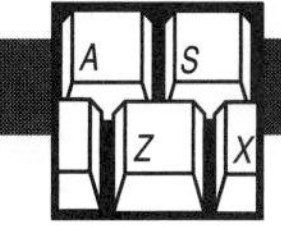

CHICKEN RATION FORMULATION

Double click on the **ChickenRation** icon. The message box in Figure 27–1 displays.

Figure 27–1

> Macros may contain viruses. It is advisable to disable macros, but if the macros are legitimate, you may lose some functionality.
> Disable Macros or Enable Macros or More Info

Click on **Enable macros**.

The message box in Figure 27–2 displays.

Figure 27–2

> Function keys F1 to F8 are set up. You may return to this location from anywhere by striking **ENTER**, then the **F1** key. Workbook by David A. Tisch. The author makes no claim for the accuracy of this application and the user is solely responsible for risk of use. You're good to go. TYPE ONLY IN THE GRAY CELLS! Note: This Workbook is made up of charts and a worksheet. The charts and worksheet are selected by clicking on the tabs at the bottom of the display. Never save the Workbook from a chart; always return to the worksheet before saving the Workbook.

Click **OK**.

Input Chicken Procedure

Click on the **Input Chicken** button. The input box in Figure 27–3 displays.

Figure 27–3

> 1. White egg layers
> 2. White egg breeders
> 3. Immature Leghorn-type white egg laying strains
> 4. Broilers
>
> ENTER THE APPROPRIATE NUMBER:

Click **OK** to select the default value of 1, White egg layers.

The input box in Figure 27–4 displays. Click **OK** to accept the predicted dry matter intake (DMI).

Figure 27–4

> Enter measured dry matter intake per 100 birds (pounds) if available or click **OK** to accept the predicted value at CX26.

Click **OK** to accept the predicted value. The message box in Figure 27–5 displays.

Figure 27–5

Enter the appropriate cell inputs in column CX.

Click **OK**.

The Cell Inputs

Enter the Cell Inputs shown in Table 27–1.

Table 27–1
Inputs for the chicken ration formulation example

3	Enter white egg layer's weight in pounds (3–3.5).
	NA
	NA
	NA
50	Enter egg mass in grams (40–70).
0	Enter daily weight change in grams. Since these are mature birds, this value should be close to 0.
50	Enter temperature (° F).
	N/A
6.3	Enter rate of egg production, eggs/week (a rate of 7 would mean one egg per hen per day).
1	Enter price ($) received for a dozen eggs.

The comment behind the cell containing the Ambient temperature label reads as shown in Figure 27–6.

Figure 27–6

Ambient temperature is used to predict energy requirement and dry matter/feed intake.

The comment behind the Rate of egg production, eggs per week label reads as shown in Figure 27–7.

Figure 27–7

Rate of egg production is also expressed as %. A rate of 100% would be the same as 7 eggs per week. This value is used here in profitability calculations only.

The notes below the input section are shown in Figure 27–8.

Figure 27–8

Notes:
1. These are nutrient targets per 100 birds.
2. The predicted DMI will decline with increasing dietary energy content.
3. Nutrient target calculations are based on a feed energy concentration of 1,462 kcal/lb. MEn, dry matter basis. This energy concentration is typical of rations based mainly on corn and soybean meal.
4. Nutrient targets for layers are based on an intake of 90 g dry matter and a 90% rate of egg production. For 100 birds, this would be 19.8 lb. of dry matter.

Strike **ENTER** and **F1**.

Select Feeds Procedure

Click on the **Select Feeds** button. The message box in Figure 27–9 displays.

Figure 27–9

Nutrient content expressed on a dry matter basis. Feedstuffs are listed first by selection status, then, within the same selection status, by decreasing protein content, then, within the same protein content, alphabetically.

Click **OK**. Select the feedstuffs listed in Table 27–2 by placing a 1 to the left of the feedstuff name.

Table 27–2
Feedstuffs to select for the chicken ration formulation example

Feedstuff
Methionine, DL
Soybean meal, 49%
Corn grain, ground
Choline chloride, 60%
Dicalcium phosphate
EDDI (Ethylenediamine dihydroiodide)
Limestone, ground
Manganous carbonate
Riboflavin
Salt (sodium chloride)
Vitamin A supplement
Vitamin B_{12} (cyanocobalamin)
Vitamin D_3 supplement
Vitamin E supplement
Vitamin K (MSB premix)
Zinc oxide

Strike **ENTER** and **F2**. Selected feedstuffs and their analyses are copied to several locations in the workbook.

Make Ration Procedure

Click on the **Make Ration** button. The message box in Figure 27–10 displays.

Figure 27–10

ENTER **POUNDS TO FEED** IN COLUMN B. TOGGLE BETWEEN NUTRIENT WEIGHTS AND CONCENTRATIONS USING THE **F5** KEY, RATION AND FEED CONTRIBUTIONS USING THE **F6** KEY. WHEN DONE STRIKE **F1**. Cell is highlighted in red if nutrient provided is poorly matched with nutrient target. The lower limit is taken as 95 to 98 percent of target, depending on the nutrient. Except for calcium and phosphorus in laying hen diets, the upper limit of acceptable mineral is taken from National Research Council (1980). *Mineral Tolerance of Domestic Animals.* National Academy Press. This publication gives safe upper limits of minerals as salts of high bioavailability. For laying hens, the upper limit of calcium is 6 percent and for available phosphorus, 1 percent. Upper limit for vitamins is from the NRC, 1987. For other nutrients, the upper limit is based on unreasonable excess and the expense of unnecessary supplementation. See Table 27–4 for specifics.

Click **OK**. The message box in Figure 27–11 displays.

Figure 27–11

The Goal Seek feature may be useful in finding the pounds of a specific feedstuff needed to reach a particular nutrient target:

1. Select the red cell highlighting the deficient nutrient
2. From the menu bar, select **Tools**, then **Goal Seek**
3. In the text box, "To Value:" enter the target to the right of the selected cell
4. Click in the text box, "By changing cell:" and then click in the gray "Pounds fed" area for the feedstuff to supply the nutrient
5. Click **OK**. You may accept the value found by clicking **OK** or reject it by clicking **Cancel**.

WARNING: Using Goal Seek to solve the unsolvable (e.g., asking it to make up an iodine shortfall with iron sulfate) may result in damage to the Workbook.

IMPORTANT: If you return to the Feedtable to remove more than one feedstuff from the selected list, you will lose your chosen amounts fed in the developing ration.

Click **OK**.

In the gray area to the right of the feedstuff name, enter the pound values shown in Table 27–3.

Table 27–3
Inclusion rates for feedstuffs in the chicken ration formulation example (unmedicated)

Methionine, DL	0.0089
Soybean meal, 49%	4.5
Corn grain, ground	15.9
Choline chloride, 60%	0.00221
Dicalcium phosphate	0.18135
EDDI (Ethylenediamine dihydroiodide)	0.00000021
Limestone, ground	1.84043
Manganous carbonate	0.00019
Riboflavin	0.000023
Salt (sodium chloride)	0.07142
Vitamin A supplement	0.04935
Vitamin B_{12} (cyanocobalamin)	0.000067
Vitamin D_3 supplement	0.00648
Vitamin E supplement	0.00791
Vitamin K (MSB premix)	0.00032
Zinc oxide	0.00029

The Nutrients Supplied *Display*

The application highlights ration nutrient levels, expressed as amount supplied per chicken per day, that fall outside the acceptable range.

The lower limit for DMI, MEn and CP is 95 percent of the target. The lower limit for all other ration nutrients is taken as 98 percent of the target. Table 27–4 shows the upper limits of the acceptable range for the various nutrients.

Table 27–4
Upper limits for ration nutrients used in the chicken ration application

	Upper Limit	Source
DMI	6% over predicted requirement	Author
MEn	6% over predicted requirement	Author
n-6 fatty acids	No upper limit	—
CP	6% over predicted requirement	Author
Arg	No upper limit	—
His	No upper limit	—
Ile	No upper limit	—
Leu	No upper limit	—
Lys	No upper limit	—
Met	No upper limit	—
Met + Cys	No upper limit	—
Phe	No upper limit	—
Phe + Tyr	No upper limit	—
Pro	No upper limit	—
Thr	No upper limit	—
Trp	No upper limit	—
Val	No upper limit	—
Ca	6% of ration DM for white egg layers and white egg breeders, 1.2% for others	Author (white egg layers/breeders) and Mineral Tolerance of Domestic Animals (1980)—poultry (others)
Pavail	1% of ration dry matter	Author
Na	0.79% of ration dry matter	Mineral Tolerance of Domestic Animals (1980)—poultry, based on limit for salt (sodium chloride)
Cl	1.33% of ration dry matter	Mineral Tolerance of Domestic Animals (1980)—poultry, based on limit for salt (sodium chloride)
K	2% of ration dry matter	Mineral Tolerance of Domestic Animals (1980)
Mg	0.3% of ration dry matter	Mineral Tolerance of Domestic Animals (1980)
Cu	300 mg/kg or ppm of ration dry matter	Mineral Tolerance of Domestic Animals (1980)—poultry
I	300 mg/kg or ppm of ration dry matter	Mineral Tolerance of Domestic Animals (1980)—poultry
Fe	1,000 mg/kg or ppm of ration dry matter	Mineral Tolerance of Domestic Animals (1980)—poultry
Mn	2,000 mg/kg or ppm of ration dry matter	Mineral Tolerance of Domestic Animals (1980)—poultry
Se	2 mg/kg or ppm of ration dry matter	Mineral Tolerance of Domestic Animals (1980)—poultry
Zn	1000 mg/kg or ppm of ration dry matter	Mineral Tolerance of Domestic Animals (1980)—poultry
Vit A	4 × the predicted requirement	NRC: Vitamin Tolerance of Animals, 1987
Vit D_3	4 × the predicted requirement	NRC: Vitamin Tolerance of Animals, 1987
Vit E	20 × the predicted requirement	NRC: Vitamin Tolerance of Animals, 1987
Vit K	1000 × the predicted requirement	NRC: Vitamin Tolerance of Animals, 1987
Biotin	10 × the predicted requirement	NRC: Vitamin Tolerance of Animals, 1987
Choline	2 × the predicted requirement	NRC: Vitamin Tolerance of Animals, 1987
Folacin	1000 × the predicted requirement	Author
Niacinavail	350 mg/kg body weight	NRC: Vitamin Tolerance of Animals, 1987
Pantothenic acid	100 × the predicted requirement	Author
Riboflavin	20 × the predicted requirement	Author
Thiamin B_1	1000 × the predicted requirement	NRC: Vitamin Tolerance of Animals, 1987
Pyridoxine B_6	50 × the predicted requirement	NRC: Vitamin Tolerance of Animals, 1987
Vitamin B_{12}	300 × the predicted requirement	NRC: Vitamin Tolerance of Animals, 1987

The *Nutrients Supplied* display shows nutrient amounts in the diet and the targeted nutrient amounts for 100 chickens as inputted. The nutrient targets for the inputted bird have been met with this ration with minimal excesses.

The cost of this ration using initial $/ton values is $2.08 per 100 chickens per day.

The first limiting amino acid: *Isoleucine*

The comment behind the cell containing this label is shown in Figure 27–12.

This is the amino acid that exists in the ration at a level that is farthest from the level required, or the amino acid whose requirement is most narrowly met.

Figure 27–12

Ca:P_{avail} *ratio:* *13.20*

The comment behind the cell containing this label is shown in Figure 27–13

Excess calcium interferes with the absorption of other minerals such as phosphorus, magnesium and zinc. A Ca-to-P ratio of 2:1 is appropriate for most poultry diets. For layers, a ratio of as high as 12 calcium to 1 available phosphorus may be acceptable (Poultry NRC, 1994).

Figure 27–13

Nutrient Concentration *Display (strike* `F5`*)*

The nutrients in the ration and the predicted nutrient targets are expressed in terms of concentration. That is, the nutrients provided by the ration are divided by the amount of ration dry matter and the nutrient targets are divided by the target amount of ration dry matter. Concentration units include percent, milligrams per kilogram or parts per million, calories per pound, and international units per pound.

Electrolyte balance: *182*

The comment behind the cell containing this label is shown in Figure 27–14

The electrolyte balance is used to assess the impact of the ration's mineral content on the body's efforts to regulate blood pH. It is calculated as mEq of excess cations: (Na + K − Cl)/kg of *dry matter* diet. The optimal electrolyte balance in the diet for chicks has been suggested to be 250, as fed (Mongin, 1981) or 278, *dry matter.* The electrolyte balance is of no value if requirements for Na, K and Cl have not been satisfied.

Figure 27–14

Strike **F5**.

The Feedstuff Contributions *Display (strike* `F6`*)*

Shown here are the nutrients contributed by each feedstuff in the ration. This display is useful in troubleshooting problems with nutrient excesses.

Strike **F6**.

Predicted Performance *Display (strike* `F7`*)*

The predicted performance is shown in Table 27–5. Entries may be made in the gray areas to determine the effect on profitability of changes in price per dozen eggs and rate of egg production.

Table 27–5
The predicted performance display for the chicken example

The calculations shown below represent income over feed cost for layers at varying price per dozen eggs and rates of egg production.

Income over feed cost is calculated as follows:

Price per dozen eggs − [12 × ration cost/(eggs per week/7)]

Enter values below to calculate projected income over feed cost (layers only).

$2.00 "What-if" $ per dozen.

5.0 "What-if" rate of egg production (eggs per hen per week).

$0.021 Current ration cost (per hen):

	Price per Dozen Eggs ($)	Rate of Egg Production (eggs per week)	Income over Feed Cost, $ per Dozen
Using Inputted $ per dozen and inputted rate of egg production:	$1.00	6.3	$0.723
Using "what-if" $ per dozen and "what-if" rate of egg production:	$2.00	5	$1.651

Strike **ENTER** and **F1**.

The Graphic *Display*

At the bottom of the home display are tabs. The current tab selected is the Worksheet tab. Other tabs are graphs based on the current ration.

DMI
Click on this tab to display a graph titled Nutrient Status: Dry Matter Intake

ME&Linoleic
Click on this tab to display a graph titled Nutrient Status: Metabolizable Energy (nitrogen-corrected) & Linoleic Acid

CP&aa
Click on this tab to display a graph titled Nutrient Status: Crude Protein & Amino Acids

Macro
Click on this tab to display a graph titled Nutrient Status: Macrominerals

Micro
Click on this tab to display a graph titled Nutrient Status: Microminerals

Vit-H_2O
Click on this tab to display a graph titled Nutrient Status: Water-Soluble Vitamins

Vit-Fat

Click on this tab to display a graph titled Nutrient Status: Fat-Soluble Vitamins

Income

Click on this tab to display a graph titled Income Over Feed Cost at Current and "What if" Price Per Dozen Eggs & Rates of Production

Click on the **Worksheet** tab.

Making a Medicated Complete Feed

Acquiring Drug Concentration and Dosage Information

Blend Feedstuffs Procedure

Click on the **Blend Feedstufs** button. The message box in Figure 27–15 displays.

Figure 27–15

YOU MUST HAVE ALREADY SELECTED THE FEEDS YOU WANT TO BLEND. When your analysis is acceptable, strike **ENTER** and **F3** to name and file the blend.

Click **OK**.

In the gray area to the left of the feedstuff name, the amounts entered in the ration are entered as the amounts to blend. Note that in the blue column to the left of the entered amounts, the feedstuff amounts are converted to an equivalent amount in a ton of mix. Feed mills prefer to have the formula expressed on a per-ton basis because the capacity of their mixer(s) is expressed in tons.

We want to use erythromycin as an additive for its ability to increase egg production.

Click on the **Additives** button for help with including additive premixes in blends. Click on the **FDA** button to go to the FDA-approved Web site.

In the search box, type **erythromycin**. This is a drug approved for use in laying hens. Click on the **Search** button.

Find Gallimycin®-100P.

Click on the **Browser View** for this product. We will use the 2.2% concentration.

Though this is a Category II (withdrawal period required), for some uses, no withdrawal is required in laying hens. This product is a Type A medicated article (see Chapter 13). Under CFR indications, find the chicken, layer information. The approved level for all classes of chickens is 4.6 to 18.5 g erythromycin thiocyanate per ton. We will use a 18.5 g/ton level.

The concentration of this additive is expressed as a percentage. The dose of this additive is expressed in grams per ton.

Determining the Appropriate Amount of Additive Premix to Make 1 Ton of Medicated Complete Feed

Close the Web site. Look for a Type that matches our additive's concentration and dose. The match is found at Type #5. In the gray areas, enter the 2.2 concentration value and the 18.5 dose value. The calculated pounds of additive premix containing 2.2 percent additive to add to one ton of finished feed to deliver the dose of 18.5 g is 1.85.

Entering the Additive Premix into the Feed Table and Selecting It

Scroll up and click on the **`Feed Table`** button to go to the feed table to enter and select the additive premix. Enter the name Gallimycin in an empty row under FEEDNAME. Enter a cost, if known. Use 20,000 for this example. Enter a DM percent of 99. Select this feedstuff along with the other components of the grain. Strike **`ENTER`** and **`F2`**.

Blending the Additive Premix into the Complete Feed

Blend Feedstuffs Procedure

Click on the **`Blend Feedstuffs`** button.

Make sure that the feedstuff amounts are entered as shown in Table 27–6.

Table 27–6
Inclusion rates for feedstuffs in the LayerMedicated blend (as yet undetermined amount of drug)

Feedstuff	Amount
Methionine, DL	0.0089
Soybean meal, 49%	4.5
Corn grain, ground	15.9
Choline chloride, 60%	0.00221
Dicalcium phosphate	0.18135
EDDI (Ethylenediamine dihydroiodide)	0.00000021
Limestone, ground	1.84043
Manganous carbonate	0.00019
Riboflavin	0.000023
Salt (sodium chloride)	0.07142
Vitamin A supplement	0.04935
Vitamin B_{12} (cyanocobalamin)	0.000067
Vitamin D_3 supplement	0.00648
Vitamin E supplement	0.00791
Vitamin K (MSB premix)	0.00032
Zinc oxide	0.00029
Gallimycin[1]	?

[1]Gallimycin®-100P, a product of Cross Vetpharm Group, Ltd.

Enter an amount of Gallimycin in the blend that converts to a value of 1.85 lb./ton or less. A value that works is 0.02 (Table 27–7).

Methionine, DL	0.0089
Soybean meal, 49%	4.5
Corn grain, ground	15.9
Choline chloride, 60%	0.00221
Dicalcium phosphate	0.18135
EDDI (Ethylenediamine dihydroiodide)	0.00000021
Limestone, ground	1.84043
Manganous carbonate	0.00019
Riboflavin	0.000023
Salt (sodium chloride)	0.07142
Vitamin A supplement	0.04935
Vitamin B_{12} (cyanocobalamin)	0.000067
Vitamin D_3 supplement	0.00648
Vitamin E supplement	0.00791
Vitamin K (MSB premix)	0.00032
Zinc oxide	0.00029
Gallimycin	0.02

Table 27–7
Inclusion rates for feedstuffs and drug in LayerMedicated blend

Strike **ENTER** and **F3**. The input box in Figure 27–16 displays.

ENTER THE NAME OF THE BLEND (names may not be composed of only numbers):

Figure 27–16

Type the name LayerMedicated and click **OK**.

The message box in Figure 27–17 displays.

The new blend has been filed at the bottom of the feed table.

Figure 27–17

Click **OK**.

View Blends Procedure

Click on the **View Blends** button. The message box in Figure 27–18 displays.

Cursor right to view the blends. Cursor down for more nutrients. DO NOT TYPE IN THE BLUE AREAS.

Figure 27–18

Click **OK**. Confirm that the formula and analysis of the LayerMedicated blend has been filed.

Strike **F1**.

Using the Medicated Complete Feed in the Ration

Select Feeds *Procedure*

Click on the **Select Feeds** button and select only the LayerMedicated feed (row 200). Unselect all other feedstuffs by entering a 0 to the left of the feedstuff name.
Strike **ENTER** and **F2**.

Make Ration *Procedure*

Click on the **Make Ration** button. Enter the amount of LayerMedicated blend that is the sum of its components in the balanced ration (Table 27–8). That amount is calculated as:

$$0.0089 + 4.5 + 15.9 + 0.00221 + 0.18135 + 0.00000021 + 1.84043 + 0.00019 + 0.000023 + 0.07142 + 0.04935 + 0.000067 + 0.00648 + 0.00791 + 0.00032 + 0.00029 + 0.02 = 22.59 \text{ lb.}$$

This value is recorded at View Blends under Formula, as entered. The ration is balanced as it was when the components of the LayerMedicated were fed unmixed. The cost of the ration is now $2.28 per 100 birds.

Table 27–8
Feeding rate for LayerMedicated blend in the chicken ration formulation example

LayerMedicated	22.59

Strike **ENTER** and **F1**.

Print Ration or Blend Procedure

Click on the **Print Ration or Blend** button. The input box in Figure 27–19 displays.

Figure 27–19

Are you printing a chicken ration evaluation or a blend formula and analysis? (**1**-RATION, **2**-BLEND)

Click **OK** to accept the default value of 1, a ration. A two-page printout will be produced by the machine's default printer.
Click on the **Print Ration or Blend** button. Type the number 2 to print a blend. Click **OK**. The message box in Figure 27–20 displays.

Figure 27–20

Click on the green number above the blend you want to print and press **F4**. Scroll right to see additional blends.

Click **OK**.
Find the LayerMedicated blend, click on the green number above it, and strike **F4**. A one-page printout will be produced by the machine's default printer.

ACTIVITIES AND WHAT-IFS

In the Forms folder on the companion CD to this text is a ChickenInput.doc file that may be used to collect the necessary inputs for use of the ChickenRation.xls file. This form may be printed out and used during on-farm visits to assist in ration evaluation activities.

1. Remake the ration described in Table 27–8 for the chicken described in this chapter. Strike **F7**. At the inputted egg selling price of $1 per dozen, what is the income over feed cost per dozen eggs produced? If the egg price went down to $0.95, what would be the new income over feed cost per dozen eggs produced? With the price per dozen back at $1, what would be the new income over feed cost per dozen eggs produced if the rate of egg production declined to 6?
2. Remake the ration described in Table 27–7 for the chicken described in this chapter. Strike **F6**. Make a list of each vitamin required by the chicken and give the top three sources for each of them as provided in this ration.
3. Remake the ration described in Table 27–7 for the chicken described in this chapter. Try replacing the soybean meal 49% in this ration with two other nonsoybean oilseed meals listed in the feed table. Explain why soybean meal 49% is the most commonly used protein supplement in chicken nutrition.

REFERENCES

Mongin, P. 1981. Recent advances in dietary cation-anion balance: applications in poultry. *Procedings of the Nutrition Society. 40*, 285–294.

National Research Council. (1980). *Mineral Tolerance of Domestic Animals.* Washington DC: National Academy Press.

National Research Council. (1987). *Vitamin Tolerance of Animals.* Washington DC: National Academy Press.

National Research Council. (1994). *Nutrient Requirements of Poultry,* 9th revised edition. Washington DC: National Academy Press.

CHAPTER 28
FEEDING DOGS

There are two things for which animals are to be envied: They know nothing of future evils, or of what people say about them.

VOLTAIRE

The nutritional phases in the dog production cycle, illustrated in Figure 28–1, include:

Bitch: gestation, lactation, maintenance/anestrous
Puppy: suckling/creep, early growth, late growth

A sample growth rate is plotted in Figure 28–2.

THE BITCH

Gestation

During pregnancy, the bitch should be fed a balanced ration split into at least two feedings or the diet should be made available free choice.

Lactation

During lactation, the bitch should be fed a balanced diet, offered free choice. The use of snacks or table scraps is undesirable as this may unbalance the diet at this critical time. Offering the nursing puppies creep feed reduces the nutritional demands on the bitch for milk production. Free-choice water should always be provided.

Maintenance/Anestrous

Bitches that are in their anestrous period should be fed rations formulated to achieve, and then maintain, the desired body condition. Likewise, pets that are not used for breeding purposes or work should be fed a maintenance diet. The biggest challenge for owners of these dogs may be to keep the animals from gaining excess weight. The issue of obesity is discussed later in this chapter.

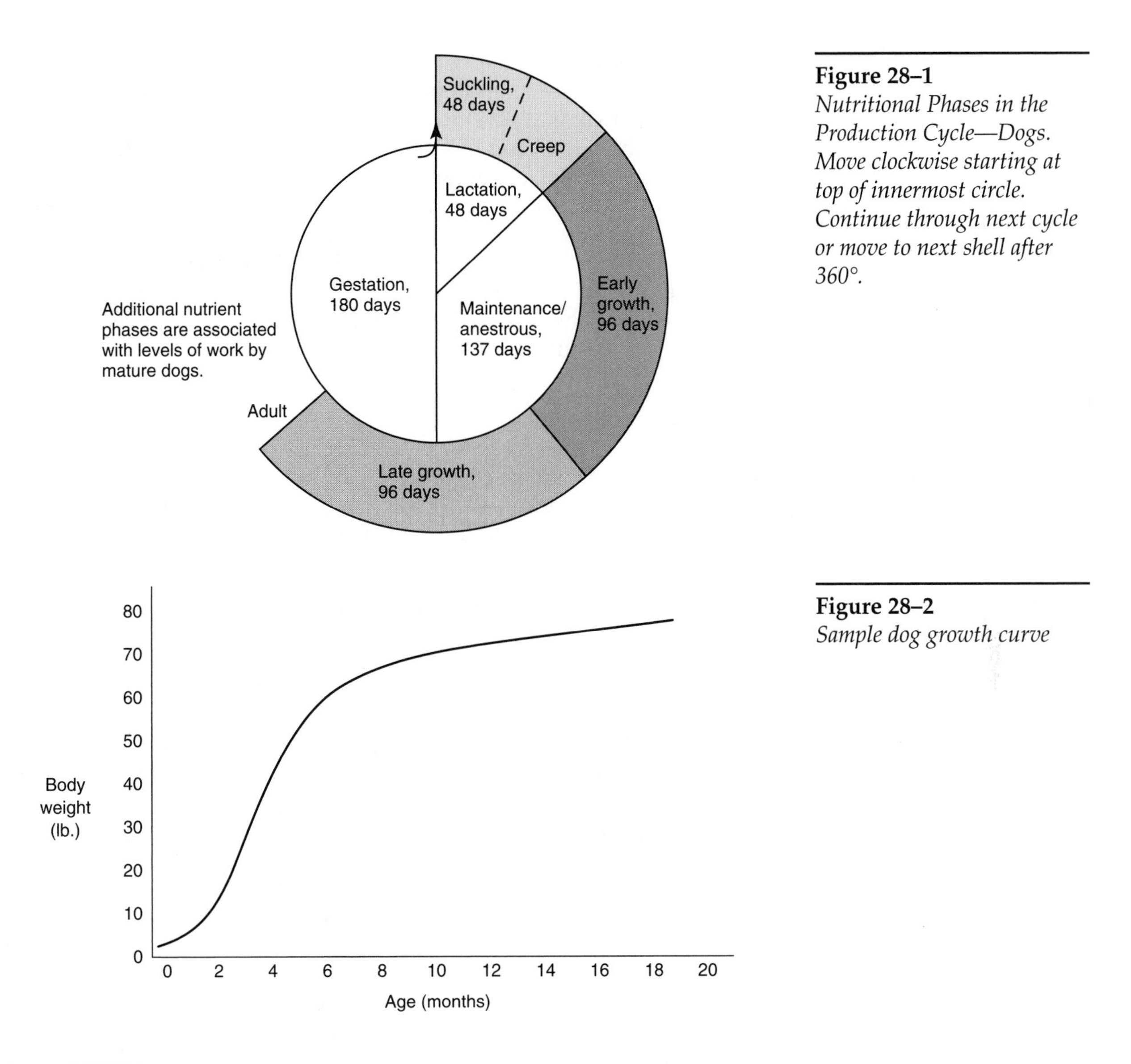

Figure 28–1
Nutritional Phases in the Production Cycle—Dogs. Move clockwise starting at top of innermost circle. Continue through next cycle or move to next shell after 360°.

Figure 28–2
Sample dog growth curve

THE PUPPY

Suckling/Creep Feeding

The structure of the bitch's placenta is described as endotheliolchorial. This type of placenta does not allow sufficient antibody immunoglobulins to pass from the bitch into the blood of the fetal puppies. The neonatal puppy acquires these antibody immunoglobulins through the bitch's first milk or colostrum. Colostrum should be consumed by the neonatal puppy as soon as possible. Colostrum is discussed in Chapter 20.

Puppies should be offered creep feed after the second week of life. A moist feed (canned, semimoist, or a dry dog food with water added) will be more attractive to them than a dry one. The bitch whose puppies have started eating creep feed by 3 weeks will be less stressed nutritionally than one whose puppies depend solely on milk until weaning. Weaning generally takes place at 6 to 8 weeks of age.

Growth

After the puppy is weaned, it should be hand fed 3 times a day or fed free choice. At 6 months of age, the number of feedings used for hand-fed dogs can be reduced to twice daily.

Growing dogs should not be fed to achieve maximum gains. Rapid gains have been associated with improper skeletal development, especially in large breeds. Improper skeletal development during early growth may lead to osteochondrosis in the puppy and hip displasia in the adult dog. Nutrients that are especially important in moderating the rate of gain include energy, calcium, and phosphorus. Proper diet formulation will reduce excesses of these nutrients to help ensure proper skeletal development.

As the growing dog approaches adulthood, its nutrient requirements can be met with a diet of reduced nutrient density. The ration for late growth should be reformulated to be less concentrated in energy to prevent the dog in this phase from accreting excess body fat.

ADULT

Adult dogs must be fed a diet that will support maintenance. For most pets, this is the only level of nutrition that should be fed. Nutrient levels beyond maintenance will be necessary if the dog is especially active or is used for work.

DOG FEEDING AND NUTRITION ISSUES

Association of American Feed Control Officials

As described in Chapter 1, the Association of American Feed Control Officials (AAFCO) is an association of government regulatory officials. The purposes of the AAFCO include regulation of production, labeling, distribution, and sale of animal feeds. The AAFCO membership works together to promote uniformity and consistency in laws pertaining to the feed industry, and to develop standards, definitions, and policies to facilitate the enforcement of such laws. Because the AAFCO did not accept the NRC publications on nutrient requirements of dogs (1985) and cats (1986) as the definitive source of minimum nutrient requirement information, the definitive source has become the AAFCO *Official Publication* (2003).

The primary concern that the AAFCO had with the NRC publication was that nutrient requirements were established for dogs and cats using purified diets. These requirement values, then, were derived from experiments using feedstuffs whose nutrients were 100 percent bioavailable. The NRC requirement values could be directly applied to commercially produced pet foods only if pet food manufacturers knew the bioavailability of each nutrient in each feedstuff included in their formulations. In other areas of livestock nutrition, researchers are actively involved in making such determinations, and in fact, the dairy NRC (2001) publication includes software with the capability to account for bioavailability of each mineral nutrient in each feedstuff. The companion application to this text uses dog NRC (1985) mineral requirement values in the determination of mineral requirements and dairy NRC (2001) mineral source bioavailability values. There are undoubtedly differences in how effectively the dog (a monogastric) and the bovine (a ruminant) are able to digest and absorb minerals from different feed sources. However, the author of this text prefers to use mineral bioavailability values determined for another species over safety factors that deliberately exaggerate mineral requirements.

The AAFCO alternative to the NRC values involves the use of safety factors or overages applied to the actual nutrient requirements. The application of inflated nutrient requirements increases the likelihood that minimum requirements will be met in formulated rations, regardless of the bioavailability of nutrients in the feedstuffs used. It is worth noting that the application of such a safety factor to the phosphorus requirement in dairy cattle is widely viewed as responsible for the phosphorus pollution problems in many areas of the country. Problems with phosphorus pollution have been one of the factors motivating legislators to fund research to determine nutrient bioavailability in rations for dairy cattle. A comparison of the AAFCO and NRC requirement values for dogs is shown in Table 28–1.

Table 28–1
Nutrient requirements for dogs: a comparison between the AAFCO, 2003 and NRC, 1985

Nutrient	AAFCO	NRC[1]
Crude protein	22	—
Arginine	0.62	0.50
Histidine	0.22	0.18
Isoleucine	0.45	0.36
Leucine	0.72	0.58
Lysine	0.77	0.51
Methionine-cystine	0.53	0.39
Phenylalanine-tyrosine	0.89	0.72
Threonine	0.58	0.47
Trytophan	0.20	0.15
Valine	0.48	0.39
Crude fat	8	5
Linoleic acid	1	1
Calcium	1	0.59
Phosphorus	0.8	0.44
Potassium	0.6	0.44
Sodium	0.3	0.06
Chloride	0.45	0.09
Magnesium	0.04	0.04
Iron	80	31.9
Copper	7.3	2.9
Manganese	5	5.1
Zinc	120	35.6
Iodine	1.5	0.59
Selenium	0.11	0.11
Vitamin A	5,000	3,710
Vitamin D	500	404
Vitamin E	50	22
Thiamine	1	1
Riboflavin	2.2	2.5
Pantothenic acid	10	9.9
Niacin	11.4	11.0
Pyridoxine	1	1.1
Folic acid	0.18	0.2
Vitamin B_{12}	0.022	.026
Choline	1,200	1,250

[1]Values are for growing dogs, dry matter basis.
Adapted from Payne, 1965.

Because the AAFCO is now the accepted authority on nutrient requirements for dogs and cats, pet food manufacturers reference the AAFCO standards in making claims of nutritional adequacy. To be able to claim that the pet food meets the AAFCO standards for "complete and balanced nutrition," pet-food manufacturers must state on the product label which of two accepted methods was used to substantiate that claim. The first method, known as the "calculation method," involves simply formulating a mix that contains the targeted range of nutrient levels published by the AAFCO. The second method involves performing AAFCO-sanctioned feeding trials on the feed formulation.

The AAFCO requires that pet food labels contain the following information.

1. A nutritional adequacy statement describing if the food is complete and balanced or if it is a snack or treat
2. A statement of which of the two legal methods was used if making the claim of complete and balanced
3. A life-stage statement if it is a complete and balanced diet, describing life stages for which the product is intended
4. Feeding directions, if described as complete and balanced
5. Brand name, if any, and product name
6. A purpose statement and animal for which the food is intended
7. A list of ingredients in descending order of predominance by weight
8. Name and address of the manufacturer or distributor
9. A quantity statement of net weight or count
10. A guaranteed analysis

The guaranteed analysis must include, at a minimum, the following analysis values, as fed basis: the minimum percentage of crude protein and crude fat, and the maximum percentage of crude fiber and moisture. In the naming of ingredients, the label must use official AAFCO feed ingredient names, if available. Such names may be of specifically identified ingredients or names of general classifications of ingredient origin such as forage products, grain products, or animal protein products.

Feeding Programs

The feeding program employed should be the one that keeps the dog in the proper body condition. Dogs fed dry food free choice will eat 10 to 13 meals per day (Mugford & Thorne, 1980). For most adult dogs, a single feeding is an acceptable way to accommodate this feeding behavior. Self-feeding is the best feeding program for very active dogs. Some dogs may not adjust to a self-feeding program and will have to be limit fed. Water should always be available.

The companion application to this text gives predicted dry matter intake (DMI) values based on both information presented in the dog NRC (1985) and on empirical evidence. However, given the variety of factors that affect dog dry matter/feed intake, users are encouraged to enter measured DMI rather than using the predicted value. This is the case whether dogs are limit fed or are fed free choice.

The Role of Meat

Dogs are carnivores, but this does not mean that they can thrive on an all meat diet or even that meat is essential in their diet (Hill, Burrows, Ellison, & Bauer, 2001). Nondomesticated carnivores devour not only the flesh of their prey but also the entrails, which are usually filled with digested vegetation. A dog eating nothing but meat will be forced to make needed glucose through gluconeogenesis in the liver and kidney. During gluconeogenesis, glucose is made from the glucogenic amino acids (all amino acids except lysine and leucine are at least partially glucogenic) in protein and glycerol in fat. Pregnant and lactating bitches fed all-meat diets may become hypoglycemic (Romsos, Palmer, Muiruri, & Bennink, 1981), indicating that gluconeogenesis alone may not be sufficient to meet these animals' glucose needs. However, some studies have suggested that if the diet contains a large protein excess, the kidney and liver may be supplied with enough amino acids that gluconeogenesis can supply the dog's total glucose needs (Blaza, Booles, & Burger, 1989; Kienzle & Meyer, 1989).

Dog Diets Compared to Human Diets

As long as the ration is balanced for nutrients, there will be no adverse effects on health or performance of the dog, no matter how many consecutive days the same ration is fed. The same would be the case for people, but the nature of meal time as a social event in human society makes food variety important.

A nutritionally balanced dog food is all that is necessary to provide the dog with complete nutrition. Feeding dogs scraps from the table will usually result in messy stools, occasional vomiting, and begging behavior. In addition, chicken or turkey bones can splinter and puncture a dog's throat, stomach, or intestine. Chop bones can block the intestines and possibly result in death.

The nutritional requirements of dogs are different from those of people. Whereas people require vitamin C, for example, dogs are able to make this vitamin in their tissues.

Chocolate

Chocolate contains theobromine, which is toxic to dogs and cats. Theobromine is found in greatest concentration in dark, unsweetened chocolate products such as baker's chocolate. Theobromine toxicity will affect the heart, central nervous system, and kidney. Early signs of toxicity include nausea and vomiting, diarrhea, and increased urination.

Dog Feeding and Public Health

Feeding a dog table scraps, not following a regular feeding schedule, or not providing a well-balanced diet will turn the dog into a forager. A forager frequently vomits, has messy stools, prowls the neighborhood for food, and may turn feral and begin to run in packs. Such a dog is a public health concern.

Feed Form

There are three categories of dog food that vary based on dry matter content. Dry dog food is about 88 to 90 percent dry matter (10 to 12 percent moisture). Semimoist dog food is 65 to 75 percent dry matter (25 to 35 percent moisture). Canned dog food is 22 to 26 percent dry matter (74 to 78 percent moisture).

Dry and semimoist dog food is usually sold in shapes that have been produced with an extruder. The extrusion process applies heat and pressure to the feed material, and starch becomes gelatinized, which improves digestibility (see Chapter 13). The processing activities involved in producing canned dog food also improve digestibility.

Dogs fed moist and semimoist dog foods are prone to dental plaque. Owners of dogs fed these types of dog foods should pay particular attention to the dental health of their animals.

Dog Body Condition

A proper feeding schedule, a balanced ration, and daily exercise are essential to keeping the dog in proper body condition and to maximize its longevity.

Energy

The majority of the energy in most commercial dog foods will come from non-fiber carbohydrate (NFC), primarily starch. Fat may also contribute a significant amount of energy. Dogs involved in sprints, such as racing greyhounds, will utilize dietary and stored carbohydrates to supply their energy needs, but will not make use of dietary fat as a source of energy for the short-duration race. Dogs involved in endurance racing, such as sled dogs, will make use of both carbohydrate and fat sources of energy during the long race.

Obesity Dogs that consume diets containing more energy than they can use become fat. This is because animals evolved in an energy-scarce environment and they needed to be able to store energy in times of plenty to survive times of scarcity. Body fat is an excellent storage medium because 1 lb. of fat contains at least 2.25 times as much energy as 1 lb. of other organic molecules.

Most pets are not fed in an energy-scarce environment and storage of excess energy in body fat serves no useful purpose. Obesity among dogs and cats is a significant problem, leading to increased risk of the development of serious health problems including diabetes and pulmonary and heart disease. There are dog foods sold specifically to help achieve weight loss. These dog foods generally use two weight-loss strategies. They may contain increased fiber, which imparts a sense of satiety without providing much energy. Drawbacks associated with this first method are that dogs will have more voluminous stools and may have increased gas production. A second method used to achieve weight loss is to reduce the fat content of the diet. This may be the best method, although it will be necessary to limit feed the dog because it may want to eat more of this diet of lower energy density.

Social facilitation describes the phenomenon by which social animals mimic the behavior of their groupmates. Dogs in multiple-dog households usually eat more food due to the effects of social facilitation: when one dog heads over to the food bowl, the other will do the same whether it is hungry or not. In multiple-dog households, it is best to provide a separate food bowl in a separate location for each dog to help prevent social facilitation behavior from contributing to obesity.

The companion application to this text estimates energy requirement using values in Table 28–2. An activity adjustment is applied to the energy requirement as described in Chapter 29.

Carbohydrate

The content of milk sugar (lactose) in the bitch's milk is 8 percent compared to 27 percent in cow's milk. This difference may lead to incomplete digestion and

Table 28–2
Calculated metabolizable energy requirements of dogs in various physiological states

Physiological State	Energy Requirement (kcal/kg $BW^{0.67}$)[1]
Start of weaning (3 weeks)	400
End of weaning (6 weeks)	375
Early growth	353
Late growth	225
Adult	159
Pregnancy, late	225
Lactation	560

[1]$BW^{0.67}$ is body weight in kilograms raised to the 0.67 power. This is a calculation of body surface area, which is used here and in the dog NRC (1985) to determine the energy requirement. See Chapter 5 for a discussion of issues concerning how the energy requirement is determined.

diarrhea in dogs and puppies that are fed foods including cow-milk products. Digestion may be improved by treatment of foods containing cow-milk products with lactase before feeding.

Dogs do not make the enzymes necessary to digest fiber. However, fiber is usually included in commercial dog foods at about 3 percent, dry matter basis, to help improve digestive health and stool formation. Higher levels of fiber may be used to reduce caloric density in diets for overweight dogs.

Protein

A diet containing as little as 6 percent of a high-quality, highly digestible protein has been shown to meet the protein requirements of adult dogs (Ullrey, 1995). Most commercial dog foods, however, will contain considerably more than this amount. The AAFCO (see Chapter 1) has adopted a value of 22 percent, dry matter basis, for the minimum crude protein content of dog food intended for growth and reproduction, and a value of 18 percent, dry matter basis, for adults at maintenance. The companion application to this text does not consider a minimum requirement for crude protein, but does consider a minimum requirement for each essential amino acid using calculations from the dog NRC (1985).

Dogs and other domestic animals that consume protein in excess of their requirements must process the excess in the liver for excretion by the kidneys. Healthy liver and kidneys are not damaged by excess protein consumption (Bovee, 1991).

Fat

Fat is included in dogs' diets to improve palatability, increase caloric density, and enhance the absorption of the fat-soluble vitamins. Dogs do not have a requirement for fat, but they do have a dietary requirement for the omega-6 fatty acid (linoleic acid) that is found in most vegetable oils, poultry and pork fat, and to a lesser extent in beef fat and butter fat (Table 8-2b).

Vitamins

No requirement values for biotin and vitamin K have been established for dogs, although it is believed that dogs have dietary requirements for these vitamins. Some of the need may be met through synthetic activity by microbes inhabiting the digestive tract. The companion application to this text calculates the ration totals for these vitamins but presents no requirement value.

END-OF-CHAPTER QUESTIONS

1. How would dogs fed only meat satisfy their metabolic need for glucose?
2. What percentage of water is found in each of the following dog food types: semimoist, dry, and canned?
3. Describe the options established by the AAFCO for substantiating a claim of nutritional adequacy in dog food.
4. Describe two strategies applied in dog food formulations to encourage weight loss.
5. Name three health problems in dogs caused by obesity.
6. What is a safety factor as it relates to dog nutrition?
7. What term is used to describe the behavior of a dog that is stimulated to eat by the sight of another dog eating?
8. What difference exists between bovine milk and bitch milk that may result in diarrhea in puppies fed bovine milk products?
9. Why is it not usually desirable to feed the growing dog a diet that results in maximal rate of gain?
10. Describe the reasons why the AAFCO did not find the dog NRC (1985) publication acceptable as the definitive source of nutrient requirement information for dogs.

REFERENCES

Association of American Feed Control Officials. (2003). *AACO Official Publication.* West Lafayette, IN.

Blaza, S. E., Booles, D., & Burger, I. H. (1989). Is carbohydrate essential for pregnancy and lactation in dogs? In I. H. Burger and J. P. W. Rivers (Eds.), *Nutrition of the dog and cat* (pp. 229–242). Cambridge, England: Cambridge University Press.

Bovee, K. C. (1991). Influence of dietary protein on renal function in dogs. *Journal of Nutrition. 121,* S128.

Hill, R. C., Burrows, C. F., Ellison, G. W., & Bauer, J. E. (2001). The effect of texturized vegetable protein from soy on nutrient digestibility compared to beef in cannulated dogs. *Journal of Animal Science. 79,* 2167–2171.

Kienzle, E., & Meyer, H. (1989). The effects of carbohydrate-free diets containing different levels of protein on reproduction in the bitch. In I. H. Burger and J. P. W. Rivers (Editors), *Nutrition of the dog and cat* (pp. 243–257). Cambridge, England: Cambridge University Press.

Mugford, R. A., & Thorne, C. (1980). Comparative studies of meal patterns in pet and laboratory housed dogs and cats. In R. S. Anderson (Editor), *Nutrition of the dog and cat* (pp. 3–14). Oxford: Pergamon Press.

National Research Council. (1985). *Nutrient Requirements of Dogs, Revised.* Washington, D.C.: National Academy Press.

National Research Council. (1986). *Nutrient Requirements of Cats, Revised.* Washington, D.C.: National Academy Press

National Research Council. 2001. *Nutrient requirements of dairy cattle.* (7th revised edition.). Washington, DC: National Academy Press.

Payne, P. R. 1965. Assessment of the protein values of diets in relation to the requirements of the growing dog. *Canine and Feline Nutritional Requirements.* O. Graham-Jones, (Editor), London: Pergamon Press.

Romsos, D. R., Palmer, H. J., Muiruri, K. L., & Bennink, M. R. (1981). Influence of a low-carbohydrate diet on performance of pregnant and lactating dogs. *Journal of Nutrition. 111,* 678.

Ullrey, D. E. (1995). Dogs and cats. In W. G. Pond, D. C. Church, & K. R. Pond (Editors.), *Basic animal nutrition and feeding* (pp. 531–545). New York: John Wiley & Sons.

CHAPTER 29

DOG RATION FORMULATION

TERMINOLOGY

Workbook A workbook is the spreadsheet program file including all its worksheets.

Worksheet A worksheet is the same as a spreadsheet. There may be more than one worksheet in a workbook.

Spreadsheet A spreadsheet is the same as a worksheet.

Cell A cell is a location within a worksheet or spreadsheet.

Comment A comment is a note that appears when the mouse pointer moves over the cell. A red triangle in the upper right corner of a cell indicates that it contains a comment. Comments are added to help explain the function and operations of workbooks.

Input box An input box is a programming technique that prompts the workbook user to type information. After typing the information in the input box, the user clicks **OK** or strikes **ENTER** to enter the typed information.

Message box A message box is a programming technique that displays a message. The message box disappears after the user clicks **OK** or strikes **ENTER**.

EXCEL SETTINGS

Security: Click on the **Tools** menu, **Options** command, **Security** tab, **Macro Security** button, **Medium** setting.

Screen resolution: This application was developed for a screen resolution of 1024 × 768. If the screen resolution on your machine needs to be changed, see Microsoft Excel Help, "Change the screen resolution" for instructions.

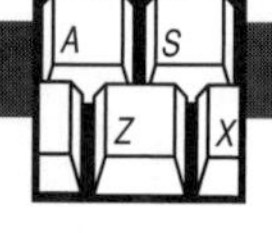

HANDS-ON EXERCISES

DOG RATION FORMULATION

Double click on the **DogRation** icon. The message box in Figure 29–1 displays.

Figure 29–1

Macros may contain viruses. It is advisable to disable macros, but if the macros are legitimate, you may lose some functionality.
Disable Macros or Enable Macros or More Info

Click on **Enable macros**.

The message box in Figure 29–2 displays.

Figure 29–2

Function keys F1 to F8 are set up. You may return to this location from anywhere by striking **ENTER**, then the **F1** key. Workbook by David A. Tisch. The author makes no claim for the accuracy of this application and the user is solely responsible for risk of use. You're good to go. TYPE ONLY IN THE GRAY CELLS! Note: This Workbook is made up of charts and a worksheet. The charts and worksheet are selected by clicking on the tabs at the bottom of the display. Never save the Workbook from a chart; always return to the worksheet before saving the Workbook.

Click **OK**.

Input Dog Procedure

Click on the **Input Dog** button. The input box shown in Figure 29–3 displays.

Figure 29–3

1. Growing dog, start of weaning (3 weeks)
2. Growing dog, end of weaning (6 weeks)
3. Early growth
4. Late growth
5. Adult, maintenance
6. Pregnancy (late)
7. Lactation

ENTER THE APPROPRIATE NUMBER:

Click **OK** to choose the default input of #5, the Adult dog at maintenance. The input box shown in Figure 29–4 displays.

Figure 29–4

Enter measured DMI (pounds) if available or click OK to accept the predicted value at CX26

Click **OK** to accept the predicted dry matter intake (DMI). The message box shown in Figure 29–5 displays.

Enter the appropriate cell inputs in column CX.

Figure 29–5

Click **OK**.

The Cell Inputs

Enter the Cell Inputs shown in Table 29-1.

Table 29–1
Inputs for the dog ration formulation example

50	Dog's weight (lb.) (2–150)
2	Dog's activity level (1–lowest to 6–highest)
10	Dry dog food percent moisture (10–12)
97	Dry dog food density (g/cup) (70–120)
33	Semi-moist dog food percent moisture (25–35)
142	Semi-moist dog food density (g/cup) (120–160)
78	Canned dog food percent moisture (74–78)
170	Canned dog food density (g/cup) (a 6-oz can is 150–180 g net)

The comment behind the cell containing the Dog's activity level label is shown in Figure 29–6.

Use the following guidelines:
1. Low activity (apartment pet or older, inactive dog: high risk of weight gain)
2. Moderate activity (outside run, sleeps inside)
3. Athlete (well-conditioned racing greyhound)
4. High-energy lifestyle (sleeps outside)
5. Herding/hunting dog
6. Working sled dog

Figure 29–6

Strike **ENTER** and **F1**.

Select Feeds Procedure

Click on the **Select Feeds** button. The message box in Figure 29–7 displays.

Nutrient content expressed on a dry matter basis. Feedstuffs are listed first by selection status; then, within the same selection status, by decreasing protein content; then, within the same protein content, alphabetically.

Figure 29–7

Click **OK**.

Explore the table. Note the nutrients listed as column headings. Note also that the table ends at row 200. You select feedstuffs for use in making two different products: (1) a blend to be mixed and sold bagged or bulk, and (2) the ration to be fed directly to the animal.

Select the feedstuffs in Table 29–2 by placing a 1 in the column to the left of the feedstuff name. All unselected feedstuffs should have a 0 value in the column to the left of the feedstuff name. If you wish to group feedstuffs but not select them, you would place a value between 0 and 1 to the left of the feedstuff name.

Table 29–2
Feedstuffs to select for the dog ration formulation example

Meat & bone meal
Soybean meal, 49%
Wheat bran
Whey, dehydrated
Corn grain, ground
Bone meal, steamed
Calcium pantothenate
Choline chloride, 60%
Copper sulfate
Corn oil
EDDI (Ethylenediamine dihydroiodide)
Magnesium oxide
Manganous carbonate
Niacin
Riboflavin
Vitamin A supplement
Vit B_{12} (cyanocobalamin)
Vitamin D supplement
Vitamin E supplement
Zinc oxide

Strike **ENTER** and **F2**.

Selected feedstuffs and their analyses are copied to several locations in the workbook.

Make Ration Procedure

Click on the **Make Ration** button. The message box shown in Figure 29–8 displays.

Figure 29–8

ENTER POUNDS TO FEED IN COLUMN B. TOGGLE BETWEEN NUTRIENT WEIGHTS AND CONCENTRATIONS USING THE **F5** KEY, RATION AND FEED CONTRIBUTIONS USING THE **F6** KEY AND FEEDING DIRECTIONS USING THE **F7** KEY. WHEN DONE STRIKE **F1**. Cell is highlighted in red if nutrient provided is poorly matched with nutrient target. The lower limit is taken as 94 to 98 percent of target, depending on the nutrient. The upper limit of acceptable mineral is taken from the dog NRC, 1985. Where data is not available, the upper limit is usually taken from the swine value found in National Research Council. 1980. *Mineral Tolerance of Domestic Animals.* National Academy Press. For calcium, phosphorus and copper, upper limits of 3 percent, 2 percent and 50 mg/kg are used, respectively. The upper limit of vitamin is taken from the NRC, 1987, *Vitamin Tolerance of Animals.* For other nutrients, the upper limit is based on unreasonable excess and the expense of unnecessary supplementation. See Table 29–4 for specifics.

Click **OK**. The message box shown in Figure 29–9 displays.

Figure 29–9

The Goal Seek feature may be useful in finding the pounds of a specific feedstuff needed to reach a particular nutrient target:

1. Select the red cell highlighting the deficient nutrient
2. From the menu bar, select **Tools**, then **Goal Seek**
3. In the text box, "To Value:" enter the target to the right of the selected cell
4. Click in the text box, "By changing cell:" and then click in the gray "Pounds fed" area for the feedstuff to supply the nutrient
5. Click **OK**. You may accept the value found by clicking **OK** or reject it by clicking **Cancel**.

WARNING: Using Goal Seek to solve the unsolvable (e.g., asking it to make up an iodine shortfall with iron sulfate) may result in damage to the Workbook.

IMPORTANT: If you return to the Feedtable to remove more than one feedstuff from the selected list, you will lose your chosen amounts fed in the developing ration.

Click **OK**.

In the gray area to the right of the feedstuff name, enter the pound values shown in Table 29–3.

Table 29–3
Inclusion rates for feedstuffs in the dog ration formulation example

Meat & bone meal	0.025
Soybean meal, 49%	0.044
Wheat bran	0.09
Whey, dehydrated	0.025
Corn grain, ground	0.36
Bone meal, steamed	0.05
Calcium pantothenate	0.0000042
Choline chloride, 60%	0.00084
Copper sulfate	0.00021
Corn oil	0.03
EDDI (Ethylenediamine dihydroiodide)	0.00000068
Magnesium oxide	0.0011
Manganous carbonate	0.0011
Niacin	0.0000065
Riboflavin	0.0000013
Vitamin A supplement	0.00043
Vitamin B_{12} (cyanocobalamin)	0.000013
Vitamin D supplement	0.000063
Vitamin E supplement	0.00022
Zinc oxide	0.0002

The Nutrients Supplied *Display*

The application highlights ration nutrient levels, expressed as amount supplied per dog per day, that fall outside the acceptable range.

The lower limit for DMI and MEdog is 94 percent of the target. There is no lower limit for CP because the target is met with amino acids. It is assumed that nonessential amino acid requirements will be met when essential amino acid targets are met. The lower limit for all other nutrients is 98 percent of the target. Table 29–4 shows the upper limits of the acceptable range for the various nutrients.

Table 29–4
Upper limits for ration nutrients used in the dog ration application

	Upper Limit	Source
DMI	6% over the predicted requirement	Author
MEdog	20% of requirement	Author
Fat	4 × the predicted requirement	Author
n-6 fatty acids	No upper limit	—
Crude protein	8% greater than requirement[1]	Author
All essential amino acids	No upper limit	—
Calcium	2.05% of diet dry matter	NRC: Nutrient Requirements of Dogs, 1985
Phosphorus, available	1.44% of diet dry matter	NRC: Nutrient Requirements of Dogs, 1985
Sodium	3.1% of diet dry matter	NRC: Mineral Tolerance of Domestic Animals, 1980—swine, based on limit for salt (sodium chloride)
Chloride	4.9% of diet dry matter	NRC: Mineral Tolerance of Domestic Animals, 1980—swine, based on limit for salt (sodium chloride)
Potassium	2% of diet dry matter	NRC: Mineral Tolerance of Domestic Animals, 1980
Magnesium	0.3% of diet dry matter	NRC: Mineral Tolerance of Domestic Animals, 1980
Copper	250 mg/kg of diet dry matter	NRC: Mineral Tolerance of Domestic Animals, 1980—swine
Iodine	400 mg/kg of diet dry matter	NRC: Mineral Tolerance of Domestic Animals, 1980—swine
Iron	3,000 mg/kg of diet dry matter	NRC: Mineral Tolerance of Domestic Animals, 1980—swine
Manganese	400 mg/kg of diet dry matter	NRC: Mineral Tolerance of Domestic Animals, 1980—swine
Selenium	2 mg/kg of diet dry matter	NRC: Mineral Tolerance of Domestic Animals, 1980—swine
Zinc	1,000 mg/kg of diet dry matter	NRC: Mineral Tolerance of Domestic Animals, 1980—swine
Vitamin A	Growing: 300,000 IU/kg body weight Gestating: 125,000 IU/kg body weight	NRC: Nutrient Requirements of Dogs, 1985
Vitamin D	4 × the requirement	NRC: Vitamin Tolerance of Animals, 1987
Vitamin E	20 × the requirement	NRC: Vitamin Tolerance of Animals, 1987
Choline	3 × the requirement	NRC: Vitamin Tolerance of Animals, 1987
Folacin	1,000 × the requirement	Author
Niacinavail	350 mg/kg body weight	NRC: Vitamin Tolerance of Animals, 1987
Pantothenic acid	100 × the requirement	Author
Riboflavin	20 × the requirement	Author
Thiamin B_1	1,000 × the requirement	NRC: Vitamin Tolerance of Animals, 1987
Pyridoxine B_6	50 × the requirement	NRC: Vitamin Tolerance of Animals, 1987
Vitamin B_{12}	300 × the requirement	NRC: Vitamin Tolerance of Animals, 1987

DMI: dry matter intake; IU: international unit.
[1]Upper limit for CP is 8% greater than either the AAFCO recommendation or the total protein computed from the NRC's total required essential and non-essential amino acids, whichever is higher.

Energy, protein, amino acids, minerals, and vitamin targets have been met with minimal excesses. Note that all dog nutrient targets are based on the assumption of 100 percent nutrient bioavailability. Feedstuff mineral bioavailabilities are taken from the dairy NRC (2001), and are shown in Table 29-5.

Table 29–5
Feedstuff mineral bioavailabilities[1]

Mineral	Bioavailability[2]
Calcium (Ca)	0.6
Phosphorus (P)	see note[3] below
Sodium (Na)	0.9
Chloride (Cl)	0.9
Potassium (K)	0.9
Magnesium (Mg)	0.16
Copper (Cu)	0.04
Iodine (I)	0.85
Iron (Fe)	0.1
Manganese (Mn)	0.0075
Selenium (Se)	1.0
Zinc (Zn)	0.15

[1]From dairy NRC (2001) and used in the dog ration formulation Workbook
[2]Value shown is grams bioavailable in each gram consumed
[3]Application uses inputted value of phosphorus bioavailability for each feedstuff

The cost of this ration using initial $/ton values is $0.09 per dog per day.

First limiting amino acid: *isoleucine*

The comment behind the cell containing this label is shown in Figure 29–10.

Figure 29–10

This is the amino acid that exists in the ration at a level that is farthest from the level required, or the amino acid whose requirement is most narrowly met.

Ca:P_{total} *ratio:* *1.07*

The comment behind the cell containing this label is shown in Figure 29–11.

Figure 29–11

When the calcium-to-phosphorous ratio is based on total phosphorus, the optimal ratio is given by the NRC as between 1.2:1 and 1.4:1. The AAFCO recommendation for dogs is to target the ratio at 1:1 with a maximum of 2:1. Excess calcium reduces phosphorus absorption.

Ca:P_{avail} *ratio:* *1.56*

The comment behind the cell containing this label is shown in Figure 29–12.

Figure 29–12

When the calcium-to-phosphorous ratio is based on available phosphorus, the recommendation in swine (no recommendation is available specifically for dogs) is to have 2 to 3 times as much calcium as phosphorus. Excess calcium reduces phosphorus absorption.

The Nutrient Concentration *Display (strike* `F5`*)*

The nutrients in the ration and the predicted nutrient targets are expressed in terms of concentration. That is, the nutrients provided by the ration are divided by the amount of ration dry matter and the nutrient targets are divided by the target amount of ration dry matter. Concentration units include percent, milligrams per kilogram or parts per million, calories per pound, and international units per pound.

Electrolyte balance: *341*

The comment behind the cell containing this label is shown in Figure 29–13.

Figure 29–13

The electrolyte balance is used to assess the impact of the ration's mineral content on the body's efforts to regulate blood pH through urinary excretion. It is calculated as mEq of excess cations: (Na + K − Cl)/kg of diet dry matter. No specific recommendations have been published for dogs. However, a lower electrolyte balance creates a more acidic urine and this is useful in helping to prevent urolithiasis.

Strike **F5**.

The Feedstuff Contributions *Display (strike* `F6`*)*

Shown here are the nutrients contributed by each feedstuff in the ration. This display is useful in trying to troubleshoot problems with nutrient excesses.

Strike **F6**.

Feeding Directions *Display (strike* `F7`*)*

The feeding directions are shown in Table 29–6.

Table 29–6
Feeding directions in the dog ration example[1]

	Inputted Percent Moisture	Inputted Density (g/cup or can)	Weight to Feed (lb. as fed)	Volume to Feed (cups or cans)
Dry	10.0	97.0	0.6	3.0
Semi-moist	33.0	142.0	0.9	2.7
Canned	78.0	170.0	2.6	6.9

[1]The calculations assume a single blended feed is fed to the dry matter requirement. Amounts to feed are expressed in units of weight and volume, varying by food type—dry, semi-moist, and canned.

Strike **F1**.

The Graphic *Display*

At the bottom of the home display are tabs. The current tab selected is the Worksheet tab. Other tabs are graphs based on the current ration.

DMI

Click on this tab to display a graph titled Nutrient Status: Dry Matter Intake

E&F

Click on this tab to display a graph titled Nutrient Status: Energy, Fat & Linoleic Acid

AA

Click on this tab to display a graph titled Nutrient Status: Amino Acids

CP

Click on this tab to display a graph titled Nutrient Status: Crude Protein

Min

Click on this tab to display a graph titled Nutrient Status: Minerals (bioavailable basis)

Vit

Click on this tab to display a graph titled Nutrient Status: Vitamins
Click on the **Worksheet** tab.

Blend Feedstuffs Procedure

Click on the **Blend Feedstuffs** button. The message box shown in Figure 29–14 is shown.

Figure 29–14

YOU MUST HAVE ALREADY SELECTED THE FEEDSTUFFS YOU WANT TO BLEND. When your analysis is acceptable, strike **ENTER** and **F3** to name and file the blend.

Click **OK**.

The amounts to blend are the same as the amounts to feed (Table 29–3). These amounts have been copied from the Make Ration section. Note that to the left of the amounts entered is a column that has converted these amounts to a pounds-per-ton basis. Feed mill mixer capacities are rated in tons so formulas to be mixed should be expressed on a pounds-per-ton basis.

Strike **ENTER** and **F3**. The input box in Figure 29–15 displays.

Figure 29–15

ENTER THE NAME OF THE BLEND (names may not be composed of only numbers):

Name the blend AdultMaint and click **OK**. The message box in Figure 29–16 displays.

Figure 29–16

The new blend has been filed at the bottom of the feed table.

Click **OK**.

View Blends Procedure

Click on the **View Blends** button. The message box in Figure 29–17 displays.

Cursor right to view the blends. Cursor down for more nutrients. DO NOT TYPE IN THE BLUE AREAS.

Figure 29–17

Click **OK**. Confirm that the AdultMaint formula and analysis have been filed.
Strike **F1**.

Using the Blended Feed in the Balanced Ration

Select Feeds *Procedure*

Click on the **Select Feeds** button. Unselect all feedstuffs that were in the AdultMaint by entering a 0 to the left of the feedstuff name. Select the AdultMaint at the bottom of the feed table (row 200) (Table 29–7).

AdultMaint

Table 29–7
Feedstuff to select for the dog ration formulation example

Strike **ENTER** and **F2**.

Make Ration *Procedure*

Click on the **Make Ration** button. Enter the ration shown in Table 29–8.

AdultMaint	0.63

Table 29–8
Feeding rate for the feedstuff in the dog ration formulation example

The amount of the AdultMaint to feed is the total amount of its component ingredients in the balanced ration. That value is:

$$0.025 + 0.044 + 0.09 + 0.025 + 0.36 + 0.05 + 0.0000042 + 0.00084 + 0.00021 + 0.03 + 0.00000068 + 0.0011 + 0.0011 + 0.0000065 + 0.0000013 + 0.00043 + 0.000013 + 0.000063 + 0.00022 + 0.0002 = 0.63.$$

This value is recorded at View Blends under Formula, as entered. The ration is balanced as it was when the components of AdultMaint were fed unmixed.

Strike **ENTER** and **F1**.

Print Ration or Blend Procedure

Make sure your name is entered at cell C1. Click on the **Print Ration or Blend** button. The input box in Figure 29–18 displays.

Are you printing a dog ration evaluation or a blend formula and analysis? (1-RATION, 2-BLEND):

Figure 29–18

Click **OK** to accept the default value of 1, a ration. A two-page printout will be produced by the machine's default printer.

Click on the **Print Ration or Blend** button. Type the number 2 to print a blend. Click **OK**. The message box in Figure 29–19 displays.

Figure 29–19

Click on the green number above the blend you want to print and press **F4**. Scroll right to see additional blends.

Click **OK**.

Find the AdultMaint blend, click on the green number above it, and strike **F4**. A one-page printout will be produced by the machine's default printer.

ACTIVITIES & WHAT-IFS

In the Forms folder on the companion CD to this text is a DogInput.doc file that may be used to collect the necessary inputs for use of the DogRation.xls file. This form may be printed out and used during kennel visits to assist in ration evaluation activities.

1. Remake the ration described in Table 29–7 for the dog described in this chapter. Strike **F7**. How many cups of this dog-food formulation would need to be fed to meet this dog's nutrient requirements if the formulation was fed as dry dog food? How many cups if fed as semi-moist dog food? How many cans if canned dog food? Discuss the advantages and disadvantages of feeding dog-food formulations in the different forms.
2. Remake the ration described in Table 29–7 for the dog described in this chapter. Input an adult dog at a low level of activity. Use the formulation made in this chapter for a dog at moderate activity for this dog at low activity. Citing your results, discuss the consequences and the likely impact on the low-activity dog.
3. Remake the ration described in Table 29–3 for the dog described in this chapter. Remove the meat and bone meal from this formulation and note the deficiencies that appear. Discuss the feasibility of formulating nutritionally complete dog foods without the inclusion of animal products.

REFERENCES

National Research Council. (1980). *Mineral Tolerance of Domestic Animals.* Washington DC: National Academy Press.

National Research Council. (1985). *Nutrient Requirements of Dogs,* revised. Washington DC: National Academy Press.

National Research Council. (1987). *Vitamin Tolerance of Animals.* Washington DC: National Academy Press.

National Research Council. (2001). *Nutrient Requirements of Dairy Cattle,* 7th revised edition. Washington DC: National Academy Press.

CHAPTER 30

FEEDING CATS

Dogs have masters; cats have staff.

ANONYMOUS

The nutritional phases in the cat production cycle, illustrated in Figure 30–1, include:

Queen: gestation, lactation, maintenance/anestrous
Kitten: suckling/creep, growth

A sample growth rate is plotted in Figure 30–2.

THE QUEEN

Gestation

Pregnancy normally lasts about 65 days. As with all domestic animals, the early portion of the pregnancy does not significantly increase the need for nutrients beyond a maintenance level. However, both the NRC (1986) and AAFCO (2003) recommend a higher level of nutrition throughout pregnancy. During pregnancy, the queen should be fed a balanced ration offered free choice. At about 2 days before parturition, food intake will decline.

Lactation

During lactation, the queen should be fed a balanced diet, offered free choice. Free choice water should also be provided. The lactating queen will usually lose body weight even when offered a palatable and nutritionally balanced diet free choice.

Maintenance/Anestrous

Queens that are in their anestrous period should be fed rations formulated to achieve, and then maintain, the desired body condition.

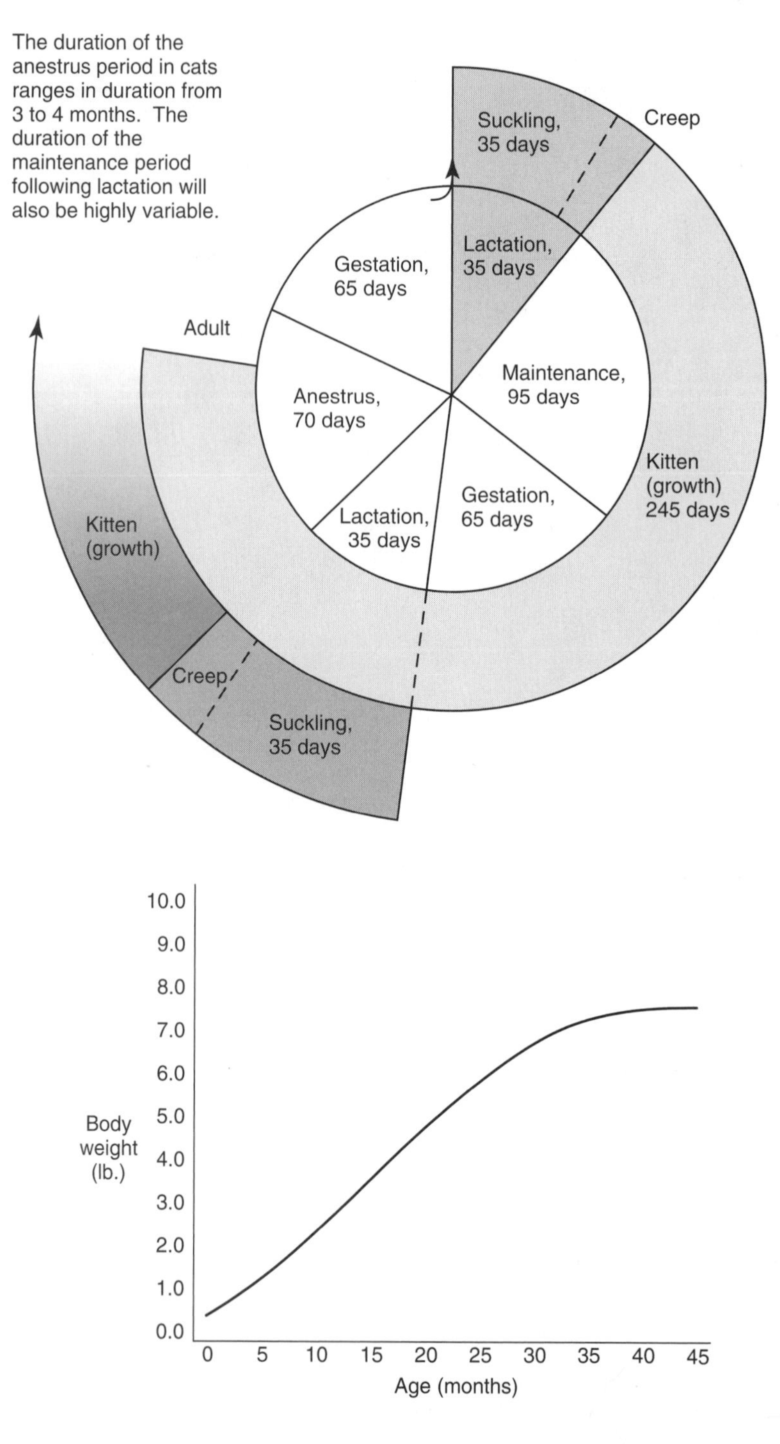

Figure 30–1
Nutritional Phases in the Production Cycle: Cats. Move clockwise starting at top of innermost circle. Continue through next cycle or move to next shell as indicated.

Figure 30–2
Growth curve for average short-haired, domestic kittens. Male cats will weigh approximately 15% higher than the average and females will weigh approximately 15% lower than the average. Adapted from Loveridge, 1987.

THE KITTEN

Suckling/Creep Feeding

Like the bitch, the structure of the queen's placenta is described as endotheliolchorial. This type of placenta does not allow sufficient antibody immunoglobulins to pass from the queen into the blood of the fetal kittens. The neonatal kitten acquires these antibody immunoglobulins through the queen's first milk or colostrum. Consumption of colostrum is important for the health of the newborn kitten. It is critical that kittens suckle as soon as possible after birth to acquire the antibodies present in colostrum. Colostrum is discussed in Chapter 20.

At about 3 weeks of age, formulated kitten food may be offered to the kittens. Weaning will normally begin to occur by 5 weeks of age. When kittens are

1. The cat has a much higher protein requirement than the dog.
2. Arginine is used in the urea cycle during the conversion of excess protein to urea. Because of the need to process excess protein, the cat will die within hours after consuming an arginine-free diet. Diets formulated with usual ingredients will contain an adequate amount of arginine.
3. Cats require taurine in the diet but dogs do not.
4. Cats cannot convert the fatty acid linoleic acid to arachidonic acid as can the dog. Therefore, the cat must consume preformed arachidonic acid.
5. Cats cannot convert beta carotene into vitamin A as can the dog. Therefore, cats must consume preformed vitamin A which is present only in animal tissues.
6. Cats cannot convert the amino acid tryptophan to the B vitamin niacin as can dogs. Therefore, dietary niacin is the only source of this B vitamin for the cat.

Figure 30–3
The major differences in nutritional requirements of cats and dogs

eating food and drinking water, the queen's intake of food should be gradually restricted and milk production will decrease. After milk production has ceased, food for the queen should be gradually adjusted to attain proper body condition and then adjusted to maintain that body condition.

Weaned Kittens

After the kitten is weaned, it is best to allow it free access to food. The digestive tract of the kitten is small and has limited storage capacity. A small kitten will be unable to consume the nutrients it requires to support growth if it has restricted access to food. If limit feeding is necessary due to overeating, feed at least four meals until 12 weeks of age, three meals until 24 weeks of age, and two meals daily thereafter. If the kitten vomits following a meal or its stomach feels taut, meal size may have to be reduced and feeding frequency increased. No matter what the feeding system, it is important that fresh food be added to the food bowl only after it has been cleaned of uneaten food.

ADULT

Growth in cats levels out at 40 to 50 weeks of age. At this point, they may be fed a maintenance diet unless they are used in breeding. Especially active cats require additional energy, and it is usually acceptable to provide this by simply feeding more of the maintenance diet rather than reformulating the ration to increase energy density. Free-choice feeding is best if the cat does not overeat because cats have been observed to eat 10 to 17 meals daily (Morris & Rogers, 1989). If overeating is a problem, the ration should be split into at least two meals daily.

CAT FEEDING AND NUTRITION ISSUES

Differences between Cats and Dogs

Dogs and cats are very different in their nutritional requirements, and must be fed appropriately formulated diets. Figure 30–3 lists the major differences in the nutritional requirements of cats and dogs.

Energy

The energy values of many feedstuffs that are commonly fed to cats have not been experimentally determined. For this reason, the energy values determined for other species are routinely used in formulating rations for cats. This situation, in addition to the fact that the same variables that affect the energy requirements of other species (age, activity, body condition, the thermal environment, and environmental stressors) also affect cats, point out the necessity to monitor cat body condition and make dietary adjustments as necessary.

Metabolizable energy (ME) is used in cat nutrition in discussions of feedstuff energy value and the cat's energy requirement. In the cat NRC (1978), the energy value of feedstuffs fed to cats was determined using the following assumptions of ME value: protein 4 kcal/g, carbohydrate 4 kcal/g, fat 9 kcal/g. This method of calculating the energy value of feedstuffs has been shown to overestimate the true energy value by 20 to 30 percent (Kendall, Smith, & Holme, 1982; Kendall, Burger, & Smith, 1985).

In the cat NRC (1986), the energy value of feedstuffs fed to cats has been taken from contemporary values determined for swine. The swine values are in vivo values acquired from metabolism studies. Where cat in vivo ME values are available, a comparison with swine values shows reasonable agreement according to the cat NRC (1986).

Among the various breeds of domestic cats, mature body weights range from about 4 lb. to 14 lb. The authors of the cat NRC (1986) have suggested that this relatively narrow range makes it unnecessary to scale energy requirements to exponents of body weight. Energy requirements are, therefore, expressed per unit of body weight (Chapter 5). The energy requirements for growth, expressed per unit of body weight, decline rapidly with age and level off at 40 to 50 weeks.

Fat

Dietary fat is not only a source of energy, but also is a source of essential fatty acids and the fat soluble vitamins. Fat type and content will affect a ration's shelf life and palatability.

As with many other domestic animals, cats have a dietary requirement for the n-6 fatty acid linoleic acid, 18:2(n-6). However, cats may have additional fatty acid requirements. The enzyme used to convert one type of n-6 fatty acid to another (delta-6-desaturase) may not be present in sufficient quantity in the cat under certain physiological conditions, so cats may have a dietary requirement for arachidonic acid, 20:4(n-6) (Pawlosky & Salem, 1996). Arachidonic acid is found only in animal fats. The fact that the delta-6-desaturase enzyme is also involved in metabolism of n-3 fatty acids suggests that the cat's requirement for fatty acids may need further study. Some researchers have suggested that maintenance of the ratio of n-6 to n-3 fatty acids between 5:1 and 10:1 may provide health benefits for cats (Schick, Schick, & Reinhart, 1996; Wander, Hall, Gradin, Du, Jewell, 1997).

Among all domestic animals, cats have the greatest ability to tolerate diets of high fat density. In one study, cats were successfully maintained on a diet containing 64 percent fat (Humphreys & Scott, 1962). However, in a study where yellow grease was used as the fat source, cats preferred diets containing 25 percent fat over those containing either 10 percent or 50 percent (Kane, Morris, & Rogers, 1981). It is likely that different fat types affect diet palatability differently. Cats showed a preference for diets made from tallow over those made from either chicken fat or butter (Kane, Morris, & Rogers, 1981). In the same study, cats showed no preference between diets using tallow and partially hydrogenated vegetable fat as the fat source.

Carbohydrate

Although cats do not have a dietary requirement for carbohydrate, they do have a metabolic requirement for the carbohydrate glucose. Because both feral cats and domestic cats consuming usual cat diets ingest little carbohydrate, these animals are in a similar position to ruminants in that they must manufacture the glucose they require. The metabolic requirement is met through gluconeogenesis occurring in the liver and kidneys. Gluconeogenesis involves compounds called glucogenic precursors that can be used to synthesize glucose. The high

Table 30–1
Apparent starch digestibilities (percent) of starch added at 35 percent of dry matter in cat diets

Coarsely ground corn	79	Coarsely ground wheat	92
Finely ground corn	94	Finely ground wheat	97
Coarsely ground and cooked corn	88	Coarsely ground and cooked wheat	96

From Pencovic & Morris (1975).

protein and fat densities of usual cat diets ensure that cats consuming these diets have adequate glucogenic precursors (glucogenic amino acids and glycerol, respectively). Interestingly, studies have shown that the starch from ground corn and wheat is well utilized by the cat (Table 30–1) and so most commercial cat food formulations include these sources of carbohydrate.

Fiber

None of the domestic animals discussed in this text are capable of manufacturing the enzymes that digest fiber. However, cats, like all animals, harbor microorganisms in their digestive tracts that do make these enzymes. Dietary fiber has a generally positive effect on these microorganisms and on the digestive processes of their host. The bacteria in the lower small intestine and colon of the cat ferment fiber to volatile fatty acids (acetic, propionic, and butyric acid). These acids are utilized by the epithelial cells of the digestive tract as an energy source. The presence of these acids in the intestinal lumen increases blood flow (Howard, Kerley, Mann, Sunvold, & Reinhart, 1997). Blood brings with it recycled urea that becomes available to the bacteria. The bacteria hydrolyze this urea and incorporate its nitrogen into their own cells before being passed in the feces. Feeding fermentable fiber to cats results in less of the excess nitrogen circulating as urea being excreted through the renal system, and more being excreted through the digestive system (Takahashi, Benno, & Mitsuoka, 1980). This situation may be helpful in managing certain renal diseases in cats.

Dietary fiber may also be helpful in preventing and managing other disorders in cats including diarrhea, obesity, and diabetes. Although it is generally agreed that the inclusion of small amounts of fiber in cat diets is helpful in maintaining normal function of the digestive tract, more research on the role of fiber in the cat is needed before specific recommendations as to dietary levels can be made. High levels of dietary fiber, as might occur in formulations designed to promote weight loss in obese cats, may result in an increased requirement for amino acids. The increased requirement would be due to an increased need to replace intestinal epithelium because the elimination of fiber promotes sloughing of intestinal epithelium. The companion application to this text calculates neutral detergent fiber (NDF) content as the ration for the cat is developed.

Protein

There are several reasons for the high protein requirement in cats.

1. Amino acids are catabolized (broken down) at a higher rate in the cat's liver than in the liver of other domestic animals. Even when fed a low-protein diet, amino acids will be catabolized at a high rate. To meet the cat's requirement for intact essential amino acids, therefore, it will be necessary to supply more than what is anticipated being deposited in the tissues or otherwise used for metabolism.
2. Cats produce a unique product called *felinine* that is made from the sulfur-containing amino acids. Although the function of felinine is not clear, it is excreted in the urine of all cats and may function in territorial marking.
3. Cats use the sulfur-containing amino acids to make the enzymes necessary for phospholipid synthesis. Cats have a large need for phospholipids because these molecules are used in absorption and metabolism of dietary fat. Of all domestic animal diets, the diet of the cat is usually highest in fat content.

4. *Taurine* is an amino acid that is found only in animal tissues and it is especially rich in shellfish. Given the cat's predatory nature, its natural diet would be rich in taurine. For this reason, the evolutionary history of the cat does not involve the establishment of a metabolic pathway for taurine synthesis. Among domestic animals, cats are unique in their requirement for this amino acid.

Taurine

Taurine, like methionine and cystine, is a sulfur-containing amino acid. All domestic animals have a metabolic requirement for taurine: it is used in bile-acid conjugation, retinal and heart muscle function, and at least in the cat, reproductive function. Except for the cat, all domestic animals are able to synthesize all the taurine needed given an adequate supply of methionine and cystine. Taurine is not found in plant tissues so the taurine requirement for cats must be met with high levels of animal products or taurine supplementation. The nature of the canning process results in some taurine destruction, so the AAFCO cat requirements for taurine are elevated if the diet is to be canned as compared to dry or semimoist diets. This adjustment is also made in the companion application to this text.

Vitamins

Unlike other domestic animals, cats have a requirement for preformed vitamin A. Whereas other domestic animals have the enzymes in the intestinal mucosa to convert beta carotene to vitamin A when needed, cats lack these enzymes. The richest sources of preformed vitamin A are animal products such as shellfish, kidney, liver, and most dairy products. Given the predatory nature of the cat, its natural diet would be rich in preformed vitamin A. For this reason, the evolutionary history of the cat did not involve the establishment of an ability to convert beta carotene to vitamin A.

Unlike beta carotene, for which the conversion to vitamin A is dependent on the metabolic need for vitamin A, preformed vitamin A consumed in excess may result in toxicity. Vitamin A toxicity in cats affects bone metabolism and may lead to deforming cervical spondylosis, which results in malformations of the cervical (neck) and thoracic (rib area) vertebrae.

The situation with niacin appears to be similar to that of arachidonic acid, taurine, and vitamin A with regard to the cat's dietary requirement. Animal tissues tend to be rich in niacin, so the natural diet of the cat would contain more than enough niacin to meet its need. For this reason, cats have not established the metabolic pathway, present in other domestic animals, to be able to convert the amino acid tryptophan to niacin.

Feline Lower Urinary Tract Disease

Feline lower urinary tract disease (FLUTD) is a syndrome that involves the development of uroliths or stones in the urinary tract of the cat. It is the same syndrome that was formerly identified as feline urological syndrome (FUS). Male and female cats are at equal risk.

The pH of the cat's urine may be a factor contributing to the high incidence of urolithiasis in cats. The cat's natural carnivorous diet results in a high intake of protein, much higher than the need for amino acids to build cat proteins. The cat's metabolism is, therefore, designed to degrade and dispose of a large portion of ingested proteins. The degraded proteins lead to the production of acids that must be eliminated by the urinary tract. The cat is naturally adapted to producing acid urine.

Rations comprised primarily of cereal grains contain large amounts of base-forming cations such as potassium salts that will result in more alkaline urine. This appears to be a contributing factor to the formation of uroliths in the cat. Excess magnesium, ammonium, and phosphate ions that have been filtered out of the blood at the kidney are more likely to crystallize in an alkaline filtrate than in an acid one. Historically, the most common stones or uroliths seen in cats were composed of magnesium ammonium phosphate (struvite).

Most commercial cat foods now include a source of anions that act as acidifying agents to help prevent the formation of struvite uroliths. Care must be exercised in acidifying the diet because diets containing excessive anions force the parathyroid hormone to mobilize available cations. The primary cation available is calcium in bone. A prolonged period of excessive anion intake in the cat may result in excessive decalcification and could result in weakened bones. Ironically, acidifying the diet may also create conditions that promote the precipitation of calcium oxalate crystals and the formation of uroliths from calcium oxalate.

Cat diets should be formulated to keep urine pH within a range that supports good urinary tract health. Tetrick (2004) suggests a target range of pH 6.0 to 6.4. The companion application to this text calculates the electrolyte balance in formulated rations as milliequivalents of excess cations: (Na + K − Cl). Although no specific recommendations have been suggested with regard to the electrolyte balance in cat diets, lower values are associated with lower-pH urine and higher values are associated with higher-pH urine. Thus, the electrolyte balance could be useful as a tool in preventing urolithiasis and FLUTD.

The formation of mineral crystals in the urinary tract is normal. Problem uroliths occur when these crystals are not voided in the urine when they are small. Water consumption and frequent urination are, therefore, more important than dietary manipulation with regard to prevention of FLUTD in cats.

Flavors

Flavors are often used in the pet food industry to enhance food palatability. Digest, a by-product of the meat processing industry, is often used in cat food formulations to enhance palatability. Sucrose, however, does not enhance the palatability of cat diets (Pfaffmann, 1955; Carpenter, 1956; Zotterman, 1956). Additional information on feed flavors is found in Chapter 3.

Feed Form

As described in Chapter 28, there are three basic forms of foods marketed for cats and dogs: dry, semimoist, and canned. Each may provide adequate nutrition but dry cat food is usually less expensive and may contribute to maintenance of oral health.

Requirements

As with the dog NRC (1985), the cat NRC (1986) gives nutrient requirements as determined in experiments feeding purified diets. The consequences of this approach are discussed in Chapter 28. With few exceptions, the cat ration application accompanying this text uses the cat nutrient requirements as published by the AAFCO in their *Official Publication* (2003). The exceptions include DMI, ME, crude protein, and myoinositol. For these exceptions, requirement values are taken from the cat NRC (1986). It should be noted that the cat NRC does not suggest that cats have a requirement for myoinositol, but it does give an dietary inclusion rate for myoinositol that has been used in nutrition studies. The AAFCO requirement values include overages or safety factors necessary to account for the lower digestibility of some products in commercial cat foods, and expected nutrient losses during processing and storage. A comparison of the requirement data offered by these two sources is found in Table 30–2.

Table 30–2
Nutrient requirements for cats—a comparison between the AAFCO, 2003 and NRC, 1986[1]

Nutrient	AAFCO	NRC
Crude protein	30	24
Arginine	1.25	1
Histidine	0.31	0.3
Isoleucine	0.52	0.5
Leucine	1.25	1.2
Lysine	1.2	0.8
Methionine-cystine	1.1	0.75
Phenylalanine-tyrosine	0.88	0.85
Taurine (extruded)	0.1	0.04
Taurine (canned)	0.2	0.04
Threonine	0.73	0.7
Trytophan	0.25	0.15
Valine	0.62	0.6
Crude fat	9	—
Linoleic acid	0.5	0.5
Arachidonic acid	0.02	0.02
Calcium	1	0.8
Phosphorus	0.8	0.6
Potassium	0.6	0.4
Sodium	0.2	0.05
Chloride	0.3	0.19
Magnesium	0.08	0.04
Iron (mg/kg)	80	80
Copper, extruded (mg/kg)	15	5
Copper, canned (mg/kg)	5	5
Manganese (mg/kg)	7.5	5
Zinc (mg/kg)	75	50
Iodine, mg/kg	0.35	0.35
Selenium (mg/kg)	0.1	0.1
Vitamin A	9,000	3,333
Vitamin D	750	500
Vitamin E	30	30
Vitamin K (mg/kg)	0.1	0.1
Thiamine (mg/kg)	5	5
Riboflavin (mg/kg)	5	4
Pantothenic acid (mg/kg)	5	5
Niacin (mg/kg)	60	40
Pyridoxine (mg/kg)	4	4
Folic acid (mg/kg)	0.8	0.8
Biotin (mg/kg)	0.07	0.07
Vitamin B_{12} (mg/kg)	0.02	0.02
Choline (mg/kg)	2,400	2,400

[1]Values are for growing kittens, dry matter basis

END-OF-CHAPTER QUESTIONS

1. Describe the urea cycle as it relates to the cat's dietary requirement for arginine.
2. What is taurine? How does food type (dry, semimoist, or canned) affect the amount of taurine fortification recommended by the AAFCO?
3. Given that the cat is a carnivore, how does it meets its metabolic need for glucose?
4. Give two examples of omega-6 fatty acids required by cats.
5. Name two types of uroliths that cause problems in cat urinary tracts. How does urinary pH affect formation of each type? What impact does a reduced dietary electrolyte balance have on urinary pH?
6. Give three reasons to expect that cats would have higher dietary protein requirements than other domestic animals.
7. How does the fact that the cat is a carnivore explain how it is different from other domestic animals in terms of dietary requirement for preformed vitamin A, niacin, and arachidonic acid?
8. How does the cat benefit from dietary fiber?
9. Describe deforming cervical spondylosis in cats. What causes deforming cervical spondylosis?
10. Which source recommends higher levels of nutrients for growing kittens: AAFCO (2003) or NRC (1986)?

REFERENCES

Association of American Feed Control Officials. (2003). Official Publication. West Lafayette, IN

Carpenter, J. A. (1956). Species differences in taste preferences. Journal of Comparative and Physiological Psychology. 49, 139.

Howard, M. D., Kerley, M. S., Mann, F. A., Sunvold, G. D., & Reinhart, G. A. (1997). Dietary fiber sources alter colonic blood flow and epithelial cell proliferation of dogs. Journal of Animal Science. 75(suppl. 1), 170.

Humphreys, E. R., & Scott, P. P. (1962). The addition of herring and vegetable oils to the diet of cats. Proceedings of the Nutrition Society. 21, xviii.

Kane, E. J., Morris, J. G., & Rogers, Q. R. (1981). Acceptability and digestibility by adult cats of diets made with various sources and levels of fat. Journal of Animal Science. 53, 1516.

Kendall, P. T., Burger, I. H., & Smith, P. M. (1985). Methods of metabolizable energy estimation in cat foods. Feline Practice. 15, 38.

Kendall, P. T, Smith, P. M., & Holme, D. W. (1982). Factors affecting digestibility and in vivo energy content of cat foods. Journal of Small Animal Practice. 25, 538.

Loveridge, G. G. (1987). Some factors affecting kitten growth. Journal of the Institute of Animal Technology. 38, 9–18.

Morris, J. G., & Rogers, G. R. (1989). Comparative aspects of nutrition and metabolism of dogs and cats. In I. H. Burger & J. P. W. Rivers (Editors.), Nutrition of the dog and cat (pp. 35–66). Cambridge, UK: Cambridge University Press.

National Research Council. (1978). Nutrient requirements of cats. Washington, DC: National Academy of Sciences.

National Research Council. (1985). Nutrient requirements of dogs. Washington, DC: National Academy of Sciences.

National Research Council. (1986). Nutrient requirements of cats. Washington, DC: National Academy of Sciences.

Pawlosky, R. J., & Salem, N. (1996). Is dietary arachidonic acid necessary for feline reproduction? Journal of Nutrition. 126, 1081S–1085S

Pencovic, T. A. & Morris, J. G. (1975). Corn and wheat starch utilization by the cat. Journal of Animal Science. 41, 325. (Abstr. 318).

Pfaffmann, C. (1955). Gustatory nerve impulses in rat, cat, and rabbit. Journal of Neurophysiology. 18, 429.

Schick, M. P., Schick, R. P., & Reinhart, G. A. (1996). The role of polyunsaturated fatty acids in the canine epidermis: Normal structural and functional components, inflammatory disease state components, and as therapeutic dietary components. In Recent advances in canine and feline nutritional research. Proceedings IAMS International Nutrition Sympposium. Wilmington, Ohio: Orange Frazer Press.

Takahashi, M., Benno, Y. Mitsuoka, T. (1980). Utilization of ammonia nitrogen by intestinal bacteria isolated from pigs. Applied and Environmental Microbiology 39, 30–35.

Tetrick, M. A. (2004). The role of diet in managing feline lower urinary tract diseases. Dayton, OH: The Iams Company.

Wander, R. C., Hall, J. A., Gradin, J. L., Du, S- H., & Jewel, D. E. The ratio of dietary (n-6) to (n-3) fatty acids influences immune system function, eicosanoid metabolism, lipid peroxidation and vitamin E status in aged dogs. Journal of Nutrition 127, 1198–1205.

Zotterman, Y. (1956). Species differences in water taste. Acta Physiologica Scandinavica. 37, 60.

CHAPTER 31

CAT RATION FORMULATION

TERMINOLOGY

Workbook A workbook is the spreadsheet program file including all its worksheets.

Worksheet A worksheet is the same as a spreadsheet. There may be more than one worksheet in a workbook.

Spreadsheet A spreadsheet is the same as a worksheet.

Cell A cell is a location within a worksheet or spreadsheet.

Comment A comment is a note that appears when the mouse pointer moves over the cell. A red triangle in the upper right corner of a cell indicates that it contains a comment. Comments are added to help explain the function and operations of workbooks.

Input box An input box is a programming technique that prompts the workbook user to type information. After typing the information in the input box, the user clicks **OK** or strikes **ENTER** to enter the typed information.

Message box A message box is a programming technique that displays a message. The message box disappears after the user clicks **OK** or strikes **ENTER**.

EXCEL SETTINGS

Security: Click on the **Tools** menu, **Options** command, **Security** tab, **Macro Security** button, **Medium** setting.

Screen resolution: This application was developed for a screen resolution of 1024 × 768. If the screen resolution on your machine needs to be changed, see Microsoft Excel Help, "Change the screen resolution" for instructions.

HANDS-ON EXERCISES

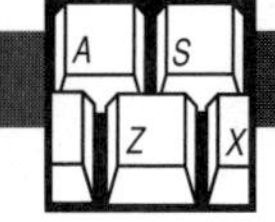

CAT RATION FORMULATION

Double click on the **CatRation** icon. The message box in Figure 31–1 displays.

> Macros may contain viruses. It is advisable to disable macros, but if the macros are legitimate, you may lose some functionality.
> Disable Macros or Enable Macros or More Info

Figure 31–1

Click on **Enable macros**. The message box in Figure 31–2 displays.

> Function keys F1 to F8 are set up. You may return to this location from anywhere by striking **ENTER,** then the **F1** key. Workbook by David A. Tisch. The author makes no claim for the accuracy of this application and the user is solely responsible for risk of use. You're good to go. TYPE ONLY IN THE GRAY CELLS! Note: This Workbook is made up of charts and a worksheet. The charts and worksheet are selected by clicking on the tabs at the bottom of the display. Never save the Workbook from a chart; always return to the worksheet before saving the Workbook.

Figure 31–2

Click **OK**.

Input Cat Procedure

Click the **Input Cat** button. The input box shown in Figure 31–3 displays.

> 1. Adult cat
> 2. Kitten
>
> ENTER THE APPROPRIATE NUMBER:

Figure 31–3

Enter the value 2 to select a kitten and click **OK**. The message box shown in Figure 31–4 displays.

> Enter measured DMI (pounds) if available or click OK to accept the predicted value at CY16:

Figure 31–4

Click **OK** to accept the predicted dry matter intake (DMI). The message box in Figure 31–5 displays.

> Enter the appropriate Cell Inputs in column CX.

Figure 31–5

Click **OK**.

The Cell Inputs

Enter the cell inputs shown in Table 31-1.

Table 31–1
Inputs for the kitten ration formulation example

	N/A
	N/A
	N/A
3.5	Kitten's weight in pounds (2–9)
10	Kitten's age in weeks (10–40)
3	Food type (1-dry, 2-semimoist, 3-canned)

The comment behind the cell containing the Kitten's weight label is shown in Figure 31–6.

Figure 31–6
Body weight (pounds)

Age (wks)	Male	Female
10	2.4	2.0
20	5.5	4.2
30	7.7	6.0
40	8.8	6.6

The comment behind the cell containing the Food type label is shown in Figure 31–7.

Figure 31–7

The food type is important, not so much because of the water content, but because energy content is assumed to vary as follows:

Food Type	Water Content (%)	ME, kcal/lb., Dry Basis
Dry	90	1,610
Semimoist	70	1,946
Canned	25	1,996

This application uses assumed energy density to predict DMI. Nutrient requirements are based on an assumed energy density of 1,814 kcal ME/lb., dry basis.

Strike **ENTER** and **F1**.

Select Feeds Procedure

Click on the **Select Feeds** button. The message box in Figure 31–8 displays.

Nutrient content expressed on a dry matter basis. Feedstuffs are listed first by selection status, then, within the same selection status, by decreasing protein content, then, within the same protein content, alphabetically.

Figure 31–8

Click **OK**.

Explore the table. Note the nutrients listed as column headings. Note also that the table ends at row 200. You select feedstuffs for use in making two different products: (1) a blend to be mixed and sold bagged or bulk, and (2) a ration to be fed directly to the animal.

Select the feedstuffs in Table 31–2 by placing a 1 in the column to the left of the feedstuff name. All unselected feedstuffs should have a 0 value in the column to the left of the feedstuff name. If you wish to group feedstuffs but not select them, you would place a value between 0 and 1 to the left of the feedstuff name.

Table 31–2
Feedstuffs and blend to select for the kitten ration formulation example

Feedstuff
Fish meal, menhaden
Cattle livers, fresh
Taurine
Meat meal
Methionine, DL
Crab meal
Corn grain, ground
Choline chloride, 60%
Cod liver oil
EDDI (Ethylenediamine dihydroiodide)
Folate (folacin, folic acid)
Inositol
Niacin
Potassium chloride
Tallow
Thiamin hydrochloride
Vitamin E supplement
Vitamin K supplement (MSB premix)
Zinc sulfate (mono-)

Strike **ENTER** and **F2**.

Selected feedstuffs and their analyses are copied to several locations in the Workbook.

Make Ration Procedure

Click on the **Make Ration** button. The message box shown in Figure 31–9 displays.

Figure 31–9

ENTER POUNDS TO FEED IN COLUMN B. TOGGLE BETWEEN NUTRIENT WEIGHTS AND CONCENTRATIONS USING THE **F5** KEY, RATION AND FEED CONTRIBUTIONS USING THE **F6** KEY. WHEN DONE STRIKE **F1**. Cell is highlighted in red if nutrient provided is poorly matched with nutrient target. The lower limit is taken as 96 to 100 percent of target, depending on the nutrient. The source of upper limits of acceptable mineral is the National Research Council, 1986, *Nutrient Requirements of Cats*; the NRC, 1987, *Vitamin Tolerance of Animals*; and the author, based on probable adverse effects on health. For other nutrients, the upper limit is based on unreasonable excess and the expense of unnecessary supplementation. See Table 31–4 for specifics.

Click **OK**.

The message box in Figure 31–10 displays.

Figure 31–10

The Goal Seek feature may be useful in finding the pounds of a specific feedstuff needed to reach a particular nutrient target:
1. Select the red cell highlighting the deficient nutrient
2. From the menu bar, select **Tools**, then **Goal Seek**
3. In the text box, "To Value:" enter the target to the right of the selected cell
4. Click in the text box, "By changing cell:" and then click in the gray "Pounds fed" area for the feedstuff to supply the nutrient
5. Click **OK**. You may accept the value found by clicking **OK** or reject it by clicking **Cancel**.

WARNING: Using Goal Seek to solve the unsolvable (e.g., asking it to make up an iodine shortfall with iron sulfate) may result in damage to the Workbook.
IMPORTANT: If you return to the Feedtable to remove more than one feedstuff from the selected list, you will lose your chosen amounts fed in the developing ration.

Click **OK**.

In the gray area to the right of the feedstuff name, enter the pound values shown in Table 31–3.

Fish meal, menhaden	0.03
Cattle livers, fresh	0.031
Taurine	0.0005
Meat meal	0.03
Methionine, DL	0.00072
Crab meal	0.03
Corn grain, ground	0.05
Choline chloride, 60%	0.0006
Cod liver oil	0.003
EDDI (Ethylenediamine dihydroiodide)	0.00000005
Folate (folacin, folic acid)	0.00000027
Inositol	0.0003
Niacin	0.000004
Potassium chloride	0.00029
Tallow	0.06
Thiamin hydrochloride	0.00015
Vitamin E supplement	0.000073
Vitamin K supplement (MSB premix)	0.0000012
Zinc sulfate (mono-)	0.0000098

Table 31–3
Inclusion rates for feedstuffs and blend in the kitten ration formulation

The Nutrients Supplied *Display*

The application highlights ration nutrient levels, expressed as amount supplied per feline per day, that fall outside the acceptable range.

The lower limit for DMI, metabolizable energy (ME), and crude protein is 96 percent of the target. The lower limit for the amino acids and essential fatty acids is 100 percent of the target. The lower limit for minerals and vitamins is 96 percent of the target. Table 31–4 shows the upper limits of the acceptable range for the various nutrients.

Table 31–4
Upper limits for ration nutrients used in the feline ration application

	Upper Limit	Source
DMI	4% over the predicted requirement	Author
Crude protein	50% of DMI	Author
Arginine	No upper limit	—
Histidine	No upper limit	—
Isoleucine	No upper limit	—
Leucine	No upper limit	—
Lysine	No upper limit	—
Methionine + Cystine	No upper limit	—
Phenylalanine + Tyrosine	No upper limit	—
Threonine	No upper limit	—
Tryptophan	No upper limit	—
Valine	No upper limit	—
Taurine	No upper limit	—
ME	25% over the predicted requirement	Author
Fat	No upper limit	—
n-6 fatty acids	No upper limit	—
Arachidonic acid	No upper limit	—
Acid detergent fiber	No upper limit	—
Calcium	10 × the predicted requirement	Author
Phosphorus	10 × the predicted requirement	Author
Sodium	3 × the predicted requirement	Author
Chlorine	10 × the predicted requirement	Author
Potassium	10 × the predicted requirement	Author
Magnesium	8.75 × the predicted requirement	NRC: Nutrient Requirements of Cats, 1986
Copper	10 × the predicted requirement	Author
Iodine	10 × the predicted requirement	Author
Iron	100 × the predicted requirement	Author
Manganese	10 × the predicted requirement	Author
Zinc	500 mg/kg	Author
Selenium	50 × the predicted requirement	NRC: Nutrient Requirements of Cats, 1986
Riboflavin	20 × the predicted requirement	Author
Pantothenic acid	100 × the predicted requirement	Author
Niacin	350 mg/kg body weight	NRC: Vitamin Tolerance of Animals, 1987
Vitamin B_{12}	300 × the predicted requirement	NRC: Vitamin Tolerance of Animals, 1987
Choline	10 × the predicted requirement	Author
Biotin	10 × the requirement	NRC: Vitamin Tolerance of Animals, 1987
Folacin	1000 × the predicted requirement	Author
Thiamin B_1	1000 × the predicted requirement	NRC: Vitamin Tolerance of Animals, 1987
Pyridoxine B_6	50 × the predicted requirement	NRC: Vitamin Tolerance of Animals, 1987
Myoinositol	25 × the predicted requirement	Author
Vitamin A	3,333 IU/kg body weight	NRC: Nutrient Requirements of Cats, 1986
Vitamin D	4 × the predicted requirement	NRC: Vitamin Tolerance of Animals, 1987
Vitamin E	20 × the predicted requirement	Vitamin Tolerance of Animals, 1987
Vitamin K	1,000 × the predicted requirement	Vitamin Tolerance of Animals, 1987

The *Nutrients Supplied* display shows nutrient amounts in the diet and nutrient targets for the inputted kitten. All nutrient levels appear to be within acceptable ranges as established by the application.

The cost of this ration using initial $/ton values is $0.27 per head per day.

1st limiting amino acid: tryptophan

The comment behind the cell containing this label is shown in Figure 31–11.

This is the amino acid that exists in the ration at a level that is farthest from the level required, or the amino acid whose requirement is most narrowly met.

Figure 31–11

Ca:P: 3.0

The comment behind the cell containing this label is shown in Figure 31–12.

The ideal calcium to phosphorus ratio for cats appears to be between 0.9:1 and 1.1:1. However, assuming some of the phosphorus will be in the relatively unavailable form of phytate, the acceptable ratio here is taken as between 0.9:1 and 2.0:1.

Figure 31–12

Electrolyte balance: 157

The comment behind the cell containing this label is found in Figure 31–13.

The electrolyte balance is used to assess the impact of the ration's mineral content on the body's efforts to regulate blood pH through urinary excretion. It is calculated as mEq of excess cations: (Na + K − Cl)/kg of diet dry matter. No specific recommendations have been published for cats. However, a lower electrolyte balance creates a more acidic urine and this is useful in helping to prevent urolithiasis, a urinary tract problem common in cats. Note that the electrolyte balance does not consider all possible compounds that would affect the pH of the cat's urine. For example, dl-methionine is a common urine acidifier for cats.

Figure 31–13

The Nutrient Concentration *Display (strike* `F5`*)*

The nutrients in the ration and the predicted nutrient targets are expressed in terms of concentration. That is, the nutrients provided by the ration are divided by the amount of ration dry matter, and the nutrient targets are divided by the target amount of ration dry matter. Concentration units include percent, milligrams per kilogram or parts per million, calories per pound, and international units per pound.

Strike **F5**.

The Feedstuff Contributions *Display (strike* `F6`*)*

Shown here are the nutrients contributed by each feedstuff in the ration. This display is useful in troubleshooting problems with nutrient excesses.

Strike **F1**.

The Graphic *Display*

At the bottom of the display are tabs. The current tab selected is the Worksheet tab. Other tabs are graphs based on the current ration.

DMI

Click on this tab to display a graph titled Nutrient Status: Dry Matter Intake

CP&AA

Click on this tab to display a graph titled Nutrient Status: Crude Protein and Amino Acids

E&FattyAcids

Click on this tab to display a graph titled Nutrient Status: Metabolizable Energy and Essential Fatty Acids

Macro

Click on this tab to display a graph titled Nutrient Status: Macrominerals or Major Minerals

Micro

Click on this tab to display a graph titled Nutrient Status: Microminerals or Trace Minerals

Vit-water

Click on this tab to display a graph titled Nutrient Status: Water Soluble Vitamins

Vit-fat

Click on this tab to display a graph titled Nutrient Status: Fat Soluble Vitamins

Click on the **Worksheet** tab.

Blend Feedstuffs Procedure

Click on the **Blend Feedstuffs** button. The message box in Figure 31–14 displays.

Figure 31–14

YOU MUST HAVE ALREADY SELECTED THE FEEDSTUFFS YOU WANT TO BLEND. When your analysis is acceptable, strike **ENTER** and **F3** to name and file the blend.

Click **OK**.

The amounts to blend are the same as the amounts to feed (Table 31–3). These amounts have been copied from the Make Ration section. Note that to the left of the amounts entered is a column that has converted these amounts to a pounds-per-ton basis. Feed mill mixer capacities are rated in tons so formulas to be mixed should be expressed on a pounds-per-ton basis.

Strike **ENTER** and **F3**. The input box in Figure 31–15 displays.

Figure 31–15

ENTER THE NAME OF THE BLEND (names may not be composed of only numbers):

Name the blend **Kittenfood**. Click **OK**. The message box in Figure 31–16 displays

> The new blend has been filed at the bottom of the feed table.

Figure 31–16

Click **OK**.

View Blends Procedure

Click on the **View Blends** button. The message box in Figure 31–17 displays.

> Cursor right to view the blends. Cursor down for more nutrients. DO NOT TYPE IN THE BLUE AREAS.

Figure 31-17

Click **OK**. Confirm that the formula and analysis of the Kittenfood blend has been filed.
Strike **F1**.

Using the Blended Feed in the Balanced Ration

Select Feeds *Procedure*

Click on the **Select Feeds** button and select only the Kittenfood (row 200) (Table 31-5). Unselect all other feedstuffs by entering a **0** to the left of the feedstuff name.

Kittenfood

Table 31–5
Feedstuff to select for the cat ration formulation example

Strike **ENTER** and **F2**.

Make Ration *Procedure*

Click on the **Make Ration** button. Enter **0.24** as the amount of Kittenfood to feed (Table 31–6).

Kittenfood	0.24

Table 31–6
Amount of Kittenfood blend to feed

The amount of Kittenfood to feed is the total amount of its component ingredients in the balanced ration. That value is:

$$0.03 + 0.031 + 0.0005 + 0.03 + 0.00072 + 0.03 + 0.05 + 0.0006 + 0.003 + 0.00000005 + 0.00000027 + 0.0003 + 0.000004 + 0.00029 + 0.06 + 0.00015 + 0.000073 + 0.0000012 + 0.0000098 = 0.24.$$

This value is recorded at View Blends under Formula, as entered. The ration is balanced as it was when the components of Kittenfood were fed unmixed.

Strike **ENTER** and **F1**.

Print Ration or Blend Procedure

Make sure your name is entered at cell C1. Click on the **Print Ration or Blend** button.

The input box shown in Figure 31–18 displays.

Figure 31–18

Are you printing a cat ration evaluation or a blend formula and analysis? (1-RATION, 2-BLEND):

Click **OK** to accept the default input of 1, a ration. A two-page printout will be produced by the machine's default printer.
Click on the **Print Ration or Blend** button. Type the number 2 to print a blend. Click **OK**. The message box in Figure 31–19 displays.

Figure 31–19

Click on the green number above the blend you want to print and press **F4**. Scroll right to see additional blends.

Click **OK**.
Find the Kittenfood blend, click on the green number above it, and strike **F4**. A one-page printout will be produced by the machine's default printer.

ACTIVITIES AND WHAT-IFS

In the Forms folder on the companion CD to this text is a CatInput.doc file that may be used to collect the necessary inputs for use of the CatRation.xls file. This form may be printed out and used during kennel visits to assist in ration evaluation activities.

1. Remake the ration described in Table 31–3 for the kitten described in this chapter. What type of feedstuff is the source of the omega-6 fatty acid, arachidonic acid, 20:4(n-6) in this ration? What is the other omega-6 fatty acid required by the cat?
2. Remake the ration described in Table 31–3 for the kitten described in this chapter. What percent fat is in this ration? How many different sources of fat are in this ration? Which feedstuff provides the lion's share of the metabolizable energy in this ration? Discuss the importance of fat in the cat's diet relative to that of other domestic animals.
3. Remake the ration described in Table 31–3 for the kitten described in this chapter. What is the calcium-to-phosphorus ratio in this kitten ration? What is the ideal range for this ratio? Which feedstuff contributes the greatest amount of calcium to this ration? Discuss concerns regarding the calcium-to-phosphorus ratio as seen in this kitten ration.

REFERENCES

National Research Council (1986). *Nutrient Requirements of Cats*, Revised Edition. Washington, DC: National Academy Press.

National Research Council (1987). *Vitamin Tolerance of Animals*. Washington DC: National Academy Press.

CHAPTER 32

FEEDING RABBITS

Cows catch no rabbits.

MISSISSIPPI PROVERB

The nutritional phases in the rabbit production cycle, illustrated in Figure 32–1, include:

Doe: gestation, lactation, maintenance
Young rabbit: suckling, growth

A sample growth rate is plotted in Figure 32–2.

THE DOE AND BUCK

Mature, open, dry does and herd bucks in good body condition and not in service should be fed a diet that contains only enough nutrients to support a maintenance level of performance.

Does and bucks are selected for replacement stock from the growing rabbits at about 12 weeks of age or when the meat animals are ready for market. The replacement doe will be ready to breed at 5 to 6 months of age, and the replacement buck will be ready to use at 6 to 7 months.

Nutrient requirements for rabbits will vary with the breed, age, body weight, and weather conditions. Through observation and managed feeding, the proper body condition can be maintained for maximum performance. Feeding pregnant does a diet high in energy and low in fiber may result in higher bunny mortality at birth when compared to diets lower in energy and higher in fiber. However, the litter size at weaning of does fed a high-energy, low-fiber diet may be greater than that of does fed a lower-energy, higher-fiber diet because of the increased energy reserves in these bunnies at parturition (Xiccato, Bernardini, Castellini, Dalle Zotte, Queaque, & Trocino, 1999). Further research is needed to determine optimal fiber and energy density in rabbit diets.

The gestation length of the doe is 30 to 32 days. During the first half of gestation, the doe in good body condition should be fed a maintenance diet.

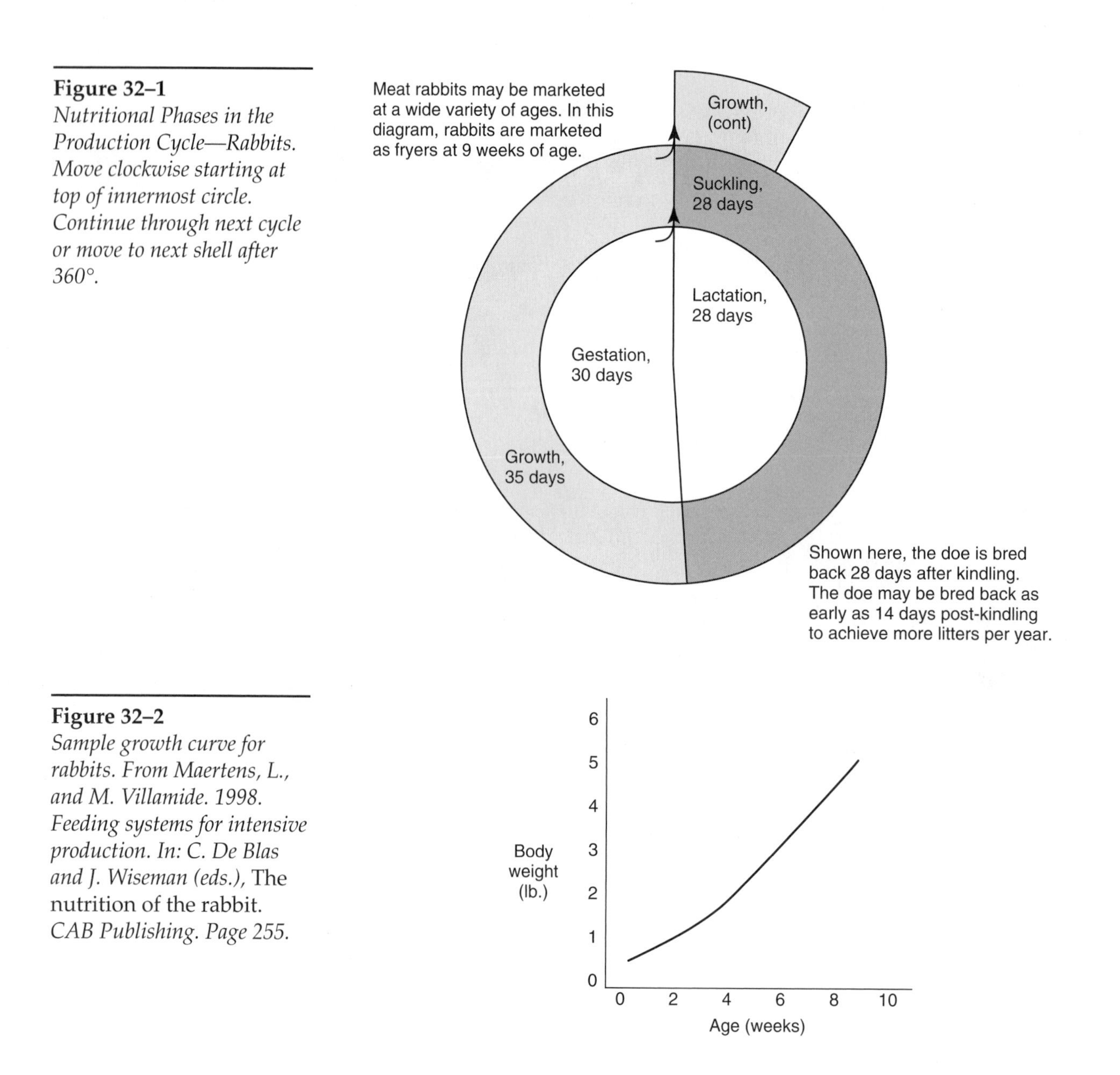

Figure 32–1
Nutritional Phases in the Production Cycle—Rabbits. Move clockwise starting at top of innermost circle. Continue through next cycle or move to next shell after 360°.

Figure 32–2
Sample growth curve for rabbits. From Maertens, L., and M. Villamide. 1998. Feeding systems for intensive production. In: C. De Blas and J. Wiseman (eds.), The nutrition of the rabbit. *CAB Publishing. Page 255.*

During the second half of gestation, the doe has an increased need for nutrients. This increased need corresponds with a time that the doe's appetite is often declining. Ideally, the diet fed the late-gestation doe is more nutrient dense than that fed to the early-gestation doe.

One or two days before kindling (parturition), the doe may go off feed. The doe should be offered small meals at this time and leftover feed should be removed. The doe should not be disturbed more than necessary during this period. Lactating does and their litters should be on full feed up until weaning time.

In rabbit does, ovulation is induced by coitus. The doe is receptive to mating immediately after parturition and so the doe may become pregnant shortly after kindling, leading to a situation where gestation and lactation commence at nearly the same time. This situation may lead to increased fetal mortality (Fortun, Prunier, & Lebas, 1993), possibly because the nutrient requirements of fetuses are not satisfied. Because milk production and nutrient requirements are directly proportional to litter size (Szendro & Maertens, 2001), lactating does should be fed free choice. Milk production peaks at about 3 weeks. Milk energy content varies with the stage of lactation—the large quantity of milk produced

Assumptions
1. The kit is weaned at 4 weeks at a weight of 1.1 lb. and requires a 56-day feeding period to reach market weight of 4 lb.
2. The rabbit averages 0.25 lb. (4.0 oz.) feed consumption per day after weaning.

Calculation
56 days × 0.25 lb. feed/day = 14 lb.

Figure 32–3
Feed calculations in the production of a 4-lb. fryer

in early lactation is lower in fat content than the reduced quantity produced in late lactation (Maertens, 2002).

THE YOUNG RABBIT

The structure of the doe's placenta is described as hemochorial. Unlike the placenta in livestock discussed thus far, the rabbit placenta does allow a significant amount of antibody immunoglobulins to pass from the blood of the doe, through the placenta, and into the blood of the fetuses. Therefore, neonatal rabbits are born with passive immunity.

Neonatal rabbits are called bunnies or kits. The litter size is usually 8 to 10 kits. The kits consume exclusively milk until about 3 weeks, when the doe's milk production begins to decline.

At about 3 weeks of age, the young rabbits come out of the nest box and begin to take solid food. Creep feeding is not practiced in the rabbit industry. Instead, the young rabbits eat the same feed as the lactating doe. The feed formulation available to both is sometimes changed after 3 weeks: initially, it is formulated with the lactating doe in mind but after 3 weeks it may be formulated specifically to address the needs of the growing rabbit. Under proper management, young rabbits from good commercial stock will weigh about 1 lb. at 4 weeks of age and can be weaned at this time.

Weaned rabbits are fed for a variable amount of time depending on the desired market weight. These animals should be on full feed. Meat rabbits less than 14 weeks old weigh approximately 4 to 6 lb. and are termed *fryers.* Meat rabbits older than 14 weeks and greater than 6 lb. are termed *roasters.*

To produce a 4-pound fryer requires approximately 14 lb. of feed. See Figure 32–3.

RABBIT FEEDING AND NUTRITION ISSUES

The Rabbit Digestive Tract

Among the other types of livestock, the rabbit's digestive tract anatomy and physiology most closely resembles that of the horse. Like the horse, the domestic rabbit is an herbivore. Also like the horse, the rabbit has a large cecum, which enables it to more effectively degrade fibrous material in forages.

Although rabbits do not have a metabolic need for fiber, a certain amount of fiber is necessary to regulate the passage rate of feed material through the digestive tract and to maintain digestive health. High levels of indigestible fiber increase the passage rate of all feed material through the gut, resulting in reduced digestive efficiency and perhaps leading to digestive upset. Low levels of fiber reduce gut motility, which could also lead to digestive upset.

The digestive system of the rabbit is unique among livestock in how it processes fiber. Fibrous feed material is effectively segregated from nonfibrous material in the colon through reverse peristalsis. Reverse peristalsis propels the nonfibrous material into the cecum, leaving the more fibrous material in the colon. The net result is that the material in the cecum is relatively rich in nutrients compared to that which was left in the colon. The microbial action in the cecum results in the production of soft pellets, which are passed as cecotropes. The material left in the colon is approximated by the acid detergent fiber (ADF) fraction and will be passed out the anus as hard pellets. The companion application to this text calculates the ADF in formulated rabbit diets. It uses a targeted a range of 15 to 25 percent for ADF, dry matter basis. Diets containing levels either above or below this range may result in digestive upset.

Cecotrophy

Rabbits produce two kinds of feces—soft and hard. *Coprophagy,* or (more appropriately), *cecotrophy* refers to consumption of the soft fecal pellets, also called cecal pellets, cecotropes or "night feces," directly from the anus. This practice is sometimes described as pseudorumination because it allows the rabbit to benefit in the same way a ruminant does from the microbial activity in the digestive tract. Cecal pellets make up 7 to 12 percent of a rabbit's total dry matter intake (DMI) (Fraga, Perez de Ayala, Carabano, & de Blas, 1991) and cecotrophy supplies the rabbit with B vitamins and essential amino acids, and permits further digestion of fiber and other nutrients by a second passage through the digestive tract. The dry matter target calculated in the companion application to this text represents the feed intake only and does not include dry matter consumed in cecal pellets.

Rabbits on high-forage diets will usually consume all of their cecal pellets. Rabbits fed diets low in fiber and excessively rich in energy and protein will not consume all their cecal pellets and these may become smeared over the rabbit and its cage.

Feeding Management

The nutrient requirements of the rabbit are greatest during lactation, less during growth and pregnancy, and least during maintenance. Ration formulation is necessary in order to meet the nutrient requirements of the rabbit in the various physiological phases. However, small rabbitries will often use a single feed formulation to feed all rabbits in all phases. The advantage of using a single formulation is that the feeding activity is that simplified and larger quantities of a single formulation can be purchased, which will usually result in cost savings. The small rabbitry must weigh the convenience and potential cost savings of a single formulation against the resulting performance loss when making feed-purchase decisions.

Most commercial rabbitries feed pelleted feed exclusively. High-moisture feedstuffs such as leafy vegetables and lawn clippings may cause digestive disturbances and the nutritional value of such feedstuffs may be difficult to quantify. The use of commercially prepared pelleted rabbit diets, though more expensive, is less labor intensive and makes it possible to more accurately balance rabbit rations for required nutrients. The size of the pellet should be matched to what the rabbit will eat. Poor pellet quality as evidenced by the quantity of fines (broken pellets) will result in reduced feed palatability. See Chapter 13 for a discussion of factors affecting pellet quality.

Feed consumption is affected by environmental factors, particularly the temperature. In warm weather, rabbits will consume less feed during the day and more at night.

Salt

If rabbits are not fed a complete feed, salt may be provided free choice by attaching a trace mineralized salt spool to the wall of the hutch.

Watering and Feeding Equipment

Rabbits must have an ample supply of clean, fresh water available at all times. Open dishes of water in the cages should be avoided since these are prone to spreading disease unless sanitized daily. Automatic watering systems are used in the larger operations. It is important that waterers and feeders be sufficiently separated to avoid contamination and spoilage. Feeders should be designed to prevent rabbits from scratching out feed.

Enteritis

Enteritis describes inflammation of the tissues of the digestive tract. In rabbits, the most common cause of enteritis is enterotoxemia in which bacteria from the genera *Clostridium* and *Escherichia* produce toxins in the lower digestive tract. Enterotoxemia is discussed in Chapter 20. Weanling rabbits are especially susceptible to enterotoxemia. Prior to 4 weeks of age, the young rabbit's stomach environment has not yet reached the very acid pH of 2 found in the adult. This makes it more likely that ingested pathogens will survive passage through the stomach, arriving at the intestine where they may cause enterotoxemia.

Enterotoxemia may also occur if the rabbit is fed a high-starch diet. As is the case with the stomach of the weanling rabbit, the weanling's pancreas may not be fully mature. The pancreatic enzymes that are responsible for starch digestion may not be present in the small intestine at a sufficient level for optimal starch digestion. Undigested starch may pass into the colon and cecum. "Starch overload" of the hindgut results in an explosion of microbial activity. Starch overload disrupts the normal resident microbial populations and it is thought that this disruption leads to the release of bacterial toxins causing enterotoxemia. Starch overload is more likely to occur when feeding starch sources, such as corn grain, that are more resistant to intestinal digestion.

Fur Block

Rabbits clean themselves by licking their coats. This behavior results in the ingestion of hair. In some cases, the ingested hair may collect in the stomach and form a hair ball or "fur block" that interferes with digestion. It has been suggested that a dietary fiber deficiency results in increased hair eating activity. The companion application to this text uses a 15% ADF (dry matter basis) minimum requirement. Another suggested cause for fur block is inadequate DMI.

Urolithiasis

Urolithiasis is a disorder common in older pet rabbits in which stones or uroliths form in the urinary tract. The etiology of this disease is somewhat different in rabbits as compared to other livestock. Unlike other livestock, rabbits excrete most of the excess dietary calcium in the urine rather than in the feces. Rabbit urine is often cloudy due to the excretion of calcium in crystals of calcium carbonate. If calcium carbonate crystals are not passed promptly, they may grow, eventually occluding the urinary tract, resulting in urolithiasis. Because rabbits are commonly fed alfalfa products, which are high in calcium, prevention presumably involves finding an alternative forage. Ensuring that rabbits have free-choice access to water will also help encourage passage of calcium carbonate crystals before they become a problem.

END-OF-CHAPTER QUESTIONS

1. Define the following terms as they relate to rabbit digestion and nutrition: coprophagy, cecotrophy, and pseudorumination.
2. What are the differences in composition and formation between the hard and soft fecal pellets produced by the rabbit?
3. What is fur block in rabbits? Give one nutrient that may, when deficient, lead to fur block. Explain the association.
4. What usually happens to the doe's dry matter/feed intake as the time of kindling approaches?
5. Give two classes of rabbits that should be offered feed free choice (should be on full feed).
6. Why are mixed feeds fed in pellet form at commercial rabbitries?
7. What is starch overload? How can starch overload lead to enterotoxemia in rabbits?
8. How might feeding alfalfa contribute to urolithiasis in rabbits?
9. Are newborn kits dependent upon colostrum to acquire antibody immunglobulins? Why or why not?
10. Explain why pregnant does are especially likely to show increased fetal mortality if not fed a balanced ration throughout gestation.

REFERENCES

Fortun, L., Prunier, A., & Lebas, F. (1993). Effects of lactation on fetal survival and development in rabbit does mated shortly after parturition. *Journal of Animal Science 71*, 1882–1886.

Fraga, M. J., Perez de Ayala, P., Carabano, R., & de Blas, C. J. (1991). Effect of type of fiber on the rate of passage and on the contribution of soft feces to nutrient intake of finishing rabbits. *Journal of Animal Science 69*, 1566–1574.

Maertens, L. (2002). Feeding Rabbits. In R. O. Kellems & D. C. Church (Editors), *Livestock Feeds & Feeding*, 5th edition. Upper Saddle River, New Jersey: Prentice Hall.

Szendro, Z., & Maertens, L. (2001). Maternal effect during pregnancy and lactation in rabbits (a review). *Acta Agraria Kaposvariensis* 5, 1–21.

Xiccato, G., Bernardini, M., Castellini, C., Dalle Zotte, A., Queaque, P. I., & Trocino, A. (1999). Effect of postweaning feeding on the performance and energy balance of female rabbits at different physiological states. *Journal of Animal Science 77*, 416–426.

CHAPTER 33

Rabbit Ration Formulation

TERMINOLOGY

Workbook A workbook is the spreadsheet program file including all its worksheets.

Worksheet A worksheet is the same as a spreadsheet. There may be more than one worksheet in a workbook.

Spreadsheet A spreadsheet is the same as a worksheet.

Cell A cell is a location within a worksheet or spreadsheet.

Comment A comment is a note that appears when the mouse pointer moves over the cell. A red triangle in the upper right corner of a cell indicates that it contains a comment. Comments are added to help explain the function and operations of workbooks.

Input box An input box is a programming technique that prompts the workbook user to type information. After typing the information in the input box, the user clicks **OK** or strikes **ENTER** to enter the typed information.

Message box A message box is a programming technique that displays a message. The message box disappears after the user clicks **OK** or strikes **ENTER**.

EXCEL SETTINGS

Security: Click on the **Tools** menu, **Options** command, **Security** tab, **Macro Security** button, **Medium** setting.

Screen resolution: This application was developed for a screen resolution of 1024 × 768. If the screen resolution on your machine needs to be changed, see Microsoft Excel Help, "Change the screen resolution" for instructions.

HANDS-ON EXERCISES

RABBIT RATION FORMULATION

Double click on the **RabbitRation** icon. The message box in Figure 33–1 displays.

Figure 33–1

Macros may contain viruses. It is advisable to disable macros, but if the macros are legitimate, you may lose some functionality.
Disable Macros or Enable Macros or More Info

Click on **Enable macros**.

The message box in Figure 33–2 displays.

Figure 33–2

Function keys F1 to F8 are set up. You may return to this location from anywhere by striking **ENTER**, then the **F1** key. Workbook by David A. Tisch. The author makes no claim for the accuracy of this application and the user is solely responsible for risk of use. You're good to go. TYPE ONLY IN THE GRAY CELLS! Note: This Workbook is made up of charts and a worksheet. The charts and worksheet are selected by clicking on the tabs at the bottom of the display. Never save the Workbook from a chart; always return to the worksheet before saving the Workbook.

Click **OK**.

Input Rabbit Procedure

Click on the **Input Rabbit** button. The input box shown in Figure 33–3 displays.

Figure 33–3

1. Growing
2. Gestating or bucks in service
3. Lactating
4. Maintenance (dry does, bucks not in service)
5. Angora or other long-hair rabbits

ENTER THE APPROPRIATE NUMBER:

Click **OK** to choose the default input of #1, a growing rabbit. The input box shown in Figure 33–4 displays.

Figure 33–4

Enter measured DMI per rabbit (pounds) if available, or click **OK** to accept the predicted value at CX26:

Click **OK** to accept the predicted dry matter intake. The message box shown in Figure 33–5 displays.

Enter the appropriate cell inputs in column CX.

Figure 33–5

Click **OK**.

The Cell Inputs

Enter the cell inputs shown in Table 33-1.

Table 33–1
Inputs for the rabbit ration formulation example

5	Rabbit age in weeks (4–42)
70	Temperature (° F)
yes	Is there significant relief from the heat at night? (yes/no)
	N/A
	N/A
	N/A
	N/A
	N/A
	N/A
	N/A
	N/A
	N/A

The comment behind the cell containing the "Rabbit age in weeks (4–42)" label is shown in Figure 33–6.

Meat rabbits are generally marketed by 8 weeks of age. No additional nutrients are included with age inputs of greater than 10 weeks. Ration amount and energy density may need to be adjusted with older rabbits depending on body condition.

Figure 33–6

The comment behind the cell containing the "Temperature in degrees F" label is shown in Figure 33–7.

Temperature is used to predict energy requirement and dry matter/feed intake.

Figure 33–7

Strike **ENTER** and **F1**.

Select Feeds Procedure

Click on the **Select Feeds** button. The message box in Figure 33–8 displays.

Figure 33–8

Nutrient content expressed on a dry matter basis. Feedstuffs are listed first by selection status; then, within the same selection status, by decreasing protein content; then, within the same protein content, alphabetically.

Click **OK**.

Explore the table. Note the nutrients listed as column headings. Note also that the table ends at row 200. You select feedstuffs for use in making two different products: (1) a blend to be mixed and sold bagged or bulk and (2) a ration to be fed directly to the animal.

Select the feedstuffs in Table 33–2 by placing a 1 in the column to the left of the feedstuff name. All unselected feedstuffs should have a 0 value in the column to the left of the feedstuff name. If you wish to group feedstuffs but not select them, you would place a value between 0 and 1 to the left of the feedstuff name.

Table 33–2
Feedstuffs to select for the rabbit ration formulation example

Feedstuff
Methionine, DL
Soybean meal, 44%
Alfalfa meal, 17%
Wheat middlings
Oat grain
Molasses, cane
Calcium pantothenate
Niacin
Potassium iodide
Pyridoxine HCl (B_6)
Riboflavin
Salt (sodium chloride)
Vitamin D supplement
Vitamin E supplement
Vitamin K (MSB premix)
Zinc oxide

Strike **ENTER** and **F2**.

Selected feedstuffs and their analyses are copied to several locations in the Workbook.

Make Ration Procedure

Click on the **Make Ration** button. The message box shown in Figure 33–9 displays.

Figure 33–9

ENTER POUNDS TO FEED IN COLUMN B. TOGGLE BETWEEN NUTRIENT WEIGHTS AND CONCENTRATIONS USING THE **F5** KEY, RATION AND FEED CONTRIBUTIONS USING THE **F6** KEY. WHEN DONE STRIKE **F1**. Cell is highlighted in red if nutrient provided is poorly matched with nutrient target. The lower limit is taken as 95 to 98 percent of target, depending on the nutrient. Where available, the upper limit of acceptable mineral is taken from the National Research Council, 1980. *Mineral Tolerance of Domestic Animals.* National Academy Press. This publication gives safe upper limits of minerals as salts of high bioavailability. The upper limit for vitamins is taken from the NRC, 1987. Where no information is available, the upper limit is based on unreasonable excess and the expense of unnecessary supplementation. See Table 33–4 for specifics.

Click **OK**. The message box shown in Figure 33–10 displays.

Figure 33–10

The Goal Seek feature may be useful in finding the pounds of a specific feedstuff needed to reach a particular nutrient target:

1. Select the red cell highlighting the deficient nutrient
2. From the menu bar, select **Tools**, then **Goal Seek**
3. In the text box, "To Value:" enter the target to the right of the selected cell
4. Click in the text box, "By changing cell:" and then click in the gray "Pounds fed" area for the feedstuff to supply the nutrient
5. Click **OK**. You may accept the value found by clicking **OK** or reject it by clicking **Cancel**.

WARNING: Using Goal Seek to solve the unsolvable (e.g., asking it to make up an iodine shortfall with iron sulfate) may result in damage to the Workbook.
IMPORTANT: If you return to the Feedtable to remove more than one feedstuff from the selected list, you will lose your chosen amounts fed in the developing ration.

Click **OK**.

In the gray area to the right of the feedstuff name, enter the pound values shown in Table 33–3.

Table 33–3
Inclusion rates for feedstuffs in the rabbit ration formulation example

Methionine, DL	0.0004
Soybean meal, 44%	0.04
Alfalfa meal, 17%	0.17
Wheat middlings	0.04
Oat grain	0.05
Molasses, cane	0.01
Calcium pantothenate	0.0000076
Niacin	0.000053
Potassium iodide	0.000000078
Pyridoxine HCl (B_6)	0.000013
Riboflavin	0.0000028
Salt (sodium chloride)	0.003
Vitamin D supplement	0.00023
Vitamin E supplement	0.000013
Vitamin K (MSB premix)	0.000038
Zinc oxide	0.0000073

The Nutrients Supplied *Display*

The application highlights ration nutrient levels, expressed as amount supplied per rabbit per day, that fall outside the acceptable range.

The lower limit for dry matter intake and DE is 95 percent of the target. The lower limit for all other nutrients is 98 percent of the target. Table 33–4 shows the upper limits of the acceptable range for the various nutrients.

Table 33–4
Upper limits for ration nutrients used in the rabbit ration application

	Upper Limit	Source
DMI	6% over predicted requirement	Author
DE	6% over predicted requirement	Author
Fat	No upper limit	—
ADF	No upper limit	—
CP	30% over predicted requirement	Author
Arg	No upper limit	—
Gly+Ser	No upper limit	—
His	No upper limit	—
Ile	No upper limit	—
Leu	No upper limit	—
Lys	No upper limit	—
Met	No upper limit	—
Met+Cys	No upper limit	—
Phe	No upper limit	—
Phe+Tyr	No upper limit	—
Pro	No upper limit	—
Thr	No upper limit	—

	Upper Limit	Source
Trp	No upper limit	—
Val	No upper limit	—
Ca	4.5% of ration dry matter	Rabbit NRC, 1977
Ptotal	1% of ration dry matter	Rabbit NRC, 1977
Pavail	The upper limit has not been established	—
Na	1.2% of ration dry matter	NRC: Mineral Tolerance of Domestic Animals, 1980—rabbit; based on limit for salt (sodium chloride)
Cl	2% of ration dry matter	NRC:Mineral Tolerance of Domestic Animals, 1980—rabbit; based on limit for salt (sodium chloride)
K	3% of ration dry matter	NRC:Mineral Tolerance of Domestic Animals, 1980
Mg	0.3% of ration dry matter	NRC:Mineral Tolerance of Domestic Animals, 1980
Cu	300 mg/kg or ppm of ration dry matter	Author
I	300 mg/kg or ppm of ration dry matter	Mineral Tolerance of Domestic Animals, 1980—value used for poultry
Fe	500 mg/kg or ppm of ration dry matter	NRC:Mineral Tolerance of Domestic Animals, 1980
Mn	400 mg/kg or ppm of ration dry matter	NRC:Mineral Tolerance of Domestic Animals, 1980
Co	10 mg/kg or ppm of ration dry matter	NRC:Mineral Tolerance of Domestic Animals, 1980
Zn	500 mg/kg or ppm of ration dry matter	NRC:Mineral Tolerance of Domestic Animals, 1980—poultry
Vitamin A	4 × the predicted requirement	NRC: Vitamin Tolerance of Animals, 1987
Carotene	No upper limit	—
Vitamin D	4 × the predicted requirement	NRC: Vitamin Tolerance of Animals, 1987
Vitamin E	20 × the predicted requirement	NRC: Vitamin Tolerance of Animals, 1987
Vitamin K	1,000 × the predicted requirement	NRC: Vitamin Tolerance of Animals, 1987
Choline	10 × the predicted requirement	Author
Folacin	No upper limit	NRC: Vitamin Tolerance of Animals, 1987
$Niacin_{total}$	No upper limit	—
$Niacin_{avail}$	350 mg/kg body weight	NRC: Vitamin Tolerance of Animals, 1987
Pantothenic acid	100 × the predicted requirement	Author
Riboflavin	20 × the predicted requirement	Author
Thiamin B_1	1,000 × the predicted requirement	NRC: Vitamin Tolerance of Animals, 1987
Pyridoxine B_6	50 × the predicted requirement	NRC: Vitamin Tolerance of Animals, 1987
Vitamin B_{12}	No upper limit	NRC: Vitamin Tolerance of Animals, 1987

DMI: dry matter intake; DE: digestible energy.

Energy, protein, amino acids, minerals, and vitamin targets have been met with minimal excesses.

The cost of this ration using initial $/ton values is $0.03 per rabbit per day.

First limiting amino acid: *Arg (Arginine)*

The comment behind the cell containing this label is shown in Figure 33–11.

This is the amino acid that exists in the ration at a level that is farthest from the level required, or the amino acid whose requirement is most narrowly met.

Figure 33–11

$Ca:P_{total}$ ratio: *2.33*

The comment behind the cell containing this label is shown in Figure 33–12.

Figure 33–12

In some species, excess calcium interferes with the absorption of other minerals such as phosphorus, magnesium and zinc. However, with rabbits a calcium-to-phosphorous ratio as high as 12 calcium to 1 phosphorus did not depress growth and resulted in normal bone ash (rabbit NRC, 1977).

$Ca:P_{avail}$ ratio: *3.33*

The comment behind the cell containing this label is shown in Figure 33–13.

Figure 33–13

Though there is no information specifically for rabbits, in swine nutrition, when the calcium-to-phosphorous ratio is based on available phosphorus, the recommendation is to have two to three times as much calcium as phosphorus.

The Nutrient Concentration *Display (strike* `F5`*)*

The nutrients in the ration and the predicted nutrient targets are expressed in terms of concentration. That is, the nutrients provided by the ration are divided by the amount of ration dry matter and the nutrient targets are divided by the target amount of ration dry matter. Concentration units include percent, milligrams per kilogram or parts per million, calories per pound, and international units per pound.

Electrolyte balance: *435*

The comment behind the cell containing this label is shown in Figure 33–14.

Figure 33–14

The electrolyte balance is used to assess the impact of the ration's mineral content on the body's efforts to regulate blood pH through urinary excretion. High intake of mineral cations has been associated with urolithiasis in rabbits. The electrolyte balance is calculated here as mEq of excess cations: (Na + K − Cl)/kg of diet *dry matter.* Because rabbits are unusual in that they excrete calcium through urinary rather than fecal means, this formula may not be suited to rabbits. The ideal balance among these electrolytes has not been defined for rabbits. The optimal electrolyte balance in the diet for pigs has been suggested to be 250, as fed (Austic & Calvert, 1981) or 278, *dry matter.* However, in pigs, optimal growth has been found to occur over the range of 0 to 667 mEq/kg of *dry matter* diet. The electrolyte balance is of no value if requirements for Na, K and Cl have not been satisfied.

Strike **`F5`**.

The Feedstuff Contributions *Display (strike* `F6`*)*

Shown here are the nutrients contributed by each feedstuff in the ration. This display is useful in trying to troubleshoot problems with nutrient excesses.

Strike **F1**.

The Graphic *Display*

At the bottom of the home display are tabs. The current tab selected is the Worksheet tab. Other tabs are graphs based on the current ration.

DMI&DE&NDF

Click on this tab to display a graph titled Nutrient Status: Dry Matter Intake, DE and NDF.

CP&aa

Click on this tab to display a graph titled Nutrient Status: Crude Protein and Amino Acids.

Macro

Click on this tab to display a graph titled Nutrient Status: Macrominerals.

Micro

Click on this tab to display a graph titled Nutrient Status: Microminerals.

Vit-H20

Click on this tab to display a graph titled Nutrient Status: Water-Soluble Vitamins.

Vit-Fat

Click on this tab to display a graph titled Nutrient Status: Fat-Soluble Vitamins.

Click on the **Worksheet** tab.

Blend Feedstuffs Procedure

Click on the **Blend Feedstuffs** button. The message box shown in Figure 33–15 is shown.

YOU MUST HAVE ALREADY SELECTED THE FEEDSTUFFS YOU WANT TO BLEND. When your analysis is acceptable, strike **ENTER** and **F3** to name and file the blend.

Figure 33–15

Click **OK**.

The amounts to blend are the same as the amounts to feed (Table 33–3). These amounts have been copied from the Make Ration section. Note that to the left of the amounts entered is a column that has converted these amounts to a pounds-per-ton basis. Feed mill mixer capacities are rated in tons so formulas to be mixed should be expressed on a pounds-per-ton basis.

Strike **ENTER** and **F3**. The input box in Figure 33–16 displays.

Figure 31–16

ENTER THE NAME OF THE BLEND (names may not be composed of only numbers):

Name the blend RabbitGrower and click **OK**. The message box in Figure 33–17 displays.

Figure 33–17

The new blend has been filed at the bottom of the feed table.

Click **OK**.

View Blends Procedure

Click on the **View Blends** button. The message box in Figure 33–18 displays.

Figure 33–18

Cursor right to view the blends. Cursor down for more nutrients. DO NOT TYPE IN THE BLUE AREAS.

Click **OK**. Confirm that the RabbitGrower formula and analysis have been filed.
Strike **F1**.

Using the Blended Feed in the Balanced Ration

Select Feeds *Procedure*

Click on the **Select Feeds** button. Unselect all feedstuffs that were in the RabbitGrower by entering a 0 to the left of the feedstuff name. Select the RabbitGrower at the bottom of the feed table (row 200) (Table 33–5).

Table 33–5
Feedstuff to select for rabbit ration formulation example

RabbitGrower

Strike **ENTER** and **F2**.

Make Ration *Procedure*

Click on the **Make Ration** button. Enter the ration shown in Table 33–6.

Table 33–6
Inclusion rate for feedstuff in rabbit ration formulation example

RabbitGrower	0.314

The amount of the RabbitGrower to feed is the total amount of its component ingredients in the balanced ration. That value is:

$$0.0004 + 0.04 + 0.17 + 0.04 + 0.05 + 0.01 + 0.0000076 + 0.000053 + 0.000000078 + 0.000013 + 0.0000028 + 0.003 + 0.00023 + 0.000013 + 0.000038 + 0.0000073 = 0.314.$$

This value is recorded at View Blends under Formula, as entered. The ration is balanced as it was when the components of RabbitGrower were fed unmixed.

Strike **ENTER** and **F1**.

Print Ration or Blend Procedure

Make sure your name is entered at cell C1. Click on the **Print Ration or Blend** button. The input box in Figure 33–19 displays.

Are you printing a rabbit ration evaluation or a blend formula and analysis? (1-RATION, 2-BLEND):

Figure 33–19

Click **OK** to accept the default value of 1, a ration. A two-page printout will be produced by the machine's default printer.
Click on the **Print Ration or Blend** button. Type the number 2 to print a blend. Click **OK**. The message box in Figure 33–20 displays.

Click on the green number above the blend you want to print and press **F4**. Scroll right to see additional blends.

Figure 33–20

Click **OK**.
Find the RabbitGrower blend, click on the green number above it, and strike **F4**. A one-page printout will be produced by the machine's default printer.

ACTIVITIES AND WHAT-IFS

In the Forms folder on the companion CD to this text is a RabbitInput.doc file that may be used to collect the necessary inputs for use of the RabbiRation.xls file. This form may be printed out and used during rabbitry visits to assist in ration evaluation activities.

1. To ensure pellet quality if no pellet binder is to be included in the formulation, feed mills often include at least 400 lb. of wheat middlings per ton. View the RabbitGrower formula at View Blends. What can you say about the need for inclusion of a pellet binder in this formulation?
2. The vitamin A value in this ration is in excess of the rabbit's requirements. What is the source of the vitamin A value in this ration? What would you say about the potential for vitamin A toxicity in the rabbit fed this ration?
3. Remake the ration described in Table 33–3 for the rabbit described in this chapter. What is the cost of feeding a growing rabbit this ration for 1 day? What are the top three feedstuffs contributing to the cost of this ration?

REFERENCES

Austic, R. E. & Calvert, C. C. (1981). Nutritional interrelationships of electrolytes and amino acids. *Federation Proceedings. 40*, 63–67.

National Research Council. (1977). *Nutrient Requirements of Rabbits*, 2nd revised edition. Washington DC: National Academy Press.

National Research Council. (1980). *Mineral Tolerance of Domestic Animals.* Washington DC: National Academy Press.

National Research Council. (1987). *Vitamin Tolerance of Animals.* Washington DC: National Academy Press.

CHAPTER 34

FEEDING FISH

Fish farming is a practical application to food production of limnology and freshwater biology.

C. F. HICKLING, 1968

CHANNEL CATFISH

The nutritional phases in the channel catfish production cycle, illustrated in Figure 34–1, include:

Broodstock: one ration is fed throughout the year
Food fish: incubator/sac fry, fry, fingerling/stocker, growout.

A sample growth rate is plotted in Figure 34–2.

Broodstock

One formulation is fed throughout the year, the amount varying with the body weight and water temperature.

Winter Season

Broodstock should be kept in good condition throughout the year. Fish fed through the winter will have reduced dry matter/feed intake due to lowered water temperature. However, it is recommended that channel catfish be fed throughout the winter to maintain body weight and to ensure a good start in the spring. This rate will be determined by the amount of feed fish will eat over a 15-minute feeding period. Winter feeding will generally be 1 percent of body weight. When the water temperature is below 65° F, catfish should be fed 1 percent of body weight every other day (Lewis & Shelton, n.d.).

Food Fish

Incubator/Sac Fry

Egg laying and subsequent egg fertilization in fish is termed *spawning*. The jelly-like mass of fertilized eggs is known as a *spawn*. The spawn is incubated

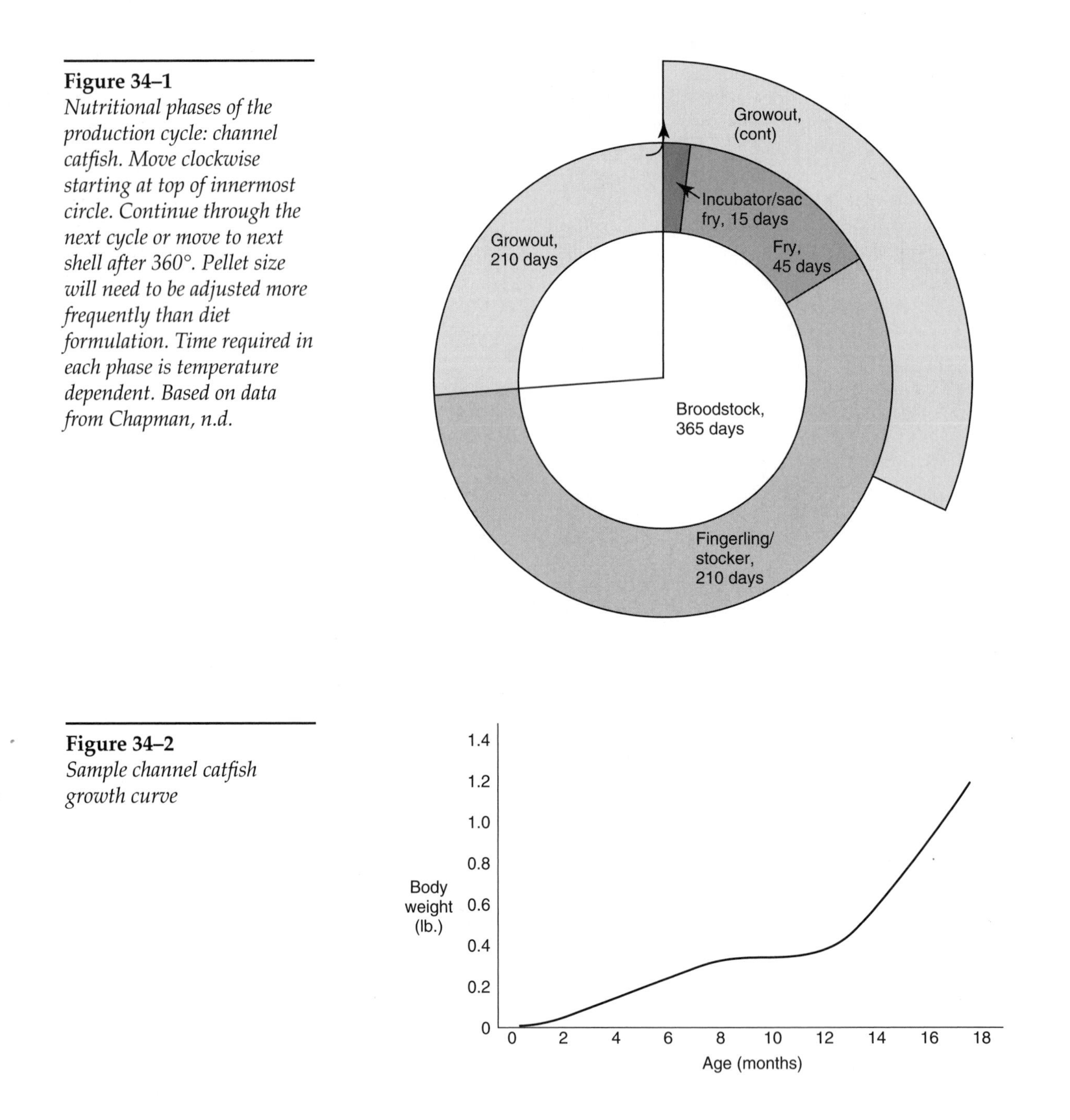

Figure 34–1
Nutritional phases of the production cycle: channel catfish. Move clockwise starting at top of innermost circle. Continue through the next cycle or move to next shell after 360°. Pellet size will need to be adjusted more frequently than diet formulation. Time required in each phase is temperature dependent. Based on data from Chapman, n.d.

Figure 34–2
Sample channel catfish growth curve

until hatching. Initially, the hatchlings, or fry, acquire nutrition from the egg sac. During this time, they are termed *sac fry.* In channel catfish, spawning through the egg sac nutrition period spans 10 to 20 days. Following this period, "swim up" occurs and this is when formulated diets begin to play a role in food fish production.

Fry

The fry are fed until they are about 2 inches long, at which time they are called *fingerlings.*

Fingerling/Stocker

Longer fingerlings are called *stockers.* Hatcheries sell catfish fingerlings and stockers to growers who grow them out and sell them as food fish.

Growout

The amount of feed needed is calculated as a percentage of fish body weight. The range for channel catfish feeding at the optimal growing temperatures of 75° to 85° F is 1.2 to 3.0 percent of body weight (Lewis, n.d.). In the growout phase, fish have increasing appetites and the nutrient density of the diet may be reduced over that of the fingerling and stocker diets. With proper management, channel catfish producers can expect conversions of 1.5 to 2.0 lb. of feed to 1 lb. of fish gain (Lewis, n.d.).

Winter Season The winter season corresponds to a nongrowing season due to the impact of low water temperature on dry matter/feed intake. The guidelines for winter feeding given for catfish broodstock also apply to the growout phase.

RAINBOW TROUT

The nutritional phases in the rainbow trout production cycle, illustrated in Figure 34–3, include:

Broodstock: one ration is fed throughout the year
Food fish: incubator/sac fry, fry, fingerling, growout

A sample growth rate is plotted in Figure 34–4.

Broodstock

One formulation is fed throughout the year, the amount varying with the body weight and water temperature.

Food Fish

Incubator/Sac Fry

Egg laying and subsequent egg fertilization in fish is termed *spawning*. The jelly-like mass of fertilized eggs is known as a *spawn*. The spawn is incubated until hatching. Initially, the hatchlings, or fry, acquire nutrition from the egg sac. During this time, they are termed *sac fry*. In rainbow trout, spawning through the egg sac nutrition period spans 60 to 70 days. Following this period, "swim up" occurs and this is when formulated diets begin to play a role in food fish production.

Fry

The fry are fed until they are about two inches long, at which time they are called *fingerlings*.

Fingerling

Hatcheries sell trout fingerlings to growers who raise them in ponds, tanks, or raceways, and sell them as food fish.

Growout

The amount of feed needed is calculated as percentage of fish body weight. The range for rainbow trout feeding at the optimal growing temperatures of 55° to 65° F is 1.5 to 6.0 (or more) percent of body weight (Hinshaw, 1999). In the growout phase, fish have increasing appetites and the nutrient density of the diet may be reduced over that of the fingerling diet. Properly managed rainbow trout farms can expect conversions of 1.2 to 2.0 lb. of feed to 1 lb. of fish gain (Cooperative State Research Service and Extension Service, n.d.).

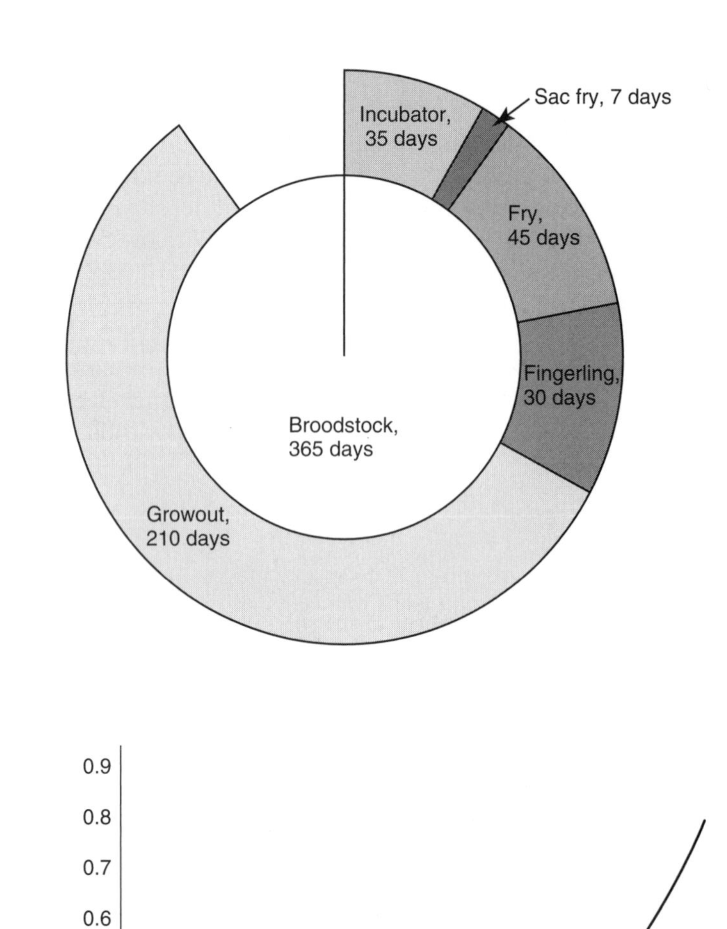

Figure 34–3
Nutritional phases of the production cycle: rainbow trout. Move clockwise starting at top of innermost circle. Continue through the next cycle or move to next shell after 360°. Pellet size will need to be adjusted more frequently than diet formulation. Time required in each phase is temperature dependent. Based on data from Clear Springs Foods (2004), Hinshaw (1999), and Klontz (1991).

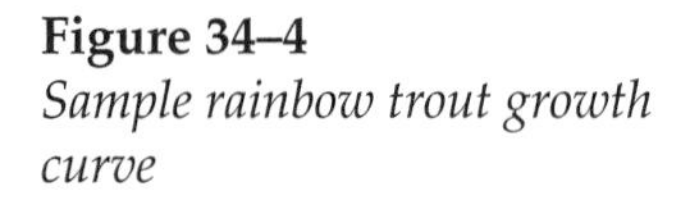

Figure 34–4
Sample rainbow trout growth curve

FISH FEEDING AND NUTRITION ISSUES

Differences between the Nutrient Needs of Fish and Those of Terrestrial Livestock

Because the body weight of fish is supported by water, they do not have to expend energy to resist the force of gravity. Because fish are poikilothermic, they do not use feed energy to maintain body temperature. Because fish excrete nitrogenous waste as ammonia, they do not have to make energy-expensive conversions to urea or uric acid. For these reasons, fish have a low maintenance energy requirement relative to terrestrial livestock (Chapter 5).

The ratio of the requirements for digestible energy (DE) to crude protein is lower for fish (8 to 10 kcal of DE per gram of crude protein) when compared to terrestrial livestock (15 to 20 kcal of DE per gram of crude protein) (Lovell,

Table 34–1
Mineral absorption sites in most commercial fish

Mineral	Absorption Site—Digestive Tract[1]	Absorption Site—Gills, Skin, and/or Oral Epithelium[1]
Ca		X
P	X	
Na		X
Cl	x	X
K	x	X
Mg	x	X
Cu		X
I	x	X
Fe	X	x
Mn	X	x
Zn	X	x
Se	x	X

[1]X indicates primary site of absorption, x indicates secondary site of absorption

1991). The higher ratio for terrestrial livestock is due to the fact that most of the feed consumed by terrestrial livestock goes toward meeting the energy requirement. Because of the diluting effect of these energy sources, the concentration of nonenergy nutrients required by land dwellers is much lower than it is for fish. With the exception of the carnivorous cat, crude protein concentrations in balanced rations for terrestrial livestock typically range from 8 to 22 percent, dry matter basis. Balanced fish rations contain 32 to 47 percent crude protein, dry matter basis and some commercial fish feeds are as high as 58 percent crude protein. The high protein concentration required by fish is, at least in part, the result of a low energy requirement.

The high level of protein required by fish is also due to the fact that fish use protein, as well as fat and carbohydrate, to meet energy needs. An amino acid can only be used for energy after the amino acid has been deaminated. Deamination produces a carbohydrate-like fragment and ammonia. For most livestock, deamination must be followed by reactions converting ammonia to urea or uric acid, reducing the net energy that may be derived from the amino acid. Because fish excrete ammonia, the net amount of usable energy derived from amino acids is higher than it is for terrestrial livestock.

The impact of interactions between water temperature and water quality on the nutritional requirements of fish have not been well studied, but are probably more significant than the impact of the environment on nutritional requirements of terrestrial livestock. For this reason, the success of the nutritional program for aquatic species is more dependent on the design of the production system than it is for terrestrial livestock.

Another difference between fish and terrestrial livestock relates to mineral nutrition. For some minerals, the primary absorption route for fish is through the gills, and dietary supplementation of these minerals is unnecessary (Table 34–1).

Of all domestic animal species discussed in this text, fish alone require myo-inositol and ascorbic acid (vitamin C). Deficiency symptoms of these vitamins in fish have been described in Chapter 10. The labile character of vitamin C and the challenge of preserving it through the pelleting/extruding processes have been described in Chapters 10 and 13. Fish rations must also be formulated to include a source of the omega-3 fatty acids that are described in Chapter 8.

Table 34–2
Pellet size specifications[1]

Description	Specifications
Starter or #0	Through U.S. sieve #30 (0.595 mm) over U.S. sieve #40 (0.420 mm)
#1	Through U.S. sieve #20 (0.841 mm) over U.S. sieve #30 (0.595 mm)
#2	Through U.S. sieve #16 (1.19 mm) over U.S. sieve #20 (0.841 mm)
#3	Through U.S. sieve #12 (1.68 mm) over U.S. sieve #16 (1.19 mm)
#4	Through U.S. sieve #8 (2.38 mm) over U.S. sieve #12 (1.68 mm)
#5	Through U.S. sieve #6 (3.36 mm) over U.S. sieve #8 (2.38 mm)

[1]Sizes larger than #4 are usually identified by the size of the pellet.

National Research Council

As with the dog and cat NRC publications (1985 and 1986, respectively), the fish NRC (1993) publication gives requirement information that was determined using purified diets. This means that the requirement values have been determined using feedstuffs of nearly 100 percent bioavailability. The companion application to this text uses the NRC requirement values as published. Because the usual feedstuffs used in commercial diets will not contain nutrients of 100 percent bioavailability, it should be expected that the evaluation of commercial feed formulations will show nutrient levels to be somewhat higher than predicted nutrient targets.

The Role of Pelleted Feed in Fish Feeding

Fish should be fed a pelleted feed. The meal form of feed coming out of the feed mill's mixer would not be located by the fish and would simply pollute the water. Pellets may be made using a pellet mill, an extruder, or an expander. For a description of these types of equipment, see Chapter 13.

In both trout and catfish, the fry are fed a small crumble feed initially. A crumble is a broken pellet that has been graded to achieve the desired particle size. In feeding fish, the pellet size will need to be adjusted frequently to correspond to the fish size. Pellet sizes are presented in Table 34–2. Table 34–3 gives fish feed size recommendations for specific fish, as well as appropriate protein and fat densities and feeding rates, expressed as a percentage of fish body weight.

Amount to Feed

A count of the fish should be made at least monthly and feeding rates adjusted accordingly. Feeding rates should also be adjusted based on water temperature. Fish in warmer water will eat more feed than fish in cooler water. The optimum temperature range for growing rainbow trout is 55° to 65° F. The optimum temperature range for growing channel catfish is 75° to 85° F. Feeding-rate adjustments should be made using predicted feeding rates from tables and formulas and through the observation of feeding behavior. Feed should be restricted when fish are sick or before fish are to be transported.

Overfeeding

One of the most serious problems in fish nutrition is overfeeding. The uneaten feed poisons the environment for the fish. Specifically, decaying proteins are converted to ammonia, nitrite, and nitrate by bacteria. Fish in water containing high levels of these compounds have decreased immune system function, leaving the fish more susceptible to disease. The toxic effect of nitrogenous compounds on fish health has been reviewed (Tomasso, 1994).

Table 34–3
Amount to feed rainbow trout (expressed as % BW, as fed basis) based on body size and water temperature. Recommended fish feed size, ration protein, and ration fat guidelines are also given.

		Fish Size in Centimeters (inches)										
		2.5–5 (1–2)	5–7.5 (2–3)	7.5–10 (3–4)	10–12.5 (4–5)	12.5–15 (5–6)	15–18 (6–7)	18–20 (7–8)	20–23 (8–9)	23–25 (9–10)	25–28 (10–11)	>28 (>11)
		Fish Weight in Grams										
		0.5	3	8	15	25	50	75	110	150	200	250
		Fish Feed Size in Millimeters (pellet size in crumble number and inches)										
		1 (#1)	1.5 (#2)	2 (#3)	3 (#4)	3 (1/8)	4 (5/32)	4.5 (5/32)	5 (5/32)	5 (3/16)	5 (3/16)	6 (1/4)
		Percent Protein in Feed (percent fat)										
		52 (15)	52 (15)	48 (15)	46 (15)	40 (15)	39 (14)	38 (13)	37 (12)	37 (11)	36 (10)	36 (10)
		Percent of Body Weight to Feed, As Fed Values										
Water Temp. °C	**°F**											
5	42	2.79	2.07	1.94	1.40	0.90	0.86	0.84	0.78	0.72	0.66	0.60
7	45	3.33	2.34	2.16	1.58	1.35	1.14	1.03	0.95	0.88	0.81	0.73
9	48	3.87	2.61	2.43	1.71	1.60	1.34	1.21	1.13	1.04	0.95	0.87
10	51	4.77	3.15	3.06	2.07	1.85	1.55	1.40	1.30	1.20	1.10	1.00
12	54	5.13	3.69	3.49	2.34	2.19	1.84	1.66	1.54	1.42	1.30	1.18
14	57	6.39	4.32	3.96	2.97	2.53	2.12	1.91	1.78	1.64	1.51	1.37
16	60	6.84	4.98	4.43	3.29	2.87	2.40	2.17	2.02	1.86	1.71	1.55

To minimize the amount of feed that goes uneaten, fish should not be fed more than they will eat in 15 minutes (Lewis & Shelton, n.d.). This assumes that the feeder is able to observe the fish eating. In trout feeding, observation of feeding behavior is facilitated by the use of floating feed. Floating feed is produced using an extrusion process (see Chapter 13). Extruded feed is expensive, and for this reason, some fish feeders use a combination of floating (10 to 20 percent) and sinking (80 to 90 percent) feed.

Uneaten Feed Because uneaten feed is of special concern in the aquaculture industries, factors that affect fish consumption are of special concern. These include the following:

1. Formulation
   a. Fish meal is generally considered the gold standard for fish feedstuff palatability. This is one reason that salmonid grower formulations typically contain 30 to 55 percent fish meal (Hardy, 1999). Catfish formulations contain less, and some researchers have questioned the necessity of including any animal or fish protein source in catfish rations (Robinson, Li, & Manning, 2003)

2. Amount of feed distributed
   a. Too much feed in too small an area will result in poor feed utilization. One recommendation is to feed no more than the fish will eat over a 15-minute feeding session, not to exceed 35 lb. of feed per acre per day (Lewis, n.d.).
3. Distance of feed from fish
   a. Feed for fry should be distributed over at least two thirds of the water surface; for larger fish, feed can be blown out over the water or fed from demand feeders spaced at 25- to 30-foot intervals along tank walls.
4. Floating versus sinking feed
   a. Some fish will have difficulty finding feed that has sunk.
5. Crowding
6. Pellet hardness
   a. Excessively hard pellets may be taken into the mouth and spit back out.
7. Fines
   a. If the pellet quality is poor, the feed may fall apart when being handled or on contact with the water. Broken pellets are called *fines.* Fish generally do not eat fines.
8. Pellet size
   a. Pellets must be sized to fit the fish. Pellets should be no greater than 20 percent of the size of the mouth opening of the fish (Dabrowski & Bardega, 1984).
9. Water temperature
   a. Cooler temperatures decrease intake.
10. Fish health
11. Chemical composition of the water

Protein Content of Fish Rations

One of the primary focuses of research in fish nutrition is protein. The usual protein concentration recommendations for channel catfish and rainbow trout are 32 to 35 percent and 38 to 47 percent (dry matter basis), respectively. Fish use protein as a source of energy and essential amino acids, as well as a nonspecific source of nitrogen with which to make the nonessential amino acids. Recognizing that each of these three items can be delivered through compounds other than intact protein, it is possible to meet fish nutrient requirements more accurately by balancing rations for energy and amino acids rather than crude protein. Doing so would reduce the crude protein content of fish diets. Depending on the cost of supplemental amino acids, this may also reduce feed cost. Lower feed protein content will also mean that uneaten feed will be less polluting in the water.

The Role of Fish Meal

In the trout, salmon, and shrimp industries, farm-raised crops are grown on feeds containing large amounts of fish meal. This is because fish meal for these species resembles the natural diet in palatability and nutrient content. However, fish meal is produced from captured fish, and traditional capture fisheries landings are not increasing and may be declining worldwide. This, in fact, is a factor contributing to the growth in market share of farm-raised fish.

Because of the uncertainty of future supply, fish producers and fish nutritionists are looking for feedstuffs to replace fish meal in fish diets without sacrificing performance. Feedstuffs that are being investigated include seafood-processing waste, rendered animal by-products, gluten meals, and oilseed meals.

Fish meals made directly from seafood processing waste are usually high in bone content. Bone content is directly proportional to ash content, and this type of feedstuff may contain 25 percent or more ash. Fish meals containing in excess of 25 percent ash are unacceptable in fish diets.

Rendered animal by-products include meat and bone meal, blood meal, poultry by-product meal, and feather meal.

Gluten meals from corn and wheat have been successfully used to replace up to 40 percent of the fish meal in trout formulations. However, corn gluten meal imparts a sometimes-undesirable yellow color to the flesh of trout. This yellow color can be masked by including pigment additives in the formulation.

Oilseed meals such as soybean meal are the most economical protein sources available to the livestock industry. However, due to nutritional and palatability concerns, soybean meal is usually limited to 25 percent of the ration for trout, salmon, and shrimp. It is possible to produce isolates from the soybean that have improved nutritional value for aquaculture species. These soy isolates can be successfully fed at higher rates but higher cost may limit their use.

Fat Content of Fish Rations

Unlike other types of livestock, channel catfish and rainbow trout have a dietary requirement for the omega-3 fatty acid linolenic acid. The only practical source of omega-3 for fish diets is fish meal and fish oils. A deficiency of this essential fatty acid may result in reduced growth rate and increased mortality.

The total fat in commercial fish feeds varies widely among the manufacturers. Excess energy consumption by fish, whether from protein, carbohydrate, or fat, produces the same result as it does in terrestrial livestock: fatty carcasses.

Types of Feed Delivery Systems

Fish may be hand-fed, fed using blower feeders mounted on trucks, fed using timers, or fed using demand feeders. Hand feeding is possible when feeding fish in relatively small ponds, tanks, or raceways. The advantage to hand feeding is that the feeder is able to observe the fish during eating activity. This observation can be used to make feeding adjustments and to assess fish health. Blower feeders mounted on trucks are able to effectively distribute a large amount of feed over a large surface area. Timers and demand feeders are used to eliminate the labor of feeding. Timers can be installed on various feed-delivery systems. Demand feeders make it possible for fish to feed themselves. With demand feeders, a hopper containing the feed is located over the water. From the hopper, a rod extends into the water. When fish hit the rod, the hopper opens for a moment, dropping a small amount of feed into the water. A fish greater than 5 inches can be trained to hit the rod to release feed. However, other aquatic species may also learn to use the demand feeder, leading to considerable feed wastage.

Whatever the feed delivery system, it is important to recognize that fish become accustomed to regularity and for this reason, fish should be fed at the same time each day.

END-OF-CHAPTER QUESTIONS

1. Which will need to be changed more frequently in the production of food fish from fry: feed formulation or pellet size?
2. How does water temperature affect feed consumption by fish?
3. What fish constituent is the primary source of ash in fish meal? How is ash content used to assess the value of fish meal as an ingredient in fish-feed formulations?

4. Explain the current interest in finding alternatives to fish meal in fish feed formulations.
5. What type of feed processing equipment is used to produce a floating fish feed? What nutrients may be destroyed by the conditions created by this equipment?
6. Name two nutrients for which fish are the only domestic animals that have a known dietary requirement.
7. Give five mineral nutrients for which the primary absorption site by fish is the gills.
8. What is the single nutritional consequence of the following characteristics of fish when compared to terrestrial livestock? (a) fish are poikilothermic; (b) fish body weight is supported by water; (c) fish excrete nitrogenous waste as ammonia. Use your answer to explain why the fish dietary protein requirement, expressed as a percentage of total feed, is higher for fish than it is for most terrestrial animals.
9. Describe the system used for identifying pellet sizes used in fish feeding.
10. Discuss the consequences of overfeeding as related to fish health.

REFERENCES

Chapman, F. A. (n.d.) Farm-raised Channel Catfish. University of Florida, Institute of Food and Agricultural Sciences. Retrieved 7/11/2005 from: http:edis.ifas.ufl.edu/topic.html

Clear Springs Foods. (2004). Acquaculture. Retrieved 7/11/2005 from: http://tommytrout/company/acquaculture.htm

Cooperative State Research Service and Extension Service, U.S. Department of Agriculture, under Special Project No. 87-EXCA-3-0836. "Trout Production." (n.d.). Retrieved 10/30/2003 from: http://aquanic.org/publicat/state/ga/trout.htm

Dabrowski, K., & Bardega, R. (1984). Mouth size and predicted food size preferences of larvae of three cyprinid fish species. *Aquaculture 40,* 41–46.

Hardy, R. W. (1999). Aquaculture's rapid growth requirements for alternate protein sources. *Feed Management, 50,* 1.

Hickling, C. F. (1968). *The Farming of Fish.* NY: Pergamon Press.

Hinshaw, J. M. (1999). Trout production feeds and feeding methods. Southern Regional Aquaculture Center. SRAC Publication No. 223.

Klontz, G. W. (1991). Manual for Rainbow Trout Production on the Family-Owned Farm. Murray, Utah: Nelson & Sons, Inc.

Lewis, G. W. (n.d.). Channel Catfish Production in Ponds. Warnell School of Forest Resources. The University of Georgia. Retrieved 8/7/2004 from: http://www.forestry.uga.edu/warnell/service/library/b0948/index.html

Lewis, G. W. & Shelton, J. L. (n.d.). Channel Catfish Production. Warnell School of Forest Resources. The University of Georgia. Retrieved 3/10/2005 from: http://warnell.forestry.uga.edu/warnell/service/library/ats-3/index.html

Lovell, R. T. (1991). Nutrition of aquaculture species. *Journal of Animal Science 69,* 4193–4200.

National Research Council. (1985). *Nutrient requirements of dogs.* Washington, DC: National Academy Press.

National Research Council. (1986). *Nutrient requirements of cats.* Revised edition. Washington, DC: National Academy Press.

National Research Council. (1993). *Nutrient requirements of fish.* Washington, DC: National Academy Press.

Robinson, E. H., Li, M. H., & and Manning, B. B. (2003). Evolving catfish diets. *Feed Management, 54,* 1.

Tomasso, J. R. (1994). Toxicity of nitrogenous wastes to aquaculture animals. *Reviews in Fisheries Science. 2,* 291–314.

CHAPTER 35

FISH RATION FORMULATION

TERMINOLOGY

Workbook A workbook is the spreadsheet program file including all its worksheets.

Worksheet A worksheet is the same as a spreadsheet. There may be more than one worksheet in a workbook.

Spreadsheet A spreadsheet is the same as a worksheet.

Cell A cell is a location within a worksheet or spreadsheet.

Comment A comment is a note that appears when the mouse pointer moves over the cell. A red triangle in the upper right corner of a cell indicates that it contains a comment. Comments are added to help explain the function and operations of workbooks.

Input box An input box is a programming technique that prompts the workbook user to type information. After typing the information in the input box, the user clicks **OK** or strikes **ENTER** to enter the typed information.

Message box A message box is a programming technique that displays a message. The message box disappears after the user clicks **OK** or strikes **ENTER**.

EXCEL SETTINGS

Security: Click on the **Tools** menu, **Options** command, **Security** tab, **Macro Security** button, **Medium** setting.

Screen resolution: This application was developed for a screen resolution of 1024 × 768. If the screen resolution on your machine needs to be changed, see Microsoft Excel Help, "Change the screen resolution" for instructions.

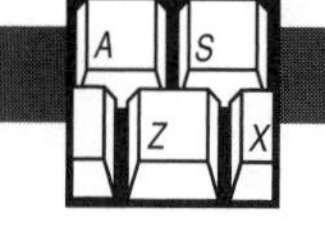

HANDS-ON EXERCISES

FISH RATION FORMULATION

Double click on the **FishRation** icon. The message box in Figure 35–1 displays.

Figure 35–1

> Macros may contain viruses. It is advisable to disable macros, but if the macros are legitimate, you may lose some functionality.
> Disable Macros or Enable Macros or More Info

Click on **Enable macros**.

The message box in Figure 35–2 displays.

Figure 35–2

> Function Keys F1 to F8 are set up. You may return to this location from anywhere by striking the **F1** key. Workbook by David A. Tisch. The author makes no claim for the accuracy of this application and the user is solely responsible for risk of use. You're good to go. TYPE ONLY IN THE GRAY CELLS! Note: This Workbook is made up of charts and a worksheet. The charts and worksheet are selected by clicking on the tabs at the bottom of the display. Never save the Workbook from a chart; always return to the worksheet before saving the Workbook.

Click **OK**.

Input Fish Procedure

Click on the **Input Fish** button. The input box shown in Figure 35–3 displays.

Figure 35–3

> 1. Rainbow trout
> 2. Channel catfish
> ENTER THE APPROPRIATE NUMBER:

Click **OK** to select the default entry of 1, Rainbow Trout. The input box in Figure 35–4 displays.

Figure 35–4

> Enter measured DMI (pounds per 1,000 fish) if available or click **OK** to accept the predicted value at CY16:

Click **OK** to accept the predicted dry matter intake (DMI). The message box in Figure 35–5 displays.

Enter the appropriate cell inputs in column CX.

Figure 35–5

Click **OK**.

The Cell Inputs

Enter the cell inputs shown in Table 35–1.

50	Trout's water temperature (° F)
150	Trout's mass in grams (0.1–575)
1	Water quality adjustment factor (0.5 for poor to 1.0 for good)
	N/A
	N/A

Table 35–1
Inputs for the fish ration formulation example

The message behind the cell containing the Trout's water temperature label is shown in Figure 35–6.

To convert from ° C to ° F:

$$°F = [°C \times 9/5] + 32$$

This application allows temperature inputs of 41° to 60° F for rainbow trout. Water temperature is assumed to impact only the fish's DMI. DMI is directly proportional to water temperature. To support a given growth rate, therefore, nutrient concentration would need to increase with decreasing water temperature. However, in this application, the required nutrient concentrations remain constant even as water temperature changes.

Figure 35–6

The comment behind the cell containing the Water quality label is shown in Figure 35–7.

The primary effect of poor water quality is probably a reduction in the fish's DMI. To support a given growth rate, therefore, nutrient concentration would need to increase with decreasing water quality. However, this application assumes that growth rate will decline with decreasing water quality and therefore required nutrient concentrations remain constant even as water quality changes.

Figure 35–7

Strike **ENTER** and **F1**.

Select Feeds Procedure

Click on the **Select Feeds** button. The message box in Figure 35–8 displays.

Figure 35–8

Nutrient content expressed on a dry matter basis. Feedstuffs are listed first by selection status, then, within the same selection status, by decreasing protein content, then, within the same protein content, alphabetically.

Click **OK**.

Explore the table. Note the nutrients listed as column headings. Note also that the table ends at row 200. You select feedstuffs for use in making two different products: (1) a blend to be mixed and sold bagged or bulk and (2) a ration to be fed directly to the fish. Select the feedstuffs listed in Table 35–2 by placing a 1 to the left of the feedstuff name. All unselected feedstuffs should have a 0 value in the column to the left of the feedstuff name. If you wish to group feedstuffs but not select them, you would place a value between 0 and 1 to the left of the feedstuff name.

Table 35–2
Feedstuffs to select for the fish ration formulation example

Fish meal, menhaden
Corn gluten meal
Soybean meal, 49%
Wheat middlings
Whey, dried
Calcium pantothenate
Copper sulfate
EDDI (Ethylenediamine dihydroiodide)
Fish oil, menhaden
Folate (folacin, folic acid)
Inositol
Iron sulfate
Manganese sulfate (mono-)
Vitamin A supplement
Vitamin C supplement
Vitamin D_3 supplement
Vitamin E supplement

Strike **ENTER** and **F2**.

Selected feedstuffs and their analyses are copied to several locations in the workbook.

Make Ration Procedure

Click on the **Make Ration** button. The message box in Figure 35–9 displays.

ENTER POUNDS TO FEED IN COLUMN B. TOGGLE BETWEEN NUTRIENT WEIGHTS AND CONCENTRATIONS USING THE **F5** KEY, RATION AND FEED CONTRIBUTIONS USING THE **F6** KEY. WHEN DONE STRIKE **F1**. Cell is highlighted in red if nutrient provided is poorly matched with nutrient target. The lower limit is taken as 96 to 100 percent of target, depending on the nutrient. The upper limit of acceptable mineral is taken from values in the fish NRC, 1993. Where no values were given, the upper limit was taken as 3 times the required level except in cases of phosphorus and manganese, for which higher limits were used. For other nutrients, the upper limit is based on unreasonable excess and the expense of unnecessary supplementation. See Table 35–4 for specifics.

Figure 35–9

Click **OK**. The message box in Figure 35–10 displays.

The Goal Seek feature may be useful in finding the pounds of a specific feedstuff needed to reach a particular nutrient target:
1. Select the red cell highlighting the deficient nutrient
2. From the menu bar, select **Tools**, then **Goal Seek**
3. In the text box, "To Value:" enter the target to the right of the selected cell
4. Click in the text box, "By changing cell:" and then click in the gray "Pounds fed" area for the feedstuff to supply the nutrient
5. Click **OK**. You may accept the value found by clicking **OK** or reject it by clicking **Cancel**.

WARNING: Using Goal Seek to solve the unsolvable (e.g., asking it to make up an iodine shortfall with iron sulfate) may result in damage to the Workbook.
IMPORTANT: If you return to the Feedtable to remove more than one feedstuff from the selected list, you will lose your chosen amounts fed in the developing ration.

Figure 35–10

Click **OK**.

In the gray area to the right of the feedstuff name, enter the pound values shown in Table 35–3.

Table 35–3
Inclusion rates for feedstuffs in the fish ration formulation example

Fish meal, menhaden	1.2
Corn gluten meal	0.5
Soybean meal, 49%	0.3
Wheat middlings	1.0
Whey, dried	0.49
Calcium pantothenate	0.00007
Copper sulfate	0.00108
EDDI (Ethylenediamine dihydroiodide)	0.0000065
Fish oil, menhaden	0.35
Folate (folacin, folic acid)	0.0000055
Inositol	0.012
Iron sulfate	0.007
Manganese sulfate (mono-)	0.025
Vitamin A supplement	0.00154
Vitamin C supplement	0.00049
Vitamin D_3 supplement	0.00192
Vitamin E supplement	0.00232

The Nutrients Supplied *Display*

The application highlights ration nutrient levels, expressed as amount supplied per 1,000 fish per day, that fall outside the acceptable range.

The lower limit for DMI, crude protein, fat, DE, P, and the trace minerals and vitamins is taken as 96 percent of the target. The lower limit for the amino acids and for n-3 fatty acid is taken as 100 percent of the target. There is no lower limit for calcium, chlorine, magnesium, potassium, and sodium since fish can acquire these minerals through their environment. Table 35–4 shows the upper limits of the acceptable range for the various nutrients.

Table 35–4
Upper limits for ration nutrients used in the fish ration application

	Upper Limit	Source
DMI	4% over the requirement	Author
Crude protein	25% over the requirement	Author
Arginine	No upper limit	—
Histidine	No upper limit	—
Isoleucine	No upper limit	—
Leucine	No upper limit	—
Lysine	No upper limit	—
Methionine & Cystine	No upper limit	—
Phenylalanine & Tyrosine	No upper limit	—
Threonine	No upper limit	—
Tryptophan	No upper limit	—
Valine	No upper limit	—
Digestible Energy	25% over the requirement	Author
Fat	1.3 × the predicted requirement for trout; 2 × the predicted requirement for catfish	Author

	Upper Limit	Source
n-3 fatty acid	No upper limit	—
n-6 fatty acid	No upper limit	—
Calcium	3% of diet dry matter	Author
Phosphorus	4.5% of diet dry matter	Author
Sodium	1.8% of diet dry matter	Author
Chloride	2.7% of diet dry matter	Author
Potassium	2.1% of diet dry matter	Author
Magnesium	0.32% of diet dry matter	NRC: Nutrient Requirements of Fish, 1993
Copper	730 mg/kg or ppm	NRC: Nutrient Requirements of Fish, 1993
Iodine	13.5 mg/kg or ppm	Author
Iron	1380 mg/kg or ppm	NRC: Nutrient Requirements of Fish, 1993
Manganese	100 mg/kg or ppm for rainbow trout, 40 mg/kg or ppm for channel catfish	Author
Zinc	1000 mg/kg or ppm	NRC: Nutrient Requirements of Fish, 1993
Selenium	13 mg/kg or ppm for rainbow trout, 15 mg/kg or ppm for channel catfish	NRC: Nutrient Requirements of Fish, 1993
Riboflavin B_2	20 × the predicted requirement	Author
Pantothenic acid	100 × the predicted requirement	Author
Niacin	350 mg/kg body weight	NRC: Vitamin Tolerance of Animals, 1987
B_{12}	300 × the predicted requirement for trout, no upper limit for catfish	NRC: Vitamin Tolerance of Animals, 1987
Choline	10 × the predicted requirement	Author
Biotin	10 × the predicted requirement for trout, no upper limit for catfish	NRC: Vitamin Tolerance of Animals, 1987
Folacin	1,000 × the predicted requirement	Author
Thiamin B_1	1,000 × the predicted requirement	NRC: Vitamin Tolerance of Animals, 1987
Pyridoxine B_6	50 × the predicted requirement	NRC: Vitamin Tolerance of Animals, 1987
Myoinositol	25 × the predicted requirement for trout, no upper limit for catfish	Author
Vitamin C	1.11% of feed dry matter	NRC: Vitamin Tolerance of Animals, 1987
Vitamin A	4 × the predicted requirement	NRC: Vitamin Tolerance of Animals, 1987
Vitamin D_3	4 × the predicted requirement	NRC: Vitamin Tolerance of Animals, 1987
Vitamin E	20 × the predicted requirement	NRC: Vitamin Tolerance of Animals, 1987
Vitamin K	No upper limit	NRC: Vitamin Tolerance of Animals, 1987

The *Nutrients Supplied* display shows nutrients supplied in the diet and nutrient targets for 1,000 of the inputted fish. All nutrient levels appear to be within acceptable ranges as established by the application.

The cost of this ration using the initial $/ton values is $2.11 per 1,000 fish per day.

First limiting amino acid: *Lysine*

The comment behind the cell containing this label is shown in Figure 35–11.

Figure 35–11

This is the amino acid that exists in the ration at a level that is farthest from the level required, or the amino acid whose requirement is most narrowly met.

	Provided	Required
mg CP/kcal DE	101	103

The comment behind the cell containing this label is shown in Figure 35–12.

Figure 35–12

Milligrams crude protein/kilocalories digestible energy. In trout nutrition it has been suggested that feeds with ratios greater than 105 may contain excess protein; feeds with ratios less than 92 may contain excess fat. In catfish nutrition it has been suggested that feeds with ratios greater than 97 may contain excess protein; feeds with ratios less than 81 may contain excess fat (fish NRC, 1993). This ratio is an attempt to establish a relationship between these two nutrients. Effective formulations may have ratios that fall outside these ranges.

	Provided	Required
kcal DE/g CP	10	10

The comment behind the cell containing this label is shown in Figure 35–13.

Figure 35–13

Kilocalories digestible energy/grams crude protein. In fish nutrition, this ratio usually falls between 8 and 10. In livestock and poultry nutrition, this ratio usually falls between 15 and 20.

Nutrient Concentration *Display (strike `F5`)*

The nutrients in the ration and the predicted nutrient targets are expressed in terms of concentration. That is, the nutrients provided by the ration are divided by the amount of ration dry matter and the nutrient targets are divided by the target amount of ration dry matter. Concentration units include percent, milligrams per kilogram or parts per million, calories per pound, and international units per pound.

Ratio of n-6:n-3 fatty acids: 0.07

The comment behind the cell containing this label is found in Figure 35–14.

Figure 35–14

The ratio of n-6:n-3 fatty acids is presented here because it is discussed in the literature, though no specific targets have been suggested for fish.

Strike **`F5`**.

Feedstuff Contributions *Display (strike* `F6`*)*

Shown here are the nutrients contributed by each feedstuff in the ration. This display is useful in trying to troubleshoot problems with nutrient excesses.

Strike **ENTER** and **F1**.

The Graphic *Display*

At the bottom of the home display are tabs. The current tab selected is the Worksheet tab. Other tabs are graphs based on the current ration.

DMI

Click on this tab to display a graph titled Nutrient Status: Dry Matter Intake

EfatCP

Click on this tab to display a graph titled Nutrient Status: Digestible Energy, Fat & Crude Protein

AA

Click on this tab to display a graph titled Nutrient Status: Amino Acids

Macro

Click on this tab to display a graph titled Nutrient Status: Macrominerals or Major Minerals

Micro

Click on this tab to display a graph titled Nutrient Status: Microminerals or Trace Minerals

Vit-water

Click on this tab to display a graph titled Nutrient Status: Water Soluble Vitamins

Vit-fat

Click on this tab to display a graph titled Nutrient Status: Fat Soluble Vitamins

Click on the **Worksheet** tab.

Blend Feedstuffs Procedure

Click on the **Blend Feedstuffs** button. The message box in Figure 35–15 displays.

YOU MUST HAVE ALREADY SELECTED THE FEEDS YOU WANT TO BLEND. When your analysis is acceptable, strike **ENTER** and **F3** to name and file the blend.

Figure 35–15

Click **OK**.

The amounts to blend are the same as the amounts to feed (Table 35-3). These amounts have been copied from the Make Ration section. Note that the blend formula is converted to a pounds-per-ton basis in the column to the left of the entered values. Feed mill mixer capacity is rated in tons so formulas to be mixed should be expressed in pounds-per-ton.

Strike **ENTER** and **F3**. The input box in Figure 35–16 displays.

Figure 35–16

ENTER THE NAME OF THE BLEND (names may not be composed of only numbers):

Name the blend TroutGrower and click **OK**. The message box in Figure 35–17 displays.

Figure 35–17

The new blend has been filed at the bottom of the feed table.

Click **OK**.

View Blends Procedure

Click on the **View Blends** button. The message box in Figure 35–18 displays.

Figure 35–18

Cursor right to view the blends. Cursor down for more nutrients. DO NOT TYPE IN THE BLUE AREAS.

Click **OK**. Confirm that the TroutGrower formula and analysis has been filed.
Strike **F1**.

Using the Blended Feed in the Balanced Ration

Select Feeds *Procedure*

Click on the **Select Feeds** button. Enter a value of 1 to the left of only TroutGrower (Table 35–5) located at row 200. Unselect all other feedstuffs by placing a 0 to the left of their names.

Table 35–5
Feedstuff to select for the fish ration formulation example

TroutGrower

Strike **ENTER** and **F2**.

Make Ration *Procedure*

Click on the **Make Ration** button.
Enter the ration shown in Table 35–6.

Table 35–6
Feeding rate for the feedstuff in the fish ration formulation example

TroutGrower	3.89

The amount of TroutGrower to feed is the total amount of its component ingredients in the balanced ration. That value is:

$$1.2 + 0.5 + 0.3 + 1.0 + 0.49 + 0.00007 + 0.00108 + 0.0000065 + 0.35 + 0.0000055 + 0.012 + 0.007 + 0.025 + 0.00154 + 0.00049 + 0.00192 + 0.00232 = 3.89.$$

This value is recorded at View Blends under Formula, as entered. The ration remains balanced as it was when the components of TroutGrower were fed unmixed. This feed may be processed into a sinking or floating pellet.

Strike **ENTER** and **F1**.

Print Ration or Blend Procedure

Make sure your name is entered at cell C1. Click on the **Print Ration or Blend** button. The input box in Figure 35–19 displays.

Are you printing a fish ration evaluation or a blend formula and analysis? (**1**-RATION, **2**-BLEND):

Figure 35–19

Click **OK** to accept the default input of 1, a ration. A two-page printout will be produced by the machine's default printer.

Click on the **Print Ration or Blend** button. Type the number 2 to print a blend. Click **OK**. The message box in Figure 35–20 displays.

Click on the green number above the blend you want to print and press **F4**. Scroll right to see additional blends.

Figure 35–20

Click **OK**.

Find the TroutGrower blend, click on the green number above it, and strike **F4**. A one-page printout will be produced by the machine's default printer.

ACTIVITIES AND WHAT-IFS

In the Forms folder on the companion CD to this text is a FishInput.doc file that may be used to collect the necessary inputs for use of the FishRation.xls file. This form may be printed out and used during on-farm visits to assist in ration evaluation activities.

1. At Input Fish, change the water temperature from 50° to 55° F. What effect does this change have on the ration balance? Explain these effects.
2. Remake the ration described in Table 35–3 for the trout described in this chapter. Replace the 1.2 lb. of fish meal menhaden with 1.2 lb. of additional soybean meal 49% to total 1.5 lb. of soybean meal. What are the primary deficiencies of soybean meal 49% when compared to fish meal menhaden?
3. Remake the ration described in Table 35–3 for the trout described in this chapter. Again, click the **Input Fish** button. What is the ideal pellet size for this trout? What is the amount to feed these fish at this water temperature expressed as a percent of body weight? By varying the inputs of fish weight and water temperature, find the values necessary to complete the following table.

	Fish weight (g)									
	50	100	150	200	250	300	350	400	450	500
Ideal pellet size										
% weight to feed in 50° F water										
% weight to feed in 55° F water										

REFERENCES

National Research Council. (1993). *Nutrient Requirements of Fish.* Washington DC: National Academy Press.

National Research Council. (1987). *Vitamin Tolerance of Animals.* Washington DC: National Academy Press.

INDEX

D

Y

Z